2D Materials for Nanoelectronics

Series in Materials Science and Engineering

Series in Materials Science and Engineering

2D Materials for Nanoelectronics

Edited by

Michel Houssa
IMEC, Leuven, Belgium

Athanasios Dimoulas
NCSR-Demokritos, Athens, Greece

Alessandro Molle
CNR IMM, Agrate Brianza, Italy

CRC Press
Taylor & Francis Group
Boca Raton London New York

CRC Press is an imprint of the
Taylor & Francis Group, an **informa** business

Cover Image: Illustration of silicene transistor device. Courtesy of Jo Wozniak, UT-Austin.

CRC Press
Taylor & Francis Group
6000 Broken Sound Parkway NW, Suite 300
Boca Raton, FL 33487-2742

First issued in paperback 2020

© 2016 by Taylor & Francis Group, LLC
CRC Press is an imprint of Taylor & Francis Group, an Informa business

No claim to original U.S. Government works

ISBN-13: 978-1-4987-0417-5 (hbk)
ISBN-13: 978-0-367-78303-7 (pbk)

Library of Congress Cataloging-in-Publication Data

Names: Houssa, Michel, editor. | Dimoulas, A. (Athanasios), editor. | Molle, Alessandro, editor.
Title: 2D materials for nanoelectronics / edited by Michel Houssa, Athanasios Dimoulas, and Alessandro Molle.
Other titles: Two D materials for nanoelectronics | Two dimensional materials for nanoelectronics | Series in materials science and engineering.
Description: Boca Raton, FL : CRC Press, Taylor & Francis Group, [2016] | "2016 | Series: Series in materials science and engineering | Includes bibliographical references and index.
Identifiers: LCCN 2015042966| ISBN 9781498704175 (hardcover ; alk. paper) | ISBN 1498704174 (hardcover ; alk. paper)
Subjects: LCSH: Nanostructured materials. | Graphene. | Nanoelectronics.
Classification: LCC TA418.9.N35 A15 2016 | DDC 621.381028/4--dc23
LC record available at http://lccn.loc.gov/2015042966

Visit the Taylor & Francis Web site at
http://www.taylorandfrancis.com

and the CRC Press Web site at
http://www.crcpress.com

Contents

SECTION I Graphene

Contents

Contents

Preface

THE SUCCESS OF THE SEMICONDUCTOR INDUSTRY relies on the continuous improvement of the performance of integrated circuits, accompanied by the reduction of their production costs. This has been achieved so far by reducing the size of the basic building block of integrated circuits: the metal-oxide–semiconductor field effect transistor (MOSFET). During the past 10–15 years, new materials were introduced to enable the reduction of MOSFET dimensions, like high-k gate dielectrics, metal gates, low-k dielectrics and copper interconnects. More recently, new devices architectures, based on multiple gates (FinFETs), replaced the conventional planar transistors. To continue device scaling, alternative materials to silicon, like germanium and III–V compounds, with high carrier mobilities, could be used as the channel in future devices.

However, further reduction in MOSFET dimensions will soon lead to a tremendous rise in power consumption as well as limited gain in the performances of integrated circuits. In this respect, two-dimensional (2D) materials such as graphene, transition metal dichalcogenides, and novel materials such as silicene, germanene, stannene and phosphorene, offer the possibility to downscale the channel thickness at the atomic level, which could lead to much improved electrostatic control of the device and suppression of the so-called short channel effects.

The isolation of graphene by Andre Geim and Konstantin Novoselov in 2004 opened the door to research on 2D materials. This monolayer of carbon atoms, arranged in a hexagonal network, has triggered considerable interest in the scientific community, due to its unique structural, mechanical and electronic properties. Graphene could be used in future nanoelectronic devices, taking advantage of its extremely high carrier mobilities and ambipolar behaviour. The absence of an energy gap in graphene, however, is problematic for logic applications, which require a sufficiently large ratio between the off and on-state current of the transistors. Such applications would thus require the use of other 2D materials, with complementary properties. Among these materials, semiconducting transition metal dichalcogenides are gaining a lot of interest, due to the possibility to tune their energy band gap through their thickness, composition and applied mechanical strain. Other 2D materials based on group-IV elements, such as silicene, germanene and stannene, were theoretically predicted, and silicene and germanene were recently successfully grown on various metallic substrates. Phosphorene, a 2D layer of phosphorous atoms, has also gained interest and has been recently integrated into functional transistors.

This book aims at giving an overview of the recent theoretical and experimental progress achieved with regard to these various 2D materials. Section I is devoted to research on graphene, first discussing its predicted properties from first principles simulations, and next highlighting progress in its growth on SiC substrates and device integration, issues with metal contacts for electrical characterisation and device fabrication, its potential use in very high-frequency analogue applications, as well as its high electric field and thermal transport properties. Section II discusses recent progress in transition metal dichalcogenides, both theoretically and experimentally. Their structural and electronic properties, predicted from ab initio calculations, are first reviewed. Insight in their physico-chemical characterisation is next highlighted, and their electrical properties and potential integration in different nanoelectronic devices are discussed, as in Schottky barrier transistors, optoelectronic devices such as photodetectors and light emitters, as well as their potential application in flexible electronics. Section III describes new 'exotic' 2D materials, such as silicene, germanene, stannene and phosphorene. The properties of silicene and germanene, predicted from first-principles calculations, are presented, followed by the discussion of recent experimental work pertaining to the growth and characterisation of silicene and germanene on different substrates. The properties of stannene, which is predicted to be a 2D topological insulator, are discussed in the next chapter. Phosphorene is next described, highlighting recent experimental work on its growth, characterisation and integration into field-effect transistors. Finally, the last chapter discusses the potential of combining various 2D materials into the so-called van der Waals heterostructures for future nanoelectronic devices.

This book benefited enormously from the contributions of internationally recognised researchers, who have paved the way to the progress in the fundamental understanding, growth, characterisation and device integration of these 2D materials. We are very grateful for their willingness to work with us on this project, the time and effort they spent on it, as well as their excellent contributions.

Michel Houssa
Athanasios Dimoulas
Alessandro Molle

Editors

Michel Houssa earned his master's and PhD in physics at the University of Liège, Belgium, in 1993 and 1996, respectively. He is a professor in the Department of Physics and Astronomy at the University of Leuven (Belgium). His current research interests focus on the first principles modeling of various materials including semi-conductor/oxide interfaces, two-dimensional materials (silicene, germanene, transition metal dichalcogenides) and their heterostructures. He has authored or co-authored about 350 publications, including 10 book chapters and 8 invited review articles, and he has given about 50 invited talks and seminars. He has been co-organiser of several international symposia and conferences, including the *Symposium on Semiconductors, Dielectrics, and Metals for Nanoelectronics of the Electrochemical Society (ECS)* and the *IEEE Semiconductor Interface Specialists Conference*. He is an ECS Fellow and a senior member of IEEE.

Athanasios Dimoulas earned a PhD in applied physics at the University of Crete in Greece in 1991 on MBE heteroepitaxial growth and char-acterisation of GaAs and related compounds on Si. Dr. Dimoulas was Human Capital and Mobility Fellow of the EU at the University of Groningen in Holland until 1994, a Research Fellow at the California Institute of Technology (Cal Tech), Chemical Engineering, Pasadena, USA, until 1996 and research associate at the University of Maryland at College Park (UMCP), USA, until February 1999. In addition, he was a visiting research scientist at IBM Zurich Research Laboratory, Switzerland, in 2006 and 2007. Since 1999, he is research direc-tor and head of the Epitaxy and Surface Science Laboratory at the National Center for Scientific Research Demokritos, Athens, Greece. He has coordi-nated several European-funded projects in the areas of advanced CMOS, the last being DUALLOGIC, a flagship CMOS project in FP7 and most recently the EU project 2D nanolattices on silicene and other 2D crystal channels for post-CMOS applications. Also, he has received the prestigious ERC (IDEAS) Advanced Investigator Grant 2011 – Smartate dealing with graphene and topological insulators at the gate of MOS devices for low power electronics. He has authored or co-authored more than 120 technical presentations in refereed journals, including 3 monographs in a Springer book. In addition, he has made more than 60 presentations in conferences, including 35 invited conferences, tutorials and summer schools. He is co-editor of a Springer book and guest editor in three special volumes of international journals. He has organised relevant MRS and E-MRS symposia in 2005, 2003, 2009, 2010 and 2013. He was the general chair of INFOS 2007 conference and served in the steering committee of INFOS and ESSDERC/ESSCIRC conferences. He has chaired the TPC committee of ESSDERC/ESSCIRC 2007 and Process Technology subcommittee of IEDM 2012. His expertise includes MBE growth of 2D selenide semiconductors and dielectric materials, VCD growth of graphene, nanodevice processing by optical and e-beam lithography and materials characterisation and device electrical characterisation.

Alessandro Molle is a research scientist at the CNR-IMM, MDM Laboratory in Agrate Brianza, Italy, where he leads the research group on low-dimensional materials and devices. He graduated in physics from the University of Genoa, Italy in 2001 and obtained his PhD degree in materials science from the same university in 2005 with a long-term beamtime project at the Synchrotron Radiation Facility (ESRF) in Grenoble, France. In 2005, he joined the MDM National Laboratory (now belonging to the CNR-IMM) as a postdoctoral fellow focusing his research on the molecular beam epitaxy of high-k oxide films on high-mobility semiconductors for logic applications in the framework of the EU project "ET4US" (FP7) and of two bilateral collaboration projects with IMEC, Belgium. From 2011 to 2014, he was the principal investigator at the CNR-IMM in the EU FP7 project "2D-Nanolattices" (FET-Open) on the synthesis of novel two-dimensional (2D) materials such as silicene. He is currently involved in the integration of silicene and other 2D elementary materials into transistors and the synthesis and structural characterization of 2D materials beyond graphene. He co-chaired an academic course on surfaces and interfaces for master's students at the University of Milan Bicocca, Italy. He was the principal organizer of two symposia at the E-MRS Spring Meeting at Strasbourg in 2010 and 2013 on the future of post-Si CMOS devices. He was the guest editor of the respective special issues of proceeding papers that appeared in *Microelectronic Engineering* (Elsevier) in 2011 and *Applied Surface Science* (Elsevier) in 2014. He is the co-author of 66 internationally peer-reviewed articles in international journals and has received 16 invitations in international conferences, workshops, and lectures.

Contributors

Rafik Addou
Department of Materials Science
and Engineering
University of Texas at Dallas
Richardson, Texas

Valery Afanas'ev
Department of Physics and
Astronomy
University of Leuven
Leuven, Belgium

Deji Akinwande
Department of Electrical and
Computer Engineering
University of Texas at Austin
Austin, Texas

Joerg Appenzeller
Department of Electrical and
Computer Engineering
Purdue University
West Lafayette, Indiana

Michele Buscema
Kavli Institute of Nanoscience
Delft University of Technology
Delft, the Netherlands

Cinzia Casiraghi
School of Chemistry
University of Manchester
Manchester, United Kingdom

Andres Castellanos-Gomez
Instituto Madrileño de Estudios
Avanzados en Nanociencia
Madrid, Spain

Hsiao-Yu Chang
Department of Electrical and
Computer Engineering
University of Texas at Austin
Austin, Texas

Saptarshi Das
Center for Nanoscale Material
Argonne National Laboratory
Lemont, Illinois

Suman Datta
Department of Electrical
Engineering
Pennsylvania State University
University Park, Pennsylvania

Yexin Deng
Department of Electrical and
Computer Engineering
Purdue University
West Lafayette, Indiana

Athanasios Dimoulas
Institute of Nanoscience and
Nanotechnology
NCSR-Demokritos
Athens, Greece

Contributors

Vincent E. Dorgan
Department of Electrical and
    Computer Engineering
University of Illinois
Urbana-Champaign, Illinois

Yuchen Du
Department of Electrical and
    Computer Engineering
Purdue University
West Lafayette, Indiana

Massimo Fischetti
Department of Materials Science
    and Engineering
University of Texas at Dallas
Richardson, Texas

Christopher L. Hinkle
Department of Materials Science
    and Engineering
University of Texas at Dallas
Richardson, Texas

Matthew Hollander
Department of Electrical
    Engineering
Pennsylvania State University
University Park, Pennsylvania

Michel Houssa
Department of Physics and
    Astronomy
University of Leuven
Leuven, Belgium

Joachim Knoch
Institute of Semiconductor
    Electronics
RWTH Aachen University
Aachen, Germany

Ortwin Leenaerts
Department of Physics
University of Antwerp
Antwerp, Belgium

Max C. Lemme
Institute of Graphene-Based
    Nanotechnology
University of Siegen
Siegen, Germany

Zuanyi Li
Department of Electrical Engineering
Stanford University
Stanford, California

Han Liu
Department of Electrical and
    Computer Engineering
Purdue University
West Lafayette, Indiana

Zhe Luo
Department of Electrical and
    Computer Engineering
Purdue University
West Lafayette, Indiana

Massoud Ramezani Masir
Department of Physics
University of Antwerp
Antwerp, Belgium

Stephen McDonnell
Department of Materials Science
    and Engineering
University of Texas at Dallas
Richardson, Texas

Alessandro Molle
CNR-IMM
Laboratorio MDM
Agrate Brianza, Italy

Thomas Mueller
Institute of Photonics
Vienna University of Technology
Vienna, Austria

Kosuke Nagashio
Department of Materials Engineering
University of Tokyo
Tokyo, Japan

Bart Partoens
Department of Physics
University of Antwerp
Antwerp, Belgium

François M. Peeters
Department of Physics
University of Antwerp
Antwerp, Belgium

Eric Pop
Department of
    Electrical Engineering
Stanford University
Stanford, California

Joshua Robinson
Department of Materials Science
    and Engineering
Pennsylvania State University
University Park, Pennsylvania

Emilio Scalise
Max-Planck-Institut für
    Eisenforschung
Düsseldorf, Germany

Frank Schwierz
Institut für Mikro- und
    Nanoelektronik
Technische Universität Ilmenau
Ilmenau, Germany

Andrey Y. Serov
Department of Electrical and
    Computer Engineering
University of Illinois
Urbana-Champaign, Illinois

Gary A. Steele
Kavli Institute of Nanoscience
Delft University of Technology
Delft, the Netherlands

André Stesmans
Department of Physics and
    Astronomy
University of Leuven
Leuven, Belgium

Ana Suarez Negreira
Department of Materials Science
    and Engineering
University of Texas at Dallas
Richardson, Texas

Akira Toriumi
Department of Materials
    Engineering
University of Tokyo
Tokyo, Japan

Dimitra Tsoutsou
Institute of Nanoscience and
    Nanotechnology
NCSR-Demokritos
Athens, Greece

William Vandenberghe
Department of Materials Science
    and Engineering
University of Texas at Dallas
Richardson, Texas

Herre S.J. van der Zant
Kavli Institute of Nanoscience
Delft University of Technology
Delft, the Netherlands

Robert M. Wallace
Department of Materials Science
    and Engineering
University of Texas at Dallas
Richardson, Texas

Freddie Withers
School of Physics and Astronomy
University of Manchester
Manchester, United Kingdom

Contributors

Xianfan Xu
Department of Electrical and
  Computer Engineering
Purdue University
West Lafayette, Indiana

Peide D. Ye
Department of Electrical and
  Computer Engineering
Purdue University
West Lafayette, Indiana

Feng Zhang
Department of Electrical and
  Computer Engineering
Purdue University
West Lafayette, Indiana

Weinan Zhu
Department of Electrical and
  Computer Engineering
University of Texas at Austin
Austin, Texas

SECTION I

Graphene

1

Theory of the Structural, Electronic and Transport Properties of Graphene

Massoud Ramezani Masir, Ortwin Leenaerts,
Bart Partoens and François M. Peeters

Contents

2D Materials for Nanoelectronics edited by Michel Houssa, Athanasios Dimoulas and Alessandro Molle © 2016 CRC Press/Taylor & Francis Group, LLC. ISBN: 978-1-4987-0417-5.

1.1 Carbon Atom and Its Allotropes

There is no material as important as carbon to our life. It is in the food we eat, the clothes we wear, the cosmetics we use and the gasoline that fuels our cars. Its name comes from the Latin word *carbo* for *coal* and *charcoal*. Carbon compounds resist extreme conditions due to their strong and stable bonds. Carbon is also an important element that can form long chains and rings of atoms, which form the basis structure of many compounds in living cells, such as deoxyribonucleic acid (DNA).

Carbon is one of the most versatile elements in terms of the number of compounds it can form. This capability is due to the diverse types of bonds it can form and the number of different elements it can join in bonding. Carbon belongs to group *IVa* in the periodic table and has four electrons available to form covalent bonds. Its ground-state configuration is $1s^2 2s^2 2p^2$, with the $2p$ orbitals ($2p_x$, $2p_y$ and $2p_z$) roughly 4 eV higher in energy than the $2s$ orbital, as shown in **Figure 1.1**. In the presence of other atoms, such as H, O or other C atoms, these orbitals can hybridise in sp, sp^2 or sp^3 configurations in order to have covalent bonds. In the sp^2 hybridisation, three equivalent sp^2-hybrid orbitals are formed which adopt a trigonal planar geometry and one

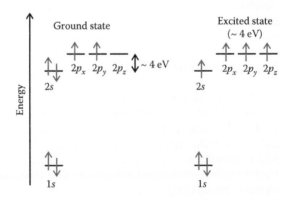

FIGURE 1.1 Electronic configuration for carbon in the ground state (left) and in an excited state (right). (Adapted from M. O. Goerbig, *Rev. Mod. Phys.* 83, 1193, 2011.)

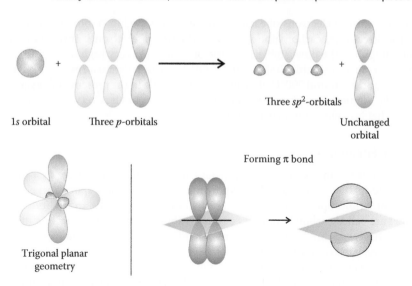

1s orbital   Three p-orbitals   Three sp^2-orbitals   Unchanged orbital

Forming π bond

Trigonal planar geometry

FIGURE 1.2 sp^2 hybridisation and formation of a π bond from two p orbitals.

unchanged p orbital lies at right angles to the plane of the hybrid orbitals, as shown schematically in **Figure 1.2**.

Atomic carbon has a very short lifetime and can only be stabilised in various multiatomic structures. Some well-known allotropes of carbon are graphite, diamond and fullerenes which also include buckyballs, nanotubes, nanobuds and nanofibers. In the next section, we briefly introduce diamond, graphite and fullerenes as three well-known allotropes of carbon.

1.1.1 Diamond

Diamond is the most famous allotrope of carbon. Each carbon atom in diamond is bonded to four other partner carbon atoms in a tetrahedral configuration with equivalent sp^3 bonds in which the 2s orbital mixes with the three 2p orbitals to form four sp^3 hybrids. A tetrahedron is composed of four triangular faces, three of which meet at each vertex. The tetrahedral arrangement of atoms is the source of many of diamond's properties. Diamond has a low electrical conductivity, the highest thermal conductivity, is highly transparent, is an excellent insulator due to its large band gap in its electronic spectrum and is among the hardest materials known. Industrially, it can be used for cutting, drilling, grinding and polishing. There is no natural substance that one can use to cut or scratch a diamond, except diamond itself.

1.1.2 Graphite

Graphite is one of the most common allotropes of carbon. In graphite, each carbon atom is covalently bonded to three other carbon atoms resulting in a trigonal geometry. Each carbon atom in graphite is sp^2 hybridised. Three out

5

of four valence electrons of each carbon atom are used in σ–bond formation with three other carbon atoms, whereas the fourth electron is involved in π bonding. It is the most stable allotrope of carbon. At very high temperatures and pressures (roughly 2000°C and 5 GPa), it can be transformed into diamond and at around 700°C, it forms carbon dioxide in the presence of oxygen. Unlike diamond, graphite is a very good conductor and it is very soft, a property that is exploited in pencils for writing.

1.1.3 Fullerenes

Fullerenes are molecules build up from carbon atoms, which were first discovered in 1985. They have different forms, such as a hollow sphere, an ellipsoid or a tube. Spherical fullerenes are also called buckyballs and cylindrical ones are called carbon nanotubes or buckytubes. From the structural point of view, fullerenes are similar to graphite, made of hexagonal carbon rings, but they may also contain pentagonal or heptagonal rings. In the past decades, the discovery of nanotubes and fullerenes attracted much attention among physicists and chemists. In 1996, the Nobel Prize in Chemistry was awarded jointly to Robert F. Curl Jr., Sir Harold W. Kroto and Richard E. Smalley 'for their discovery of fullerenes'.

1.2 Graphene

Graphene is an allotrope of carbon, whose structure is a single layer of a hexagonal arrangement of sp^2-bonded carbon atoms [2]. In fact, graphite is nothing else than van der Waals coupled layers of graphene. This name was first given in Reference 3 to describe a single layer of graphene as one of the components of graphite.

For a long time, one was convinced that two-dimensional (2D) crystals cannot exist because they cannot withstand thermal fluctuations [4–8]. Several efforts in the past to create atomically thin films of other materials failed since the films became unstable and tended to separate and *clump up* rather than form perfect layers. It came as a big surprise when in 2004, a group of scientists from the University of Manchester led by Andre Geim and Konstantin Novoselov succeeded in extracting stable 2D crystals, including a stable monolayer of carbon atoms. Nowadays, the stability of graphene is explained by postulating small out-of-plane corrugations [9,10]. Another part of the explanation is that a 2D crystal is never a completely freestanding system. It is always in contact with a substrate or clamped at the edges, which stabilises the structure.

The method used by the Manchester group to extract a single layer of graphite was simple. Graphite consists of several graphene layers coupled to each other by weak van der Waals bonding. They extracted a single layer of graphene using micromechanical exfoliation, simply called the scotch tape technique. This technique consists of repeated peeling of multilayered graphite of a highly ordered pyrolytic graphite crystal with a cellophane tape followed by pressing the tape on a Si/SiO$_2$ substrate to deposit the graphene samples. The sample sizes produced in this way are on the order of several square micrometer. In fact, if you write with a pencil and you look at the results under a microscope, you may also find single-layer sheets.

After extracting a single layer of graphene from graphite, the Manchester researchers transferred the graphene sheet onto thin SiO$_2$ on a silicon wafer in order to isolate it electrically. Since SiO$_2$ interacts weakly with graphene, this process can provide nearly charge-neutral graphene layers. Doped silicon under the SiO$_2$ substrate allows one to use it as a *back gate* electrode to change the charge density in the graphene layer. This device is nothing else than a field-effect transistor and led to the first observation of the anomalous Hall effect in graphene by Geim and Novoselov [11] and by Philip Kim and Yuanbo Zhang [12] in 2005. These experiments demonstrated the theoretically predicted π Berry's phase of massless Dirac fermions in graphene.

In 1947, Wallace [13] studied a single layer of graphite theoretically. His purpose was to understand the electronic properties of three-dimensional (3D) graphite. Later in 1984, Semenoff [14] considered graphene in a magnetic field as an analogue of (2 + 1)-dimensional electrodynamics. Semenoff obtained the Landau levels (LLs) of this system and found a LL precisely at the Dirac point, that is, with zero energy. This level is responsible for the anomalous integer quantum Hall effect [11,12].

Geim and Novoselov received several awards for their pioneering research on graphene, including the Mott medal for Geim in 2007 for the *discovery of a new class of materials – free-standing 2D crystals – in particular graphene,* the 2008 EuroPhysics Prize for *discovering and isolating a single free-standing atomic layer of carbon (graphene) and elucidating its remarkable electronic properties* and the Körber Prize for Geim in 2009 for *developing the first 2D crystals made of carbon atoms.* On 5 October 2010, the Nobel Prize in Physics was awarded to Andre Geim and Konstantin Novoselov from the University of Manchester *for groundbreaking experiments regarding the 2D material graphene.*

1.3 Electronic Structure

The electronic properties of graphene are well described by the electronic structure of a perfect, flat, freestanding and infinite graphene crystal in vacuum. This is an idealisation in several ways since every 'real' system has defects and ripples and is supported in some way or the other (and of course, it is finite and not in vacuum). Such an idealised case can be treated theoretically and can therefore provide a lot of insight. The results of several theoretical approximations for the electronic band structure and the density of states (DOS) of graphene are shown in **Figure 1.3**. These were calculated with density functional theory (DFT), first-neighbour tight binding (TB) involving only p_z atomic orbitals and the Dirac–Weyl equation (i.e. the Dirac equation for massless fermions), respectively. More information on these theoretical calculations is given in the following sections.

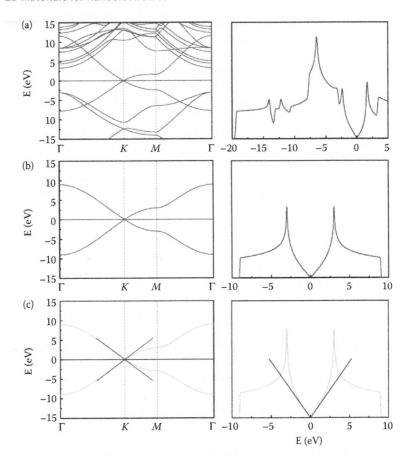

FIGURE 1.3 Electronic band structure (left) and the DOS (right) for intrinsic graphene, calculated with different theoretical models: (a) DFT, (b) NN TB and (c) Dirac theory (QED) for massless fermions. The Fermi energy is set to zero.

The DFT calculations are the most accurate ones, but the TB results give a qualitatively correct description of the valence and conduction bands. For small excitations (<2 eV) from the Fermi level of the neutral system, a good description is provided by a linear spectrum that can be deduced from the 2D Dirac–Weyl Hamiltonian $H \approx \hbar v_F \sigma \cdot \mathbf{p}$, where v_F denotes the Fermi velocity (a constant) and σ represents the 2D vector of Pauli matrices (σ_x, σ_y). The Fermi velocity is isotropic for small excitations and can be measured or calculated to be $v_F \approx 10^6$ ms^{-1}. This is about 300 times smaller than the velocity of light in vacuum. The point in the energy spectrum that coincides with the Fermi level of the neutral system is called the Dirac point, but this name is also used for the corresponding point in reciprocal space (usually the K point) of the hexagonal lattice. The DOS is zero at this Dirac point but at the same time, there is no gap in the band structure. Therefore, graphene is called a zero-gap semiconductor or semimetal.

1.3.1 Electronic Structure from DFT

In condensed-matter physics and chemistry, DFT is the preferred calculation tool for many electronic properties of crystalline materials. Although the usual implementations of DFT have some serious shortcomings, most notably the band-gap problem, the electronic properties of graphene are nicely reproduced and correspond well to the available experimental data.

In **Figure 1.4a**, the projected bands and DOS of graphene are shown. These were calculated within the DFT formalism making use of the generalised gradient approximation. The electronic band structure of graphene has a very characteristic appearance with linearly crossing bands at the Fermi level. The densities of the electron states in the different bands are pictured in **Figure 1.4b**. The states close to the Fermi level are built from p_z orbitals that form π bonds. The lower bands are made from s, p_x and p_y orbitals that show substantial hybridisation (sp^2). These hybridised orbitals form in-plane σ bonds among each other and do not interact with the p_z orbitals due to their different symmetry: the sp^2 orbitals are even with respect to reflection over the graphene plane, whereas the p_z orbitals are odd.

Some peculiar electron states are observed at higher energies. The states in the second and third conduction bands are nearly free-electron states which correspond to electrons that are weakly bound above and below the graphene sheet at a distance of about 2 and 2.5 Å. The densities of these electron states are shown in **Figure 1.4b**. The next states correspond to the free-electron states which form a continuum starting approximately 4.5 eV (the work function of graphene) above the Dirac point. The continuum is only visible as a discrete collection of parabolic energy bands, because of the periodic

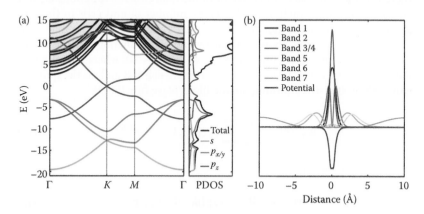

FIGURE 1.4 (a) The (projected) electronic band structure (left) and DOS (right) of graphene. The free-electron continuum is indicated by a yellow shade in the band structure and the Fermi level is taken as the origin of the energy scale. (b) The densities of the electron states of different bands at the Γ point in the direction normal to the graphene layer. The bands are numbered from lower to higher energy as they appear in (a). The local electrostatic (pseudo) potential is also shown by a dashed line.

boundary conditions used in the DFT calculations, but is indicated by the yellow area in **Figure 1.4a**.

1.3.2 Electronic Structure from TB

Further insight into the electronic structure of graphene can be gained from a qualitative TB model. The TB model is used to calculate the electronic band structure using an approximate set of wave functions based upon a super-position of wave functions for isolated atoms located at each atomic site. The method is closely related to the linear combination of atomic orbitals method. Graphene is composed of carbon atoms arranged in a hexagonal lattice with each carbon atom covalently bonded to three other carbon atoms. Core electron orbitals do not overlap much with the orbitals of adjacent atoms, which results in electrons localised at the atomic positions. The TB model describes the band structures using only a few parameters whose values are chosen such that it reproduces the experimental band structure or the outcome of first-principles calculations.

The electrons are assumed to be localised around the atomic positions. The probability to find an electron on an adjacent atom is small and we can expand the Bloch wave function of the crystal in a linear combination of local Wannier functions:

$$\psi_{k,n}(\mathbf{r}) = \frac{1}{\sqrt{N}} \sum_{\mathbf{R}} \Phi_n(\mathbf{R}, \mathbf{r}) e^{i\mathbf{k}\cdot\mathbf{R}}, \tag{1.1}$$

where $\mathbf{R}$ is the real-space lattice vector of the C atoms. The functions $\Phi_n(\mathbf{R}, \mathbf{r})$ are called Wannier functions and N is the number of unit cells in the crystal. To develop a useful model, several approximations must be made. The Wannier functions are approximated by the eigenfunctions of the atomic Hamiltonian, that is, the atomic orbitals $\phi_{\mu,s}(\mathbf{r} - \mathbf{t}_\ell - \mathbf{R})$, where $\mathbf{t}_\ell$ is the position vector of the atom l inside the primitive unit cell at $\mathbf{R}$ and s is the spin state of the μth orbital. The resulting on-site Bloch wave function is

$$\psi_{k,n}(\mathbf{r}) = \frac{1}{\sqrt{N}} \sum_{\mathbf{R}} \phi_{\mu,s}(\mathbf{r} - \mathbf{t}_\ell - \mathbf{R}) e^{i\mathbf{k}\cdot\mathbf{R}}, \tag{1.2}$$

which obeys the Bloch theorem

$$\psi_{k,j}(\mathbf{r} + \mathbf{R}') = \exp(i\mathbf{k}\cdot\mathbf{R}')\psi_{k,j}(\mathbf{r}). \tag{1.3}$$

In order to find the Hamiltonian matrix elements, we start with the Schrödinger equation:

$$H\Psi_k(\mathbf{r}) = E_k\Psi_k(\mathbf{r}). \tag{1.4}$$

The crystal wave function $\Psi_k(\mathbf{r})$ can be expanded on the basis of the on-site Bloch wave functions,

$$\Psi_{\mathbf{k}}(\mathbf{r}) = \sum_i c_{\mathbf{k},i} \psi_{\mathbf{k},i}(\mathbf{r}). \tag{1.5}$$

Substituting **Equation 1.5** in **Equation 1.4** and using the orthogonality of the Bloch wave functions, we obtain

$$\sum_{i,j} c_{\mathbf{k},j}^* c_{\mathbf{k},i} \left[\psi_{\mathbf{k},j}^*(\mathbf{r}) H \psi_{\mathbf{k},i}(\mathbf{r}) - E_{\mathbf{k}} \psi_{\mathbf{k},j}^*(\mathbf{r}) \psi_{\mathbf{k},i}(\mathbf{r}) \right] = 0. \tag{1.6}$$

The Hamiltonian matrix elements are defined as

$$H_{i,j}(\mathbf{k}) = \frac{1}{N} \sum_{\mathbf{R},\mathbf{R}'} e^{i\mathbf{k}\cdot(\mathbf{R}-\mathbf{R}')} \int d\mathbf{r} \phi_j^*(\mathbf{r}-\mathbf{R}') H \phi_i(\mathbf{r}-\mathbf{R}), \tag{1.7}$$

and the overlap matrix elements are

$$S_{i,j}(\mathbf{k}) = \frac{1}{N} \sum_{\mathbf{R},\mathbf{R}'} e^{i\mathbf{k}\cdot(\mathbf{R}-\mathbf{R}')} \int d\mathbf{r} \phi_j^*(\mathbf{r}-\mathbf{R}') \phi_i(\mathbf{r}-\mathbf{R}). \tag{1.8}$$

Owing to the fact that the atomic orbitals centred at the different sites are not orthogonal, the overlap integrals are in general non-zero but small. This non-orthogonality is useful to obtain the electronic spectrum over a wide range of wave vector space. Substituting **Equations 1.7** and **1.8** in **Equation 1.6**, we find the energy spectrum for fixed **k** as

$$E_{\mathbf{k}} = \frac{\sum_{i,j} H_{i,j}(\mathbf{k}) c_{\mathbf{k},j}^* c_{\mathbf{k},i}}{\sum_{i,j} S_{i,j}(\mathbf{k}) c_{\mathbf{k},j}^* c_{\mathbf{k},i}}. \tag{1.9}$$

Minimising the energy using $\partial E_{\mathbf{k}}/\partial c_{\mathbf{k},i}^* = 0$, we obtain the following secular equation:

$$\sum_i \left[H_{i,j}(\mathbf{k}) - E_{\mathbf{k}} S_{i,j}(\mathbf{k}) \right] c_{\mathbf{k},i} = 0. \tag{1.10}$$

1.3.3 TB Model for Monolayer Graphene

Now, we consider a monolayer of graphene as shown in **Figure 1.5** with a unit cell that contains two atoms denoted by A and B. The positions of type A atoms can be generated using a linear combination of basis vectors:

$$A(n_1, n_2) = n_1 \mathbf{a}_1 + n_2 \mathbf{a}_2, \tag{1.11}$$

with lattice vectors $\mathbf{a}_1$ and $\mathbf{a}_2$ given by

$$\mathbf{a}_1 = \begin{pmatrix} \sqrt{3}/2 \\ -1/2 \end{pmatrix} a, \quad \mathbf{a}_2 = \begin{pmatrix} \sqrt{3}/2 \\ 1/2 \end{pmatrix} a, \tag{1.12}$$

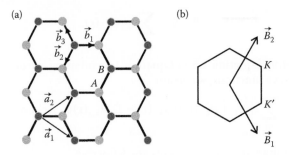

FIGURE 1.5 (a) Monolayer graphene with the two atoms of the two sublattices in different colours. (b) The corresponding Brillouin zone.

where $a = \sqrt{3}a_0 = 0.246$ nm, a_0 is the carbon–carbon atom distance and n_1 and n_2 are two arbitrary integers. The A sublattice is connected to the B sublattice by the three sublattice vectors:

$$\mathbf{b}_1 = \begin{pmatrix} 1/\sqrt{3} \\ 0 \end{pmatrix} a, \quad \mathbf{b}_2 = \begin{pmatrix} -1/(2\sqrt{3}) \\ 1/2 \end{pmatrix} a, \quad \mathbf{b}_3 = \begin{pmatrix} -1/(2\sqrt{3}) \\ -1/2 \end{pmatrix} a. \quad (1.13)$$

The position of the atoms in the B sublattice is given by $\mathbf{B}(m_1, m_2) = m_1\mathbf{a}_1 + m_2\mathbf{a}_2 + \mathbf{b}_1$, where m_1 and m_2 are the two integer numbers. The six second nearest neighbours (NNs) are given by

$$\mathbf{R} = \{\mathbf{a}_1, \mathbf{a}_2, \mathbf{a}_1 - \mathbf{a}_2, \mathbf{a}_2 - \mathbf{a}_1, -\mathbf{a}_1, -\mathbf{a}_2\}. \quad (1.14)$$

The reciprocal lattice of graphene is also a 2D triangular lattice with reciprocal lattice vectors given by

$$\mathbf{B}_1 = \frac{2\pi}{a}\frac{1}{\sqrt{3}}\begin{pmatrix} 1 \\ -\sqrt{3} \end{pmatrix}, \quad \mathbf{B}_2 = \frac{2\pi}{a}\frac{1}{\sqrt{3}}\begin{pmatrix} 1 \\ \sqrt{3} \end{pmatrix}, \quad (1.15)$$

and the hexagonal Brillouin zone is shown in **Figure 1.5b**. Two different corners of the Brillouin zone are defined by the wave vectors:

$$K = \frac{1}{3}\mathbf{B}_1 + \frac{2}{3}\mathbf{B}_2 = \frac{2\pi}{a}\begin{pmatrix} 1/\sqrt{3} \\ 1/3 \end{pmatrix},$$

$$K' = \frac{2}{3}\mathbf{B}_1 + \frac{1}{3}\mathbf{B}_2 = \frac{2\pi}{a}\begin{pmatrix} 1/\sqrt{3} \\ -1/3 \end{pmatrix}. \quad (1.16)$$

These points are inequivalent because they cannot be connected by the reciprocal lattice vectors of **Equation 1.15**. The degeneracy of the high-symmetry K and K' points, called *valley degeneracy*, is a consequence of time inversion symmetry. We can reach the remaining (and equivalent) K and K' points through a rotation over an angle $\pm 2\pi/3$. There are two atoms per unit cell in the graphene lattice, which forms two triangular sublattices A and B and the Bloch wave functions can be built up by a linear combination of the two sublattice Bloch wave functions:

$$\Psi_{\mathbf{k}}(\mathbf{r}) = a_{\mathbf{k}} \Psi_{\mathbf{k}}^{(A)}(\mathbf{r}) + b_{\mathbf{k}} \Psi_{\mathbf{k}}^{(B)}(\mathbf{r}), \tag{1.17}$$

where $a_{\mathbf{k}}$ and $b_{\mathbf{k}}$ are complex functions of the quasi-momentum $\mathbf{k}$. They can be considered as the amplitudes of the sublattice pseudospin. As shown in **Figure 1.6**, the pseudospin *up* state corresponds to the electron density localised at sublattice A and the pseudospin *down* state at sublattice B. Limiting ourselves to the nearest-neighbour approximation, the hopping amplitude is given by

$$
\begin{aligned}
H^{AB}(\mathbf{k}) &= \frac{1}{N} \sum_{\mathbf{R},\mathbf{R}'} e^{i\mathbf{k}\cdot(\mathbf{R}-\mathbf{R}')} \int d\mathbf{r} \phi_{\sigma}^{*}(\mathbf{r} - \mathbf{R}') H \phi_{\sigma'}(\mathbf{r} - \mathbf{R}) \\
&\approx \sum_{m=1}^{3} e^{i\mathbf{k}\cdot\mathbf{b}_m} \int d\mathbf{r} \phi_{\sigma}^{*}(\mathbf{r}) H \phi_{\sigma'}(\mathbf{r} - \mathbf{b}_m) \\
&= \sum_{m=1}^{3} t_m e^{i\mathbf{k}\cdot\mathbf{b}_m} \\
&= (H^{BA}(\mathbf{k}))^{*}, \tag{1.18}
\end{aligned}
$$

where σ corresponds to the different atomic orbitals. The overlap parameters between orbitals on the NN sites are given by

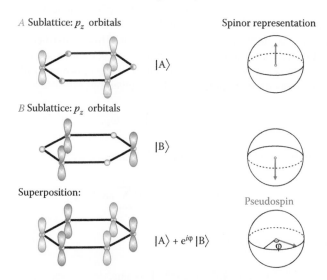

A Sublattice: p_z orbitals

$|A\rangle$

Spinor representation

B Sublattice: p_z orbitals

$|B\rangle$

Superposition:

$|A\rangle + e^{i\varphi} |B\rangle$

Pseudospin

FIGURE 1.6 Pseudospin representation.

$$S^{AB}(\mathbf{k}) \approx \sum_{m=1}^{3} e^{i\mathbf{k}\cdot\mathbf{b}_m} \int d\mathbf{r}\phi_\sigma^*(\mathbf{r})\phi_{\sigma'}(\mathbf{r} - \mathbf{b}_m)$$

$$= \sum_{m=1}^{3} s_m e^{i\mathbf{k}\cdot\mathbf{b}_m}$$

$$= (S^{BA}(\mathbf{k}))^*, \tag{1.19}$$

and the on-site overlap is

$$S^{AA}(\mathbf{k}) = S^{BB}(\mathbf{k}) \approx \int d\mathbf{r}\phi_\sigma^*(\mathbf{r})\phi_{\sigma'}(\mathbf{r}) = \delta_{\sigma\sigma'}. \tag{1.20}$$

We also consider the next-nearest neighbour (NNN) hopping which connects the same sublattice sites:

$$H^{AA}(\mathbf{k}) = H^{BB}(\mathbf{k}) = \frac{1}{N}\sum_{\mathbf{R},\mathbf{R}'} e^{i\mathbf{k}\cdot(\mathbf{R}-\mathbf{R}')} \int d\mathbf{r}\phi_\sigma^*(\mathbf{r} - \mathbf{R}')H\phi_{\sigma'}(\mathbf{r} - \mathbf{R})$$

$$\approx \int d\mathbf{r}\phi_\sigma^*(\mathbf{r})H\phi_{\sigma'}(\mathbf{r}) + \sum_{i=1}^{6} e^{i\mathbf{k}\cdot\mathbf{R}_i} \int d\mathbf{r}\phi_\sigma^*(\mathbf{r})H\phi_{\sigma'}(\mathbf{r} - \mathbf{R}_i)$$

$$= \varepsilon_\tau \delta_{\sigma\sigma'} + 2t' \sum_{i=1}^{3} \cos\mathbf{k}\cdot\mathbf{a}_i, \tag{1.21}$$

where ε_τ is the energy of the atomic orbital $\tau = s, p, d,\ldots$ and t' is the NNN-hopping parameter. For the electronic properties of graphene, we consider the particular case in which only the π bonding created by the transverse p_z orbitals of the carbon atom are taken into account. This is justified because, as shown in **Figure 1.4a**, the p_z orbitals do not interact with the other orbitals, and the low-energy physics is determined by these p_z orbitals only. The off-diagonal elements of the hopping matrix are given by

$$H_{p_z,p_z}^{AB} = \left(H_{p_z,p_z}^{BA}\right)^* = t\omega_\mathbf{k}, \tag{1.22}$$

the diagonal matrix elements including NNN are

$$H_{p_z,p_z}^{AA} = H_{p_z,p_z}^{BB} = \varepsilon_{p_z} + 2t' \sum_{i=1}^{3} \cos\mathbf{k}\cdot\mathbf{a}_i, \tag{1.23}$$

and the overlap elements are

$$S_{p_z,p_z}^{AB} = \left(S_{p_z,p_z}^{AB}\right)^* = s\omega_\mathbf{k},$$
$$S_{p_z,p_z}^{AA} = S_{p_z,p_z}^{BB} = 1. \tag{1.24}$$

The phase function ω_k is a sum over the NN phase factors:

$$\omega_k = \sum_i e^{i k \cdot b_i}. \tag{1.25}$$

Substituting the matrix elements in **Equation 1.9**, we obtain the following secular equation:

$$\det \begin{bmatrix} \varepsilon_{p_z} + 2t' \sum_{i=1}^{3} \cos k \cdot a_i - \varepsilon_k & (t - s\varepsilon_k)\omega_k \\ (t - s\varepsilon_k)\omega_k^* & \varepsilon_{p_z} + 2t' \sum_{i=1}^{3} \cos k \cdot a_i - \varepsilon_k \end{bmatrix} = 0, \tag{1.26}$$

where the two solutions of this equation correspond to the valence ($\alpha = +1$) and the conduction band ($\alpha = -1$). In case NNN interactions are neglected, they are given by

$$\varepsilon_k = \frac{\varepsilon_{p_z} + \alpha t \, |\omega(k)|}{1 + \alpha s \, |\omega(k)|}, \tag{1.27}$$

with

$$|\omega(k)| = \sqrt{3 + 2 \sum_{i \neq j} \cos k \cdot (b_i - b_j)} = \sqrt{3 + f(k)}, \tag{1.28}$$

and

$$f(k) = 2\cos(k_y a) + 4\cos\left(\frac{k_y a}{2}\right)\cos\left(\frac{\sqrt{3}k_x a}{2}\right). \tag{1.29}$$

With NNN interactions included and using the fact that $s \approx 0$, we obtain

$$\varepsilon_k \approx \varepsilon_{p_z} + \alpha t \sqrt{3 + f(k)} + t' f(k). \tag{1.30}$$

The parameters t and t' were estimated using first-principles calculations yielding the value $t \approx -3$ eV for NN hopping and $0.02t < t' < 0.2t$ for the NNN-hopping parameter. The conduction ($\alpha = +1$) and valence ($\alpha = -1$) bands (see **Figure 1.7**) touch each other in the Dirac points K (K') given by the roots of the energy dispersion $\epsilon_k = 0$. The valence band is completely filled and the conduction band is completely empty due to the fact that only one electron (with spin up or spin down) of each carbon atom contributes to the π band. When $t' = 0$, the energy dispersion has electron–hole symmetry $\varepsilon_k^\alpha = -\varepsilon_k^{-\alpha}$. The Hamiltonian has time-reversal symmetry ($H_k = H_{-k}^*$) which means that $\varepsilon_k = \varepsilon_{-k}$ and thus, a pair of Dirac points are formed that have doubly degenerate zero-energy states. This is the so-called valley degeneracy at low energies. The positions of these Dirac points are given by **Equation 1.16**.

Note that also another but equivalent way of writing is used for the TB Hamiltonian (including both nearest and NNN atoms) [15]:

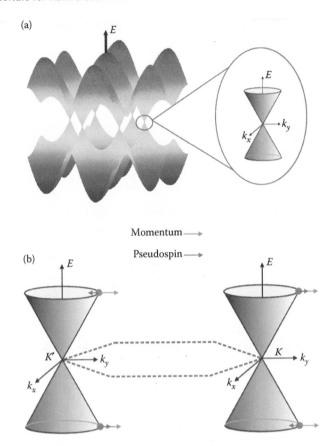

FIGURE 1.7 (a) Graphene band structure and linear spectrum close to the Dirac point. (b) Schematic representation of the helicity in the two valleys including the momentum and pseudospin direction.

$$H = -t \sum_{<i,j>,\sigma} (a^\dagger_{\sigma,i}b_{\sigma,j} + Hc) - t' \sum_{\ll i,j\gg,\sigma} (a^\dagger_{\sigma,i}a_{\sigma,j} + b^\dagger_{\sigma,i}b_{\sigma,j} + H.c.), \quad (1.31)$$

where $a_{i,\sigma}(a^\dagger_{i,\sigma})$ annihilates (creates) an electron with spin σ ($\sigma = \uparrow,\downarrow$) on site R_i on sublattice A (an equivalent definition is used for sublattice B). t and t' are the above-defined NN hopping and NNN- hopping parameters, respectively.

1.3.4 Effective Hamiltonian and Continuum Limit

From **Equations 1.21** and **1.22**, we can define an effective Hamiltonian, neglecting the NNN-hopping parameter t',

$$H = \begin{bmatrix} 0 & t\omega^*(\mathbf{k}) \\ t\omega(\mathbf{k}) & 0 \end{bmatrix}, \quad (1.32)$$

and the eigenstate of this Hamiltonian is given by

$$\Psi_k = \begin{pmatrix} a_k^\alpha \\ b_k^\alpha \end{pmatrix}, \tag{1.33}$$

and we obtain the eigenvalues given by

$$\lambda = \alpha \, |\omega(\mathbf{k})|, \tag{1.34}$$

with corresponding eigenstates:

$$\Psi_k = \frac{e^{i\mathbf{k}\cdot\mathbf{r}}}{\sqrt{2}} \begin{pmatrix} 1 \\ e^{-i\varphi} \end{pmatrix}, \tag{1.35}$$

with $\varphi = \arctan(\mathrm{Im}(\omega(\mathbf{k}))/\mathrm{Re}(\omega(\mathbf{k})))$.

1.3.4.1 Continuum Limit

Expanding the energy dispersion (**Equation 1.30**) close to the K (K') point using $\mathbf{k} = \mathbf{K} + \mathbf{q}$ with $|\mathbf{q}| \ll |\mathbf{K}|$, we find to the first order [15]:

$$\varepsilon_{\mathbf{k},\eta=\pm}^\alpha = \alpha v_F \, |\mathbf{q}| + O[(|\mathbf{q}| \, / \, |\mathbf{K}|)^2], \tag{1.36}$$

with $v_F = \sqrt{3}ta/2\hbar$. The energy ϵ_k to the first approximation is independent of the valley pseudospin η. This results in an effective Hamiltonian that is valid close to the Dirac points:

$$H = \eta\hbar v_F(q_x\sigma_x + \eta q_y\sigma_y), \tag{1.37}$$

where σ_i are the Pauli matrices and $\eta = \pm$ corresponds to the individual $K(+)$ and $K'(-)$ points. This Hamiltonian is simply the 2D Dirac–Weyl (massless Dirac) Hamiltonian with $c \to v_F$. It is useful to use the four-spinor representation with the effective low-energy Hamiltonian as

$$H_q^\eta = \hbar v_F \tau^z \otimes \mathbf{q} \cdot \boldsymbol{\sigma}, \tag{1.38}$$

where

$$\tau^z \otimes \sigma = \begin{pmatrix} \sigma & 0 \\ 0 & -\sigma \end{pmatrix}. \tag{1.39}$$

The eigenstates are given by

$$\Psi_{q,\alpha}^{\eta=+} = \frac{1}{\sqrt{2}} \begin{pmatrix} 1 \\ \alpha e^{i\phi} \\ 0 \\ 0 \end{pmatrix} \quad \Psi_{q,\alpha}^{\eta=-} = \frac{1}{\sqrt{2}} \begin{pmatrix} 0 \\ 0 \\ 1 \\ -\alpha e^{i\phi} \end{pmatrix}, \tag{1.40}$$

where $\phi = \arctan(q_y/q_x)$. The Pauli matrices in **Equation 1.38** represent the sublattice pseudospin with *spin up* for one of the sublattices *A* or *B* and *spin down* for the other one. The valley isospin is described by the Pauli matrix τ^z in the Hamiltonian of **Equation 1.38**. The twofold valley degeneracy is indirectly related to the two different sublattices. The expansion of the spectrum around the Dirac point including t' up to the second order in $|\mathbf{q}|/|\mathbf{K}|$ is given by

$$E_{\pm}(\mathbf{q}) \approx 3t' \pm v_F\,|\mathbf{q}| - \left(\frac{9t'a^2}{4} \pm \frac{3ta^2}{8}\sin(3\theta_\mathbf{q}) \right)|\mathbf{q}|^2. \qquad (1.41)$$

In the presence of t', the position of the Dirac point shifts and it breaks the electron–hole symmetry. Note that up to order $(|\mathbf{q}|/|\mathbf{K}|)^2$, the dispersion depends on the direction in momentum space and has a threefold symmetry. This is the so-called *trigonal warping* of the electronic spectrum.

1.3.5 Helicity and Chirality

Mathematically, an object is chiral if it cannot be mapped to its mirror image using only rotations and translations. A helix and a Möbius strip are two famous examples of 2D chiral objects in 3D space. Other familiar objects which exhibit chiral symmetry are gloves and shoes. In particle physics, chirality is determined by whether the particle transforms in a right- or left-handed representation of the Poincaré group. Helicity is the projection of the spin of the particle onto the direction of its motion.

In graphene, we can write the helicity operator for the sublattice pseudo-spin as

$$h_\mathbf{q} = \sigma \cdot \frac{\mathbf{q}}{|\mathbf{q}|}. \qquad (1.42)$$

As shown in **Figure 1.7b**, the helicity of the electron is right-handed if the pseudospin and momentum are in the same direction and left-handed if the pseudospin and momentum are in opposite directions. For massless particles, chirality is the same as helicity. Electrons in graphene have a well-defined chirality, due to the linear spectrum close to the Dirac points. The helicity operator commutes with the Hamiltonian and has the same eigenstates with eigenvalues $h = \pm 1$:

$$\begin{aligned} h_\mathbf{q}\Psi_K &= \pm\Psi_K, \\ h_\mathbf{q}\Psi_{K'} &= \mp\Psi_{K'}. \end{aligned} \qquad (1.43)$$

It is also possible to present the Hamiltonian using the helicity operator:

$$H = \eta \hbar v_f\,|\mathbf{q}|\,h_\mathbf{q}, \qquad (1.44)$$

where η is the valley degeneracy. The band index $\alpha = \pm 1$ can be determined using $\alpha = \eta h$. The helicity eigenvalue is a good quantum number for energies close to the K and K' points.

1.3.6 Klein Tunnelling

In the presence of a slowly varying potential, chirality remains a good quantum number for an elastic scattering process since such a potential cannot mix the K and K' valleys. When intervalley scattering and the lack of symmetry between sublattices are neglected, a potential barrier shows no reflection for a normal incident electron and the electron is fully transmitted. This effect is known as Klein tunnelling or the Klein paradox [16]. An important consequence of this effect is that no charge carriers in graphene can be confined by electrostatic fields.

In order to present this problem in more detail, we start with the low-energy Dirac Hamiltonian given by **Equation 1.37**. For low energy, we can neglect intervalley scattering and the electrons in K and K' behave independently. Then, we can write the equation as $H\Psi(x,y) = E\Psi(x,y)$, which admits spinor solutions of the form:

$$\Psi(x, y) = \begin{pmatrix} \psi_{\mathrm{I}}(x, y) \\ \psi_{\mathrm{II}}(x, y) \end{pmatrix}.$$ (1.45)

We will consider two systems: (a) a potential step and (b) a potential barrier, along the x direction. Owing to the translational invariance along the y direction, we attempt solutions of the form $\Psi(x,y) = \exp(ik_y y)(\phi_1(x),\phi_2(x))^{\mathrm{T}}$ with T denoting the transpose of the row vector. Then, $\phi_1(x)$ and $\phi_2(x)$ obey the coupled first-order differential equations:

$$-i\hbar v_{\mathrm{F}}\left[d/dx + k_y\right]\phi_2 = \varepsilon\phi_1,$$ (1.46)

$$-i\hbar v_{\mathrm{F}}\left[d/dx - k_y\right]\phi_1 = \varepsilon\phi_2,$$ (1.47)

where $\varepsilon = E - V$. For a step potential $V(x) = V\Theta(x)$ (see **Figure 1.8a**), the solutions of this coupled set of differential equations in the two regions are

$$\Psi_1(x, y) = e^{ik_y y}\begin{pmatrix} e^{ik_1 x} + re^{-ik_1 x} \\ s_i\left[e^{ik_1 x + i\theta} - re^{-ik_1 x - i\theta}\right] \end{pmatrix},$$ (1.48)

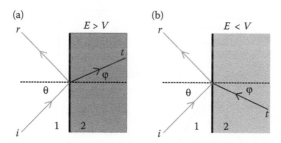

(a) E > V (b) E < V

FIGURE 1.8 (a) Schematics of the transmission through a potential step for an electron with $E > V$ or positive refraction index. (b) Same as (a) but for a negative refraction index or $E < V$.

$$\Psi_2(x, y) = e^{ik_y y} \begin{pmatrix} te^{\pm ik_2 x} \\ \pm s_i te^{\pm(ik_2 x + i\phi)} \end{pmatrix}, \tag{1.49}$$

where $s_i = \text{sgn}(E - V_i)$, $k_i = \sqrt{\varepsilon_i^2 - k_y^2}$, with $i = 1, 2$, $\tan \theta = k_y/k_1$ and $\tan \phi = k_y/k_2$. When matching the waves at $x = 0$, we obtain the transmission amplitude t and probability $T = tt^*$:

$$t_{E<V} = \frac{2\cos\theta}{e^{-i\theta} + e^{-i\phi}}, \quad T_{E<V} = \frac{2\cos^2\theta}{1 + \cos(\theta - \phi)}, \tag{1.50}$$

and the $E > V$ result is obtained by $\phi \to -\phi$. Using the fact that the momentum of the electron in the y-direction is continuous at $x = 0$, we obtain Snell's law as $\sin \theta = (1 - V/E)\sin \phi$. When $E < V$, the step potential acts like a medium with *negative refraction index* (see **Figure 1.8b**). This means that a divergent electron beam in region 1 that hits the potential step gets focussed again in region 2 [17].

Next, we consider a potential barrier with width W, as shown in **Figure 1.9**. From the point of view of optics, it is like a medium with refraction index $1 - V/E$. When we inject a wave with the incident angle θ, it splits into transmitted and reflected waves. The transmitted wave inside the barrier will resonate between the two edges at $x = 0$ and W, as shown in **Figure 1.10a**. Analogous with optical waves, the difference in the optical paths along the barrier is

$$\Delta L = \left(1 - \frac{V}{E}\right)(BC + CD) - (BN), \tag{1.51}$$

where $BC = CD = (W/\cos\phi)$ and $BN = 2W \tan \phi \sin \theta$.

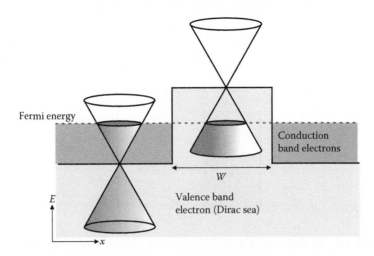

FIGURE 1.9 Illustration of the Klein tunnelling through a potential barrier.

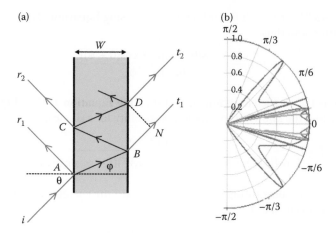

FIGURE 1.10 (a) Schematic figure to show the multiple reflections in a barrier. (b) Transmission probability for Klein tunnelling in graphene as a function of the angle for an incoming electron with an energy of 800 meV (blue), 250 meV (yellow) and 150 meV (purple). The width of the potential barrier is 100 nm and its height is 200 meV.

Using $\sin \theta = (1 - V/E)\sin \phi$, we obtain

$$\Delta L = 2(1 - V/E)W \cos\phi. \tag{1.52}$$

The total transmission is given by $T = |t_1|^2 + |t_2|^2 + 2|t_1||t_2|\cos \delta$, with the corresponding phase difference:

$$\delta = k_1\Delta L = 2k_1(1 - V/E)W \cos\phi. \tag{1.53}$$

The transmission is maximum when $|\delta| = 0, 2\pi, 4\pi,...$ and minimum when $|\delta| = \pi, 3\pi,....$

In order to obtain the total transmission, we introduce r and t to be the reflection and transmission coefficient for the potential step outside the barrier and r' and t' to be the corresponding coefficients inside the barrier. The different combinations of transmitted waves through the barrier are

$$tt', tt'r'^2e^{i\delta}, ..., tt'r'^{2(n-1)}e^{i(n-1)\delta}, ... \tag{1.54}$$

and we obtain the total transmission amplitude by summing over all terms:

$$
\begin{aligned}
t_{\text{tot}} &= tt'\left(1 + r'^2e^{i\delta} + \cdots + r'^{2(n-1)}e^{i(n-1)\delta} + \cdots\right) \\
&= \frac{tt'}{1 - r'^2e^{i\delta}}.
\end{aligned}
\tag{1.55}
$$

The corresponding transmission probability $T_{\text{tot}} = t_{\text{tot}}t_{\text{tot}}^*$ is

$$T_{\text{tot}} = \frac{1}{1 + F\sin^2(\delta/2)}, \tag{1.56}$$

here $F = 4R/(1 - R)^2$ with $R = |r'|^2$ and $T = tt'$. Using **Equations 1.48** and **1.49**, we obtain r' and thus:

$$F = \frac{[1 - \cos(\theta - \phi)][1 + \cos(\theta + \phi)]}{\cos^2 \theta \cos^2 \phi}, \tag{1.57}$$

where F is the coefficient of finesse. Substituting **Equation 1.57** in **Equation 1.56**, we find the known transmission probability for a single barrier [18]:

$$T = \frac{\cos^2 \theta \cos^2 \phi}{\left[\cos(k_2 W)\cos\theta\cos\phi\right]^2 + \sin^2(k_2 W)(1 - ss_0 \sin\theta\sin\phi)^2}, \tag{1.58}$$

with $\delta = 2k_2 W$. Substituting $k_2 W = n\pi$, we obtain the energies E at which the resonances occur (i.e. $T = 1$):

$$E = V \pm [k_y^2 + n^2\pi^2/W^2]^{1/2}. \tag{1.59}$$

For a normal incident wave with $\theta = 0$, we find $T = 1$ as expected from the Klein paradox. The angular dependence of the transmission probability is shown in **Figure 1.10b** for a potential barrier with a width of 100 nm and a height of 200 meV for three different energies of the incoming electron. Note that $T(\theta) = T(-\theta)$, and for values of $k_2 W$ satisfying the relation $k_2 W = n\pi$, with n as an integer, the barrier becomes completely transparent since $T = 1$ becomes independent of the value of θ.

1.3.7 High Mobility, Minimum Conductivity and Universal Optical Conductivity

Transport measurements in graphene show a remarkably high electron mobility at room temperature [2]. The nearly symmetric conductance measured experimentally shows that mobilities for holes and electrons should be nearly the same [2] and are almost independent of temperature [11,12], which implies that the dominant scattering mechanism is defect scattering.

In spite of the vanishing carrier density near the Dirac points, graphene shows a minimum conductivity close to the conductivity quantum $4e^2/h$ per carrier type [11]. A number of theories [19–24] have been formulated to describe the minimum conductivity but still its physical origin is not completely understood. Most theories suggest that the minimum conductivity should be $4e^2/\pi h$, which in comparison with the experimental measurements is π times smaller [2], and depends on the impurity concentration [25]. This disagreement is well known as 'the mystery of the missing π' [2]. Close to the Dirac point, graphene conducts by randomly distributed electron and hole puddles due to the presence of rippling and defects. Theory did not take into account this inhomogeneity, and possibly this can explain the higher conductivity in the experimental data (see **Figure 1.11**).

Evidence for the universality of the optical (or high-frequency) conductivity in graphene has also been observed [26,27]. These experiments show that at half-filling and small temperatures, this optical conductivity is essentially constant

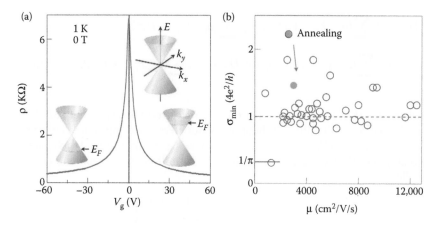

FIGURE 1.11 (a) Ambipolar electric field effect in single-layer graphene. The insets show its conical low-energy spectrum. (b) Minimum conductivity of graphene, independent of the carrier mobility. (Adapted from A. K. Geim and K. S. Novoselov, *Nat. Mater.* **6**, 183, 2007.)

and equal, up to a few percent, to $\pi e^2/2h$. Such value only depends on the von Klitzing constant and not on the material parameters, like the Fermi velocity. Theoretically, the computation of the optical conductivity in the absence of interactions gives exactly this value $\pi e^2/2h$, both in the idealised Dirac description [28,29] and even beyond the linear-spectrum approach [30]. The universality of this optical conductivity also implies that observable quantities such as graphene's optical transmittance and reflectance are also universal [31].

1.3.8 Bilayer Graphene

Bilayer graphene consists of two graphene layers on top of each other. Bilayer graphene has a zero band gap and finite DOS at the Fermi level and thus behaves like a metal. A band gap can be introduced by applying a perpendicular electric field. The ground-state structure of bilayer graphene corresponds to an AB stacking shown in **Figure 1.12**. The TB Hamiltonian for bilayer graphene can be written as provided in References 15 and 32

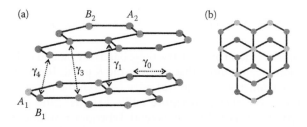

FIGURE 1.12 Lattice structure of bilayer graphene: (a) side view with indication of the hopping parameters, and (b) top view.

$$H = -\gamma_0 \sum_{m,i} (a^\dagger_{m,i,s} b_{m,i,s} + H.c.) - \gamma_1 \sum_{j,s} (a^\dagger_{1,j,s} a_{2,j,s} + H.c.)$$

$$-\gamma_4 \sum_{j,s} (a^\dagger_{1,j,s} b_{2,j,s} + a^\dagger_{2,j,s} b_{1,j,s} + H.c.) - \gamma_3 \sum_{j,s} (b^\dagger_{1,j,s} b_{2,j,s} + H.c.), \qquad (1.60)$$

where $a_{m,i,s}(b_{i,m,s})$ is the annihilation of an electron with spin s at site R_i of sublattice $A(B)$. The different hopping parameters γ_i are approximately given by $\gamma_0 = t$, $\gamma_1 \approx 0.4$ eV, $\gamma_3 \approx 0.3$ eV and $\gamma_4 = 0.04$ eV, corresponding to their values in graphite. Neglecting γ_4, we can write the Hamiltonian close to the K and K' valleys as

$$H = \begin{pmatrix} -\Delta/2 & v_3\pi & 0 & v_F\pi \\ v_3\pi^\dagger & \Delta/2 & v_F\pi & 0 \\ 0 & v_F\pi^\dagger & \Delta/2 & \xi\gamma_1 \\ v_F\pi & 0 & \xi\gamma_1 & -\Delta/2 \end{pmatrix}, \qquad (1.61)$$

where $v_F = \sqrt{3}a\gamma_0/2\hbar$ is the in-plane velocity, $v_3 = (\sqrt{3}/2)a\gamma_3/\hbar$ is an effective velocity with $v_3 = v$, $\pi = p_x + ip_y$, $\pi^\dagger = p_x - ip_y$, $\mathbf{P} = (p_x, p_y) = p(\cos\phi, \sin\phi)$ is the momentum and $\xi = \pm 1$ for the two different valleys K and K', $\Delta = \varepsilon_2 - \varepsilon_1$ is the difference between on-site energies in the two layers. Corresponding eigenstates are

$$\Psi_K = \begin{pmatrix} \psi_{A_1} \\ \psi_{B_2} \\ \psi_{A_2} \\ \psi_{B_1} \end{pmatrix}, \quad \Psi_{K'} = \begin{pmatrix} \psi_{B_2} \\ \psi_{A_1} \\ \psi_{B_1} \\ \psi_{A_2} \end{pmatrix}. \qquad (1.62)$$

From the Hamiltonian in **Equation 1.61**, one obtains the energy band as

$$\varepsilon^2_{\pm,\alpha} = \frac{\gamma^2_1}{2} + \frac{\Delta^2}{4} + \left(v^2 + \frac{v^2_3}{2}\right)p^2 + (-1)^\alpha\sqrt{\Gamma}, \qquad (1.63)$$

$$\Gamma = \frac{1}{4}(\gamma^2_1 - v^2_3 p^2)^2 + v^2_F p^2[\gamma^2_1 + \Delta^2 + v^2_3 p^2] + 2\xi\gamma_1 v_3 v^2_F p^3 \cos 3\phi. \qquad (1.64)$$

In the absence of layer asymmetry, these bands are plotted in **Figure 1.13a**. The dispersions $\varepsilon_{\pm 1}$ are those that touch at the K point. In the presence of a finite-valley asymmetry Δ, a gap will open between the upper band $\varepsilon_{+,1}$ and the lower one $\varepsilon_{-,1}$ as shown in **Figure 1.13b**. External gates can be used to place the separate layers at different potential energies, resulting in a non-zero Δ and thus an energy gap. Such an external potential will also induce different charges in both layers. This has been studied within a self-consistent Hartree approximation in a TB model in Reference 33.

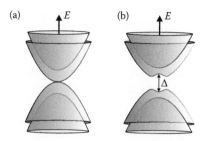

FIGURE 1.13 Electronic band of bilayer graphene for (a) $\Delta = 0$ and (b) $\Delta \neq 0$.

In order to study the transport properties of bilayer graphene, it is useful to find an effective low-energy Hamiltonian. Such an effective Hamiltonian can be obtained because γ_1 is large in comparison with γ_3 and γ_4. Using this approximation, we obtain a two-component Hamiltonian describing the effective hopping between the two sites A_1 and B_2 as

$$
H = \frac{1}{2m}\begin{pmatrix} 0 & \pi^{\dagger 2} \\ \pi^2 & 0 \end{pmatrix} + \xi v_3 \begin{pmatrix} 0 & \pi^{\dagger} \\ \pi & 0 \end{pmatrix}
$$
$$
- \xi \Delta \left[\frac{1}{2}\begin{pmatrix} 1 & 0 \\ 0 & -1 \end{pmatrix} - \frac{v_F^2}{\gamma_1^2}\begin{pmatrix} \pi^{\dagger}\pi & 0 \\ 0 & -\pi\pi^{\dagger} \end{pmatrix} \right], \tag{1.65}
$$

where $m = \gamma_1 / 2v_F^2$. This Hamiltonian is applicable when $\varepsilon < \gamma_1/4$. The eigenstates for the two different valleys K and K' are given by

$$
\Psi^K_{\xi=+1} = \begin{pmatrix} \psi_{A_1} \\ \psi_{B_2} \end{pmatrix}, \quad \Psi^{K'}_{\xi=-1} = \begin{pmatrix} \psi_{B_1} \\ \psi_{A_1} \end{pmatrix}. \tag{1.66}
$$

Note that the components are reversed for different valleys.

1.3.9 Multilayer Graphene

Continuing stacking graphene layers will finally lead to graphite. The lowest-energy stacking sequence is the AB Bernal stacking. While graphene is a zero-gap semiconductor, graphite is a semimetal with a small band overlap of ≈41 meV. The evolution of the band structure from graphene to graphite has been studied in Reference 34. While bilayer graphene has already an extremely small band overlap (<0.2 meV) if up to γ_4 is taken into account, the band overlap in AB-stacked multilayer graphene increases with an increasing number of layers. It is within 10% of the band overlap of graphite for 11 or more layers. The linear spectrum at the Dirac points is present in case of an odd number of layers [35].

Also, ABC rhombohedral stacking of graphene multilayers has been experimentally realised [36]. The low-energy spectrum is given by (only taking into account γ_0 and γ_1)

$$H = \frac{(\hbar v_F)^n}{(-\gamma_1)^{n-1}} \begin{pmatrix} 0 & \pi^{\dagger n} \\ \pi^n & 0 \end{pmatrix}, \tag{1.67}$$

with n the number of layers [37]. This is a very nice example of the tuning of the energy spectrum, that is, $E \sim k^n$, by changing the number of graphene layers.

As in bilayer graphene, the band gap can be tuned in these multilayer graphene systems by applying an external perpendicular electric field. It has been shown that this induced band gap strongly depends on the stacking order [38,39].

1.4 In the Presence of a Magnetic Field

1.4.1 Homogeneous Magnetic Field

An electron in a single graphene layer in the presence of a perpendicular magnetic field $B(x)$ that may vary along the x direction is described by the Hamiltonian:

$$H = v_F \tau_z \otimes \sigma \cdot \Pi. \tag{1.68}$$

Here, we replaced the momentum by its gauge-invariant form [40] also called the *minimal substitution*,

$$\mathbf{p} \rightarrow \Pi = \mathbf{p} + e\mathbf{A}(\mathbf{r}), \tag{1.69}$$

where $\mathbf{p}$ is the momentum operator and $\mathbf{A}(x)$ is the vector potential that generates the magnetic field $\mathbf{B} = \nabla \times \mathbf{A}$. Neither the $\mathbf{p}$ nor the $\mathbf{A}$ are gauge invariant. Adding a gradient of an arbitrary function ∇f to the vector potential $\mathbf{A}$ results in the same magnetic field. Under a gauge transformation, one has to transform $\mathbf{p} \rightarrow \mathbf{p} - e\nabla f$ in order to have a gauge invariant Π. In the context of electrons on a lattice, the transformation **Equation 1.69** is valid as long as the lattice spacing a is much smaller than the magnetic length ℓ_B,

$$\ell_B = \sqrt{\frac{eB}{\hbar}}. \tag{1.70}$$

The lattice spacing a is approximately ~ 0.142 nm and $\ell_B = 26$ nm$/\sqrt{B[T]}$ and thus, $a = \lambda_B$ is satisfied for magnetic fields as large as 10^3 T. Let us now consider such a Dirac electron in the presence of a homogeneous magnetic field B_0. Let us work in the Landau gauge, where $\mathbf{A}(x) = (0, Bx, 0)$. We can write **Equation 1.68** explicitly as

$$H = -iv_F \hbar \begin{pmatrix} 0 & \partial_x - i\partial_y + eBx/\hbar \\ \partial_x + i\partial_y - eBx/\hbar & 0 \end{pmatrix}. \tag{1.71}$$

Then, the equation $H\Psi(x,y) = E\Psi(x,y)$ admits the following solutions:

$$\Psi(x, y) = \begin{pmatrix} \psi_A(x, y) \\ \psi_B(x, y) \end{pmatrix}, \tag{1.72}$$

with $\psi_A(x,y)$, $\psi_B(x,y)$ obeying the coupled set of equations:

$$i\left[\frac{\partial}{\partial x} - i\frac{\partial}{\partial y} + \frac{eB}{\hbar}x\right]\psi_B + E\psi_A = 0, \tag{1.73}$$

$$i\left[\frac{\partial}{\partial x} + i\frac{\partial}{\partial y} - \frac{eB}{\hbar}x\right]\psi_A + E\psi_B = 0. \tag{1.74}$$

For such a gauge, $[H, p_y] = 0$, and p_y is a good quantum number. Owing to the translational invariance along the y direction, we can assume solutions of the form $\Psi(x,y) = \exp(i\,k_y y)(a(x),b(x))^T$, with T denoting the transpose of the row vector. Then, **Equations 1.73** and **1.74** take the following form:

$$-i\hbar v_F\left[\frac{d}{dx} + (k_y + x / \ell_B)\right]b = Ea, \tag{1.75}$$

$$-i\hbar v_F\left[\frac{d}{dx} - (k_y + x/\ell_B)\right]a = Eb. \tag{1.76}$$

Operating on **Equations 1.75** and **1.76** with $-i(d/dx - (k_y + x))$ gives

$$\left[\frac{d^2}{dx^2} - (k_y + x)^2 \mp 1 + \left(\frac{\ell_B^2}{\hbar^2 v_F^2}\right)E^2\right]c_{\mp} = 0, \tag{1.77}$$

where $c_- = a$ and $c_+ = b$. Solutions of **Equation 1.77** are the well-known Hermite polynomials $H(x)$. For $c_- = a$, the wave function is

$$a(x) = \exp(-z^2/4)H_{E^2/2-1}(x + k_y),$$

and the energy spectrum is

$$E_n = \pm\frac{\hbar v_F}{\ell_B}\sqrt{2(n + 1)}, \tag{1.78}$$

with the lowest level:

$$E_0 = \sqrt{2}\frac{\hbar v_F}{\ell_B}. \tag{1.79}$$

Repeating this procedure for $c_+ = b$ gives $b(x) = \exp(-z^2/4)H_{E^2/2}(x + k_y)$ with the spectrum:

$$E_n = \pm\frac{\hbar v_F}{\ell_B}\sqrt{2n} \propto \sqrt{nB}. \tag{1.80}$$

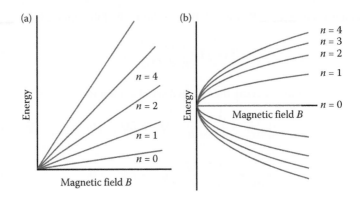

FIGURE 1.14 LL as a function of the magnetic field. (a) Non-relativistic case with $E_n \propto (n + 1/2)B$. (b) Dirac electron case with $E_n \propto \sqrt{nB}$.

Note the difference of these spectra with the one for Schrödinger electrons, $E_n = \hbar\omega_c(n + 1/2) \propto (n + 1/2)B$, which depends linearly on the magnetic field (see **Figure 1.14** for comparison). Actually, this expression for the LLs coincides with the previous one of **Equation 1.78**, except for the lowest level which is now $E_0 = 0$. In other words, the lowest LL for a Dirac electron has a twice-smaller degeneracy as compared with the other LLs. (Similar results hold for the holes.) These *relativistic* LLs in graphene have been observed experimentally using scanning tunnelling spectroscopy, which is used to provide information about the density of electrons in a sample as a function of their energy. In this method, one shines monochromatic light on the sample and measures the intensity of the transmitted light [41–43].

1.4.2 LLs in Bilayer Graphene

Consider a homogeneous magnetic field B_0 normal to the 2D plane (x, y) of bilayer graphene. Let us use the Landau gauge for the vector potential $A(x) = (0, B_0 x, 0)$ and make the change $\Pi \rightarrow p + eA$. Based on **Equation 1.61**, the one-electron Hamiltonian for bilayer graphene, in which not only γ_4 but also γ_3 is neglected, can be written as

$$H = \begin{pmatrix} V_1 & \Pi & t & 0 \\ \Pi^\dagger & V_1 & 0 & 0 \\ t & 0 & V_2 & \Pi^\dagger \\ 0 & 0 & \Pi & V_2 \end{pmatrix}, \tag{1.81}$$

where $\Pi = v_F[p_x + i(p_y + eA)]$, V_1 and V_2 are the potentials at the two layers and t is the tunnel coupling between the layers, assumed to be constant. This Hamiltonian is valid near the Dirac point K or K'. Thus, scattering between the K and K' valleys is neglected. This scattering was shown [44] to be negligible for fields below 10^4 T in single-layer graphene and this is also expected to be the case in bilayer graphene. For more details as well as the motivation why we may neglect trigonal warping, we refer to Reference 12. To simplify the notation, we introduce the length scale $\ell_B = [\hbar/eB_0]^{1/2}$ and the energy scale $E_0 = \hbar v_F/\ell_B$. Similar to the Dirac Hamiltonian, the Hamiltonian in **Equation 1.81** commutes with p_y and therefore is a conserved quantity. This allows us to write $\Psi(x,y) = \Phi(x)\exp(ik_y y)$ and solve the equation $H\Psi(x,y) = E\Psi(x,y)$ for the wave function $\Psi(x, y) = (\phi_a(x), \phi_b(x), \phi_c(x), \phi_d(x))^T \exp(ik_y y)$ with T denoting the transpose. Then, the components of $\Psi(x,y)$ obey the following set of coupled differential equations:

$$\begin{cases} -i(d/dx - (k_y + x))\phi_b + t\phi_c = (E - V_1)\phi_a, \\ -i(d/dx + (k_y + x))\phi_a = (E - V_1)\phi_b, \\ -i(d/dx + (k_y + x))\phi_d + t\phi_a = (E - V_2)\phi_c, \\ -i(d/dx - (k_y + x))\phi_c = (E - V_2)\phi_d. \end{cases} \tag{1.82}$$

Setting $V_0 = (V_1 + V_2)/2$, $\Delta V = V_1 - V_2$, $\delta = \Delta V/2$ and $\epsilon = E - V_0$, **Equations 1.81** can be decoupled. The result for ϕ_a is

$$[d^2/dz^2 - z^2/4 + \gamma_\pm/2]\phi_a = 0, \tag{1.83}$$

where $\gamma_\pm = \epsilon^2 + \delta^2 \pm [(1 - 2\delta)^2 + (\epsilon^2 - \delta^2)t^2]^{1/2}$ and $\sqrt{2}(x + k_y) = z$. The solutions of **Equation 1.83** can be written in terms of Weber functions. For an asymptotically vanishing wave function for $z \to \infty$, we define $\phi_a(z) = e^{-z^2/4}g(z)$ and substitute it in **Equation 1.83**. For $\delta = 0$ and using standard power-series procedures, we complete the solution and find the energy spectrum:

$$\epsilon_{n,\pm} = \pm\left[2n + 1 + \frac{t^2}{2} \pm \left[\frac{t^4}{4} + (2n + 1)t^2 + 1 \right]^{1/2} \right]^{1/2}, \tag{1.84}$$

where n is an integer, the LL index. For small interlayer tunnel parameter t', this becomes: $E_n = \pm(\hbar v_F/\ell_B)\sqrt{2(n + 1)}$. Note the similarity with the spectrum for single-layer graphene, $E_n = \pm(\hbar v_F/\ell_B)\sqrt{2n}$, and the difference from that for the usual electrons with a parabolic energy–momentum relation: $E_n = \hbar\omega_c(n + 1/2)$ consisting of equidistant LLs. For $t \to 0$, **Equation 1.84** reduces to that of two *decoupled* layers with LLs $E_n = \pm(\hbar v_F/\ell_B)\sqrt{2n + 1} \pm 1$.

1.4.3 Anomalous Quantum Hall Effect

In 1981, Klaus von Klitzing et al. measured the Hall resistance and the magneto resistance of a 2DEG [45]. They found that the Hall resistance exhibits plateaus that have quantised values h/ne^2, where n is an integer. In 1985, Klaus von Klitzing received the Nobel Prize for this discovery. Later in 1981, Laughlin showed [46] that edge effects and broadening of the LLs due to disorder have no influence on the accuracy of the Hall quantisation.

Graphene shows an interesting behaviour in the presence of a magnetic field and with respect to the conductivity quantisation: it displays an anomalous quantum Hall effect with the sequence of steps shifted by 1/2 with respect to the standard sequence, and with an additional factor of 4. The number of occupied states are $N = 2 \times (2n + 1)$, since for each LL index n, there are in total $2n$ states for K and K' valleys and +1 is due to the zero mode that is shared between two Dirac points. Thus, in graphene, the Hall conductivity is

$$\sigma_{xy} = \frac{I}{V_{\mathrm{H}}} = \pm 4\left(n + \frac{1}{2}\right)\frac{e^2}{h}. \qquad (1.85)$$

Then, the factor 4 is a consequence of the degeneracy of the two valley states and two spin states. This unconventional quantum Hall effect can even be measured at room temperature [2]. This anomalous behaviour is a direct result of the emergent massless Dirac electrons in graphene [11]. In a magnetic field, their spectrum has an LL with energy precisely at the Dirac point. This level is a consequence of the Atiyah–Singer index theorem and is half-filled in neutral graphene [14], leading to the +1/2 in the Hall conductivity. This result has been experimentally observed [11,12] as shown in **Figure 1.15**.

The quantum Hall effect in bilayer graphene leads to a Hall conductivity given by

$$\sigma_{xy} = \pm 4n\frac{e^2}{h}, \qquad (1.86)$$

that is, with only one of the two anomalies. The integer quantum Hall effect in bilayer graphene has been observed experimentally in Reference 47

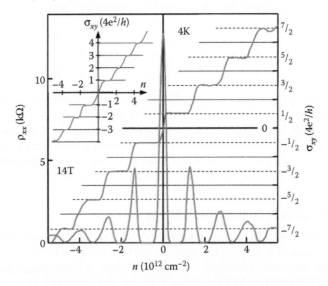

FIGURE 1.15 Quantum Hall effect (QHE) for massless Dirac fermions. Hall conductivity and longitudinal resistivity of graphene as a function of their concentration at $B = 14$ T. (Adapted from A. K. Geim and K. S. Novoselov, *Nat. Mater.* **6**, 183, 2007.)

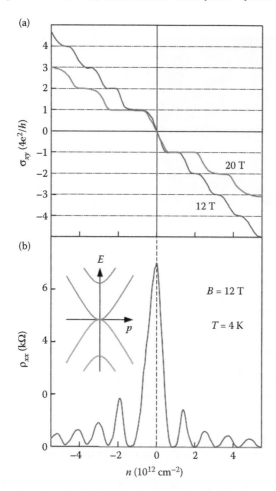

FIGURE 1.16 QHE in bilayer graphene: (a) σ_{xy} and (b) ρ_{xx} are plotted as a function of the carrier density at a fixed magnetic field and temperature. (Adapted from K. S. Novoselov et al., *Nat. Phys.* **2**, 177, 2006.)

(see **Figure 1.16**). Interestingly, concerning the second anomaly, the first plateau at $n = 0$ is absent, indicating that bilayer graphene stays metallic at the Dirac point.

1.4.4 Gauge Fields Induced by Lattice Deformation

Geometrical deformations of the graphene lattice result in local strains which can induce large pseudo-magnetic fields and produce a pseudo-quantum Hall effect [48–50]. It was reported [51] that nanobubbles grown on a Pt(111) surface can induce a pseudo-magnetic field exceeding 300 T (see **Figure 1.17**). It was shown that different graphene bubbles can be observed on different substrates [52,53] and furthermore it is possible to control the curvature by applying an

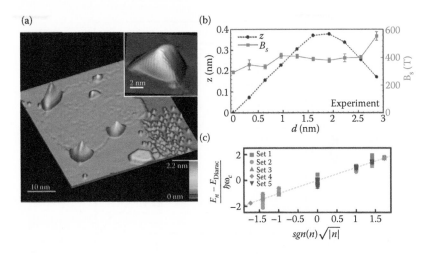

FIGURE 1.17 (a) Nanobubbles of graphene monolayer on Pt(111). (b) Experimental topographic line scan and experimentally determined B_s and deformation z. (c) Normalised peak energy versus $sgn(n)\sqrt{|n|}$. (Adapted from N. Levy et al., *Science* **329**, 544, 2010.)

external electric field [53]. With such large strain-induced pseudo-magnetic fields, it becomes possible to control the electronic properties of graphene, which is called 'strain engineering'. Below, we briefly describe the gauge field induced by an elastic strain and ripples.

1.4.4.1 Deformation
The TB Hamiltonian for electrons in graphene that include both nearest and NNN atoms can be written in the form of **Equation 1.87**:

$$H = -t \sum_{<i,j>,\sigma} (a_{\sigma,i}^\dagger b_{\sigma,j} + H.c.) - t' \sum_{<i,j>,\sigma} (a_{\sigma,i}^\dagger a_{\sigma,j} + b_{\sigma,i}^\dagger b_{\sigma,j} + H.c.). \quad (1.87)$$

By applying strain, we change the distance or angles between the p_z orbitals of the different sites. Then, the hopping parameters will change between different sites and a new term will appear in the original Hamiltonian (**1.87**):

$$H_{od} = \sum_{i,j} \left\{ \delta t_{i,j}^{(a,b)} (a_i^\dagger b_j + h.c.) + \delta t_{i,j}^{(a,a)} (a_i^\dagger a_j + b_j^\dagger b_j) \right\}. \quad (1.88)$$

A Fourier transformation of this extra term due to deformation leads to

$$
\begin{aligned}
H_{od} = \sum_{k,k'} a_k^\dagger b_{k'} \sum_{i,\delta_{ab}} \delta t_i^{(ab)} e^{i(k-k')\cdot R_i - i\delta_{aa}\cdot k'} + H.c. \\
+ (a_k^\dagger a_{k'} + b_k^\dagger b_{k'}) \sum_{i,\delta_{ab}} \delta t_i^{(aa)} e^{i(k-k')\cdot R_i - i\delta_{ab}\cdot k'},
\end{aligned} \quad (1.89)
$$

where $\delta t_{i,j}^{(a,b)}$ $\left(\delta t_{i,j}^{(a,a)}\right)$ is the change of the hopping energy between orbitals on lattice sites $\mathbf{R}_i$ and $\mathbf{R}_j$ on the same (different) sublattices (we have written $\mathbf{R}_j = \mathbf{R}_i + \delta$, where δ_{ab} is the NN vector and δ_{aa} is the NNN vector). With projection of the operators close to K and K', we find

$$a_n \simeq e^{-i\mathbf{K}\cdot\mathbf{R}_n}a_{1,n} + e^{-i\mathbf{K}'\cdot\mathbf{R}_n}a_{2,n},$$
$$b_n \simeq e^{-i\mathbf{K}\cdot\mathbf{R}_n}b_{1,n} + e^{-i\mathbf{K}'\cdot\mathbf{R}_n}b_{2,n}. \tag{1.90}$$

To prevent valley scattering, we assume δt_{ij} is smooth over the lattice-spacing scale, and does not have a Fourier component with momentum $\mathbf{K} - \mathbf{K}'$. Then, we will rewrite **Equation 1.89** in real space for the K point as

$$H_{od} = \int d^2r\{A(\mathbf{r})a_1^\dagger(\mathbf{r})b_1(\mathbf{r}) + H.c. + \phi(\mathbf{r})[a_1^\dagger(\mathbf{r})a_1(\mathbf{r}) + b_1^\dagger(\mathbf{r})b_1(\mathbf{r})]\}. \tag{1.91}$$

where

$$A(\mathbf{r}) = \sum_{\delta_{ab}} \delta t^{(ab)}(\mathbf{r})e^{-i\delta_{ab}\cdot\mathbf{K}},$$
$$\phi(\mathbf{r}) = \sum_{\delta_{aa}} \delta t^{(aa)}(\mathbf{r})e^{-i\delta_{aa}\cdot\mathbf{K}}. \tag{1.92}$$

For the other valley, we have a similar expression except we have to replace A by A^*. A is complex due to the lack of inversion symmetry for NN hopping and we have

$$A(\mathbf{r}) = A_x(\mathbf{r}) + iA_y(\mathbf{r}). \tag{1.93}$$

Due to the inversion symmetry of the two triangular sublattices that make up the honeycomb lattice, we have $\phi(\mathbf{r}) = \phi^*(\mathbf{r})$. Finally, we can rewrite **Equation 1.91** as

$$H_{od} = \int d^2r\left[\hat{\Psi}_1^\dagger(\mathbf{r})\sigma\cdot A(\mathbf{r})\hat{\Psi}_1(\mathbf{r}) + \phi(\mathbf{r})\hat{\Psi}_1^\dagger(\mathbf{r})\hat{\Psi}_1(\mathbf{r})\right]. \tag{1.94}$$

The effective magnetic field in the presence of the vector potential is given by $\mathbf{B} = (c/ev_F)\nabla \times A$. A magnetic field results in a broken time-reversal symmetry, although the original problem was time-reversal invariant. This broken time-reversal symmetry is not real since we limited our discussion to one of the Dirac cones. The sign of the strain-induced magnetic field in the two cones is opposite and the two cones are related to each other by time-reversal symmetry. As a consequence, the total system preserves time-reversal symmetry.

1.4.4.2 Elastic Strain

The elastic free energy for graphene can be written in terms of the in-plane displacement $\mathbf{u}(r) = (u_x, u_y)$ as

$$F[\mathbf{u}] = \frac{1}{2}\int d^2r\left[(B - G)\left(\sum_{i=1,2} u_{ii}\right)^2 + 2G\sum_{i,j=1,2} u_{ij}^2\right], \tag{1.95}$$

where B is the bulk modulus, G is the shear modulus and

$$u_{ij} = \frac{1}{2}\left(\frac{\partial u_i}{\partial x_j} + \frac{\partial u_j}{\partial x_i}\right), \tag{1.96}$$

is the strain tensor ($x_1 = x$ and $x_2 = y$).

There are many different types of static deformations of the honeycomb lattice that can affect the propagation of Dirac fermions. The simplest one is due to changes in the area of the unit cell due to either dilation or contraction. Changes in the unit-cell area lead to local changes in the density of electrons and, therefore, local changes in the chemical potential in the system. In this case, the effective potential is given by

$$\phi_{dp} = g(u_{xx} + u_{yy}), \tag{1.97}$$

and their effect is diagonal in the sublattice index.

The NN hopping depends on the length of the carbon bond. Hence, elastic strains that modify the relative orientation of atoms also lead to an effective gauge field, which acts on each K point separately, as first discussed in relation to carbon nanotubes. Consider two carbon atoms located in two different sublattices in the same unit cell at $\mathbf{R}_i$. The change in the local bond length can be written as

$$\delta u_i = \frac{\delta_{ab}}{a} \cdot [\mathbf{u}_A(\mathbf{R}_i) - \mathbf{u}_B(\mathbf{R}_i + \delta_{ab})]. \tag{1.98}$$

The local displacements of the atoms in the unit cell can be related to $\mathbf{u}(\mathbf{r})$ by

$$(\delta_{ab} \cdot \nabla)\mathbf{u} = \kappa^{-1}(\mathbf{u}_A - \mathbf{u}_B), \tag{1.99}$$

where κ is a dimensionless quantity that depends on microscopic details. Changes in the bond length lead to changes in the hopping amplitude:

$$t_{ij} \approx t_{ij}^0 + \frac{\partial t_{ij}}{\partial a}\delta u_i. \tag{1.100}$$

We can write

$$\delta t^{ab}(\mathbf{r}) \approx \beta \frac{\delta u^{(ab)}}{a} \quad \text{with } \beta = \frac{\partial t^{(ab)}}{\partial \ln(a)}. \tag{1.101}$$

Substituting **Equation 1.98** into **1.101** and the final result into **Equation 1.93**, we find

$$A_x^{(s)} = \frac{3}{4}\beta\kappa(u_{xx} - u_{yy}),$$

$$A_y^{(s)} = \frac{3}{2}\beta\kappa u_{xy}, \tag{1.102}$$

where $\beta/v_F \approx a^{-1} \sim 1$ Å^{-1}. In conclusion, changes in the hopping amplitude induce a vector potential A and a scalar potential ϕ in the Dirac Hamiltonian. As shown in Reference 48, this induced vector potential leads to a uniform pseudo-magnetic field, in case the applied strain field has triaxial symmetry. In case of uniaxial strain, the Fermi velocity becomes anisotropic [54]. A treatment up to the second order in the strain has shown that the above-discussed treatment up to the first order is valid up to a strain of 15% [54].

References

1. M. O. Goerbig, *Rev. Mod. Phys.* **83**, 1193, 2011.
2. A. K. Geim and K. S. Novoselov, *Nat. Mater.* **6**, 183, 2007.
3. S. Mouras, A. Hamm, D. Djurado and J. C. Cousseins, *Rev. Chim. Miner.* **24**, 572, 1987.
4. R. E. Peierls, *Helv. Phys. Acta* **7**, 81, 1934.
5. R. E. Peierls, *Ann. Inst. Henri Poincare* **5**, 177, 1935.
6. L. D. Landau, *Phys. Z. Sowjetunion* **11**, 26, 1937.
7. L. D. Landau and E. M. Lifshitz, *Statistical Physics*, Part 1, Vol. 5, 3rd ed., Butterworth-Heinemann, Oxford, 1980.
8. N. D. Mermin, *Phys. Rev.* **176**, 250, 1968.
9. J. C. Meyer, A. K. Geim, M. I. Katsnelson, K. S. Novoselov, T. J. Booth and S. Roth, *Nature* **446**, 60, 2007.
10. A. Fasolino, J. H. Los and M. I. Katsnelson, *Nat. Mater.* **6**, 858, 2007.
11. K. S. Novoselov, A. K. Geim, S. V. Morozov, D. Jiang, M. I. Katsnelson, I. V. Grigorieva, S. V. Dubonos and A. A. Firsov, *Nature* **438**, 197, 2005.
12. Y. Zhang, Y. Tan, H. L. Stormer and P. Kim, *Nature* **438**, 201, 2005.
13. P. R. Wallace, *Phys. Rev.* **71**, 622, 1947.
14. G. W. Semenoff, *Phys. Rev. Lett.* **53**, 2449, 1984.
15. A. H. Castro Neto, F. Guinea, N. M. R. Peres, K. S. Novoselov and A. K. Geim, *Rev. Mod. Phys.* **81**, 109, 2009.
16. O. Klein, *Z. Phys.* **53**, 157, 1929.
17. V. V. Cheianov, V. Fal'ko and B. L. Altshuler, *Science* **315**, 1252, 2007.
18. M. I. Katsnelson, K. S. Novoselov and A. K. Geim, *Nat. Phys.* **2**, 620, 2006.
19. E. Fradkin, *Phys. Rev. B* **33**, 3263, 1986.
20. E. V. Gorbar, V. P. Gusynin, V. A. Miransky and I. A. Shovkovy, *Phys. Rev. B* **66**, 045108, 2002.
21. M. I. Katsnelson, *Eur. Phys. J. B* **57**, 225, 2007.
22. J. Tworzydlo, B. Trauzettel, M. Titov, A. Rycerz and C. W. J. Beenakker, *Phys. Rev. Lett.* **96**, 246802, 2006.
23. K. Ziegler, *Phys. Rev. Lett.* **80**, 3113, 1998.
24. P. M. Ostrovsky, I. V. Gornyi and A. D. Mirlin, *Phys. Rev. B* **74**, 235443, 2006.
25. J. H. Chen, C. Jang, S. Xiao, M. Ishigami and M. S. Fuhrer, *Nat. Nanotechnol.* **3**, 206, 2008.
26. R. R. Nair, P. Blake, A. N. Grigorenko, K. S. Novoselov, T. J. Booth, T. Stauber, N. M. R. Peres and A. K. Geim, *Science* **320**, 1308, 2008.
27. Z. Q. Li, E. A. Henriksen, Z. Jiand, Z. Hao, M. C. Martin, P. Kim, H. L. Stormer and D. N. Basovi, *Nat. Phys.* **4**, 532, 2008.
28. T. Ando, Y. Zheng and H. Suzuura, *J. Phys. Soc. Jpn.* **71**, 1318, 2002.

29. V. P. Gusynin, S. G. Sharapov and J. P. Carbotte, *Phys. Rev. Lett.* **96**, 256802, 2006.
30. T. Stauber, N. M. R. Peres and A. K. Geim, *Phys. Rev. B* **78**, 085432, 2008.
31. A. B. Kuzmenko, E. van Heumen, F. Carbone and D. van der Marel, *Phys. Rev. Lett.* **100**, 117401, 2008.
32. E. McCann, D. S. Abergel and V. I. Fal'ko, *Solid State Commun.* **143**, 110, 2007.
33. E. McCan, *Phys. Rev. B* **74**, 161403, 2006.
34. B. Partoens and F. M. Peeters, *Phys. Rev. B* **74**, 075404, 2006.
35. B. Partoens and F. M. Peeters, *Phys. Rev. B* **75**, 193402, 2007.
36. C. J. Shih, A. Vijayaraghavan, R. Krishnan, R. Sharma, J. H. Han, M. H. Ham, Z. Jin et al. *Nat. Nanotechnol.* **6**, 439, 2011.
37. H. Min and A. H. MacDonald, *Phys. Rev. B* **77**, 155416, 2008.
38. A. A. Avetisyan, B. Partoens and F. M. Peeters, *Phys. Rev. B* **79**, 035421, 2009.
39. A. A. Avetisyan, B. Partoens and F. M. Peeters, *Phys. Rev. B* **81**, 115432, 2010.
40. J. D. Jackson, *Classical Electrodynamics*, 3rd ed., Wiley, New York, 1998.
41. G. Li and E. Y. Andrei, *Nat. Phys.* **3**, 623, 2007.
42. M. L. Sadowski, G. Martinez, M. Potemski, C. Berger and W. A. de Heer, *Phys. Rev. Lett.* **97**, 266405, 2006.
43. Z. Jiang, E. A. Henriksen, L. C. Tung, Y. J. Wang, M. E. Schwartz, M. Y. Han, P. Kim and H. L. Stormer, *Phys. Rev. Lett.* **98**, 197403, 2007.
44. S. Park and H. Sim, *Phys. Rev. B* **77**, 075433, 2008.
45. K. V. Klitzing, G. Dorda and M. Pepper, *Phys. Rev. Lett.* **45**, 494, 1980.
46. R. B. Laughlin, *Phys. Rev. Lett.* **50**, 1395, 1983.
47. K. S. Novoselov, E. McCann, S. V. Morozov, V. I. Fal'ko, M. I. Katsnelson, U. Zeitler, D. Jiang, F. Schedin and A. K. Geim, *Nat. Phys.* **2**, 177, 2006.
48. F. Guinea, M. I. Katsnelson and A. K. Geim, *Nat. Phys.* **6**, 30, 2009.
49. F. Guinea, A. K. Geim, M. I. Katsnelson and K. S. Novoselov, *Phys. Rev. B* **81**, 035408, 2010.
50. F. Guinea, M. I. Katsnelson and M. A. H. Vozmediano, *Phys. Rev. B* **77**, 075422, 2008.
51. N. Levy, S. A. Burke, K. L. Meaker, M. Panlasigui, A. Zettl, F. Guinea, A. H. C. Neto and M. F. Crommie, *Science* **329**, 544, 2010.
52. E. Stolyarova, D. Stolyarov, K. Bolotin, S. Ryu, L. Liu, K. T. Rim, M. Klima et al. *Nano Lett.* **9**, 332, 2008.
53. T. Georgiou, L. Britnell, P. Blake, R. V. Gorbachev, A. Gholinia, A. K. Geim, C. Casiraghi and K. S. Novoselov, *Appl. Phys. Lett.* **99**, 093103, 2011.
54. M. Ramezani Masir, D. Moldovan and F. M. Peeters, *Solid State Commun.* **76**, 175–176, 2013.

2

Epitaxial Graphene
Progress on Synthesis and Device Integration

Joshua Robinson, Matthew Hollander and Suman Datta

Contents

2.1 Introduction

While the initial discovery of graphene was groundbreaking in establishing the stability of two-dimensional crystals, the synthesis technique used to make this remarkable discovery has not been conducive to wafer-scale production of graphene-based electronics. The technique, referred to as exfoliation, relies on micro-mechanically cleaving two-dimensional crystals from a

2D Materials for Nanoelectronics edited by Michel Houssa, Athanasios Dimoulas and Alessandro Molle © 2016 CRC Press/Taylor & Francis Group, LLC. ISBN: 978-1-4987-0417-5.

three-dimensional graphite bulk by physically dragging highly ordered pyrolytic graphite (HOPG) across a substrate[1] or by using a scotch tape to peel individual layers of carbon from HOPG.[2] The process of exfoliation is a relatively simple and cost-effective method to produce high-quality graphene crystallites as large as 10 μm; however, this process produces bilayer and multi-layer crystallites as well. Because of the various types of crystallites produced during exfoliation, optical microscopy must be utilised to search out and identify single-layer graphene crystallites for further processing or testing. Although exfoliation produces graphene samples that exhibit excellent electronic properties, the technique is decidedly time consuming and impractical for large-scale manufacturing. In recent years, synthesis by mechanical cleavage has become more efficient, even incorporating the use of ultra-sonication to create suspensions of sub-micron graphene crystals which can be used to coat arbitrary substrates,[3] yet remains unsuitable for wafer-scale production of graphene electronic devices. Alternatively, the development and optimisation of synthesis techniques such as chemical vapour deposition (CVD) and sublimation from SiC wafers have allowed for wafer-scale synthesis of large-grained, polycrystalline graphene films and provided promise for the eventual commercialisation of graphene-based electronic technologies that require high-quality, affordable graphene substrates.

Although both CVD and Si sublimation techniques are able to produce high-quality graphene over a large area, they differ dramatically in their mechanism and methods. CVD techniques may operate through segregation of bulk-dissolved carbon at the surface or through surface decomposition of carbon-containing precursors (i.e. methane, propene, ethylene, etc.).[4] In both methods, a metal substrate at high temperature (1000°C) is used to segregate the carbon or to act as a catalytic substrate for graphene growth. Owing to the use of a metal substrate, graphene growth by CVD techniques necessitates the incorporation of a transfer step to move the graphene material from the metal substrate to a suitable supporting substrate for device fabrication. Unlike CVD-based synthesis techniques, growth of graphene by sublimation of Si from SiC requires no transfer step. Instead, the technique produces a thin carbon film on top of a semi-insulating SiC substrate that is suitable for subsequent device processing.[5] Silicon sublimation occurs at high temperatures (~1300–1600°C) and can take place either in ultra-high vacuum (UHV) or in an inert, low-pressure argon atmosphere. In the following sections, we will discuss the synthesis and integration of epitaxial graphene to form high-performance devices and circuits.

2.2 Synthesis of Epitaxial Graphene

Synthesis of epitaxial graphene from SiC substrates is accomplished through the controlled, high-temperature sublimation of silicon from the underlying SiC.[6-8] Sublimation of silicon from the SiC leaves behind a reconstructed layer of carbon atoms, which ultimately rearrange to form a layer of graphene. This process can be considered bottom up, where growth of each subsequent layer of graphene takes place below the previously grown layers, instead of above them. This is unlike most other growth processes, which can be considered top down, where growth of subsequent layers typically takes place on top of previously grown layers. The growth kinetics and quality of the resulting graphene depend on the structure of the SiC surface that is used for growth. The most common substrates are 4H-SiC and 6H-SiC hexagonal structures.[6–8] Each of the two polymorphs is composed of SiC bilayers, where each bilayer contains a plane of C atoms as well as a plane of Si atoms. 4H-SiC and 6H-SiC contain four and six SiC bilayers, respectively, and exhibit different stacking sequences.[9] The SiC crystal structures have a Si-terminated surface (0001) as well as a C-terminated surface (000$\bar{1}$). **Figure 2.1** shows the crystal structure of these two SiC polymorphs.

Before graphitisation, the SiC substrate is subjected to a H_2 etch step at elevated temperatures in order to create a regular array of atomically flat surfaces. The use of a hydrogen etch prior to graphene synthesis is effective in removing much of the residual surface damage caused by SiC-polishing techniques.[10] Importantly, the hydrogen etch also leads to significant step bunching and the formation of terraces across the SiC surface, which will be discussed later.[10] The Si sublimation growth process follows and is used to grow graphene either in UHV or in an inert argon atmosphere at about 1200°C.[11–13] The use of argon during the sublimation process can help to improve film morphology, leading to reduced pitting and much larger domain sizes than achievable using high-vacuum sublimation.[10] Growth can take place on either the Si- or C-face of the SiC substrate and the final structure of the graphene depends on which face the growth takes place. Growth on the C-face is faster than on the Si-face; yet, the graphene that grows on this face has a variety of orientations and can be considered rotationally faulted relative to the conventional Bernal stacking in graphite crystals.[5] The growth on the Si-face is significantly slower and produces Bernal stacked layers of graphene which have an orientation of 30° with respect to the underlying SiC plane.[6,14] While graphene growth on the Si-face is sensitive to growth temperature, it is found to be relatively independent of growth time. Because of this fact, graphene growth on the Si-face is often considered a self-limited process.[15] Owing to the additional control over film

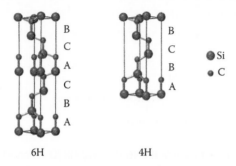

FIGURE 2.1 Crystal structure of the 6H- and 4H-SiC polymorphs showing a stacking order.

morphology provided by the slow growth speeds and well-ordered graphene stacking, Si-face growth processes are considered the most promising candidates for producing large, homogenous graphene substrates for wafer-scale device fabrication. As a result, this section will focus primarily on the growth mechanisms that drive graphene formation on the Si-face.

Throughout the graphitisation process, the Si-face goes through a sequence of surface reconstructions that ultimately result in the growth of a graphene sheet and a carbon buffer layer which exists between the graphene and the underlying SiC substrate. The series of surface reconstructions are listed below:[16]

$$(3 \times 3) \rightarrow (1 \times 1) \xrightarrow{1000°C} (\sqrt{3} \times \sqrt{3}) \xrightarrow{1080°C} (6\sqrt{3} \times 6\sqrt{3}) + \text{Graphene}$$

The process starts with a hydrogen-etched sample and the preparation of the Si-rich (3×3) phase. Heating this surface to about 950°C leads to a $(\sqrt{3} \times \sqrt{3})R30°$ phase. This phase can be considered as a 1/3 monolayer of Si adatoms atop the SiC substrate. Further annealing to 1100–1150°C sublimes more silicon from the surface and leads to the formation of a well-ordered $(6\sqrt{3} \times 6\sqrt{3})R30°$ phase, in which the SiC is covered by a carbon layer. The large surface unit cell of $(6\sqrt{3} \times 6\sqrt{3})R30°$ comes into existence only because of the difference in lattice parameters of graphene and SiC (2.46 and 3.08 Å, respectively). The side length of the $(6\sqrt{3} \times 6\sqrt{3})R30°$ is approximately 32 Å and contains about 108 carbon atoms and 108 silicon atoms per SiC bilayer, which can be seen from **Figure 2.2.** The $(6\sqrt{3} \times 6\sqrt{3})R30°$ reconstruction is equivalent in size to a (13×13) graphene unit cell. The quasi-(6×6) honeycomb hexagons are slightly varying in size during these phase transitions.[17] These phases and patterns are observed using low-energy electron diffraction.[17]

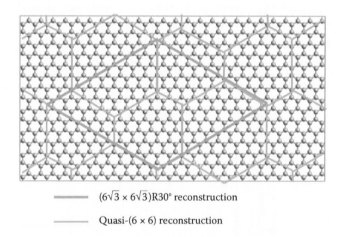

———— $(6\sqrt{3} \times 6\sqrt{3})R30°$ reconstruction

———— Quasi-(6×6) reconstruction

FIGURE 2.2 Structural model of the $(6\sqrt{3} \times 6\sqrt{3})R30°$ reconstruction in the top view showing the Si-terminated (1×1)-SiC substrate. The $(6\sqrt{3} \times 6\sqrt{3})R30°$ unit cell and the three hexagons in quasi-(6×6) periodicity are indicated. (Adapted from Riedl, C., Coletti, C. and Starke, U. *J. Phys. D Appl. Phys.* **43**, 374009, 2010.)

The $(6\sqrt{3} \times 6\sqrt{3})$R30° superstructure, which is often referred to as the carbon buffer layer, exists between the crystalline SiC substrate and the overlying graphene. It is sometimes referred to as 'zero-layer graphene' as it represents the precursor to graphitisation. It serves as a buffer layer for growth of epitaxial graphene layers. The carbon layer which sits atop the buffer layer represents the first carbon layer which can be considered graphene, although it does not actually grow on top of the buffer layer. Instead, each layer of graphene that sits atop the buffer layer is the result of the desorption of Si atoms from the underlying SiC bilayers. Desorption of Si atoms from a total of three SiC bilayers is required to form a single layer of graphene, which can be understood simply by counting the number of C atoms needed to form a single-graphene monolayer and realising that a single SiC bilayer yields only 1/3 of the required C atoms. Each new layer of graphene starts with the formation of a new $(6\sqrt{3} \times 6\sqrt{3})$R30° buffer layer, during which the previous buffer layer is released of its covalent bonding with the SiC surface and is transformed into a true graphene layer. This process results in a stacking of all graphene sheets with a 30° rotation relative to the underlying SiC lattice. Although the graphene is rotated relative to the SiC lattice, additional layers of epitaxial graphene on the Si-face are found to be Bernal stacked relative to the first layer.

Examining the C 1s peak of the x-ray photoelectron spectroscopy (XPS) spectra of the carbon buffer layer can provide a valuable insight into its structure. In the XPS spectra, the C 1s level is composed of several different peaks. Apart from the SiC bulk peak at 283.73 eV, there are two additional peaks called S1 and S2, at 284.99 and 285.60 eV, respectively. A graphene peak is expected at 284.7 eV; yet, neither S1 nor S2 are at that position. Instead, the $(6\sqrt{3} \times 6\sqrt{3})$R30° layer has two non-equivalent, non-graphene-like carbon atoms. Since this layer is partially covalently bound to the substrate, S1 originates from the sp^3-bonded carbon atoms which are simultaneously bound to one Si atom of the Si(0001) layer and three C atoms in the overlying graphene and S2 originates from the remaining sp^2-bonded carbon atoms in the buffer layer.[17] The two components have a ratio of approximately 1:2 and thus, almost 1/3 of the carbon atoms in the initial layer are bound to the SiC substrate.

2.2.1 Role of Step Edges

When discussing the growth and electrical properties of epitaxial graphene, it is important to address the impact of step edges. Currently, it remains difficult to obtain atomically flat SiC wafers and the nominally on-axis SiC substrate typically exhibits some small degree of off-angle, resulting in atomic step edges across the surface of the substrate. The aforementioned H_2 etch step is helpful in rearranging the SiC surface to achieve regions of atomically flat SiC for subsequent graphitisation according to the growth kinetics described previously. However, during the high-temperature H_2 etch step, significant step-edge bunching occurs across the SiC substrate. These step edges coalesce into large $(110n)$ inclined facets, which are separated by atomically flat (0001)

FIGURE 2.3 Cross-sectional TEM micrographs of graphene nucleated on the terrace step edge of SiC at 1325°C. Many-layer graphene is possible along the ($1\bar{1}0n$) plane and can occur before the growth of graphene on the terrace face (0001). (Adapted from Robinson, J. et al. *ACS Nano* **4**, 153–8, 2010.)

plateaus which can vary in size from less than 1 μm to greater than 10 μm. The step edges between the plateaus can grow in size to be as high as 10 nm and are important to both the ultimate electronic properties and growth of the graphene.

Although the growth of graphene on the Si-face described in the previous section is self-limited and highly controllable down to a single monolayer of graphene, the presence of step edges across the substrate surface lead to regions of multi-layer graphene interspersed between few-layer regions. This phenomenon is a result of the fact that graphene growth on the ($1\bar{1}0n$) crystal face differs from the (0001) face, where growth on the ($1\bar{1}0n$) face occurs at lower temperatures and is not self-limiting.[18]

It is thought that step edges serve as nucleation sites for the graphene film due to their high density of dangling bonds and defective nature.[18] **Figure 2.3** shows a cross-sectional transmission electron microscopy (TEM) micrograph of a step edge that shows multi-layer graphene growth along the ($1\bar{1}0n$) crystal face, while the (0001) terrace ledges show no graphene.[18] For this sample, the graphitisation temperature of 1325°C is slightly lower than the typical temperature for graphitisation of the (0001) face. The presence of multi-layer graphene at the step edge confirms the lower growth temperature and suggests that the ($1\bar{1}0n$) face may serve as a nucleation site for graphene growth across the (0001) plateau. Researchers have shown that defects at the step edges can facilitate the sublimation of Si and hence, graphitisation first occurs on the (0001) face where the step edge meets the (0001) plateau. Low- energy electron microscopy and scanning electron microscopy of partial graphitisation show that this reconstruction phase is initially concentrated at the step edges and then grows outwards from the step edge. These regions stretch out into the flat terraces of SiC and form islands which merge together to form a graphene monolayer.[19,20]

2.2.2 Electrical Properties of Epitaxial Graphene

Along the (0001) crystal face, Si-face epitaxial graphene layers are strongly bound to the substrate by means of the carbon buffer layer, which induces strong electron doping and significant spectral disorder at low energies near the Dirac point.[21,22] Away from the Dirac point (at higher energies), Si-face epitaxial graphene displays the typical linear dispersion relation of graphene.[21] Mobilities for Si-face epitaxial graphene are typically much lower than C-face epitaxial graphene, CVD graphene and exfoliated graphene, averaging values between 500 and 1500 cm^2/V/s at electron densities between 1×10^{12} and 1×10^{13} cm^{-2}.[10] The limited mobility of Si-face epitaxial graphene is likely a result of a significant remote-charged impurity scattering due to dangling bonds in the carbon buffer layer.[23]

Furthermore, several researchers have quantified the negative impact of step edges on the ultimate carrier mobility of bulk Si-face epitaxial graphene showing that step edges are a significant source of additional resistance.[23,24] Nano-probe measurements along and across the plateau and step edge show the step-edge resistance to be an order higher than along the step plateau ((0001) face).[25] A temperature-dependent transport study on epitaxial graphene indicated that step edges may be an additional source of both remote-charged impurity scattering and surface optical phonon scattering.[23] **Figure 2.4** shows the impact of step edges on degrading the measured transport properties. In this plot, measured Hall effect mobility is shown as a function of the step-edge height. For this experiment, 5×5-μm Hall crosses were used to sample the transport properties of a localised region of the epitaxial graphene on SiC. The step-edge height was measured for each individual Hall cross by summing the total distance propagated in the (0001) direction when moving from one side of the Hall cross to the other.

Besides the problem of step edges, reduced mobility of Si-face samples remains an issue for high-performance graphene devices using Si-face epitaxial graphene. Recent work to cleave Si-face epitaxial graphene from the carbon buffer layer by hydrogen intercalation has led to substantial improvements in electronic transport properties as well as reduced electron doping,[26-28] suggesting that passivation or removal of the buffer layer is integral to high-performance Si-face epitaxial graphene. These experiments are often able to improve mobility to 3000 cm^2/V/s at similar carrier densities (1×10^{12} to 1×10^{13} cm^{-2}).[27,28] In this method, the epitaxial graphene is exposed to molecular hydrogen at elevated temperatures (1000°C), whereupon the hydrogen acts to passivate the SiC substrate and is found to convert the carbon buffer layer into a layer of graphene. In this way, the technique can be used to convert a single carbon buffer layer into a monolayer of graphene atop the hydrogen-passivated SiC substrate, effectively decoupling the graphene from the underlying substrate. Transport studies on hydrogen-intercalated epitaxial graphene, also known as quasi-freestanding epitaxial graphene, indicate that the improved transport properties are mostly attributed to a reduction in a remote-charged impurity scattering, which are likely associated with the dangling bonds present in the carbon buffer layer.[23] Using this technique, quasi-freestanding epitaxial graphene radio frequency (RF) transistors have been demonstrated with improved extrinsic current gain cutoff frequency, f_T. Improved f_T, can be directly related to the increase in carrier mobility due to

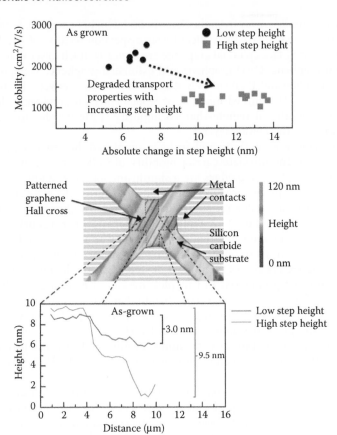

FIGURE 2.4 Effect of step-edge height on mobility shows degradation of transport properties as step-edge height increases. Here, step-edge height is measured using optical profilometry. (Adapted from Hollander, M. J. et al. *Phys. Status Solidi* **210**, 1062–70, 2013.)

the hydrogen intercalation step, where f_T is related to the mobility according to the relation:

$$f_T \approx \frac{g_m}{2\pi(C_{GS} + C_{GD})(1 + g_{ds}(R_S + R_D)) + C_{GD}g_m(R_S + R_D)}$$

In this relation, transconductance, or g_m, linearly scales with carrier mobility and thus, any increase in mobility directly leads to an increase in f_T and RF performance. **Figure 2.5** plots the current gain, H_{21}, as a function of frequency for the epitaxial graphene transistor with and without hydrogen intercalation. In this plot, f_T can be extracted by identifying the point, where H_{21} is unity. **Figure 2.5** shows that hydrogen intercalation can be an effective technique

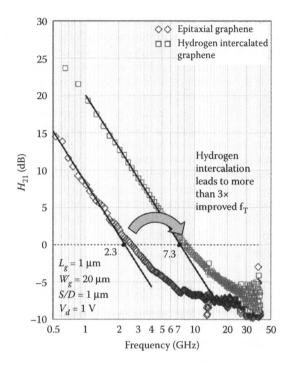

FIGURE 2.5 Plot of extrinsic current gain as a function of frequency showing how the use of a hydrogen intercalation step can lead to improved device performance through improved transport properties. (Adapted from Robinson, J. A. et al. *Nano Lett.* **11**, 3875–80, 2011.)

for increasing RF performance, indicating an increase in extrinsic f_T of more than 3× from 2.3 to 7.3 GHz.

2.3 Materials Integration with Epitaxial Graphene for Device Applications

Since its discovery, there has been a significant interest in utilising graphene for RF device applications. This interest is a result of graphene's excellent transport properties, with mobilities reported as high as 200,000 cm²/V/s reported for suspended monolayer graphene in vacuum and at low temperatures[29] and values of 10,000–40,000 cm²/V/s routinely reported for exfoliated and CVD graphene samples on various supporting substrates at room temperature.[1,30] Besides high mobilities, graphene also demonstrates an exceedingly high saturation velocity, with values of 3×10^7 cm/s measured for graphene on SiO_2 at low carrier concentrations[31] and theoretical predictions ranging from 3 to 6×10^7 cm/s for graphene and its variations.[32] These properties far surpass typical semiconducting materials and make graphene and its alternative forms attractive for super-high-frequency and extremely high-frequency operation. Specifically, the high mobility and saturation velocity of graphene suggest extremely high switching speeds for graphene-based field-effect transistors (FETs), which have the potential to yield ultra-fast RF devices into the terahertz regime. Fuelled by this interest, research groups around the world have demonstrated RF graphene transistors with operating frequencies as high as 427 GHz[33]; yet, the ultimate performance is typically limited by materials integration of contact metals and gate dielectrics, which often degrade the intrinsic transport properties of graphene and limit the extrinsic device performance. In this way, the ultimate RF performance depends on several factors, including the optimisation of charge carrier mobility through the synthesis technique or top-gate dielectric integration; minimisation of parasitic elements such as contact resistance, access resistance and parasitic capacitances and aggressive scaling of gate length and oxide thickness to reach ultra-short-channel lengths.

2.3.1 Dielectric Integration with Epitaxial Graphene

Epitaxial graphene presents several unique challenges to device optimisation for RF applications relative to other forms of graphene. Most of these challenges are a direct result of the complex growth morphology due to the presence of step edges and multi-layer graphene across them. Step edges have often been cited as a source of additional scattering, with reports of degradation of bulk mobility occurring with increasing step-edge density[34] as well as reports of conductivity anisotropy parallel and perpendicular to step edges.[35,36] In 2011, Ross et al. utilised atomic-scale measurements to directly measure the additional resistance introduced across the step edge.[25] Step resistances normalised to the absolute step height have yielded values between 14–25 Ω/mm/nm for epitaxial graphene and 19–27 Ω/mm/nm for quasi-freestanding epitaxial graphene on SiC.[11,24,37] Furthermore, significant substrate interaction between the graphene and underlying SiC as compared to other types of graphene such as exfoliated and CVD grown (i.e. carbon buffer layer, dangling bonds) can have a significant impact on transport properties.

As a result of the combined effects of step edges and substrate interactions, transport is found to be almost completely dominated by remote-charged impurity scattering for both epitaxial and quasi-freestanding epitaxial graphene.[23] This presents an interesting situation for the integration of top-gate dielectrics. In fabricating a graphene-based device, various techniques have been used to implement top-gate dielectrics, including electron-beam physical vapour deposition (EBPVD),[38,39] functionalised atomic layer deposition (ALD)[40–42] and seeded ALD (either through the use of a thin oxidised metal layer,[43,44] thin oxide layer[45] or a polymer buffer layer[46]). In addition, physical assembly techniques such as the use of thin-alumina nanoribbons[47] or self-aligned oxidised nanowires[48] have been employed, although the extent to which these techniques are scalable is not clear.

While current implementations of the graphene FET have shown great promise using these techniques, top-gate dielectrics often cause an undesirable degradation in the transport properties of the underlying graphene.[46,49,50] This degradation in carrier mobility has been mainly attributed to the presence of additional remote-charged scatterers in the dielectric and close to the graphene or the additional scattering due to remote surface optical phonons, which propagate along the surface of the dielectric and act as a time-varying potential which can dynamically couple to and scatter charge carriers in the graphene. Alternatively, calculations have shown that high dielectric constant (high-κ) materials should have the effect of suppressing charged impurity scattering in the underlying graphene leading to an improvement in transport properties.[51,52] Such an improvement has been shown variously by increasing the dielectric constant of a solvent overlayer[53,54] or by the use of an ice overlayer.[55]

Because of the dominance of remote-charged impurity scattering for epitaxial graphene samples, high-k dielectrics can be of particular benefit in preserving or even improving the as-grown transport properties.[23,45] It was found that ALD hafnia dielectrics seeded with a 2-nm hafnia seed layer deposited by EBPVD produced an average increase in Hall mobility of 73%–57%, while ALD alumina seeded with 2-nm alumina deposited by EBPVD produced an average increase of 52%–43% and dielectrics deposited by *only* EBPVD produced an average increase of ~15% regardless of composition.[45] **Figure 2.6** plots the mobilities for the EBPVD-seeded high-k dielectric- coated epitaxial graphene compared to their as-grown values.

These results contrast with attempts to integrate ALD-grown high-k dielectrics on exfoliated and CVD graphene which resulted in degraded mobility.[46,49,50] These results can be understood by considering the various competing scattering mechanisms in graphene. For the case of high-mobility exfoliated and CVD graphene, top-gate dielectric of high-k dielectrics can be difficult due to the hydrophobic nature of graphene[40] and the lack of any functional groups that might allow for the molecular absorption of the gas precursors. Chemical functionalisation or a seeding process is often used to promote ALD coverage and uniformity. These techniques often degrade the ultimate transport properties because the high-mobility samples often exhibit a low as-grown density of remote-charged impurity scatterers. By utilising an imperfect seeding technique, a significant density of additional remote-charged scatterers can be introduced. Alternatively, for the case of epitaxial graphene, the carrier mobility is already severely limited by the presence of a high density of remote-charged impurity scatterers intrinsic to the SiC substrate and due to the presence of step edges across the surface. In this situation, the integration of a high-k dielectric can be beneficial if the density of additional remote-charged impurity scatterers introduced to the system is low relative to the density present before dielectric integration by acting to screen the carriers from these impurities and thus improve the carrier mobility. To this end, Hollander et al. utilised temperature-dependent Hall effect measurements to experimentally extract an effective remote-charged impurity density before and after dielectric integration on epitaxial

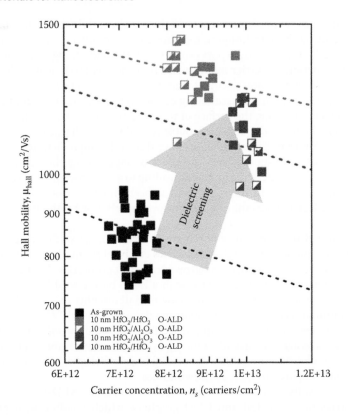

FIGURE 2.6 Measured Hall mobility as a function of carrier density for as-grown epitaxial graphene along with HfO$_2$ and Al$_2$O$_3$-coated epitaxial graphene. The dashed lines represent the measured dependencies of mobility with carrier density for the samples. The increase in mobility is attributed to dielectric screening by the high-k dielectric overlayer which acts to offset the large density of intrinsic remote-charged impurity scatterers present in as-grown epitaxial graphene. (Adapted from Hollander, M. J. et al. *Nano Lett.* **11**, 3601–7, 2011.)

graphene.[23] In their work, they showed that for epitaxial graphene samples with an as-grown impurity density above approximately 1×10^{12} cm^{-2} conventional integration of high-k top-gate dielectrics using a direct deposited seed can lead to improved mobility. Alternatively, they showed that for high-quality hydrogen-intercalated epitaxial graphene with minimal impact from step edges and, thus, the low as-grown remote-charged impurity density, the integration of high-k dielectrics led to an overall degradation in transport properties. These results suggest the presence of two regimes, a remote surface optical phonon-dominated regime at low remote-charged impurity densities and a remote-charged impurity-dominated regime at high densities, where high-k dielectrics as integrated are only beneficial for the latter case. **Figure 2.7** plots the simulated Hall mobility as a function of

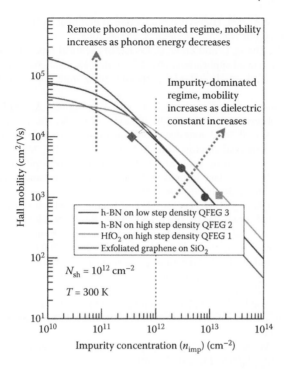

FIGURE 2.7 Simulated mobility as a function of remote-charged impurity density. As the density of scatterers increases, high-k dielectrics become beneficial (right side). When the impurity density is low, low-k dielectrics are typically better due to the high energy of their remote surface optical phonon modes. For epitaxial graphene, the impurity density is typically high, but can be lowered by reducing step edges or through hydrogen intercalation. (Adapted from Hollander, M. J. et al. *Phys. Status Solidi* **210**, 1062–70, 2013.)

impurity concentration. This plot highlights the two scattering regimes for epitaxial graphene.

These results emphasise that the overall benefit of high-k dielectrics is found to critically depend on the charged impurity density within the dielectric–graphene system. For the case of epitaxial graphene where remote-charged impurity scattering is often found to be the limiting scattering process, high-k dielectrics are particularly applicable even though dielectric integration can often lead to additional incorporation of charged impurity scatterers.

References

1. Novoselov, K. S. et al. Electric field effect in atomically thin carbon films. *Science* **306**, 666–9, 2004.
2. Geim, A. K. and Novoselov, K. S. The rise of graphene. *Nat. Mater.* **6**, 183–91, 2007.

3. Hernandez, Y. et al. High-yield production of graphene by liquid-phase exfoliation of graphite. *Nat. Nanotechnol.* **3**, 563–8, 2008.

4. Zhang, Y., Zhang, L. and Zhou, C. Review of chemical vapor deposition of graphene and related applications. *Acc. Chem. Res.* **46**, 2329–39, 2013.

5. De Heer, W. A. et al. Epitaxial graphene. *Solid State Commun.* **143**, 92–100, 2007.

6. Van Bommel, A. J., Crombeen, J. E. and Van Tooren, A. LEED and Auger electron observations of the SiC(0001) surface. *Surf. Sci.* **48**, 463–72, 1975.

7. Charrier, A. et al. Solid-state decomposition of silicon carbide for growing ultrathin heteroepitaxial graphite films. *J. Appl. Phys.* **92**, 2479, 2002.

8. Forbeaux, I., Themlin, J.-M. and Debever, J.-M. Heteroepitaxial graphite on 6H-SiC(0001): Interface formation through conduction-band electronic structure. *Phys. Rev. B* **58**, 16396–406, 1998.

9. Park, C., Cheong, B.-H., Lee, K.-H. and Chang, K. Structural and electronic properties of cubic, 2H, 4H and 6H SiC. *Phys. Rev. B* **49**, 4485–93, 1994.

10. Emtsev, K. V. et al. Towards wafer-size graphene layers by atmospheric pressure graphitization of silicon carbide. *Nat. Mater.* **8**, 203–7, 2009.

11. Riedl, C. and Starke, U. Structural properties of the graphene–SiC(0001) interface as a key for the preparation of homogeneous large-terrace graphene surfaces. *Phys. Rev. B* **76**, 245406, 2007.

12. Starke, U. and Riedl, C. Epitaxial graphene on SiC(0001) and [formula: see text]: From surface reconstructions to carbon electronics. *J. Phys. Condens. Matter* **21**, 134016, 2009.

13. De Heer, W. A. et al. Epitaxial graphene electronic structure and transport. *J. Phys. D Appl. Phys.* **43**, 374007, 2010.

14. Forbeaux, I., Themlin, J.-M., Charrier, A., Thibaudau, F. and Debever, J.-M. Solid-state graphitization mechanisms of silicon carbide 6H–SiC polar faces. *Appl. Surf. Sci.* **162–163**, 406–12, 2000.

15. Hass, J., de Heer, W. A. and Conrad, E. H. The growth and morphology of epitaxial multilayer graphene. *J. Phys. Condens. Matter* **20**, 323202, 2008.

16. Chen, W. et al. Atomic structure of the 6H–SiC(0001) nanomesh. *Surf. Sci.* **596**, 176–86, 2005.

17. Riedl, C., Coletti, C. and Starke, U. Structural and electronic properties of epitaxial graphene on SiC(0001): A review of growth, characterization, transfer doping and hydrogen intercalation. *J. Phys. D Appl. Phys.* **43**, 374009, 2010.

18. Robinson, J. et al. Nucleation of epitaxial graphene on SiC(0001). *ACS Nano* **4**, 153–8, 2010.

19. Hupalo, M., Conrad, E. and Tringides, M. Growth mechanism for epitaxial graphene on vicinal 6H-SiC(0001) surfaces: A scanning tunneling microscopy study. *Phys. Rev. B* **80**, 041401, 2009.

20. Ohta, T., Bartelt, N. C., Nie, S., Thürmer, K. and Kellogg, G. L. Role of carbon surface diffusion on the growth of epitaxial graphene on SiC. *Phys. Rev. B* **81**, 121411, 2010.

21. Zhou, S. Y. et al. Substrate-induced bandgap opening in epitaxial graphene. *Nat. Mater.* **6**, 770–5, 2007.

22. Kim, S., Ihm, J., Choi, H. and Son, Y.-W. Origin of anomalous electronic structures of epitaxial graphene on silicon carbide. *Phys. Rev. Lett.* **100**, 176802, 2008.

23. Hollander, M. J. et al. Heterogeneous integration of hexagonal boron nitride on bilayer quasi-free-standing epitaxial graphene and its impact on electrical transport properties. *Phys. Status Solidi* **210**, 1062–70, 2013.

24. Ciuk, T. et al. Step-edge-induced resistance anisotropy in quasi-free-standing bilayer chemical vapor deposition graphene on SiC. *J. Appl. Phys.* **116**, 123708, 2014.

25. Ji, S.-H. et al. Atomic-scale transport in epitaxial graphene. *Nat. Mater.* **11**, 114–9, 2012.

26. Riedl, C., Coletti, C., Iwasaki, T., Zakharov, A. A. and Starke, U. Quasi-free-standing epitaxial graphene on SiC obtained by hydrogen intercalation. *Phys. Rev. Lett.* **103**, 246804, 2009.

27. Speck, F. et al. The quasi-free-standing nature of graphene on H-saturated SiC(0001). *Appl. Phys. Lett.* **99**, 122106, 2011.

28. Robinson, J. A. et al. Epitaxial graphene transistors: Enhancing performance via hydrogen intercalation. *Nano Lett.* **11**, 3875–80, 2011.

29. Bolotin, K. I. et al. Ultrahigh electron mobility in suspended graphene. *Solid State Commun.* **146**, 351–5, 2008.

30. Chen, J.-H., Jang, C., Xiao, S., Ishigami, M. and Fuhrer, M. S. Intrinsic and extrinsic performance limits of graphene devices on SiO_2. *Nat. Nanotechnol.* **3**, 206–9, 2008.

31. Dorgan, V. E., Bae, M.-H. and Pop, E. Mobility and saturation velocity in graphene on SiO[sub 2]. *Appl. Phys. Lett.* **97**, 082112, 2010.

32. Schwierz, F. Graphene transistors: Status, prospects and problems. *Proc. IEEE* **101**, 1567–84, 2013.

33. Cheng, R. et al. High-frequency self-aligned graphene transistors with transferred gate stacks. *Proc. Natl. Acad. Sci. USA* **109**, 11588–92, 2012.

34. Robinson, J. A. et al. Effects of substrate orientation on the structural and electronic properties of epitaxial graphene on SiC(0001). *Appl. Phys. Lett.* **98**, 222109, 2011.

35. Yakes, M. K. et al. Conductance anisotropy in epitaxial graphene sheets generated by substrate interactions. *Nano Lett.* **10**, 1559–62, 2010.

36. Jouault, B. et al. Probing the electrical anisotropy of multilayer graphene on the Si face of 6H-SiC. *Phys. Rev. B* **82**, 085438, 2010.

37. Bryan, S. E., Yang, Y. and Murali, R. Conductance of epitaxial graphene nanoribbons: Influence of size effects and substrate morphology. *J. Phys. Chem. C* **115**, 10230–5, 2011.

38. Wu, Y. Q. et al. Top-gated graphene field-effect-transistors formed by decomposition of SiC. *Appl. Phys. Lett.* **92**, 092102, 2008.

39. Kedzierski, J. et al. Epitaxial graphene transistors on SiC substrates. *IEEE Trans. Electron. Devices* **55**, 2078–85, 2008.

40. Lee, B. et al. Conformal Al[sub 2]O[sub 3] dielectric layer deposited by atomic layer deposition for graphene-based nanoelectronics. *Appl. Phys. Lett.* **92**, 203102, 2008.

41. Lin, Y.-M. et al. Operation of graphene transistors at gigahertz frequencies. *Nano Lett.* **9**, 422–6, 2009.

42. Wang, X., Tabakman, S. M. and Dai, H. Atomic layer deposition of metal oxides on pristine and functionalized graphene. *J. Am. Chem. Soc.* **130**, 8152–3, 2008.

43. Kim, S. et al. Realization of a high mobility dual-gated graphene field-effect transistor with Al[sub 2]O[sub 3] dielectric. *Appl. Phys. Lett.* **94**, 062107, 2009.

44. Xu, H. et al. Quantum capacitance limited vertical scaling of graphene field-effect transistor. *ACS Nano* **5**, 2340–7, 2011.

45. Hollander, M. J. et al. Enhanced transport and transistor performance with oxide seeded high-κ gate dielectrics on wafer-scale epitaxial graphene. *Nano Lett.* **11**, 3601–7, 2011.

46. Farmer, D. B. et al. Utilization of a buffered dielectric to achieve high field-effect carrier mobility in graphene transistors. *Nano Lett.* **9**, 4474–8, 2009.
47. Liao, L. et al. High-kappa oxide nanoribbons as gate dielectrics for high mobility top-gated graphene transistors. *Proc. Natl. Acad. Sci. USA* **107**, 6711–5, 2010.
48. Liao, L. et al. High-speed graphene transistors with a self-aligned nanowire gate. *Nature* **467**, 305–8, 2010.
49. Fallahazad, B., Kim, S., Colombo, L. and Tutuc, E. Dielectric thickness dependence of carrier mobility in graphene with HfO[sub 2] top dielectric. *Appl. Phys. Lett.* **97**, 123105, 2010.
50. Robinson, J. A. et al. Epitaxial graphene materials integration: Effects of dielectric overlayers on structural and electronic properties. *ACS Nano* **4**, 2667–72, 2010.
51. Hwang, E. H. and Das Sarma, S. Dielectric function, screening and plasmons in two-dimensional graphene. *Phys. Rev. B* **75**, 205418, 2007.
52. Konar, A., Fang, T. and Jena, D. Effect of high-κ gate dielectrics on charge transport in graphene-based field effect transistors. *Phys. Rev. B* **82**, 115452, 2010.
53. Ponomarenko, L. A. et al. Effect of a high-κ environment on charge carrier mobility in graphene. *Phys. Rev. Lett.* **102**, 206603, 2009.
54. Chen, F., Xia, J., Ferry, D. K. and Tao, N. Dielectric screening enhanced performance in graphene FET. *Nano Lett.* **9**, 2571–4, 2009.
55. Jang, C. et al. Tuning the effective fine structure constant in graphene: Opposing effects of dielectric screening on short- and long-range potential scattering. *Phys. Rev. Lett.* **101**, 146805, 2008.

3

Metal Contacts to Graphene

Akira Toriumi and Kosuke Nagashio

Contents

2D Materials for Nanoelectronics edited by Michel Houssa, Athanasios Dimoulas and Alessandro Molle © 2016 CRC Press/Taylor & Francis Group, LLC. ISBN: 978-1-4987-0417-5.

3.1 Introduction

Since the pioneering research on graphene [1], many fundamental research studies have been carried out, and a number of new phenomena have been found ideal for the two-dimensional material. Electrical contacts, however, are critically important in addition to the surprisingly high electron and hole mobility from the viewpoint of the electron device applications of graphene. This is more significant in the higher carrier mobility channel materials. It is noted that an interaction of metal/graphene might be different from those of conventional metals/semiconductors as there is no energy band gap in graphene. Nevertheless, it is worth considering similarities between them to understand what occurs at the metal/graphene interface. Therefore, let us start with Schottky contacts on ordinary semiconductors.

3.1.1 Metal Contacts to Conventional Semiconductors

Metal contacts represent a key technology for any electron devices. Although the contact resistance in graphene is now recognised to be a serious issue for high-performance devices, high carrier mobility was given attention in the early stages of graphene-engineering research. The contact issue, however, is becoming increasingly more important even in conventional semiconductor devices [2].

Contact resistance is basically determined by the Schottky barrier height (SBH) ϕ_b and dopant concentration N_D in semiconductors, according to the following formula [3]:

$$Rc \propto \exp\left(\lambda \frac{\varphi_b}{\sqrt{N_D}} \right), \tag{3.1}$$

where λ is a constant depending on the effective mass and permittivity of a semiconductor. Thus, the SBH ϕ_b and dopant concentration N_D are critically important in controlling the contact resistance.

There are two extreme models for understanding SBH [4]. The first is the Schottky–Mott model and the second is the Bardeen model. The former denotes the ideal interface ($S = 1$ and pure limit), and the latter does perfect pinning ($S = 0$ and dirty limit). S is called the pinning parameter. **Figure 3.1** outlines the band alignments for forming the Schottky interface. We still have not clearly understood the origin of the Fermi-level pinning at semiconductor interfaces, which is still one of the most significant issues in solid-state electron device physics. S is generally between zero and one. There are several kinds of models to explain the Fermi-level pinning [5], and the following three have been well cited. In any case, a dipole formation at the metal/semiconductor interface changes the band alignment. The main issue is how the interface dipoles are formed.

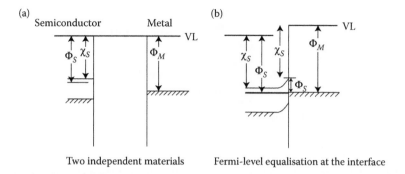

FIGURE 3.1 Band alignment of metal and semiconductor. (a) Metal is finite distance from semiconductor. Vacuum level should be aligned. (b) Metal is in contact with semiconductor, resulting in aligned Fermi levels. This view for Schottky barrier formation is called the Schottky–Mott model.

3.1.2 Metal-Induced Gap States

Let us consider an ideal metal/semiconductor interface. Virtual gap states formed by metal wave function penetration into a semiconductor have been proposed. The energy levels of such states might be found within the bulk band gap [6]. These gap states may form the interface dipole, resulting in the SBH not being determined simply by the difference between the work function of metals and the electron affinity of semiconductors. The dipole formation due to wave function penetration into vacuum is one of the origins for the metal work function in a vacuum. If the gap states density is huge, the Fermi level of the metal will be completely pinned. The states are called metal-induced gap states (MIGS) [7]. Wave function penetration depends on the energy band gap (E_g) of semiconductors as can easily be expected. The larger E_g induces less penetration and less pinning. The data reported so far have empirically followed this model. This fact is regarded as strong evidence for the MIGS model. The Fermi level is actually perfectly pinned near the valence band edge for Ge, which is a typical narrow gap semiconductor, which means that S is approximately zero and that the SBH does not change at all even if various metals make contact [8,9]. However, it is not clear whether the charge-neutrality level in the MIGS model is simply determined by the branch point of semiconductors. Although the MIGS model is still controversial both theoretically and experimentally, the idea is quite intriguing.

3.1.3 Disorder-Induced Gap States

There should be more or fewer defects at the metal/semiconductor interface because there are certainly lattice and/or bonding mismatches there. These defects should form interface states inside the semiconductor energy band gap and could capture carriers, which form interface dipoles. These defects are also modulated by interface structural disorder, and with an energy distribution of gap states (the energy spectrum of defect distribution) [10]. This view seems quite probable in actual metal/semiconductor interfaces.

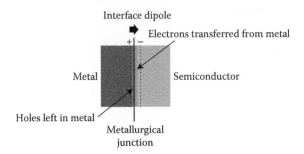

FIGURE 3.2 Schematic illustration of dipole formation at the metal/semiconductor interface. There are several models explaining how charges are transferred between the metal and semiconductor.

3.1.4 Bond Polarisation Model

As it is unrealistic to expect actual metal/semiconductor interfaces to have fixed separation between the metal and semiconductor used in analysis, more specific atomic structures of the interface should be taken into consideration. In that sense, the Fermi-level position is not pinned as usually discussed but is likely to be determined by local parameters that vary spatially at the interface [11], and the SBH that is experimentally determined merely represents the averaged position of the Fermi level with respect to the semiconductor band structure.

In any models, interface dipoles are formed by the charge transfer between metal and interface gap states and modulate the simple Schottky–Mott model significantly as schematically shown in **Figure 3.2.** We do not intend to discuss the validity or accuracy of each model, but it is worth keeping these views in mind when metal contact to graphene is considered.

3.1.5 Doping

The effect of impurity doping in semiconductors on electrical contacts is discussed. Doping concentration is quite important in achieving low-contact resistance, as already demonstrated in **Equation 3.1.** High levels of doping reduce the tunnelling distance, resulting in decreased contact resistance at the interface [3]. Thus, high-impurity doping into semiconductors is a conventional but powerful method of achieving low-contact resistance. Although this is straightforward and easy to understand intuitively, it is not evident whether the SBH might be changed in heavily doped semiconductors or not.

Since metal/graphene contacts represent the main issue that needs to be discussed in this section, we will bring our discussion of conventional metal/semiconductor interfaces to a close. However, the basic idea behind this is important for understanding metal/graphene contacts. Graphene seems to be located just between semiconductors and metals in terms of no band gap and no free carriers at the Fermi level [12]. These facts make graphene quite

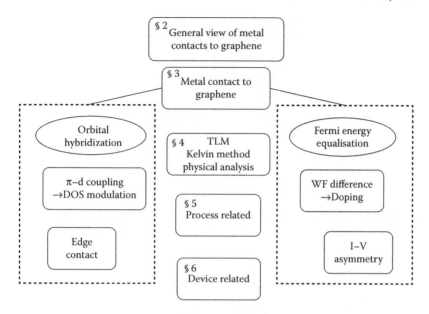

FIGURE 3.3 Framework for overlooking this chapter.

different from conventional semiconductors or metals. Graphene can generally and easily makes contact ohmic with any metal, but the absolute value of contact resistance is relatively high. These facts have been clearly identified in the graphene community. This section discusses the metal contacts to graphene, which are mainly based on our experimental results.

Figure 3.3 outlines the overall framework discussed in this section.

3.2 General View of Metal Contact to Graphene

First, let us think about an interface at which graphene is separated from metal with a finite thickness, as outlined in **Figure 3.4**. Graphene intrinsically has linear energy dispersion in the density of states (DOSs) with no energy band gap at the Dirac point. This fact makes graphene significantly different from conventional Schottky contacts in that no depletion layers are formed due to lack of an energy band gap.

The carrier transport between metal and graphene can theoretically be described with carrier transmission theory. Therefore, the effect of graphene DOSs on contact resistance is apparently critical because electrons in the metal should be transferred to empty states in graphene. In fact, there are much fewer empty states in graphene as final states than occupied states in the metal, as indicated in **Figure 3.5**. A very small number of states are available for electrons in graphene. Thus, poor graphene DOSs are intrinsically a bottleneck in achieving low-contact resistance on graphene [13]. This is the major challenge in the electron device applications of graphene. Here, it is

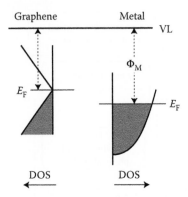

FIGURE 3.4 Comparison of energy dependence of DOSs between graphene and metal, where both are located independently without any interactions.

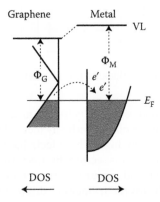

FIGURE 3.5 Fermi-level alignment where metal is in contact with graphene. Electron transfer that changes the band alignment is schematically illustrated.

worth mentioning that graphene DOSs increase with an increase in the carrier number (E_F). The injection probability is calculated by using elementary quantum mechanics, where the energy barrier height and width of the barrier should limit the transmission probability. If the energy barrier is assumed to be the vacuum in **Figure 3.4**, the separation between graphene and metal is quite important. The width between graphene and the metal depends on how the interactions occur at the metal/graphene interface.

Momentum conservation in the electron transmission process is in principle needed theoretically [14], but is not seriously considered in this section because the contact metal is polycrystalline in most cases.

3.3 Metal/Graphene Interactions

There are roughly two kinds of interactions between metals and adsorbates. The first is physical adsorption (physisorption), which is based on van der

Waals attractive interaction. A strong repulsive force at quite-near regions of the surface works between them. Therefore, physisorption is a bound state under a balance between attractive and repulsive potentials. The binding energy is generally small. The second is chemical adsorption (chemisorption), which is discussed in more detail below, because when we consider electrical contact, chemisorption interactions are more important. The former induces weaker interactions and the latter induces stronger ones.

A number of reports on metal/graphene contacts in addition to our reports have been published so far [15–23]. If we can increase the carrier density in graphene, an increase in DOSs is expected, as previously explained. This is in contrast with the decrease in tunnelling distance in conventional semiconductors. Since C–C bonds are quite stable, it is very difficult in practice to dope column III or V elements substitutionally into graphene, which is quite different from doping into Si or Ge. Therefore, it is critically important how the metal/graphene interface is controlled without impurity doping.

3.3.1 Chemical Bonding of Metal with Graphene

3.3.1.1 Electrochemical Equalisation

A metal with a large difference in the work function from graphene may transfer a number of carriers to graphene from the viewpoint of *electrochemical equalisation*, which increases the DOSs of graphene. Since the work function of graphene has been reported to be ~4.5 eV, electrons should theoretically be transferred from graphene to metal in the case of a high work function metal such as Pt or Au, which means hole doping into graphene. Therefore, even though there are no chemical interactions between them, electrochemical potential balance will be achieved, as shown in **Figure 3.5**, which corresponds to the Schottky–Mott limit discussed for conventional metal/semiconductor interfaces. Since charge transfer forms a dipole at the metal/graphene interface, the vacuum level is not a constant at the interface. Furthermore, charge transfer is not simply determined by the difference in work functions. This is due to the fact that orbital hybridisation features, discussed later, might partly be involved at the interface in real contacts.

Charge transfer influences I-V characteristics in back-gated graphene field-effect transistors (FETs). In fact, although graphene underneath the metal is in principle p-type for high work function metals, graphene in the channel should ideally be neutral. This effect can experimentally be observed in the asymmetric R_{ch}-V_{GS} characteristics, as typically plotted in **Figure 3.6**. This is understandable from the viewpoint of the formation of p–n junctions inside the graphene channel [24]. **Figure 3.7** outlines the case of Ni/graphene, in which a chemical reaction at the interface might be involved as well as asymmetric channel conductance as a function of the back-gate bias; a significant change in channel properties can be observed. In the case of electron doping, n–p–n is formed at negative V_{GS}, whereas n–n–n is formed at positive V_{GS}. This will be discussed again in Section 3.4.4 from the viewpoint of current flow at the edge of the metal/graphene interface.

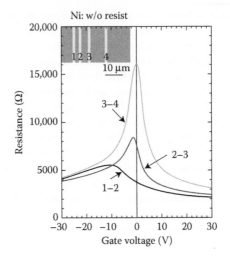

FIGURE 3.6 Gate bias dependence of channel resistance in back-gated graphene FET. Source and drain metal is Ni. This indicates a larger negative shift in Dirac point at a shorter distance between two electrodes. Asymmetric resistance characteristics as a function of back-gate bias are also clearly shown.

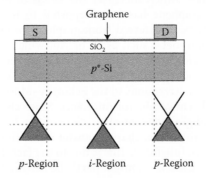

FIGURE 3.7 Schematic of back-gated graphene FET. Regions underneath the source and drain are carrier transferred to be n- or p-type, which is different from the graphene channel region. Here, effects due to substrate SiO_2 are ignored. Thus, p–n junctions are automatically formed by applying gate bias on the channel.

3.3.1.2 Orbital Hybridisation

It has been generally found that graphene has no dangling bonds on the surface, which is ideally true. However, this does not mean there is no orbital hybridisation with the metal. In fact, chemical bonding between graphene and the metal such as π–d interactions is formed. In this case, the bonding does not result from the electrochemical equalisation, discussed above, but from *orbital hybridisation*. A simple depiction of the electronic structure in

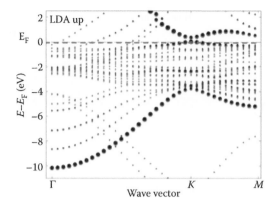

FIGURE 3.8 Theoretical results on E–k dispersion relationship around K point. Linear dispersion of graphene is destroyed by making contact with metal. (Adapted from M. Vanin et al., *Phys. Rev.* **B81**, 2010, 081408.)

graphene is no longer valid, and should be reconsidered for respective metals from the viewpoint of how the bonding orbitals are hybridised. Consequently, linear dispersion in graphene may be modified or partly destroyed, as indicated in **Figure 3.8** [25], and contact resistance should be significantly modified.

It is more realistic to consider that both electrochemical equalisation and orbital hybridisation more or less occur simultaneously. The chemical interaction strength at the metal/graphene interface for various metals is summarised from the viewpoint of the Dirac point shift in **Figure 3.9** [26]. There seems to be a trend, but it still includes large amounts of data scattering. One of the main reasons for this is that the Dirac point is determined not only by charge transfer but also by the change in electronic dispersion due to chemical bonding, as previously discussed. Another reason is that charge neutrality is also influenced by charge transfer from the substrate. Since most experiments have been carried out on SiO_2, the charge transfer from substrate SiO_2 has significantly affected the Dirac point of graphene. **Figure 3.9** presents the overall trend in the results that have so far been experimentally reported. Note that this does not always directly indicate metal/graphene interactions.

What is the intrinsic origin of different interactions depending on what metal is on graphene, apart from the effects of substrate SiO_2? Unoccupied d-states in metals seem to have something to do with the chemical bonding strength between metal and graphene. Let us think about both Au and Ni for graphene contacts, as typical examples [27]. When orbital hybridisation is concerned, both bonding and anti-bonding states of π-orbitals in graphene with d-orbitals in metal are expected. However, no energy gain is obtained by π–d hybridisation for Au with the occupied d-band because the Fermi level of Au is above the anti-bonding state, and orbital hybridisation will not provide energy gain. Finite energy gain for Ni with the unoccupied d-band is expected, on the other hand, because the Fermi level for transition metals is

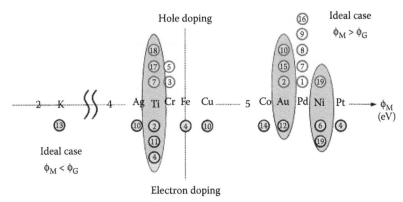

(1) Nature Nano, 6, 179 (2011)   (6) APL, 96, 253503 (2010)   (11) Sol. Stat.com, 149, 1068 (2009)   (16) I. PhysE, 40, 228 (2007)
(2) Nature Nano, 3, 486 (2008)   (7) PRB, 78, 1214028 (2008)   (12) PRB, 81, 115453 (2010)   (17) EEEE EDL, 28, 1282 (2007)
(3) Nano Lett, 9, 3430 (2009)   (8) PRB, 79, 245430 (2008)   (13) Nature Phys., 4, 377 (2008)   (18) Nature Nano, 3, 491 (2008)
(4) PRB, 80, 75406 (2009)   (9) Nano Lett, 9, 1089 (2009)   (14) APL, 93, 152104 (2008)   (19) APL, 97, 143514 (2010)
(5) Nano Lett, 9, 1742 (2009)   (10) APL, 97, 53107 (2010)   (15) PRB, 84, 33407 (2011)

FIGURE 3.9 Differences in doping types of graphene obtained experimentally in literature for various metals. Doping type is not simply determined by the work function, particularly for Ti, Au and Ni. Numbers correspond to references cited in the text.

between the bonding and anti-bonding states, as schematically outlined in **Figure 3.10**. This is the driving force in the π–d hybridisation of graphene with Ni. Therefore, the coupling between Au and graphene is mainly dominated by electrochemical equalisation, while that between Ni and graphene is dominated by orbital hybridisation. These intuitive views help us to better understand how metal interacts with graphene.

Note that strong orbital hybridisation is not necessarily better to achieve lower contact resistance on graphene because it may change the peculiar electronic structure of graphene. Therefore, there have been too many inconsistent results. As a result, understanding graphene/metal interactions is the key to achieving lower contact resistance at metal/graphene interfaces.

This understanding has recently been extended to σ–d interactions by considering the 'end-contact', outlined in **Figure 3.11.** Theoretical calculations have predicted that σ–d interactions should dramatically lower the contact resistance [28]. sp^2-based σ-orbital bonding with metal d-orbitals should intuitively be stronger than π–d interactions. Rather, low-contact resistance was demonstrated on the basis of the same concept by forming zigzag edges in the source and drain [29], as outlined in **Figure 3.12**. Furthermore, we can also see that there is a 'sweet spot' for patterned contacts in terms of the reduction in total resistance.

Edge contact has recently been further extended to side edge contacts in boron nitride (BN)-sandwiched graphene structures, as schematically shown in **Figure 3.13a**. Improved contact resistance such as 150 Ω/μm is plotted in **Figure 3.13b** [30]. sp^2 σ-orbitals can bond with the graphene edge at shorter

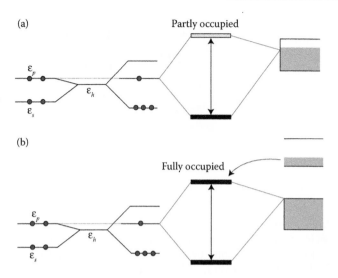

FIGURE 3.10 Schematic illustration of interaction of p-orbital of graphene with (a) Ni and (b) Au. Graphene is described by the sp² tight binding model, in which ε_s, ε_p and ε_h are s-, p- and sp² hybridized orbitals, respectively. In (a), anti-bonding state is partly occupied, while it is fully occupied in (b). In case (a), Ni, electrons can communicate between graphene and Ni, while in case (b), Au, π-d orbital hybridization is not expected.

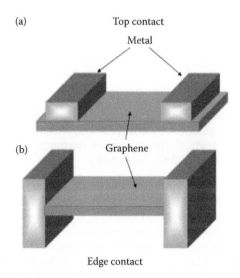

FIGURE 3.11 Schematics of (a) top contact and (b) edge contact. (Adapted from Y. Matsuda, W.-Q. Deng and W. A. Goddard III, *J. Phys. Chem.* **C 114**, 2010, 17845.)

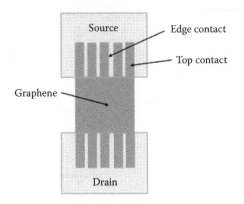

FIGURE 3.12 Image of zigzag contacts with graphene. Graphene is cut into several pieces in source and drain regions to expose edges to the metal. The number of edges increases significantly by decreasing the top-contact area. (Adapted from J. Smith et al., *ACS Nano* **7**, 2013, 3661.)

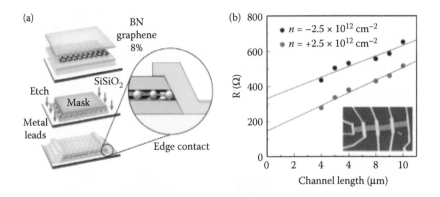

FIGURE 3.13 (a) Schematic of edge contact in BN/graphene/BN sandwich structure. (b) Two-probe resistances for both polarities of carriers in the channel as a function of channel length. Quite-low-contact resistances are shown. (Adapted from L. Wang et al., *Science* **342**, 2013, 614.)

distances, which reduces electron transmission. Interestingly, it has also been found that the oxygen-terminated graphene edge, shown in **Figure 3.14,** is more beneficial for reducing contact resistance than hydrogen- or fluorine-terminated edges due to the smaller binding distance [31]. Such orbital engineering is certainly the key to pushing graphene technology ahead.

We should mention that graphene has an atomically flat surface without dangling bonds, but the metal is polycrystalline in most cases. Therefore, metal/graphene interactions may not be uniform but rather inhomogeneous [32]. This may actually cause distributed interactions of physisorption and chemisorption on graphene. This should be further studied in the future.

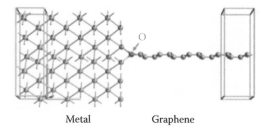

Metal   Graphene

FIGURE 3.14 Atomic image of edge contact. Theoretical results suggest that incorporation of oxygen atoms between carbon and metal would decrease contact resistance. (Adapted from L. Wang et al., *Science* **342**, 2013, 614. Supporting data.)

3.4 Electrical Characterisation of Metal Contact to Graphene

3.4.1 Current Flow Path at Metal/Graphene Interface

The specific contact resistance is generally described by a unit of (Ω/cm^2). However, current should be uniformly injected into the area to estimate the intrinsic contact resistance per given area rather than per width. In fact, current is always injected non-uniformly to graphene because of the finite resistance of the graphene layer. Therefore, the current flow path needs to be clarified to estimate contact resistance more accurately. The 'transmission line model (TLM)' can be applied to describe real contact as outlined in **Figure 3.15a and b**. Here, the specific sheet resistance of a metal (ρ_S^M $\Omega/\square$), the specific

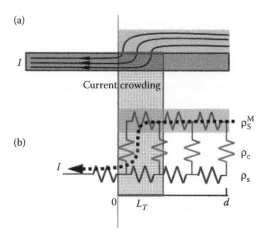

FIGURE 3.15 (a) Schematic image of current flow around the contact edge. Current crowding generally occurs at the corner region due to low metal resistance and finite contact resistance. (b) Equivalent circuit model indicating current crowding at the corner.

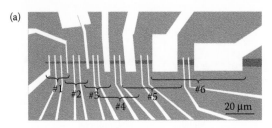

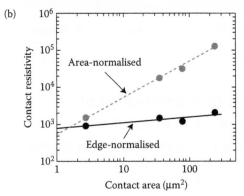

FIGURE 3.16 (a) Optical micrograph of a four-terminal contact resistance measurement device. Area of the source and drain is changed, while the potential probe has the same area. The method of contact resistivity estimation is explained in **Figure 3.19**. (b) Comparison between area-normalised and edge-normalised contact resistivity. Edge-normalised contact resistivity apparently better describes contact resistivity.

sheet resistance of graphene (ρ_s $\Omega/\square$) and the specific contact resistivity (ρ_c $\Omega/$ cm^2) determine current injection properties. This model has been well used in the descriptions of conventional semiconductors to characterise the 'current crowding' effect. Electrons in graphene FETs should preferentially flow into the metal at the contact edge when there is high graphene resistance. Because ρ_S^M is smaller than ρ_s, it is reasonable to consider that the current flows preferentially in the metal and that it is injected into graphene at the contact edge.

Figure 3.16a outlines the multi-terminal back-gated FET structure on monolayer graphene with various electrode distances. The contact resistivity is plotted in **Figure 3.16b** normalised by both the metal contact area and the edge length. These results suggest that the effective contact area is mostly determined by the edge rather than the physical contact area. This fact directly indicates that ρ_c should not be defined by the apparent contact area [33].

3.4.2 Transfer Length Method

The simplest way of estimating contact resistance is with the 'transfer length method (TLM)'. It should be distinguished from the 'transmission line model' because it is also abbreviated as TLM.

It is quite useful to more quantitatively characterise this effect in **Figure 3.15** by assuming L_T, which is approximately characterised by the relative magnitude of ρ_s and ρ_c as $L_T = \sqrt{\rho_C/\rho_S}$, where metal sheet resistance has been neglected. The underlying idea is as follows [34]. The potential distribution under the contact is not uniform but determined by both ρ_c and ρ_s as

$$V(x) = \frac{I\sqrt{\rho_S\rho_C}\cosh[(L-x)/L_T]}{W\sinh(L/L_T)}, \quad L_T = \sqrt{\frac{\rho_C}{\rho_S}}, \tag{3.2}$$

where L and W are the apparent contact length and width, I is the current flowing into the contact and x is the distance from the contact edge. The voltage drops nearly exponentially with the distance from the edge. The '$1/e$' distance of the voltage curve is defined as the transfer length, L_T. Thus, the graphene/metal contact is more accurately characterised by using both ρ_c and L_T, particularly for $L > 1.5\,L_T$ instead of edge-normalised or area-normalised ρ_c.

Furthermore, it is obvious that the channel resistance in an FET consists of contact resistance (R_{con}) and channel resistance (R_{ch}). Namely, $R_{tot} = R_{ch} + 2R_{con}$. Therefore, by measuring R_{tot} between two probes as a function of the channel length L_{ch}, $2\cdot R_{con}$ is obtained by extrapolating R_{tot} to $L_{ch} = 0$ because R_{ch} should be in proportion to L_{ch}. $2L_T$ is obtained, as shown in **Figure 3.17**, by extrapolation to zero on the x-axis. It should be noted here that R_{con} is not simply ρ_c/LW, but rather ρ_c/L_TW by taking into account the effective area, L_TW, for current injection at the contact. This method has often been used in conventional semiconductor analysis. It is quite straightforward and can easily be understood.

All the contact and channel resistances are assumed to be the same in this method for any two probes, but this is not always correct particularly for graphene because graphene conductance and the Dirac point strongly depend on

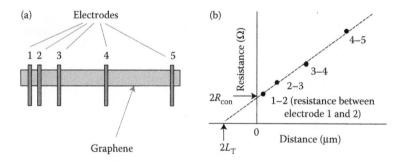

FIGURE 3.17 (a) Typical experimental setup in two-probe resistance measurement of graphene. Graphene sample with many electrodes is schematically illustrated. (b) Measured resistances are plotted as a function of distance between two electrodes. Both contact resistance and L_T (transfer-length) are obtained simultaneously by extrapolating experimental values to zero on x- and y-axes. The number of 'n-m' means the resistance between electrodes 'n' and 'm'.

how graphene is in contact with the back interface (SiO_2). Further, graphene resistance underneath the metal may depend on the back-gate voltage. This is closely related to what L_T is and what the interaction of graphene with metals is. Therefore, results should carefully be examined from the viewpoints of experimental conditions.

3.4.3 Cross-Bridge Kelvin Method

The cross-bridge Kelvin (CBK) structure for one contact has been used [34] to overcome these problems with TLM, in which three electrodes are connected to an L-shaped channel in **Figure 3.18a.** Here, the voltage is measured between contacts 2 and 3, while the current flows from 1 to 2. The electrical potential distribution is not uniform, as previously discussed. Since the voltage between contacts 2 and 3 is an average of the potential over the contact length from zero (the front edge) to L (the end edge), the measured voltage is

$$V(x) = \frac{1}{L} \int_0^L V(x) = \frac{\rho_c}{LW} I.$$

Thus, the contact resistance is described as

$$R_{con} = \frac{V}{I} = \frac{\rho_c}{LW}. \tag{3.3}$$

Thus, ρ_c with an area-normalised unit of Ω/cm^2 can be directly estimated.

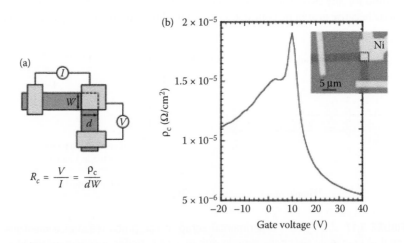

(a) $R_c = \dfrac{V}{I} = \dfrac{\rho_c}{dW}$

FIGURE 3.18 (a) CBK method to estimate contact resistivity Ω/cm^2. (b) Typical experimental results for contact resistivity estimated by the CBK method are plotted as a function of back-gate bias. Significant asymmetry is observed as a function of back-gate bias.

The results obtained for Ni/graphene contacts with this method are shown in **Figure 3.18b** [33]. Contact resistivity significantly depends on the back-gate bias, and exhibits large asymmetry due to the carrier-doping effect.

3.4.4 Four-Probe Method and Graphene Resistivity Underneath Metal

The results in **Figure 3.18b** indicate that graphene underneath the metal still behaves as graphene. The resistivity of graphene in contact with metal was analysed using simple four-probe measurements to reveal this effect more directly. Before discussing it, it is worth mentioning that the four-probe measurements in **Figure 3.19** can be used for R_{con} estimation. Since ρ_{ch} can be accurately estimated using measurements of the potential difference, both R_{ch} and R_{con} can be simultaneously estimated. Namely, $R_{tot} = 2R_{con} + R_{ch}$, where R_{ch} is $\rho_{ch} \cdot L_{ch}$. As a matter of fact, the contact resistivity in **Figure 3.16** was obtained by using this method.

The resistivity of graphene underneath the metal can also be estimated with this four-probe method. Source–drain current flows through the metal (with a length of L_M) on the graphene channel because the resistivity of metal film is much lower than that of graphene. However, with decreasing L_M, the current may flow preferentially through graphene due to finite graphene/metal contact resistance. The resistivity of a metal/graphene-stacked layer with different L_M was measured as a function of V_{BG} in the test structure outlined in **Figure 3.20a**, in which the metal electrode was Ni. The results are plotted in **Figure 3.20b** [26]. This device was fabricated by using electron beam lithography with a chemical resist. As L_M shortened, the resistivity of the metal/graphene-stacked structure clearly demonstrated ambipolar behaviour as a function of V_{GS}. More importantly, V_{BG}-dependent resistivity is direct experimental evidence of the carrier modulation of graphene underneath the Ni electrode. That is, the Fermi level of graphene is certainly changed by V_{BG}, even when graphene is in contact with Ni. This behaviour can be understood by considering the simple resistor TLM shown in **Figure 3.21a**. The current flow path was substantially changed by changing the metal length. Higher

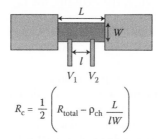

$$R_c = \frac{1}{2}\left(R_{total} - \rho_{ch}\frac{L}{lW}\right)$$

FIGURE 3.19 Four-probe method for evaluating contact resistance. Actual contact resistance can be estimated by accurately estimating intrinsic graphene resistivity with the four-probe method.

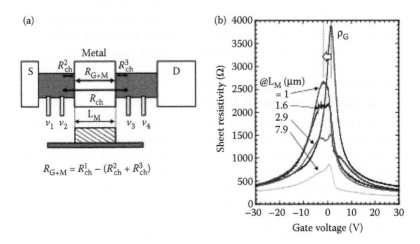

FIGURE 3.20 (a) Configuration for resistance measurements of graphene in contact with metal, which could be estimated without considering contact resistance. (b) Sheet resistance of graphene in contact with Ni indicates a large gate bias dependence. A longer metal leads to less gate bias dependence. Dirac point shift is also observed.

ratios of channel current were injected into longer metals with smaller contact resistivity [26], as plotted in **Figure 3.21b.** The results are intuitively coherent. It should be noted here that metal/graphene interactions depend on both the metal material and processes employed, as previously discussed. A resist-processed Ni contact was used in this experiment. This effect will be discussed later in Section 3.4.5.

3.4.5 Quantum Capacitance Measurements

Graphene DOSs are directly related to contact properties, and it would be quite informative to characterise graphene DOSs experimentally, as explained in Section 3.4.2. We focussed attention on quantum capacitance (C_Q) measurements for this reason because, theoretically, the C_Q of graphene has directly been described [35] as follows:

$$C_Q = e^2 DOS. \tag{3.4}$$

C_Q is regarded as energy compensation for carrier accumulation in graphene, as outlined in **Figure 3.5,** and metal/graphene interactions were directly evaluated by measuring C_Q through a modification to graphene DOSs. Thus, C_Q measurements, if possible, can be quite a powerful tool for characterising the DOSs of graphene in contact with metals.

Note that C_Q is always connected to gate capacitance in series, as outlined in the equivalent circuit of **Figure 3.22,** and this is theoretically masked by a small gate capacitance because C_Q is quite large. Since 300-nm or 90-nm-thick

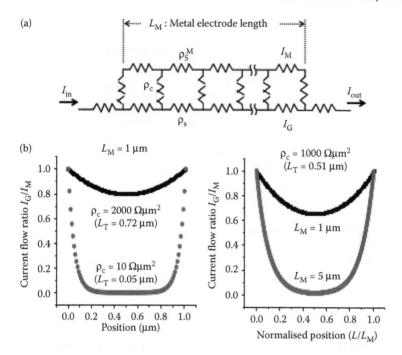

(a)

(b)

FIGURE 3.21 (a) Equivalent circuit to represent results in **Figure 3.20b**. (b) Ratio of current flow in metal to that in graphene by assuming typical parameters for L_M, L_T and ρ_c in TLM. Two kinds of calculated results are plotted. Current flow ratio varies depending on position underneath the metal. Results are in good agreement with those expected from experiments in which resist process was used.

SiO_2 has often been used for back-gated FETs in graphene research, it is impossible to detect C_Q experimentally in conventional back-gated FETs. Therefore, we need to examine graphene on a very thin SiO_2 with C_Q-comparable equivalent-oxide thickness.

Fortunately, the contrast of graphene on SiO_2 increased in a few nanometer-thick SiO_2 regions according to our calculated results on the visibility

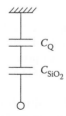

FIGURE 3.22 Equivalent circuit for quantum capacitance in back-gated graphene FET on SiO_2.

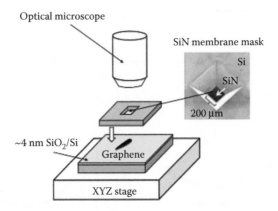

FIGURE 3.23 Metal electrode pattern formation without any resist process. SiN membrane mask on Si substrate was used for making fine patterns, which was fabricated by FIB etching. PMMA was first deposited, ad then removed in acetone, in order to measure the device with resist. The membrane mask with arbitrary patterns was adjusted on graphene underneath an optical microscope.

of graphene on SiO_2. Therefore, the exfoliated graphene layer was placed on an ~3-nm-thick SiO_2/n^+Si wafer. Although the contrast of graphene on SiO_2 was very weak, it could be detected under a microscope. Au and Ni were deposited on graphene, and then metal/graphene/SiO_2/n^+-Si stacks were fabricated. Here, Au and Ni were examples of charge transfer for the former and orbital hybridisation for the latter. As discussed later, the process prior to metal deposition critically affects metal/graphene interactions. Thus, metal patterning without a resist process was developed by fabricating very finely patterned shadow masks for this experiment [24], as outlined in **Figure 3.23.** The polymethyl methacrylate (PMMA) resist was intentionally spin coated on graphene in the resist process, and then removed with acetone before the metal was deposited. It should be experimentally noted that peripheral C_{SiO_2} should be taken into account because the gate area is not the same as that of the graphene flake shown in **Figure 3.24a.** Parallel capacitance has also been included from the equivalent circuit in **Figure 3.24b. Figure 3.25** plots the extracted C_Q of graphene in contact with Ni and Au as a function of the Fermi energy (E_F). Note that there is a significant difference in DOSs of graphene in contact with Ni with and without a resist process. Without a resist, the linear dispersion of graphene is substantially destroyed, while with the resist, it is rather moderate and comparable to that of Au near the Dirac point. The Ni contact with the resist process looks like an Au-type interaction. It clearly indicates that a nonvisible resist residue substantially affected the graphene DOS [24].

C_Q measurements directly and sensitively detected changes in the DOSs of graphene in contact with metal. The effect of the resist on contact resistance is further discussed in Section 3.5.

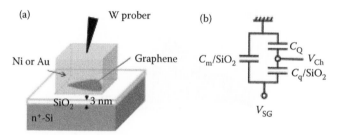

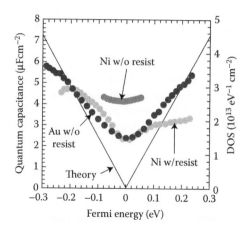

FIGURE 3.24 (a) Arrangements for quantum capacitance measurements of small flakes of graphene. Graphene can be detected under an optical microscope on 3–4-nm SiO_2 on Si substrate. Small square metal patterns were directly deposited through the membrane mask. (b) Equivalent circuit for measuring graphene quantum capacitance. Parallel capacitance derived from a simple metal–oxide–semiconductor capacitor should be taken into account for analysing this system. C_m/SiO_2 and C_g/SiO_2 represent capacitances of metal/SiO_2/n+Si with and without graphene, respectively.

FIGURE 3.25 Quantum capacitance estimated in the system in **Figure 3.24**. Three kinds of results in graphene covered by Ni with and without resist process, and by Au without resist are plotted. DOS was simply calculated by using **Equation 3.4**.

3.5 Process-Related Issues

As discussed in the previous sections, the resist process for making contact patterns has a large effect on metal/graphene interactions. A straightforward experiment is described in this section. It is known that graphene is grown on Ni (111) due to appropriate lattice matching [36]. **Figure 3.26** shows two images of the electron backscattering pattern (EBSP), in which the result on as-exfoliated graphene is compared to that on resist-processed

EBSP

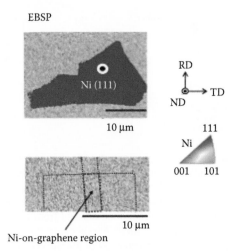

FIGURE 3.26 Comparison of EBSP patterns of deposited Ni on graphene with and without resist process. Near-epitaxial growth on as-exfoliated graphene and random orientation growth on resist-processed graphene are clearly observed.

graphene layers [26]. Nearly epitaxially grown Ni on as-exfoliated graphene and a random orientation of Ni on w/resist-processed graphene are observed. Namely, the resist residue on graphene is not easily removed with the conventional resist removal process. This fact strongly suggests that contact resistance should be dependent on how the resist residue is removed, and that graphene technology should be developed by considering the invisible resist residue.

Resist residue may relax metal/graphene interactions, while it may not increase the DOSs of graphene at the metal contact. It is clear that the resist-free interaction of graphene with Ni is much stronger than that with Au. The DOSs at the Dirac point are much higher and rather flat in the resist-free Ni/graphene case. This means that direct interaction of Ni with graphene is beneficial in terms of enhancing graphene DOSs. However, note that graphene conductivity is reduced because its peculiar dispersion of graphene is destroyed. The TLM pattern was fabricated by using a focussed ion beam (FIB) on the SiN membrane mask with the method previously described to compare the contact properties between w/resist and w/o resist processes, as outlined in **Figure 3.27**. Two-terminal resistance was measured by using the TLM patterns in **Figure 3.28a** and **b** [24]. A significantly large effect of the resist process on two-probe resistance is observed. The contact resistivity for three kinds of contacts (Au and Ni without resist and Ni with resist) was obtained from these data, as shown in **Figure 3.29**. The contact resistivity was as low as or less than 50 ($\Omega/\mu m$) for Au without a resist. This value is extremely low to the best of our knowledge [24].

Current annealing or thermal annealing has been utilised to achieve high mobility and low-contact resistance from the early stages of graphene

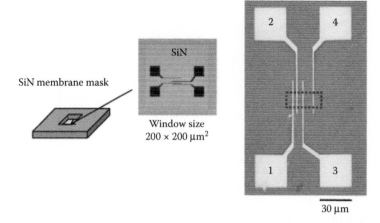

FIGURE 3.27 TLM pattern inside 200×200 μm^2 window was delineated by FIB by using the same technique as that in **Figure 3.23**. The actual metal pattern on graphene is also shown. Two-probe contact resistivity was estimated for graphene with and without resist. The pattern was aligned underneath an optical microscope.

research. This is quite reasonable by considering that contact resistance is not only determined by the distance between metal and graphene but also by modulated DOSs through metal/graphene interactions, and/or process-related residues.

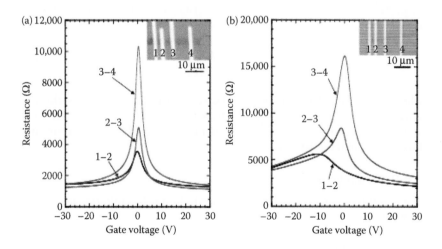

FIGURE 3.28 Two-probe resistance as a function of back-gate bias for several probe distances for Ni electrodes (a) with and (b) without resist processes. Large differences are observed depending on whether resist processes were used or not. With resist process, gate bias dependence looks quite moderate, while it is drastically degraded without resist process.

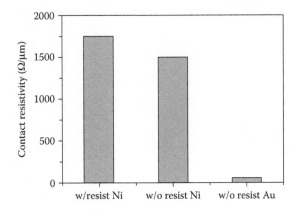

FIGURE 3.29 Two-probe contact resistivity ($\Omega/\mu m$) measured by TLM method for three cases described in **Figure 3.28**. Contact resistivity for Au without resist has quite a low value of ~50 $\Omega/\mu m$.

3.6 Effect of Contact Resistance on Graphene Devices

3.6.1 Metal–Insulation–Semiconductor Field-Effect Transistors

Contact resistance has a great effect on total transconductance in conventional semiconductor FETs, particularly those that are an ultra-short channel. Namely, when intrinsic channel conductance becomes higher, contact resistance has more severe effects as parasitic resistance. The transconductance, g_m, is described in the first-order approximation [37] as

$$g_m = \frac{g_m^0}{1 + R_S g_m^0},$$ (3.5)

where g_m^0 is the value without parasitic resistance. This is rather serious in graphene FETs. If g_m^0 is infinitely large, actual g_m approaches $1/R_s$. In other words, ultimate device performance is definitely limited by source resistance including contact resistance.

3.6.2 RF Applications

The cutoff frequency, f_T, is a performance indicator in radio frequency (RF) applications.

$$f_T \approx \frac{g_m}{2\pi} \frac{1}{(C_{GS} + C_{GD})[1 + g_D(R_S + R_D)] + C_{GD}g_m(R_S + R_D)}.$$ (3.6)

Parasitic resistance such as R_S and R_D (source and drain resistance including contact resistance) obviously degrades f_T, but a large g_D (no saturation in I_{DS}–V_{DS}) due to no energy band gap is more serious in graphene devices [37].

3.7 Conclusion

Contact properties at the metal/graphene interface are determined by metal/ graphene interactions. These interactions theoretically mean both charge transfer and the orbital hybridisation between them. Neither of these are independent but rather are related to each other. Moreover, the linear dispersion of graphene is substantially modified by contact with metal.

The carrier number in semiconductors in conventional metal/semiconductor interfaces affects the tunnelling distance, while in metal/graphene, it substantially changes the DOSs in graphene. This is the Schottky–Mott interaction for graphene. However, gap states are formed in semiconductors, while the dispersion relationship in itself is modified by the interaction by the interface states that induced dipole formation. This may correspond to the bond polarisation model or that of disorder-induced gap states in semiconductors. Although there are similarities and differences between metal/semiconductor and metal/graphene interfaces, the solutions to elucidate them atomically seem to be quite similar. One characteristic feature is that the graphene surface is always 'atomically flat' in metal/graphene. In addition, graphene properties should be investigated by taking into account the graphene/SiO_2 interface, as has been emphasised throughout this section.

The views on metal/graphene interactions that have been discussed in this section are mainly based on our experimental results. It is needless to say that although many reports on graphene have recently been published, further research is needed to solve contact problems in the engineering sense.

References

1. K. S. Novoselov, A. K. Geim, S. V. Morozov, D. Jiang, Y. Zhang, S. V. Dubonos, I. V. Grigorieva and A. A. Firsov, *Science* **306**, 2004, 666.
2. http://www.itrs.net/
3. S. M. Sze, *Physics of Semiconductor Devices*, 2nd ed. (John Wiley & Sons, New York, 1981).
4. W. A. Harrison, *Electronic Structure and the Properties of Solids: The Physics of the Chemical Bond* (W. H. Freeman and Company, San Francisco, 1980).
5. E. H. Rhoderick and R. H. Williams, *Metal–Semiconductor Contacts* (2nd ed.), Chapter 2 (Clarendon Press, Oxford, 1988).
6. V. Heine, *Phys. Rev.* **A138**, 1965. 1689.
7. W. Monch, *Electronic Properties of Semiconductor Interface* (Springer, Berlin, 2004).
8. A. Dimoulas, P. Tsipas, A. Sotiropoulos and E. K. Evangelou, *Appl. Phys. Lett.* **89**, 2006, 252110.
9. T. Nishimura, K. Kita and A. Toriumi, *Appl. Phys. Lett.* **91**, 2007, 123123.
10. H. Hasegawa and H. Ohno, *J. Vac. Sci. Technol.* **B4**, 1986, 1130.
11. R. Tung, *Appl. Phys. Rev.* **1**, 2014, 011304.
12. A. H. Castro Neto, F. Guinea, N. M. R. Peres, K. S. Novoselov and A. K. Geim, *Rev. Mod. Phys.* **81**, 2009, 109.
13. K. Nagashio, T. Nishimura, K. Kita and A. Toriumi, *Technical Digest of International Electron Device Meeting*, Baltimore, USA. p. 565, 2009.

14. E. L. Wolf, *Principles of Electron Tunneling Spectroscopy* (Oxford University Press, New York, 2012).
15. B. Huard, N. Stander, J. A. Sulpizio and D. Goldhaber-Gordon, *Phys. Rev.* **B78**, 2080, 121402(R).
16. E. J. H. Lee, K. Balasubramaniani, R. T. Weitzi, M. Burghard and K. Kern, *Nat. Nanotechnol.* **3**, 2008. 486.
17. P. Blake, R. Yanga, S. V. Morozov, F. Schedin, L. A. Ponomarenko, A. A. Zhukov, R. R. Nair, I. V. Grigorieva, K. S. Novoselov and A. K. Geim, *Solid State Commun.* **149**, 2009, 1068.
18. R. Nouchi and K. Tanigaki, *Appl. Phys. Lett.* **96**, 2010, 253503.
19. F. Xia, V. Perebeinos, Y.-M. Lin, Y. Wu and P. Avouris, *Nat. Nanotechnol.* **6**, 2011, 179.
20. A. Hsu, H. Wang, K. K. Kim, J. Kong and T. Palacios, *IEEE Electron Device Lett.* **32**, 2011, 1008.
21. D. Berdebes, T. Low, Y. Sui, J. Appenzeller and M. S. Lundstrom, *IEEE Electron Device* **58**, 2011, 3925.
22. C. Gong, D. Hinojos, W. Wang, N. Nijem, B. Shan, R. M. Wallace, K. Cho and Y. J. Chabal, *ACS Nano* **6**, 2012, 5381.
23. T. Chu and Z. Chen, *ACS Nano* **8**, 2014, 3584.
24. R. Ifuku, K. Nagashio, T. Nishimura and A. Toriumi, *Appl. Phys. Lett.* **103**, 2013, 033514.
25. M. Vanin, J. J. Mortensen, A. K. Kelkkanen, J. M. Garcia-Lastra, K. S. Thygesen and K. W. Jacobsen, *Phys. Rev.* **B81**, 2010, 081408.
26. T. Moriyama, K. Nagashio, T. Nishimura and A. Toriumi, *J. Appl. Phys.* **114**, 2013, 024503.
27. B. Hammer and J. K. Norskov, *Nature* **376**, 1995, 238.
28. Y. Matsuda, W.-Q. Deng and W. A. Goddard III, *J. Phys. Chem.* **C 114**, 2010, 17845.
29. J. Smith, A. D. Franklin, D. B. Farmer and C. D. Dimitrakopoulos, *ACS Nano* **7**, 2013, 3661.
30. L. Wang et al., *Science* **342**, 2013, 614.
31. L. Wang et al., *Science* **342**, 2013, 614. Supporting data.
32. E. Watanabe et al., *Diam. Relat. Mater.* **24**, 2012, 171.
33. K. Nagashio, T. Nishimura, K. Kita and A. Toriumi, *Appl. Phys. Lett.* **97**, 2010, 143514.
34. D. K. Schroder, *Semiconductor Material and Device Characterization*, 2nd ed. (John Wiley & Sons, New York, 1998).
35. S. Datta, *Quantum Transport: Atom to Transistor* (Cambridge University Press, Cambridge, 2005).
36. C. Oshima and A. Nagashima, *J. Phys.: Condens. Matter* **9**, 1997, 1.
37. F. Schwierz, *Nat. Nanotechnol.* **5**, 2010, 487.

4

Graphene for RF
Analogue Applications

Max C. Lemme and Frank Schwierz

Contents

4.1 Introduction

In 2004, two seminal papers on the preparation of graphene, a two-dimensional (2D) carbon-based material, put the scientific community in a frenzy [1,2]. Not only did physicists and chemists become excited by the new material, but also the extraordinary high carrier mobilities observed in graphene caught the attention of the electronic device community and raised the expectation that graphene could be the perfect channel material for fast field-effect transistors (FETs) and replace conventional semiconductors in this field [3,4]. Indeed, soon after the publication of References 1 and 2, several device

2D Materials for Nanoelectronics edited by Michel Houssa, Athanasios Dimoulas and Alessandro Molle © 2016 CRC Press/Taylor & Francis Group, LLC. ISBN: 978-1-4987-0417-5.

research groups started working intensively on graphene transistors, and in 2007, the first FET with a large-area graphene channel was demonstrated [5]. In the following, we designate such transistors as graphene field-effect transistors (GFETs). **Figure 4.1a** shows the basic structure of a GFET. Essentially, it resembles the classical metal–oxide–semiconductor FET (MOSFET) architecture and consists of an insulating substrate with the large-area graphene channel on top, the ohmic source and drain contacts, and the gate which is separated from the channel by a thin-gate dielectric.

A key feature of large-area graphene is the fact that the conduction and valence bands touch each other, and thus, it does not possess a bandgap. The missing gap has serious consequences for the operation of GFETs; most notably, it impedes proper switch-off of the transistor channel. Therefore, GFETs are not suitable as switches in digital logic circuits.

In radio-frequency (RF) circuits, on the other hand, FETs are continuously operated in the on-state and switch-off is not necessarily required. This fact, combined with the fast carrier transport in graphene, was the motivation to start research on GFETs for RF applications and already in late 2008, the first GFET capable of GHz operation was reported [7]. Since then, many groups became active in research on RF GFETs and achieved impressive results. We will show below, however, that the missing bandgap of the GFETs' channels not only prevents switch-off but also affects the on-state operation of GFETs and deteriorates their RF performance. To take advantage of graphene in RF transistors nonetheless, different strategies are conceivable:

- ✿ Trying to compensate the undesirable effects of the missing gap by exploiting other more preferable properties of graphene, most notably the high carrier mobility.

- ✿ Opening a gap in narrow graphene nanoribbons (GNRs) or bilayer graphene (BLG) and using GNRs or BLG as semiconducting channels in GNR FETs or BLG FETs.

- ✿ Developing GFETs for applications where other channel materials perform poorly or even fail, for example, for flexible electronics where the superior bendability of graphene can be exploited.

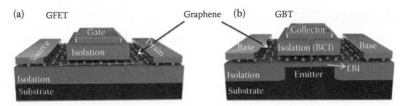

FIGURE 4.1 (a) Schematic of a GFET. Common substrates for GFETs are oxidised Si wafers, Si wafers covered with an insulator other than SiO$_2$ or SiC for GFETs with epitaxial large-area graphene channel. (b) Schematic of a GBT. Here, the charge carriers are transported vertically through the graphene. (Adapted from S. Vaziri et al., *Nanoscale*, 2015.)

❊ Elaborating alternative RF device concepts such as the vertical graphene-base transistor (GBT, **Figure 4.1b**) that do not rely on a bandgap and capitalise on the specific peculiarities of graphene.

In this chapter, we will review graphene-based transistors with respect to their suitability for RF applications.

4.2 RF Transistor Figures of Merit

In the following, we introduce three important RF transistor figures of merit (FOM) and discuss the preconditions to be met regarding the properties of the material of the active transistor region, that is, the channel in case of FETs, and the specific device design to achieve good RF transistor performance. Since so far most work on graphene transistors has been related to the FET type, mostly GFETs and, to a lesser extent, GNR FETs and BLG FETs, our main emphasis in this section is placed on FETs. We support our considerations with insights gained from the long-lasting research on III–V RF FETs. This will help us to judge the merits and problems of graphene as the channel material for RF FETs.

RF FETs are used to amplify high-frequency input signals and their amplification characteristics are described in terms of small-signal gains. Usually considered are the current gain h_{21} and the power gain, for which several definitions are used in the RF community, for example, the maximum stable gain, the maximum available gain and the unilateral power gain [8]. For the sake of simplicity, from these power gains, we select only one, the unilateral power gain U, for the following discussion. Both h_{21} and U are frequency-dependent and roll off with increasing frequency at a characteristic slope of −20 dB/dec. Closely related to these gains are the two probably most widely used FOMs of RF transistors, namely the cutoff frequency f_T and the maximum frequency of oscillation f_{max}. The cutoff frequency is the frequency at which the magnitude of h_{21} has dropped to unity (i.e. 0 dB) and at f_{max}, the unilateral power gain equals unity. Thus, f_T and f_{max} mark upper frequency limits since beyond these frequencies the transistor loses its ability to amplify. It is important to note that for the majority of RF applications simultaneously high f_T and f_{max} are desirable and that in many applications power gain and f_{max} are even more important than current gain and f_T.

A third important RF transistor FOM is the minimum noise figure NF_{min}. Every transistor generates fluctuations of voltage and current called noise. Noise is always undesirable and particularly critical for the amplification of small RF signals. A measure for the amount of noise generated in a transistor is the noise figure NF defined in **Equation 4.1** as

$$NF = \frac{P_{Si}/P_{Ni}}{P_{So}/P_{No}}, \tag{4.1}$$

where P_{Si} and P_{So} are the signal powers at the input and output and P_{Ni} and P_{No} are the noise powers at input and output, respectively. In the RF community, the noise figure is frequently not used as a dimensionless number according to **Equation 4.1** but given in units of dB as **Equation 4.2**:

$$NF\,[\text{dB}] = 10 \times \log \frac{P_{Si}/P_{Ni}}{P_{So}/P_{No}}. \tag{4.2}$$

Under optimum matching and bias conditions, the noise figure reaches a minimum called minimum noise figure NF_{min}.

When assessing the RF capabilities of FETs, it is helpful to consider the transistor small-signal equivalent circuit shown in **Figure 4.2**. Based on this equivalent circuit, approximate expressions for the three FOMs introduced above can be elaborated, from which, in turn, a good understanding of the potential merits and drawbacks of graphene-based FETs compared to other RF FET types can be achieved.

By two-port analysis, from the equivalent circuit, the following approximate expressions for f_T and f_{max} can be derived [9]:

$$f_T = \frac{g_m}{2\pi(C_{gs} + C_{gd})} \times \frac{1}{1 + g_{ds}(R_S + R_D) + (C_{gd}g_m(R_S + R_D)/(C_{gs} + C_{gd}))}, \tag{4.3}$$

$$f_{max} = \frac{g_m}{4\pi C_{gs}} \times \frac{1}{(g_{ds}(R_i + R_S + R_G) + g_m R_G(C_{gd}/C_{gs}))^{1/2}}. \tag{4.4}$$

Also based on the equivalent circuit from **Figure 4.2**, an expression for the minimum noise figure has been elaborated [10]:

$$NF_{min} = 1 + 2\pi k_f f C_{gs} \sqrt{\frac{R_G + R_S}{g_m}}, \tag{4.5}$$

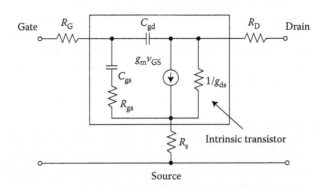

FIGURE 4.2 Small-signal equivalent FET circuit consisting of the intrinsic transistor with the gate–source capacitance C_{gs}, the charging resistance R_{gs}, the gate–drain capacitance C_{gd}, the transconductance g_m and the drain conductance g_{ds}, as well as of the extrinsic parasitic gate, source and drain resistances R_G, R_S and R_D.

where k_f is an empirical factor and f is the operating frequency of the transistor. An analysis of **Equations 4.3** through **4.5** reveals the following simple message: A good RF FET with high f_T, high f_{max} and low NF_{min} should show a transconductance as high as possible and ALL other elements of its equivalent circuit should be as small as possible.

Let us next discuss in more detail the preconditions to obtain a high transconductance, a small gate–source capacitance and a small drain conductance, the latter being particularly important for GFETs. The drain current I_D of a FET is related to the carrier sheet carrier concentration in the channel n_{sh} and the carrier drift velocity v according to

$$I_D \propto n_{sh} \, v \, W, \tag{4.6}$$

where W is the gate width. The FET's transconductance, defined as the variation of the drain current caused by a variation of the gate–source voltage V_{GS} at fixed drain–source voltage V_{DS}, can be expressed as

$$g_m = \left.\frac{dI_D}{dV_{GS}}\right|_{V_{DS}=const} \propto v W \left.\frac{dn_{sh}}{dV_{GS}}\right|_{V_{DS}=const}. \tag{4.7}$$

Equation 4.7 reveals that, to attain a high transconductance, fast carriers are needed as a prerequisite. Since in most semiconductors, electrons are faster than holes, RF FETs usually are n-channel devices, where electrons constitute the drain current. For large-area gapless graphene, however, the situation is a bit different since here electrons and holes show quite similar transport characteristics. Generally, measures for the speed of carrier transport are the mobility μ, the peak velocity v_{peak} and/or the high-field saturation velocity v_{sat}, which all should be high. It should be noted, however, that for fast RF FETs mobility and peak velocity are more relevant than saturation velocity and that one of the preconditions for a high mobility is a light carrier effective mass.

A second prerequisite for a high transconductance is an effective gate control, that is, a large variation of carrier sheet density with respect to a variation of the gate–source voltage dn_{sh}/dV_{GS}. This can be achieved by a thin gate dielectric with a high dielectric constant and a channel material with a high density of states (DOSs). The latter, however, implies a heavier carrier effective mass and thus a degraded mobility. This already shows us that it is not possible to meet all requirements on a good RF FET channel material simultaneously, but that the device engineer always faces trade-offs.

Table 4.1 shows some details of the fastest transistors from different classes of III–V high electron mobility transistors (HEMTs, a specific type of FET particularly popular in RF electronics), namely conventional GaAs HEMTs, GaAs pHEMTs (p stands for pseudomorphic), GaAs mHEMTs (m stands for metamorphic), InP HEMTs and InSb-channel HEMTs. In addition, **Table 4.2** summarises relevant properties of the channel materials of these HEMTs.

It can be seen that the transistors with InAs channels show the highest f_T and f_{max}. Obviously, the beneficial effect of the high mobility overcompensates the low DOS (indicated by the lighter effective mass) in the InAs channels.

Table 4.1   Frequency Performance and Channel Material of Different Classes of III–V HEMTs

Transistor Class	L (nm)	f_T (GHz)	f_{max} (GHz)	Channel Material	References
Conventional GaAs HEMT	100	113		GaAs	[13]
	240		151	GaAs	[14]
GaAs pHEMT	100	152		$In_{0.25}Ga_{0.75}As$	[15]
	100		290	$In_{0.25}Ga_{0.75}As$	[16]
GaAs mHEMT	20	660	1000	InAs	[17]
	40	688	800	$In_{0.7}Ga_{0.3}As$	[18]
InP HEMT	25	610	1500	InAs composite channel	[19]
	35	700	1000	InAs composite channel	[20]
InSb-channel HEMT	85	340	270	InSb	[12]
	84	280	490	InSb	[11]

Note: L: Gate length. Composite channel: three-layer channel consisting of an InAs layer sandwiched between two $In_{0.53}Ga_{0.47}As$ layers. Note that (i) InP HEMTs and GaAs mHEMTs are currently the fastest of all RF FETs and (ii) their f_T–L and f_{max}–L trend lines (for f_T trend lines, see **Figure 4.3**) indicate an f_T around 480 GHz and an f_{max} around 750 GHz for a gate length of 85 nm, that is, significantly higher compared to InSb-channel HEMTs.

Let us first consider the HEMTs with $In_xGa_{1-x}As$ channels (x is the In content). When moving from pure GaAs channels (i.e. $In_xGa_{1-x}As$ with $x = 0$) via $In_xGa_{1-x}As$ with $0 < x < 1$ to InAs (i.e. $In_xGa_{1-x}As$ with $x = 1$), we see that the channel mobility increases with increasing In content and that, by and large, the record f_T and f_{max} values for the different classes of $In_xGa_{1-x}As$-channel

Table 4.2   Room-Temperature Electron Mobility of Undoped Bulk Material μ_{bulk}, Room-Temperature Mobility in a 2DEG (Two-Dimensional Electron Gas) FET Channel μ_{2DEG}, Electron Effective Mass m_{eff} and Bandgap E_G of III–V HEMT Channels

Channel Material	μ_{bulk} (cm²/Vs)	μ_{2DEG} (cm²/Vs)	m_{eff} (m₀)	E_G (eV)
GaAs	8500	6000	0.067	1.42
$In_{0.25}Ga_{0.75}As$	16,400	7000	0.057	1.17
$In_{0.53}Ga_{0.47}As$	30,500	11,000	0.038	0.75
InAs	40,000	15,000	0.026	0.35
InSb	77,000	30,000	0.014	0.17

HEMTs follow this trend. One could argue that our comparison is not entirely fair since the conventional GaAs HEMTs and the GaAs pHEMTs from **Table 4.1** have longer gates than the GaAs mHEMTs and InP HEMTs. It has been shown, however, (i) that for conventional GaAs HEMTs and GaAs pHEMTs shorter gates do not lead to an improvement of f_T and f_{max} beyond the record values given in **Table 4.1** [8] and (ii) that for all gate lengths GaAs mHEMTs and InP HEMTs show a better f_T and f_{max} performance than conventional GaAs HEMTs and GaAs pHEMTs [8,9]. This could lead to the general conclusion that higher channel mobilities automatically lead to faster HEMTs with higher f_T and f_{max} and that the beneficial effect of a higher mobility overcompensates that of a lower DOS (indicated by the lighter effective mass).

Since InSb shows an even higher electron mobility than InAs and given the trend discussed above, one could expect that InSb-channel HEMTs outperform GaAs mHEMTs and InP HEMTs. Experiments have shown, however, that InSb-channel HEMTs, although being fast, show lower f_T and f_{max} compared to InP HEMTs and GaAs mHEMTs with similar gate length [11,12]. Several reasons may contribute to the weaker f_T-f_{max} performance of InSb-channel HEMTs: (i) the beneficial high mobility may no longer be able to compensate the deteriorating effects of the low DOS in InSb, (ii) the narrow gap of InSb ($E_G = 0.17$ eV) may deteriorate the drain current saturation (see the discussion on the drain conductance below) and (iii) the processing technology for InSb-channel HEMTs is less mature compared to that of GaAs mHEMTs and InP HEMTs. In any case, it is fair to state that a higher carrier mobility is certainly desirable but in itself does not guarantee better RF performance and is therefore not the only prerequisite for fast FETs.

We conclude that, although a high mobility is beneficial for RF FETs in general, channel materials with extremely high mobility, combined with a too light carrier effective mass and, consequently, a too low DOS may not be the optimum choice for RF FETs.

Next, we consider the gate–source capacitance C_{gs}. The overall gate–source capacitance consists of the intrinsic component $C_{gs\text{-int}}$, which is responsible for the control of the drain current and therefore, mandatorily needed for proper FET operation, and the extrinsic component $C_{gs\text{-ext}}$, which deteriorates FET performance and includes parasitic fringing and stray capacitances. The intrinsic component of C_{gs} obeys the following relation:

$$C_{gs\text{-int}} = \left| \frac{dQ_{ch}}{dV_{GS}} \right|_{V_{DS}=const} \propto LW \left. \frac{dn_{sh}}{dV_{GS}} \right|_{V_{DS}=const}, \qquad (4.8)$$

where Q_{ch} is the mobile channel charge, that is, $q \times n_{sh}$, where q is the elementary charge. Comparing **Equation 4.7** with **Equation 4.8** reveals that the only option to minimise $C_{gs\text{-int}}$ without affecting the transconductance is to use a short gate (i.e. short L), whereas the term dn_{sh}/dV_{GS} should be large. This means, in other words, that a good RF FET should show a small overall C_{gs}, a small C_{gs-int}, and a small C_{gs-ext}, BUT simultaneously a large intrinsic C_{gs} per unit area, C'_{gs-int}, given as

$$C'_{gs-int} = \frac{C_{gs-int}}{L\,W}.$$

(4.9)

An efficient gate control, that is, a large dn_{sh}/dV_{GS}, is obtained when the FET has a channel with a high DOS (i.e. heavy m_{eff}) and, most of all, a thin gate dielectric (i.e. a thin barrier between gate and channel) with a high dielectric constant ε_r. Note again the trade-off between DOS and carrier effective mass versus carrier mobility.

Finally, we take a look on the drain conductance g_{ds} which is defined as

$$g_{ds} = \frac{dI_D}{dV_{DS}}\bigg|_{V_{GS}=const},$$

(4.10)

and describes the slope of the FET output characteristics $I_D = f(V_{DS})$. A small g_{ds} as required for a good RF FET means that in the direct current (DC) operating point, the drain current should vary only little when the drain–source voltage changes, that is, the transistor needs to be operated in the regime of drain current saturation. It has been shown that a precondition for good saturation behaviour of a FET is a truly semiconducting channel, that is, a channel having a sufficiently wide bandgap [9]. Provided this requirement is fulfilled, the saturation behaviour can further be improved by a thin gate dielectric with high ε_r. As we will show, a small g_{ds} is particularly important to achieve high-power gain and high f_{max}, whereas current gain and f_T are much less affected by g_{ds}.

Table 4.3 summarises the preconditions and requirements for a good RF FET and measures to be taken to meet the requirements.

The preceding discussion leads to the conclusion that, regarding the requirements on the channel material for a good RF FET, the picture for graphene is quite mixed. While the high carrier mobility of gapless large-area graphene certainly represents a big advantage, the missing bandgap and the resulting poor drain current saturation cause serious problems. In GNRs, a sufficiently wide gap can be opened (provided the GNRs are less than about 10 nm wide and possess well-defined edges) so that drain current saturation in GNR FETs should be satisfactory. The gap opening in BLG (realistically 0.13 eV at maximum [21]) is less than the bandgap of InSb and thus, probably too narrow for a good RF FET. In addition, semiconducting GNRs and BLG suffer from a dramatically reduced mobility compared to gapless large-area graphene [9,22].

4.3 Experimental RF GFETs

We have shown that a high carrier mobility is needed for a good RF transistor. Gapless large-area graphene meets this specific requirement. In suspended

Table 4.3   Guide for Achieving Good RF FETs with High f_T, High f_{max} and Low NF_{min}

Circuit Element	Target	Requirement	Means to Meet the Requirement
g_m	Large	Fast carriers	i. Electrons instead of holes (n-channel FET)
		Efficient gate control	ii. Light m_{eff}, high μ, high v_{peak} and v_{sat}
			iii. Thin barrier with high ε_r
			iv. Large DOS, heavy m_{eff}
g_{ds}	Small	Good saturation	i. Semiconducting channel with sufficiently wide gap
			ii. Thin barrier with high ε_r
C_{gs}	Small	Small C_{gs-int} and C_{gs-ext}	Short gate length
	High C_{gs-int}	Effective gate control	i. Thin barrier with high ε_r
			ii. Large DOS, heavy m_{eff}
C_{gd}	See C_{gs}	See C_{gs}	See C_{gs}
R_{gs}	Small	Fast carriers	Light m_{eff}, high μ, high v_{peak} and v_{sat}
R_S, R_D	Small	Fast carriers	Light m_{eff}, high μ, and low R_{co}
R_G	Small	Large gate cross-section	Mushroom gate (small food print and wide head)
		Small gate finger width	Multifinger gate

Note: R_{co}: Contact resistance.

graphene, mobilities of more than 200,000 cm^2/Vs have been measured at room temperature [23] and for graphene on insulating substrates still impressive mobilities between 20,000 and 38,000 cm^2/Vs have been reported [24,25]. Therefore, it was only natural to investigate if the high carrier mobilities of graphene can be exploited for fast RF FETs and if the beneficial effects of the high mobility can overcompensate the undesirable consequences of the missing bandgap. Indeed, several groups started fabricating graphene transistors specifically designed for GHz operation [7,26,27] and research programmes on graphene RF FETs have been established [28]. Within a short period of time, the RF performance of GFETs, particularly in terms of f_T, could be improved significantly. In late 2008, the f_T record for GFET was 14.7 GHz [7], in 2009, it climbed to 50 GHz [29], in 2010, already 300 GHz have been achieved [30] and in 2012, a GFET showing an f_T of 427 GHz has been reported [31].

Figure 4.3 shows the f_T performance of GFETs together with that of the fastest III–V HEMTs (i.e. GaAs mHEMTs and InP HEMTs) and of competing

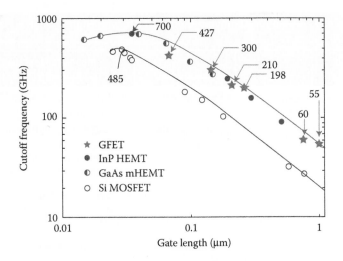

FIGURE 4.3 Cutoff frequency f_T of GFETs versus gate length, together with the best cutoff frequencies reported for InP HEMTs, GaAs mHEMTs and Si RF MOSFETs. Experimental data are shown by symbols while the lines are a guide for the eye and serve as trend lines. Stars: GFETs, data taken from References 30 through 35 and from the compilation in Reference 9. Full circles: InP HEMTs and GaAs mHEMTs, data from Reference 20 and from the compilation in Reference 9. Open circles: Si RF MOSFETs, data from the compilation in Reference 9. The numbers assigned to the symbols by arrows indicate the cutoff frequency in GHz.

nanoscale silicon RF MOSFETs. Obviously, GFETs compete well, outperform Si MOSFETs with comparable size, and show similar cutoff frequencies as the best InP HEMTs and GaAs mHEMTs down to gate lengths of about 100 nm.

However, as mentioned before, for many RF applications not the current gain and f_T, but rather the power gain and f_{max} are most important. Unfortunately, here the picture looks less promising for GFETs as can be seen from the f_{max}–f_T plot in **Figure 4.4**. Compared to the record f_{max} of InP HEMTs and GaAs mHEMTs exceeding 1 THz and that of Si MOSFETs of 420 GHz, the reported record f_{max} for GFETs is only slightly above 100 GHz [36].

When comparing the f_T–f_{max} performance of GFETs with that of other RF FETs one should bear in mind that the measured RF data of GFETs frequently undergo a de-embedding procedure that is different from the one commonly used by the RF community for RF FETs and leads to more optimistic (i.e. higher) f_T and f_{max} values [32,39].

Since the origin of the relatively poor power gain and f_{max} performance of GFETs is not obvious on first sight, with the help of the following two considerations, we discuss this issue in more detail and show that the weak saturation of the drain current and the resulting large drain conductance g_{ds} both caused by the missing bandgap of large-area graphene channels, are the main problem of RF GFETs and are predominantly responsible for their relatively low f_{max}.

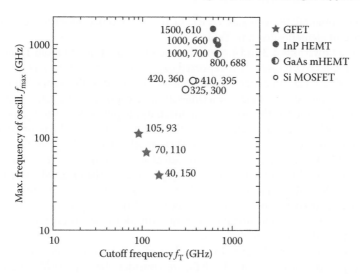

FIGURE 4.4 Maximum frequency of oscillation f_{max} versus cutoff frequency f_T of GFETs {stars [36–38] together with the best f_{max}–f_T data pairs for competing Si RF MOSFETs (open circles, taken from the compilation in Reference 39 and III–V HEMTs [InP HEMTs: full circles taken from Reference 19 and from the compilation in Reference 39, GaAs mHEMTs: half-full circles taken from Reference 17 and from the compilation in Reference 39])}. The numbers next to the symbols indicate f_{max} (first number) and f_T (second number) in GHz for the best transistors of each type. Note that only those transistors from **Figure 4.2** could be included for which both f_T and f_{max} have been reported.

Consideration 1, after Reference 9. For the intrinsic transistor (see **Figure 4.2**), the cutoff frequency f_{T-int} and the maximum frequency $f_{max-int}$ read as

$$f_{T-int} = \frac{g_m}{2\pi(C_{gs} + C_{gd})}, \tag{4.11}$$

$$f_{max-int} = \frac{g_m}{4\pi C_{gs}} \times \frac{1}{\sqrt{g_{ds} R_{gs}}}. \tag{4.12}$$

It can be seen that f_{T-int} is not affected by the drain conductance g_{ds} at all. Thus, the influence of g_{ds} on the overall f_T in **Equation 4.3** is only a weak second-order effect caused by the interaction of g_{ds} with R_S and R_D. On the other hand, already $f_{max-int}$ is seriously deteriorated by g_{ds}, according to $f_{max} \propto 1/\sqrt{g_{ds}}$. Hence, the influence of g_{ds} on f_{max} represents a fundamental and strong first-order effect.

Consideration 2, after Reference 40. Here, we analyse the mathematical structure of **Equations 4.3** and **4.4** to show that a large drain conductance has only a limited effect on f_T but is critical for f_{max}. To this end, we first consider the experimental 182-nm gate GFET T1 from Reference 30 with the equivalent circuit elements given in **Table 4.4.**

Using **Equations 4.3** and **4.4** and the circuit elements for T1 from **Table 4.4**, an f_T of 156 GHz (close to the reported experimental f_T of 168 GHz [30] and an f_{max} of 52 GHz (no experimental data are provided in Reference 30) is calculated. In past discussions, sometimes an unacceptably large gate resistance has been identified as the origin of an unsatisfying power gain and f_{max} performance of RF FETs. To check whether the gate resistance or the poor current saturation (i.e. large g_{ds}) cause the rather small calculated f_{max} of GFET T1 from **Table 4.4**, we first set g_{ds} to zero (ideal saturation) and obtain a significantly increased f_{max} of 456 GHz. Next, g_{ds} is reset to the original value from Reference 30 and the gate resistance R_G is set to zero. In this case, only a moderately improved f_{max} of 74 GHz is obtained. This clearly indicates that the large g_{ds} (caused by an insufficient current saturation) has a more serious effect on f_{max} than the large gate resistance.

To obtain further insights in the effect of the drain conductance on the f_T and f_{max} performance of RF FETs, we consider the InP HEMT T2 from **Table 4.4**. With the circuit elements from the table, an f_T of 385 GHz and an f_{max} of 1.5 THz, both close to the experimental results from Reference 41 ($f_T = 394$ GHz, $f_{max} = 1.1 \dots 1.44$ THz) are calculated. Now, we take the equivalent circuit elements from **Table 4.4** and, as a kind of a Gedanken experiment, vary the value of g_{ds} by setting it first to zero (ideal saturation), next to 2.5 mS (original value from Reference 41) and finally to 115 mS (corresponds to $g_m/g_{ds} = 0.408$, which is the g_m/g_{ds} ratio of the GFET T1) and analyse the terms in the denominators of **Equations 4.3** and **4.4**. For convenience, these two equations are given below as **Equations 4.13** and **4.14** again and the terms are designated.

$$f_T = \frac{g_m/(2\pi(C_{gs} + C_{gd}))}{1 + \underbrace{g_{ds}(R_S + R_D)}_{\text{1st}} + \underbrace{(C_{gd}\,g_m/C_{gs} + C_{gd})(R_S + R_D)}_{\text{2nd}}\,\,\,^{\text{3rd term}}}, \qquad \textbf{(4.13)}$$

$$f_{max} = \frac{g_m/(4\pi C_{gs})}{[\underbrace{g_{ds}(R_{gs} + R_S + R_G)}_{\text{1st}} + \underbrace{(g_m)/(C_{gs} + C_{gd})R_G C_{gd}}_{\text{2nd term}}]^{1/2}}. \qquad \textbf{(4.14)}$$

Table 4.4 Elements of the Equivalent Circuit of the GFET T1 and the InP HEMT T2

Transistor	g_m (mS)	C_{gs} (fF)	C_{gd} (fF)	R_{gs} (Ω)	g_{ds} (mS)	R_G (Ω)	R_S (Ω)	R_D (Ω)
T1 (GFET)	2.6	2.2	0.2	250	6.37	180	2.4	13.9
T2 (InP HEMT)	47	15	2.8	1.5	2.5	0.7	6	0.8

Source: L. Liao et al., *Nature*, vol. 467, no. 7313, pp. 305–308, 2010; R. Lai et al., *Technical digest, IEDM*, pp. 609–611, 2007.

Table 4.5   Terms in Denominator of Equation 4.13, Value of the Denominator, and Relative $f_T = f(g_{ds})/f(g_{ds} = 0)$

Cutoff Frequency f_T Using Equation 4.13					
g_{ds} (mS)	First Term	Second Term	Third Term	Denominator	Relative f_T
0	1	0	0.05	1.05	1
2.5	1	0.017	0.05	1.067	0.985
115	1	0.782	0.05	1.832	0.572

The results of our Gedanken experiment are listed in **Tables 4.5** and **4.6** and show that f_{max} reacts much more sensitively to g_{ds} than f_T due to the structure of **Equations 4.13** and **4.14**, in particular due to the fact that for small and moderate g_{ds}, the first term (unity) in **Equation 4.13** clearly dominates the denominator and limits the effect of g_{ds} on f_T.

We see that the drain conductance g_{ds} of 2.5 mS of the experimental InP HEMT results in a very limited decrease (<2%) compared to the fictive InP HEMT with ideal drain current saturation, that is, zero g_{ds}, and that even the assumption of the large g_{ds} of 115 mS still results in more than 50% of the f_T performance of the ideal HEMT. On the other hand, the effect of the moderate g_{ds} of 2.5 mS deteriorates f_{max} already by more than 50% compared to the fictive InP HEMT with ideal saturation and a g_{ds} of 115 mS leaves only 8% of the f_{max} performance. This leads us to the conclusion that the high mobility of graphene is not able to compensate the deteriorating effect of the unsatisfying drain current saturation, that is, of the missing gap, on power gain and f_{max} of GFETs. Thus, no matter how carefully the design of large-area GFETs with gapless channels is optimised, these transistors will not compare favourably with high-performance RF III–V HEMTs (and probably Si MOSFETs as well) in terms of power gain and f_{max}.

The situation could be changed by replacing the gapless graphene channel by semiconducting GNR channels. Unfortunately, experimental RF GNR FETs have not been reported yet. An overview on the theoretical work on these transistors and a summary of the predicted f_T–f_{max} performance of GNR FETs can be found in Reference 42.

Table 4.6   Terms in Denominator of Equation 4.14, Value of the Denominator, and Relative $f_{max} = f_{max}(g_{ds})/f_{max}(g_{ds} = 0)$

Maximum Frequency of Oscillation f_{max} Using Equation 4.14				
g_{ds} (mS)	First Term	Second Term	Denominator	Relative f_{max}
0	0	0.006	0.078	1
2.5	0.02	0.006	0.163	0.48
115	0.943	0.006	0.974	0.08

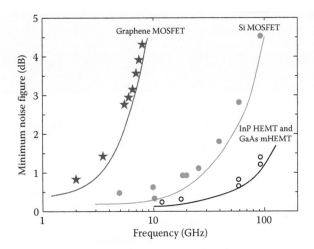

FIGURE 4.5 Minimum noise figures versus frequency reported for experimental GFETs [43], Si RF MOSFETs [44–46] and III–V HEMTs [46–50]. The lines indicate the lower limit of the noise figures reported for three transistor classes so far.

So far, only little data are available on the RF noise performance of GFETs. In Reference 43, the minimum noise figure of GFETs has been measured up to 8 GHz. **Figure 4.5** shows the experimental NF_{min} data from Reference 43 together with the best reported minimum noise figures of Si RF MOSFETs and III–V HEMTs. As can be seen, the noise performance of GFETs significantly lags behind that of the other RF FETs. With minimum noise figures below 1.5 dB up to 100 GHz, the III–V HEMTs show the best noise performance. Si RF MOSFETs with minimum noise figures below 1 dB up to 20 GHz have been reported and recently, their noise characterisation has been extended to 94 GHz, where an NF_{min} of 4.5 dB has been measured. For comparison, the GFETs from Reference 43 show comparable noise figures already at 8 GHz.

While **Figure 4.5** shows a clear trend at first sight, the gate lengths of the transistors should be taken into account for a fair comparison. In doing so, a significant difference becomes obvious: the graphene noise data stem from 1-μm gate length GFETs, while the gate lengths of the Si MOSFETs in **Figure 4.5** are in the range between 38 and 65 nm and those for the III–V HEMTs range from 50 to 150 nm. One approach to compare the noise performance of RF FETs with different gate lengths is to consider the gate length dependent noise factor M_{noise} defined as

$$M_{noise} = \frac{NF_{min}}{f \times L}, \qquad (4.15)$$

where NF_{min} is the minimum noise figure in dB, f is the frequency in GHz and L is the gate length in μm. When calculating the noise factor for the transistors from **Figure 4.5**, one obtains $M_{noise} \approx 0.5$ for the GFET, $M_{noise} = 0.5 \dots 2.5$

for the Si RF MOSFETs and $M_{noise} \approx 0.1$ for the III–V MOSFETs, that is, the GFETs are noisier than III–V HEMTs but show noise factors similar to or even lower than those of the Si MOSFETs. Regarding the minimum noise figure, the Si MOSFETs from **Figure 4.5** definitely benefit from their very short gate lengths. An improved noise performance of GFETs with shorter gate lengths can be expected, albeit not as low as those of the III–V HEMTs.

4.4 Flexible RF Electronics

So far, we have compared the performance potential of GFETs high-performance III–V and Si RF FETs or SiGe heterojunction bipolar transistors (HBTs) on common rigid substrates. There is, however, a growing interest in flexible electronics where, apart from digital circuits, also medium-performance RF transistors are needed [51]. Graphene is bendable and can easily be transferred to flexible substrates without seriously affecting the carrier mobility. For example, electron and hole mobilities of up to 8000 cm²/Vs and 6600 cm²/Vs, respectively, have been reported for chemical vapor deposition (CVD)-grown graphene transferred to polyimide [52]. This is orders of magnitude more than the mobility observed in organic semiconductors which currently are popular channel materials for flexible FETs. Since 2012, significant progress has been made in particular on research on RF GFETs on flexible substrates. **Table 4.7** highlights the state of the art in the field in mid-2015.

We note that flexible GFETs suffer from the absent off-state as much as their counterparts on rigid substrates. Their RF performance, however, is probably sufficient for many medium-performance applications, particularly when bearing in mind that the fastest of the competing organic FETs shows an f_T of less than 30 MHz [54]. Even though this transistor has a long gate length of 2 μm, it behaves much worse than flexible GFETs. A flexible graphene-based radio receiver system operating at 2.4 GHz, reported in 2014 [52] clearly demonstrates the potential of flexible GFETs.

4.5 Vertical Graphene Transistors: Performance Potential

Electronic devices with vertical carrier transport, as opposed to lateral transport as in GFETs, based on graphene and other 2D materials have received considerable attention lately. Some of these devices are based on Schottky barriers such as the 'Barristor' [55]. Others are based on quantum-mechanical

Table 4.7 State of the Art of Flexible RF GFET in Terms of f_T and f_{max}

Gate Length (nm)	f_T (GHz)	f_{max} (GHz)	References
260	198	28	[32]
500	10.7	3.7	[53]

tunnelling such as a graphene–boron nitride–graphene device [56], also suggested to operate in resonant tunnelling regime [57], the concept of the SymFET, which utilises the intrinsic symmetry of the graphene band structure [58] or hot-electron transistors with a graphene-base electrode, a GBT [59]. Here, we will focus on the projected potential of the GBT for RF applications and experimental data available to date. The device structure of the GBT is based on a graphene sheet that is sandwiched between two insulating or semiconducting barriers. These are capped by metal or doped semiconductor contacts (the emitter and the collector, **Figure 4.1b**). The carrier transport is vertical to the graphene sheet through quantum-mechanical tunnelling across the barrier between the emitter and the graphene base. There have been many implementations of similar devices in the past, including metal base hot-electron transistors [60] and III–V heterojunction devices [61,62]. However, metals need to have a certain thickness in order to conduct sufficiently, which sets a lower limit to the charge carrier transition time for vertical transport. It is therefore the combination of the high electrical conductivity and extreme thinness of graphene (at least 10 times thinner than metals) that should enable fast devices and high charge carrier transmission rates.

The operating principle of a GBT is explained in **Figure 4.6**. In the off-state, with the base electrode unbiased, the combined emitter-base insulator (EBI) and base-collector insulator (BCI) barriers prevent carrier transport despite a fixed finite collector voltage (**Figure 4.6a**). When a voltage is applied to the graphene base, the potential modulates the tunnelling barrier between the emitter and the base. Above a certain threshold, charge carriers can tunnel through the emitter-base barrier via the Fowler–Nordheim mechanism. If the materials are chosen carefully, these hot carriers can then pass through the graphene above the base-collector barrier and reach the collector electrode through ballistic transport (**Figure 4.6b**). Here, injected electrons with energies comparable to the emitter Fermi level are considered hot electrons. If the transport is truly ballistic, the carrier transit time in the GBT approaches zero. Ideally, if the base-collector barriers are low enough to suppress quantum-mechanical

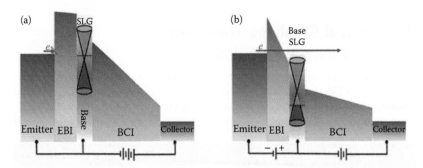

FIGURE 4.6 Simplified band diagram of the GBT in the off-state (a) and the on-state (b) in the common-emitter configuration. The energy difference between the Fermi level and the Dirac potential in the graphene represents quantum capacitance effect.

backscattering phenomena at the base-collector interface, the injected hot electrons contributing to the collector on-current would result in a current gain of 1. As an alternative concept, EBIs with low barrier heights can facilitate thermionic emission of electrons (not shown) as a result of barrier height lowering with the base potential.

The performance of GBTs strongly depends on the design parameters to maximise the current and minimise the loss mechanisms. The scattering of hot electrons in the base is already minimised due to the one-atom-thin graphene. Nevertheless, the EBI parameters need to be accurately chosen to guarantee high injection current densities, while preventing the emission of cold electrons (i.e. leakage currents) with energies comparable to the base Fermi energy via defect mediated electron transport or direct tunnelling. These cold electrons would not be able to surpass the base-collector barrier and thus lead to parasitic base currents. As a consequence, cold electron transport limits the common-base current gain or current transfer ratio α (I_c/I_E). The requirements for the BCI are likewise defined: the BCI needs to act as an electron filter, which allows the passage of the hot electrons and blocks the cold electron emission (leakage) from the base to the collector. This requires a low barrier to minimise the quantum-mechanical backscattering of hot electrons at the BCI barrier, and, simultaneously, suppress thermionic, tunnelling and defect-mediated electron transport from the base to the collector.

The potential performance of GBTs has been predicted in a zero-order estimation based on quantum-mechanical simulations in Reference 59. **Figure 4.7** shows simulated DC transfer and output characteristics for a device with an Er$_2$Ge$_3$/Ge emitter barrier ($\Phi_{EBI} = 0.2$ eV, i.e. the barrier height for electrons towards the base electrode, **Figure 4.6a**) and a compositionally graded Ti$_x$Si$_{1-x}$O$_2$ BCI. This highly idealised device shows switching over several orders of magnitude and saturating output characteristics. This early model did not

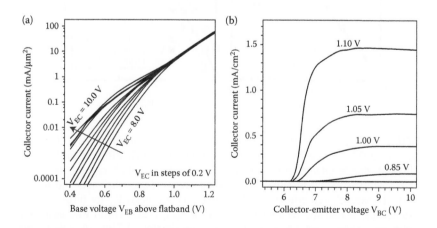

FIGURE 4.7 Simulated DC transfer (a) and output (b) characteristics of a GBT with a 3-nm EBI and an 80-nm BCI. (W. Mehr et al., Vertical graphene base transistor, *IEEE Electron Device Lett.*, 33, 691–693 © 2012 IEEE.)

include scattering effects and predicted that an EBI with a barrier of 0.4 eV or smaller and a thickness of less than 3–5 nm could lead to terahertz operation at emitter-base voltages of approximately 1 V.

An updated model has been introduced that calculates the GBT electrostatics self-consistently with the charge stored in the graphene and the electrons tunnelling through the EBI and BCI [63]. The model, therefore, accounts for the electrostatic impact of the charge travelling along the GBT, including space charge effects at high current levels that typically reduce the maximum f_T in bipolar transistors. The model calculates the I–V characteristics and it has been verified through comparison with experimental data from Reference 64. The model confirms the GBT's potential for THz operation taking into account parasitics within a fairly large design space. **Figure 4.8** shows the simulated cutoff frequency f_T as a function of the on-current density J_C of an optimised GBT with a Si emitter, a Ta_2O_5 EBI and a SiCOH BCI, as well as f_{max} as a function of the base voltage V_{BE} for different device geometries [65].

The graphene-based heterojunction transistor (GBHT) is a promising variant of the GBT proposed by Di Lecce et al. [66]. Here, the graphene base is sandwiched between an n⁺-semiconductor emitter and an n-semiconductor collector. In contrast to the hot-electron driven GBT, the carrier transport

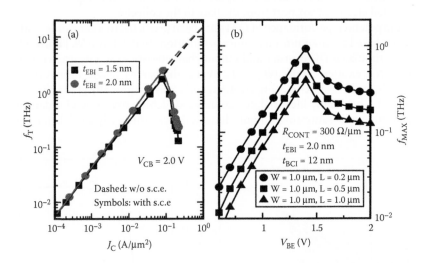

FIGURE 4.8 (a) Cutoff frequency f_T versus collector current density J_C for an optimised GBT with a highly doped silicon emitter, Ta_2O_5 EBI and SiCOH BCI. Although space charge effects in the BCI limit f_T at high current densities, THz operation is still feasible. The dashed lines indicate the modelled results without space charge effects. (b) f_{max} versus base voltage V_{BE} for different GBT geometries. A realistic contact resistance of $R_{CONT} = 300\ \Omega/\mu m$ was assumed. Optimised dimensions allow an f_{max} of approximately 1 THz. (Adapted from S. Venica et al., Graphene base transistors with optimized emitter and dielectrics, in *2014 37th International Convention on Information and Communication Technology, Electronics and Microelectronics (MIPRO)*, 2014, pp. 33–38.)

mechanism is dominated by thermionic emission over the emitter-based Schottky barrier. **Figure 4.9** shows the simplified band diagram of a silicon-based GBHT. At zero bias, the work function difference between the silicon emitter/collector and graphene forms depletion regions and causes band bending in both the emitter and collector. This leads to a triangular energy barrier between the emitter and the collector, which blocks the collector current in the off-state. This is the case even though a positive collector–emitter voltage is applied, as would be the case in typical applications (**Figure 4.9a**). In the on-state, the height and shape of the barrier is modulated by the emitter-base voltage. The base potential reduces the barrier height and electrons are injected to the base, pass through the graphene and are eventually accelerated by the electric field in the collector region (**Figure 4.9b**). The GBHT is thus very similar to an n–p–n HBT in structure and behaviour, with the graphene monolayer replacing the p-type base. The model used in Reference 66 for the GBHT simulations is the same used for the GBT in Reference 67.

The GBHT may help to overcome some of the engineering challenges encountered in GBTs, in particular, the ultra-thin tunnel barriers and low Schottky barriers. Nevertheless, the formation of a high-quality (crystalline) semiconductor layer on top of graphene presents a serious processing challenge.

A comparison through numerical simulations between an n-type Si GBHT and an n–p–n SiGe HBT structure ([69]; aggressively scaled to reach its performance limits) showed that the GBHT may deliver an f_T more than twice as high as for optimised HBTs (**Figure 4.10**, [70]). This is assuming a transparent graphene/Si interface, thanks to the monolayer thickness of the base region and the smaller emitter transit time due to the unipolar nature of the device. The fact that very few holes are present in the n-type GBHT is a key feature, which distinguishes the GBHT from the HBT.

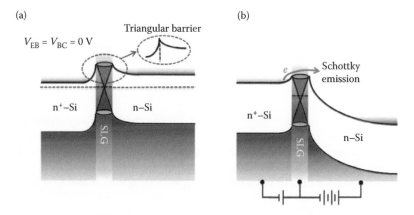

FIGURE 4.9 Simplified band diagram of the GBHT in the off-state (a) and the on-state (b) in the common-emitter configuration. The energy difference between the Fermi level and the Dirac potential in the graphene represents the quantum capacitance effect. (Adapted from S. Vaziri et al., *Solid State Commun.*, 2015.)

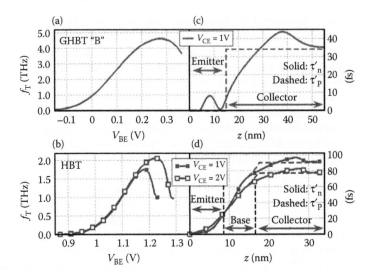

FIGURE 4.10 Simulated f_T of the GBHT (a) and an HBT (b). The GBHT has the potential to outperform the HBT's cutoff frequency by a factor of more than 2. This is confirmed by the accumulated transit times τ_p and τ_n along the devices, at the peak-f_T bias for the GBHT (c) and HBT (d).

Figure 4.10c and **d** shows the accumulated transit times $\tau'_p(z)$ and $\tau'_n(z)$ across the devices at peak f_T, defined in **Equations 4.16** and **4.17** as

$$\tau'_p(z) = \int_{z_E}^{z} \frac{d(\rho_{GR}(z') + \rho_p(z'))}{dI_C}\,dz', \tag{4.16}$$

and

$$\tau'_n(z) = \int_{z_E}^{z} \frac{d\rho_n(z)}{dI_C}\,dz'. \tag{4.17}$$

z_E is the location of the emitter contact, ρ_{GR} is the charge density on the graphene sheet (set to zero for the HBT) and $\rho_{p/n}$ are the hole/electron charge densities across the devices. The emitter, base and collector transit times for the two devices can be extracted (τ_E, τ_B, τ_C): for the GBHT, $\tau_E = 6$ fs, $\tau_B \approx 0$ fs and $\tau_C = 28$ fs; for the HBT at VCE = 1 V, $\tau_E = 25$ fs, $\tau_B = 55$ fs and $\tau_C = 10$ fs.

4.6 Vertical Graphene Transistors: Experiments

Experimentally, GBTs have been demonstrated on a chip/die scale [64,71]. In Vaziri et al., the devices composed of an n-doped silicon emitter, a 5-nm-thick thermal silicon dioxide (SiO_2) EBI tunnelling barrier, a graphene base, a 15–25-nm-thick atomic layer-deposited aluminium oxide (Al_2O_3) BCI and a metal (titanium/gold) collector. DC electrical data confirm the working principles of the GBT. **Figure 4.11a** illustrates the transfer characteristics of such a GBT at $V_B = 2$ V. The device is in its off-state before turning on at a threshold voltage of approximately 4.5 V. The device shows very low off-state leakage currents and an on/off collector current ratio of more than four orders of magnitude. **Figure 4.11b** shows temperature dependent I–V characteristics of the emitter-base tunnelling diode. For the high-field range, the dependence of the current on the temperature diminishes. This fact together with the excellent linear fit of the I–V characteristics to the Fowler–Nordheim model (inset of **Figure 4.11b**) confirm that the transport mechanism is dominated by tunnelling.

The collector current levels of the GBT in **Figure 4.11** are quite low, which would severely limit their applicability as RF devices. This can be overcome by using low-barrier dielectrics or semiconductors. Experimentally, double-layer dielectrics provide a promising route. Bilayers of a high-quality dielectric (layer 1) and a very low bandgap dielectric (layer 2) can efficiently suppress both direct tunnelling and defect-mediated leakage currents and thus make Fowler–Nordheim tunnelling (FNT) the dominant transport mechanism (**Figure 4.12a**). This can be further enhanced if the layer 2 dielectric has a very high electron affinity. Such a configuration can, in principle, result in step tunnelling (ST) [72] (**Figure 4.12b**), in which the effective barrier thickness is suddenly reduced to the thickness of the layer with the lower electron affinity (layer 1). Experimentally, this has been demonstrated with a dielectric combination of tullium silicate (TmSiO) and titanium dioxide (TiO_2). Injection diodes with this material combination showed over four orders of magnitude higher on state currents [6].

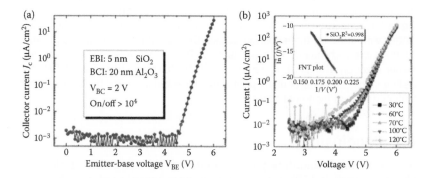

FIGURE 4.11 (a) Transfer characteristics of a GBT with 5-nm-thick SiO_2 and 20-nm-thick Al_2O_3 as the EBI and the BCI, respectively. The transfer characteristics show an on/off ratio exceeding 104. (b) Temperature dependence I–V for the injection tunnel diode with 5-nm SiO_2 as the tunnel barrier. The inset FN plot shows an excellent linear fit. (Adapted from S. Vaziri et al., *Solid State Commun.*, 2015.)

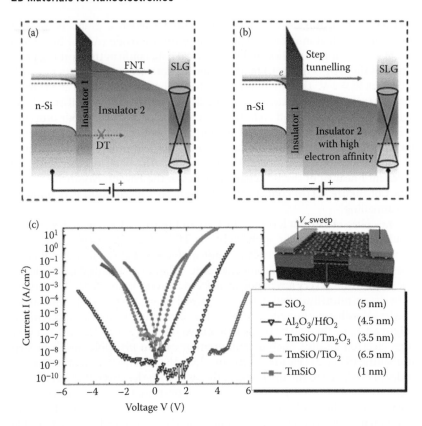

FIGURE 4.12 (a) Simplified band diagram of the GBT emitter-base injection diode in the on-state with a bilayer insulator stack showing FNT. (b) The same injection diode as in (a), but with a higher electron affinity insulator 2, resulting in the sudden onset of ST. (c) I–V characteristics of the emitter-base tunnel diodes with different tunnel barrier stacks. Devices with bilayer insulators, which combine the high quality interface layer of TmSiO with a second insulator with higher electron affinity show far superior I–V characteristics. TmSiO/TiO$_2$ tunnelling stacks show particularly promising characteristics: low threshold voltage, high current and high nonlinearity.

4.7 Conclusion

Because of the extraordinary high mobility of their large-area gapless graphene channels, GFETs have intensively been investigated and their suitability for high-performance RF applications has been examined. Indeed, GFETs with cutoff frequencies in excess of 400 GHz and maximum frequencies of oscillation slightly above 100 GHz have been demonstrated. Thus, the suitability of GFETs for GHz operation has been confirmed and in terms of the cutoff frequency f_T, they compete well with the best conventional III–V-based RF FETs and surpass the performance of Si RF FETs. Their maximum frequency of oscillation f_{max}, however, lags significantly behind that of Si and III–V RF FETs and the origin of the weak f_{max} performance is a fundamental property of large-area graphene channels – the missing gap – and cannot be overcome by optimised transistor designs and improving the processing technology. On the other hand, GFETs show promise for flexible RF applications since graphene is extremely bendable. While GFETs on flexible substrates show rather moderate f_{max}, similar to their counterparts on rigid substrates, their RF power gain and f_{max} performance is by far better than that of flexible organic RF FETs.

Vertical graphene-based transistors are considered as alternatives to the 'standard' graphene FETs. GBTs, quite similar in structure to conventional bipolar transistors, are promising devices due to their high-performance potential and their similarity to existing technology. GBTs (and GBHTs) are predicted to enable f_T and f_{max} values exceeding or approaching 1 THz, even though there is currently no experimental data to confirm the RF performance. Vertical devices are in principle also suitable for integration on flexible substrates, as demonstrated in Reference 73, where a vertical diode of graphene and WSe_2 is investigated. Another example is planar tunnel transistors with a vertical stack of 2D materials, as demonstrated by Roy et al. [74].

Finally, we note that the intensive work on graphene and graphene transistors resulted in a primarily unexpected side effect: it paved the way for 2D materials beyond graphene [39,75]. Meanwhile, several hundred of such 2D materials are known and a substantial part of them shows a bandgap and therefore is of interest for electronics. Prominent examples are the Mo- and W-based transition metal dichalcogenides and phosphorene which, due to their semiconducting nature, complement gapless large-area graphene perfectly. We are confident that 2D materials, graphene and beyond graphene, will eventually find their applications in electronics. We do not expect, however, that these materials will replace the conventional semiconductors in mainstream digital logic and in high-performance RF applications in the near-to-medium term.

References

1. K. S. Novoselov, A. K. Geim, S. V. Morozov, D. Jiang, Y. Zhang, S. V. Dubonos, I. V. Grigorieva and A. A. Firsov, Electric field effect in atomically thin carbon films, *Science*, vol. 306, pp. 666–669, 2004.
2. C. Berger, Z. M. Song, T. B. Li, X. B. Li, A. Y. Ogbazghi, R. Feng, Z. T. Dai et al., Ultrathin epitaxial graphite: 2D electron gas properties and a route toward graphene-based nanoelectronics, *J. Phys. Chem. B*, vol. 108, pp. 19912–19916, 2004.
3. A. K. Geim and K. S. Novoselov, The rise of graphene, *Nat. Mater.*, vol. 6, pp. 183–191, 2007.
4. A. K. Geim, Graphene: Status and prospects, *Science*, vol. 324, pp. 1530–1534, 2009.
5. M. C. Lemme, T. J. Echtermeyer, M. Baus and H. Kurz, A graphene field-effect device, *IEEE Electron Device Lett.*, vol. 28, no. 4, pp. 282–284, 2007.

6. S. Vaziri, M. Belete, E. Dentoni Litta, A. D. Smith, G. Lupina, M. C. Lemme and M. Ostling, Bilayer insulator tunnel barriers for graphene based vertical hot-electron transistors, *Nanoscale*, vol. 7, pp. 13096–13104, 2015.

7. I. Meric, P. Baklitskaya, P. Kim and K. Shepard, RF performance of top-gated, zero-bandgap graphene field-effect transistor, *Technical digest, IEDM*, pp. 1–4, 2008.

8. F. Schwierz and J. J. Liou, *Modern Microwave Transistors*. John Wiley & Sons, Hoboken, NJ, 2003.

9. F. Schwierz, Graphene transistors: Status, prospects, and problems, *Proc. IEEE*, vol. 101, no. 7, 1567–1584, 2013.

10. H. Fukui, Optimal noise figure of microwave GaAs MESFET's, *IEEE Trans. Electron Devices*, vol. 26, no. 7, pp. 1032–1037, 1979.

11. T. Ashley, M. T. Emeny, D. G. Hayes, K. P. Hilton, R. Jefferies, J. O. Maclean, S. J. Smith, A. W.-H. Tang, D. J. Wallis and P. J. Webber, High-performance InSb based quantum well field effect transistors for low-power dissipation applications, *Technical digest, IEDM*, pp. 1–4, 2009.

12. T. Ashley, L. Buckle, M. T. Emeny, M. Fearn, D. G. Hayes, K. P. Hilton, R. Jefferies et al., Indium antimonide based technology for RF applications, in *IEEE Compound Semiconductor Integrated Circuit Symposium, 2006. CSIC 2006*, pp. 121–124, 2006.

13. A. N. Lepore, H. M. Levy, R. C. Tiberio, P. J. Tasker, H. Lee, E. D. Wolf, L. F. Eastman and E. Kohn, 0.1 μm gate length MODFETs with unity current gain cutoff frequency above 110 GHz, *Electron. Lett.*, vol. 24, no. 6, p. 364, 1988.

14. I. Hanyu, S. Asai, M. Nunokawa, K. Joshin, Y. Hirachi, S. Ohmura and Y. Aoki, Super low-noise HEMTs with a T-shaped WSi$_x$ gate, *Electron. Lett.*, vol. 24, no. 21, pp. 1327–1328, 1988.

15. L. D. Nguyen, P. J. Tasker, D. C. Radulescu and L. F. Eastman, Characterization of ultra-high-speed pseudomorphic AlGaAs/InGaAs (on GaAs) MODFETs, *IEEE Trans. Electron Devices*, vol. 36, no. 10, pp. 2243–2248, 1989.

16. K. L. Tan, R. M. Dia, D. C. Streit, T. Lin, T. Q. Trinh, A. C. Han, P. H. Liu, P.-M. D. Chow and H. C. Yen, 94-GHz 0.1 μm T-gate low-noise pseudomorphic InGaAs HEMTs, *IEEE Electron Device Lett.*, vol. 11, no. 12, pp. 585–587, 1990.

17. T. Merkle, A. Leuther, S. Koch, I. Kallfass, A. Tessmann, S. Wagner, H. Massler, M. Schlechtweg and O. Ambacher, Backside process free broadband amplifier MMICs at D-band and H-band in 20 nm mHEMT technology, in *2014 IEEE Compound Semiconductor Integrated Circuit Symposium (CSICs)*, 2014, pp. 1–4.

18. D.-H. Kim, B. Brar and J. A. del Alamo, f_T = 688 GHz and f_{max} = 800 GHz in L_g = 40 nm In$_{0.7}$Ga$_{0.3}$As MHEMTs with g$_{m\_max}$ > 2.7 mS/μm, *Technical digest, IEDM*, pp. 13.6.1–13.6.4, 2011.

19. X. Mei, W. Yoshida, M. Lange, J. Lee, J. Zhou, P.-H. Liu, K. Leong et al., First demonstration of amplification at 1 THz using 25-nm InP high electron mobility transistor process, *IEEE Electron Device Lett.*, vol. 36, no. 4, pp. 327–329, 2015.

20. S. Sarkozy, M. Vukovic, J. G. Padilla, J. Chang, G. Tseng, P. Tran, P. Yocom, W. Yamasaki, K. M. K. H. Leong and W. Lee, Demonstration of a G-band transceiver for future space crosslinks, *IEEE Trans. THz Sci. Technol.*, vol. 3, no. 5, pp. 675–681, 2013.

21. G. Iannaccone, G. Fiori, M. Macucci, P. Michetti, M. Cheli, A. Betti and P. Marconcini, Perspectives of graphene nanoelectronics: Probing technological options with modeling, *Technical digest, IEDM*, pp. 1–4, 2009.

22. G. Fiori and G. Iannaccone, Multiscale modeling for graphene-based nanoscale transistors, *Proc. IEEE*, vol. 101, no. 7, pp. 1653–1669, 2013.
23. K. I. Bolotin, K. J. Sikes, Z. Jiang, M. Klima, G. Fudenberg, J. Hone, P. Kim and H. L. Stormer, Ultrahigh electron mobility in suspended graphene, *Solid State Commun.*, vol. 146, pp. 351–355, 2008.
24. J. Y. Tan, A. Avsar, J. Balakrishnan, G. K. W. Koon, T. Taychatanapat, E. C. T. O'Farrell, K. Watanabe et al., Electronic transport in graphene-based hetero-structures, *Appl. Phys. Lett.*, vol. 104, no. 18, p. 183504, 2014.
25. Y. Wang, B.-C. Huang, M. Zhang, C. Miao, Y.-H. Xie and J. Woo, High performance graphene FETs with self-aligned buried gates fabricated on scalable patterned Ni-catalyzed graphene, in *2011 Symposium on VLSI Technology (VLSIT)*, pp. 116–117, 2011.
26. J. S. Moon, D. Curtis, M. Hu, D. Wong, C. McGuire, P. M. Campbell, G. Jernigan et al., Epitaxial-graphene RF field-effect transistors on Si-face 6H-SiC substrates, *IEEE Electron Device Lett.*, vol. 30, pp. 650–652, 2009.
27. Y.-M. Lin, K. A. Jenkins, A. Valdes-Garcia, J. P. Small, D. B. Farmer and P. Avouris, Operation of graphene transistors at gigahertz frequencies, *Nano Lett.*, vol. 9, pp. 422–426, 2009.
28. J. Albrecht, Overview of DARPA carbon electronics for RF applications program (CERA), *Presented at the Device Research Conference*, pp. 207–208, 2010.
29. Y.-M. Lin, H.-Y. Chiu, K. A. Jenkins, D. B. Farmer, P. Avouris and A. Valdes-Garcia, Dual-gate graphene FETs with f_T of 50 GHz, *Electron Device Lett. IEEE*, vol. 31, pp. 68–70, 2010.
30. L. Liao, Y.-C. Lin, M. Bao, R. Cheng, J. Bai, Y. Liu, Y. Qu, K. L. Wang, Y. Huang and X. Duan, High-speed graphene transistors with a self-aligned nanowire gate, *Nature*, vol. 467, no. 7313, pp. 305–308, 2010.
31. R. Cheng, J. Bai, L. Liao, H. Zhou, Y. Chen, L. Liu, Y.-C. Lin, S. Jiang, Y. Huang and X. Duan, High-frequency self-aligned graphene transistors with transferred gate stacks, *Proc. Natl. Acad. Sci.*, vol. 109, no. 29, pp. 11588–11592, 2012.
32. N. Petrone, I. Meric, T. Chari, K. L. Shepard and J. Hone, Graphene field-effect transistors for radio-frequency flexible electronics, *IEEE J. Electron Devices Soc.*, vol. 3, no. 1, pp. 44–48, 2015.
33. Y.-M. Lin, K. Jenkins, D. Farmer, A. Valdes-Garcia, P. Avouris, C.-Y. Sung, H.-Y. Chiu and B. Ek, Development of graphene FETs for high frequency electronics, *Technical digest, IEDM*, pp. 1–4, 2009.
34. Y.-M. Lin, D. B. Farmer, K. A. Jenkins, Y. Wu, J. L. Tedesco, R. L. Myers-Ward, C. R. Eddy, D. K. Gaskill, C. Dimitrakopoulos and P. Avouris, Enhanced performance in epitaxial graphene FETs with optimized channel morphology, *IEEE Electron Device Lett.*, vol. 32, no. 10, pp. 1343–1345, 2011.
35. J. Lee, H.-J. Chung, J. Lee, J. Shin, J. Heo, H. Yang, S.-H. Lee et al. RF performance of pre-patterned locally-embedded-back-gate graphene device, *Technical digest, IEDM*, pp. 23.5.1–23.5.4, 2010.
36. Z. H. Feng, C. Yu, J. Li, Q. B. Liu, Z. Z. He, X. B. Song, J. J. Wang and S. J. Cai, An ultra clean self-aligned process for high maximum oscillation frequency graphene transistors, *Carbon*, vol. 75, pp. 249–254, 2014.
37. Y. Wu, K. A. Jenkins, A. Valdes-Garcia, D. B. Farmer, Y. Zhu, A. A. Bol, C. Dimitrakopoulos et al., State-of-the-art graphene high-frequency electronics, *Nano Lett.*, vol. 12, no. 6, pp. 3062–3067, 2012.
38. Z. Guo, R. Dong, P. S. Chakraborty, N. Lourenco, J. Palmer, Y. Hu, M. Ruan et al., Record maximum oscillation frequency in C-face epitaxial graphene transistors, *Nano Lett.*, vol. 13, no. 3, pp. 942–947, 2013.

39. F. Schwierz, J. Pezoldt and R. Granzner, Two-dimensional materials and their prospects in transistor electronics, *Nanoscale*, vol. 7, no. 18, pp. 8261–8283, 2015.

40. F. Schwierz, Graphene transistors 2011, in *2011 International Symposium on VLSI Technology, Systems and Applications (VLSI-TSA)*, pp. 1–2, 2011.

41. R. Lai, X. B. Mei, W. R. Deal, W. Yoshida, Y. M. Kim, P. H. Liu, J. Lee et al., Sub 50 nm InP HEMT device with F_{max} greater than 1 THz, *Technical digest, IEDM*, pp. 609–611, 2007.

42. F. Schwierz, Graphene electronics, in A. Chen, J. Hutchby, V. Zhirnov and G. Bourianoff, eds., *Emerging Nanoelectronic Devices*, John Wiley & Sons, Hoboken, NJ, pp. 298–314, 2014.

43. M. Tanzid, M. A. Andersson, J. Sun and J. Stake, Microwave noise characterization of graphene field effect transistors, *Appl. Phys. Lett.*, vol. 104, no. 1, p. 013502, 2014.

44. J.-O. Plouchart, J. Kim, J. Gross, R. Trzcinski and K. Wu, Scalability of SOI CMOS technology and circuit to millimeter wave performance, in *IEEE Compound Semiconductor Integrated Circuit Symposium, 2005. CSIC '05*, p. 4, 2005.

45. L. Poulain, N. Waldhoff, D. Gloria, F. Danneville and G. Dambrine, Small signal and HF noise performance of 45 nm CMOS technology in mmW range, in *2011 IEEE Radio Frequency Integrated Circuits Symposium (RFIC)*, pp. 1–4, 2011.

46. P. C. Chao, A. J. Tessmer, K.-H. G. Duh, P. Ho, M.-Y. Kao, P. M. Smith, J. M. Ballingall, S.-M. J. Liu and A. A. Jabra, W-band low-noise InAlAs/InGaAs lattice-matched HEMTs, *IEEE Electron Device Lett.*, vol. 11, no. 1, pp. 59–62, 1990.

47. K. H. G. Duh, P. C. Chao, S. M. J. Liu, P. Ho, M. Y. Kao and J. M. Ballingall, A super low-noise 0.1 μm T-gate InAlAs–InGaAs–InP HEMT, *IEEE Microw. Guid. Wave Lett.*, vol. 1, no. 5, pp. 114–116, 1991.

48. P. M. Smith, InP-based HEMTs for microwave and millimeter-wave applications, Indium Phosphide and Related Materials. Conference Proceedings, Seventh International Conference on May 1995, pp. 68–72, 1995.

49. C. S. Whelan, P. F. Marsh, S. M. Lardizabal, W. E. Hoke, R. A. McTaggart and T. E. Kazior, Low noise and power metamorphic HEMT devices and circuits with $X = 30\%$ to 60% In$_x$GaAs channels on GaAs substrates, *Proc. IPRM*, pp. 337–340, 2000.

50. B. O. Lim, M. K. Lee, T. J. Baek, M. Han, S. C. Kim and J.-K. Rhee, 50-nm T-gate InAlAs/InGaAs metamorphic HEMTs with low noise and high f_T characteristics, *IEEE Electron Device Lett.*, vol. 28, no. 7, pp. 546–548, 2007.

51. D. Akinwande, N. Petrone and J. Hone, Two-dimensional flexible nanoelectronics, *Nat. Commun.*, 5, 2014, Article number: 5678.

52. M. N. Yogeesh, K. Parish, J. Lee, L. Tao and D. Akinwande, Towards the design and fabrication of graphene based flexible GHz radio receiver systems, in *Microwave Symposium (IMS), 2014 IEEE MTT-S International*, pp. 1–4, 2014.

53. N. Petrone, I. Meric, J. Hone and K. L. Shepard, Graphene field-effect transistors with gigahertz-frequency power gain on flexible substrates, *Nano Lett.*, vol. 13, no. 1, pp. 121–125, 2013.

54. M. Kitamura and Y. Arakawa, High current-gain cutoff frequencies above 10 MHz in n-channel C60 and p-channel pentacene thin-film transistors, *Jpn. J. Appl. Phys.*, vol. 50, no. 1S2, p. 01BC01, 2011.

55. H. Yang, J. Heo, S. Park, H. J. Song, D. H. Seo, K.-E. Byun, P. Kim, I. Yoo, H.-J. Chung and K. Kim, Graphene barristor, a triode device with a gate-controlled Schottky barrier, *Science*, vol. 336, pp. 1140–1143, 2012.

56. L. Britnell, R. V. Gorbachev, R. Jalil, B. D. Belle, F. Schedin, A. Mishchenko, T. Georgiou et al., Field-effect tunneling transistor based on vertical graphene heterostructures, *Science*, vol. 335, pp. 947–950, 2012.

57. L. Britnell, R. V. Gorbachev, A. K. Geim, L. A. Ponomarenko, A. Mishchenko, M. T. Greenaway, T. M. Fromhold, K. S. Novoselov and L. Eaves, Resonant tunnelling and negative differential conductance in graphene transistors, *Nat. Commun.*, vol. 4, p. 1794, 2013.

58. P. Zhao, R. M. Feenstra, G. Gu and D. Jena, SymFET: A proposed symmetric graphene tunneling field effect transistor, in *2012 70th Annual Device Research Conference (DRC)*, pp. 33–34, 2012.

59. W. Mehr, J. Dabrowski, J. C. Scheytt, G. Lippert, Y.-H. Xie, M. C. Lemme, M. Ostling and G. Lupina, Vertical graphene base transistor, *IEEE Electron Device Lett.*, vol. 33, pp. 691–693, 2012.

60. C. A. Mead, Operation of tunnel-emission devices, *J. Appl. Phys.*, vol. 32, pp. 646–652, 1961.

61. M. Heiblum, D. C. Thomas, C. M. Knoedler and M. I. Nathan, Tunneling hot-electron transfer amplifier: A hot-electron GaAs device with current gain, *Appl. Phys. Lett.*, vol. 47, pp. 1105–1107, 1985.

62. K. Seo, M. Heiblum, C. M. Knoedler, J. E. Oh, J. Pamulapati and P. Bhattacharya, High-gain pseudomorphic InGaAs base ballistic hot-electron device, *IEEE Electron Device Lett.*, vol. 10, pp. 73–75, 1989.

63. S. Venica, F. Driussi, P. Palestri, D. Esseni, S. Vaziri and L. Selmi, Simulation of DC and RF performance of the graphene base transistor, *IEEE Trans. Electron Devices*, vol. 61, no. 7, pp. 2570–2576, 2014.

64. S. Vaziri, G. Lupina, C. Henkel, A. D. Smith, M. Östling, J. Dabrowski, G. Lippert, W. Mehr and M. C. Lemme, A graphene-based hot electron transistor, *Nano Lett.*, vol. 13, no. 4, pp. 1435–1439, 2013.

65. S. Venica, F. Driussi, P. Palestri and L. Selmi, Graphene base transistors with optimized emitter and dielectrics, in *2014 37th International Convention on Information and Communication Technology, Electronics and Microelectronics (MIPRO)*, pp. 33–38, 2014.

66. V. Di Lecce, R. Grassi, A. Gnudi, E. Gnani, S. Reggiani and G. Baccarani, Graphene-base heterojunction transistor: An attractive device for terahertz operation, *IEEE Trans. Electron Devices*, vol. 60, no. 12, pp. 4263–4268, 2013.

67. V. Di Lecce, R. Grassi, A. Gnudi, E. Gnani, S. Reggiani and G. Baccarani, Graphene base transistors: A simulation study of DC and small-signal operation, *IEEE Trans. Electron Devices*, vol. 60, no. 10, pp. 3584–3591, 2013.

68. S. Vaziri, A. D. Smith, M. Ostling, G. Lupina, J. Dabrowski, G. Lippert, F. Driussi et al., Going ballistic: Graphene hot electron transistors, *Solid State Commun.*, vol. 224, pp. 64–75, 2015. doi:10.1016/j.ssc.2015.08.012.

69. M. Schroter, G. Wedel, B. Heinemann, C. Jungemann, J. Krause, P. Chevalier and A. Chantre, Physical and electrical performance limits of high-speed SiGeC HBTs – Part I: Vertical scaling, *IEEE Trans. Electron Devices*, vol. 58, no. 11, pp. 3687–3696, 2011.

70. V. Di Lecce, A. Gnudi, E. Gnani, S. Reggiani and G. Baccarani, Graphene-base heterojunction transistors for post-CMOS high-speed applications: Hopes and challenges, presented at the *Device Research Conference, DRC*, Columbus, OH, 2015.

71. C. Zeng, E. B. Song, M. Wang, S. Lee, C. M. Torres, J. Tang, B. H. Weiller and K. L. Wang, Vertical graphene-base hot-electron transistor, *Nano Lett.*, vol. 13, no. 6, pp. 2370–2375, 2013.

72. N. Alimardani and J. F. Conley Jr, Step tunneling enhanced asymmetry in asymmetric electrode metal-insulator-insulator-metal tunnel diodes, *Appl. Phys. Lett.*, vol. 102, no. 14, p. 143501, 2013.

73. T. Georgiou, R. Jalil, B. D. Belle, L. Britnell, R. V. Gorbachev, S. V. Morozov, Y.-J. Kim et al., Vertical field-effect transistor based on graphene-WS$_2$ heterostructures for flexible and transparent electronics, *Nat. Nanotechnol.*, vol. 8, no. 2, pp. 100–103, 2013.

74. T. Roy, M. Tosun, J. S. Kang, A. B. Sachid, S. B. Desai, M. Hettick, C. C. Hu and A. Javey, Field-effect transistors built from all two-dimensional material components, *ACS Nano*, vol. 8, no. 6, pp. 6259–6264, 2014.

75. G. Fiori, F. Bonaccorso, G. Iannaccone, T. Palacios, D. Neumaier, A. Seabaugh, S. K. Banerjee and L. Colombo, Electronics based on two-dimensional materials, *Nat. Nanotechnol.*, vol. 9, no. 10, pp. 768–779, 2014.

5

High-Field and Thermal
Transport in Graphene

Zuanyi Li, Vincent E. Dorgan,
Andrey Y. Serov and Eric Pop

Contents

2D Materials for Nanoelectronics edited by Michel Houssa, Athanasios Dimoulas and Alessandro Molle © 2016 CRC Press/Taylor & Francis Group, LLC. ISBN: 978-1-4987-0417-5.

5.1 Introduction

The rise of interest in two-dimensional (2D) materials started with graphene (Geim and Novoselov, 2007), a one-atomic layer of carbon atoms arranged into a honeycomb lattice with an interatomic distance of 1.42 Å, as shown in **Figure 5.1**. Carbon atoms in graphene are bonded through orbitals with sp^2 hybridisation, which leads to a strong bond and excellent mechanical strength. A single atomic layer (monolayer) of graphene is nearly 98% transparent to visible light, enabling the design of graphene-based transparent electrodes (Kim et al., 2009a). Graphene is flexible and shows great electrical characteristics under mechanical strain, which also enables the design of flexible electronic applications (Lee et al., 2008). In addition to outstanding mechanical, electrical and thermal properties, which we will cover later, graphene-based technology benefits from the abundance, nontoxicity and biocompatibility of carbon.

There are several ways to obtain graphene. The simplest method is mechanical exfoliation from graphite using adhesive (e.g. Scotch™) tape, which led to the discovery of the field-effect properties of graphene in 2004 (Novoselov et al., 2004) and eventually to the Nobel Prize in Physics in 2010 (Geim, 2011, Novoselov, 2011). Although this method produces small samples (e.g. ~10 μm) of excellent quality, its industrial applications are limited due to lack of process scalability. Graphene can also be grown by epitaxy on SiC substrate (Kedzierski et al., 2008). Some difficulties associated with this process include the relatively expensive and small size of SiC wafers, plus the difficulty in obtaining undoped graphene samples by this method. The most common way to obtain graphene these days is the chemical vapour deposition (CVD) process using a catalyst such as Cu (Li et al., 2009). After growth, the graphene is coated with a polymer such as poly(methyl methacrylate) or poly(bisphenol A carbonate) and transferred to the substrate of interest (Wood et al., 2015). A challenge of the CVD process is that due to somewhat random character of the growth, it leads to polycrystalline graphene layers with grain boundaries and line defects (Huang et al., 2011).

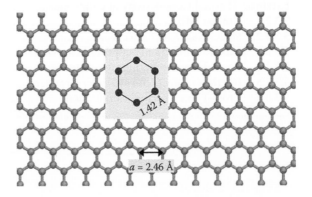

FIGURE 5.1 Structure of graphene showing hexagonally arranged carbon atoms with spacing 1.42 Å and lattice constant $a = 2.46$ Å.

As mentioned earlier, graphene has outstanding electrical and thermal properties stemming from its electron and phonon energy dispersions, shown in **Figure 5.2**. Unlike most semiconductors or metals, graphene has a linear electron energy band structure near the Fermi level, that is, $E = \hbar v_F |\mathbf{k}|$, where $\hbar$ is the reduced Planck constant and $v_F \approx 10^6$ m/s is the Fermi velocity of charge carriers, that is, their ballistic velocity between collisions (Castro Neto et al., 2009). The linear dispersion also indicates that the density of states (DOS) for electrons in graphene is linearly proportional to their energy (Fang et al., 2007):

$$\text{DOS}(E) = \frac{2}{\pi(\hbar v_F)^2}|E| \tag{5.1}$$

where factors of 2 are included for valley degeneracy (K and K′) and spin degeneracy, respectively.

As shown in **Figure 5.2a**, graphene has a zero band gap, with the point where conduction band meets valence band called the Dirac point. There are two equivalent zero-gap points at the K and K' locations in the Brillouin zone (BZ) of graphene. The minimum carrier density is observed when the Fermi level is at the Dirac point, and in theory is limited only by the thermally generated carrier density in thermal equilibrium (Fang et al., 2007):

$$n = p = n_i = \frac{\pi}{6}\left(\frac{k_B T}{\hbar v_F}\right)^2, \tag{5.2}$$

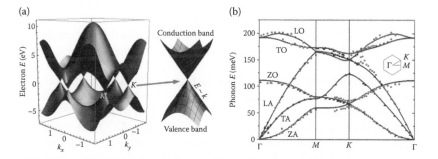

FIGURE 5.2 (a) Graphene electron energy band structure with zoom-in to show linear dispersion near the K (or K') point of the BZ. The energy bands deviate from linear behaviour at energies greater than ~1 eV (Serov et al., 2014). (Biro, L. P., Nemes-Incze, P. and Lambin, P. 2012. Graphene: Nanoscale processing and recent applications. *Nanoscale*, 4, 1824–1839. Reproduced by permission of The Royal Society of Chemistry.) (b) Phonon energy dispersion along high-symmetry lines for monolayer graphene obtained from *ab initio* calculations, where red circles (Yanagisawa et al., 2005) and blue triangles (Mohr et al., 2007) are experimental data plotted for comparison. The inset shows the graphene BZ.

where k_B is the Boltzmann constant and T is the temperature. However, in reality, various ionised impurities can induce local changes in the Dirac point, so-called 'potential energy puddles', which also increase the minimum carrier density (Zhang et al., 2009, Li et al., 2011a).

The fact that we cannot effectively decrease the carrier density means that graphene field-effect transistors (FETs) are difficult to switch off, making it challenging to utilise graphene for applications in digital electronics. Although a band gap can be introduced by quantum confinement, that is, patterning graphene into nanoribbons (GNRs), the GNR width corresponding to a reasonable band gap of ~0.5 eV is around just a few nanometers (Son et al., 2006, Han et al., 2007, Li et al., 2008, Huang et al., 2009), making the edges and associated edge roughness a significant part of the device. In reality, the mobility in narrow GNRs is typically lower than in pristine graphene even for 10–20-nm-wide devices (Jiao et al., 2009, Behnam et al., 2012). High mobility and good current density in large-scale graphene can be utilised in analogue applications and interconnects (Lin et al., 2010, Behnam et al., 2012, Wu et al., 2012), where off-state leakage is less important, while high on-current and transconductance (gain) are crucial. These applications as well as potential high-performance digital electronics require careful thermal management, where graphene can also play a role due to high thermal conductivity (k) (Balandin, 2011, Pop et al., 2012, Xu et al., 2014b).

The thermal properties of graphene can be better understood by inspecting the phonon dispersion, as shown in **Figure 5.2b**. Phonons, that is, the quantised lattice vibrations of single-layer graphene (SLG) have six branches corresponding to two atoms in the elementary cell: three acoustic modes (transverse TA, longitudinal LA and flexural ZA) and three optical modes (transverse TO, longitudinal LO and flexural ZO). Transverse and longitudinal acoustic modes have high sound velocity $v_{LA} \approx 21$ km/s and $v_{TA} \approx 14$ km/s (Pop et al., 2012), which leads to a strong contribution to thermal conductivity. Flexural ZA and ZO modes correspond to out-of-plane vibrations. Unlike linear LA and TA modes, ZA has a quadratic dependence of frequency ω on wave vector q: $\omega \approx \alpha q^2$, which leads to the high DOS and big contribution to the thermal conductivity for *suspended* samples (Lindsay et al., 2010), which could exceed 2000 W/m/K at room temperature (Balandin, 2011, Pop et al., 2012, Xu et al., 2014b). For samples on a substrate (e.g. SiO_2), ZA modes are suppressed by the interaction with the substrate (Seol et al., 2010, Ong and Pop, 2011), which leads to lower but still great thermal conductivity of supported samples $k \approx 600$ W/m/K (Seol et al., 2010). The thermal transport in graphene will be discussed in detail in Section 5.3.

Now, we first return to the electrical properties of graphene. Suspended samples with exfoliated graphene allow probing of intrinsic electrical properties and exhibit charge carrier mobility as high as ~100,000 cm^2/V/s albeit at low temperature (~5 K) and low carrier density $n \sim 10^{10}$–10^{11} cm^{-2} (Bolotin et al., 2008). Such a high low-field mobility in suspended graphene can be explained by weak electron–phonon coupling, especially at lower carrier densities due to relatively high energy of optical phonons ($\hbar\omega_{OP} \approx 200$ meV) and intervalley acoustic phonons ($\hbar\omega_{AC,i} \approx 140$ meV) (Balandin, 2011,

Pop et al., 2012). As the energy of these phonons is much higher than $k_B T \approx 26$ meV at room temperature, the number of carriers which can *emit* these phonons is very low for equilibrium Fermi–Dirac distribution. On the other hand, the probability to *absorb* a high-energy phonon is limited by the nature of the Bose–Einstein distribution for phonons. The reasoning provided here is valid only in near-equilibrium conditions (low field), as it is much easier to emit optical phonons for non-equilibrium high-energy so-called 'hot' carriers (Serov et al., 2014).

Although suspended graphene devices exhibit outstanding electrical properties, their utilisation for realistic electronics is very limited. First, the manufacturing of suspended devices is a challenging process with low throughput (Bolotin et al., 2008, Dorgan et al., 2013). Second, it is difficult to modulate charge carrier density in suspended samples because of very weak capacitive coupling between graphene and the gate through the air or vacuum gap. Therefore, in order to make devices and circuits, graphene is most commonly placed on substrates compatible with conventional CMOS (complementary metal–oxide–semiconductor) processing, for example, Si/SiO_2, Al_2O_3 or HfO_2. However, when graphene is placed on a substrate, its electrical characteristics degrade quite significantly, with the low-field mobility typically less than 10,000 $cm^2/V/s$. Such steep degradation of the low-filed mobility is related to various substrate-induced scattering mechanisms such as ionised impurity scattering and surface phonon scattering (Adam et al., 2009, Zhu et al., 2009, Perebeinos and Avouris, 2010, Ong and Fischetti, 2012c, Serov et al., 2014).

As graphene could be targeted towards interconnects and analogue applications, the understanding of electrical transport at high electric fields can be beneficial for device analysis and optimisation. High-field transport in graphene has been studied both theoretically and experimentally. Experimental analysis is usually performed on a four-point structure or in multi-finger Hall configuration to exclude the effect of the contact resistance (Barreiro et al., 2009, DaSilva et al., 2010, Dorgan et al., 2010). It is usually observed that the current in graphene tends to saturate at higher electric field but does not reach 'true' saturation with output conductance $g_d = dI_d/dV_d$ nearly zero (Barreiro et al., 2009, DaSilva et al., 2010, Dorgan et al., 2010). However, despite the lack of current saturation, it is possible to observe velocity saturation, albeit accompanied by the rising device temperature due to self-heating (Dorgan et al., 2010).

From the theoretical side, the velocity saturation in graphene is not an easy problem to solve, mostly due to the complicated nature of interaction of charge carriers in graphene with the substrate. The first theoretical studies of velocity saturation in graphene focussed on intrinsic graphene behaviour, therefore, mostly neglecting the substrate effect (Akturk and Goldsman, 2008, Shishir and Ferry, 2009). The tool of choice in this case is usually the Monte-Carlo method in momentum space, which naturally provides a simulation scheme for multiple scattering mechanisms (Jacoboni and Reggiani, 1983). The hydrodynamic model can also be used for modelling of high-field effects and has some advantages (Bistritzer and MacDonald, 2009, Svintsov et al., 2012, Serov et al., 2014).

Earlier theoretical studies on intrinsic graphene exhibit quite different results for velocity saturation: for example, one study demonstrated carrier-density-dependent saturation velocity v_{sat} in the range between $0.3v_F$ and $0.45v_F$ using the ensemble Monte-Carlo method (Shishir and Ferry, 2009). Another study showed almost no carrier-density dependence in velocity saturation with much lower bound for v_{sat} around $0.1v_F$ (Bistritzer and MacDonald, 2009). The main difference between the two models lies in values for deformation potentials for acoustic and optical phonons employed.

Later investigations (Li et al., 2010, 2011b, Perebeinos and Avouris, 2010), which incorporated the role of the substrate, pointed to the importance of interfacial substrate phonon modes and self-heating. However, the treatment of the substrate phonons was first performed with relatively simplistic models including no screening or static screening. A Monte-Carlo study involving substrate phonons (with static screening, but no substrate impurities) showed the velocity saturation between $0.4v_F$ and $0.6v_F$ (Li et al., 2010). The same group performed a similar study with self-heating taken into account and found carrier-density-dependent velocity saturation with v_{sat} ranging between $0.15v_F$ and $0.5v_F$ (Li et al., 2011b). Similar results were shown using Monte-Carlo methods by another group, where substrate phonons were treated without screening yielding very high values for v_{sat} between $0.4v_F$ and $0.8v_F$ depending on the substrate material (between $0.4v_F$ and $0.5v_F$ for intrinsic graphene) (Perebeinos and Avouris, 2010). However, a recent detailed pseudo-potential-based Monte-Carlo study (with dynamic screening theory but without self-heating) pointed to lower numbers about $0.2v_F$ to $0.3v_F$ at room temperature (Fischetti et al., 2013). A more comprehensive study (Serov et al., 2014) using a hydrodynamic model considered self-heating and thermal coupling to the substrate, scattering with ionised impurities, graphene phonons, dynamically screened interfacial plasmon–phonon (IPP) modes and various substrates. It found that the high-field behaviour is determined by scattering with IPP modes and a small contribution of graphene phonons, and the drift velocity was benchmarked with available experimental data (Dorgan et al., 2010).

High-field effects in graphene are important not only in static (DC) regimes, but can also lead to a variety of interesting physical transient effects, especially at higher frequencies (Sekwao and Leburton, 2013). These effects render interesting features of electron–electron and electron–phonon interactions in zero band gap graphene. Up to this point, we have been discussing the characterisation of material properties of graphene usually performed experimentally in specialised test structures or in simulation with complicated methods such as the Monte Carlo or hydrodynamic model. A simpler characterisation usually involves a lumped compact model and basic extraction of transport coefficients such as mobility and contact resistance on a simple three- or four-terminal metal–oxide–semiconductor field-effect transistor (MOSFET) device. In order to bring together material properties and real device geometry, other methods such as the drift-diffusion model can be used to assist with device analysis. For instance, current saturation in graphene devices is an important topic because it is related to the analogue transistor gain, with better saturation resulting in higher gain ($= g_m/g_d$, where $g_m = \partial I_D/\partial V_{GS}$ is the transconductance and g_d is the output conductance defined earlier). However, current saturation is not only caused by velocity saturation but also can be assisted electrostatically similar to the pinch-off effect in silicon MOSFETs, which can be captured by the Poisson equation in the drift-diffusion scheme. Other physical effects occurring alongside current saturation are band-to-band carrier generation and impact ionisation (Winzer et al., 2010, Girdhar and Leburton, 2011, Pirro et al., 2012). Despite the drift-diffusion method proving its merit for bulk semiconductors, relatively fewer efforts had assessed the applicability of drift-diffusion techniques to graphene (Ancona, 2010, Bae et al., 2011, Jimenez, 2011, Islam et al., 2013). In the next section, we will mainly discuss high-field electronic transport in graphene.

5.2 High-Field Electronic Transport in Graphene

5.2.1 Practical Device Operation and High-Field Transport

Understanding high-field electronic transport in a material like graphene is not only important from a scientific point of view, but is also essential for achieving practical applications. This can be further understood by comparing the long-channel model versus the short-channel model for a typical n-channel FETs. The basic long-channel NMOSFET equation for calculating drain current (I_D) is given by (Muller et al., 2003)

$$I_D = \mu C_{ox} \frac{W}{L}\left[\left(V_{GS} - V_T - \frac{1}{2}V_{DS}\right)V_{DS}\right], \tag{5.3}$$

where μ is the carrier mobility, L and W are the channel length and width, respectively, C_{ox} is the gate oxide capacitance per unit area, V_{GS} is the gate voltage, V_T is the threshold voltage and V_{DS} is the drain voltage, all with reference to the grounded source terminal. Using the long-channel model, when $V_{DS} > V_{GS} - V_T$ the inversion layer at the drain is effectively 'pinched-off' as shown in **Figure 5.3**, and I_D no longer rises with an increase in V_D. This occurs when the gate-to-*drain* voltage is smaller than the threshold voltage ($V_{GD} = V_{GS} - V_{DS} < V_T$) and can no longer support sufficient inversion charge at the drain end of the channel. The classical long-channel saturation current (I_{Dsat}) is defined by the drain voltage at the pinch-off point ($V_{DS} = V_{GS} - V_T$) and given by

$$I_{Dsat} = \mu C_{ox} \frac{W}{2L}(V_{GS} - V_T)^2, \tag{5.4}$$

where the classical I_{Dsat} increases quadratically with the gate voltage overdrive ($V_{GS} - V_T$) and is inversely proportional to L.

The long-channel analysis, which assumes carrier mobility independent of lateral field, is no longer applicable at higher electric fields present in short-channel transistors. These higher fields result in the drift velocity (v_d) of carriers in short-channel devices approaching a limiting value known as the saturation velocity (v_{sat}). This leads to current saturation occurring in

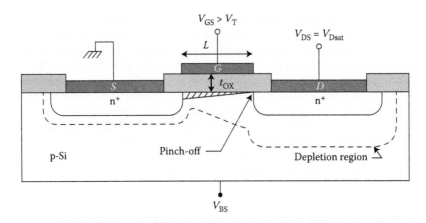

FIGURE 5.3 Schematic of a typical long-channel n-type MOSFET, shown in this case at the onset of saturation such that the pinch-off point is at the drain side of the channel. (Adapted from Taur, Y. and Ning, T. H. 2009. *Fundamentals of Modern VLSI Devices*. Cambridge University Press, New York. Copyright 2009.)

a short-channel transistor at much lower voltages than one would predict if using the long-channel model (Taur et al., 1993, Taur and Ning, 2009). Consequently, for short-channel MOSFETs, the saturation current is given by

$$I_{Dsat} \approx v_{sat} W C_{ox} (V_{GS} - V_T - V_{Dsat}),\tag{5.5}$$

if one assumes velocity saturation along the entire channel length, where V_{Dsat} is the drain voltage at the onset of current saturation. For shorter and shorter gate lengths ($L \to 0$), we can estimate the maximum drain current as

$$I_{Dmax} = v_{sat} W C_{ox} (V_{GS} - V_T),\tag{5.6}$$

where I_{Dmax} increases linearly, not quadratically, with overdrive voltage $(V_{GS} - V_T)$ and the term $C_{ox}(V_G - V_T)$ is an approximation of the 2D carrier density (n) in the channel. **Figure 5.4** shows a comparison of the high-field behaviour for long- and short-channel devices discussed here, and further emphasises that with increased scaling and higher electric fields in modern transistors, improved understanding of high-field transport (i.e. the energy dissipation mechanisms that determine v_{sat} for a given material) is critical for enhancing practical device operation. The velocity saturation in a typical semi-conductor with *parabolic* energy bands (like Si) can be simply estimated as (Lundstrom, 2000)

$$v_{sat} \approx \left(\frac{\hbar \omega_{OP}}{m^*} \right)^{1/2},\tag{5.7}$$

where $\hbar \omega_{OP}$ is the optical phonon energy (responsible for the majority of inelastic scattering) in the material and m^* is the conductivity effective mass.

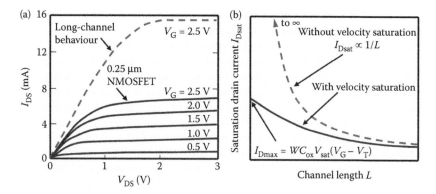

FIGURE 5.4 (a) Current–voltage characteristics for a Si NMOSFET with $L = 0.25\ \mu m$ and $W = 9.5\ \mu m$. The solid lines are experimental curves, whereas the dashed line is the long-channel approximation with velocity saturation effects ignored. (Adapted from Taur, Y. et al. 1993. *Solid-State Electron.*, 36, 1085–1087.) (b) Model predictions for MOSFET saturation current versus channel length without velocity saturation (dashed) and with velocity saturation (solid). (Adapted from Muller, R. S. et al. 2003. *Device Electronics for Integrated Circuits.* John Wiley & Sons, New York.)

$v_{sat} \approx 10^7$ cm/s in Si for both electrons and holes (Jacoboni et al., 1977). However, as we will see, this expression does not hold for graphene, where the energy bands near the Fermi level have a *linear* dispersion relationship.

Detailed knowledge of the coupling of high-field transport with self-heating is also necessary when discussing practical device operation, since the ability to effectively remove heat from integrated circuits is a limiting factor for future scaling. At high fields, the charge carriers (e.g. electrons in conduction band) accelerate and gain energy, or 'heat up'. Mechanisms that may limit electron transport include electron scattering with other electrons, phonons, interfaces, defects and impurities. These scattering events not only determine v_{sat}, but also when electrons scatter with phonons, electrons can lose energy to the lattice and effectively raise the temperature of the lattice (i.e. Joule heating or self-heating) (Pop et al., 2005, Pop and Goodson, 2006, Pop et al., 2006b). Consequently, it is important to account for self-heating effects when investigating high-field transport, as the electronic properties of a material may vary drastically with temperature. For example, the saturation velocity in Si shows a slight decrease with rising temperature due to an increase in phonon scattering (Jacoboni et al., 1977).

5.2.2 High-Field Transport in Graphene

For high-field transport in graphene, our discussion here is primarily concerned with monolayer graphene under steady-state transport. First, we focus on simulation results for ideal graphene and then include analysis of substrate effects and experimental data.

A detailed review of the current theoretical understanding of electron transport in graphene has recently been presented by Fischetti et al. (2013). Here, we summarise some of the key points from their study, as well as other theoretical works, as they pertain to our discussion of high-field transport. We initially give attention to phonon-limited transport in ideal graphene, as shown in **Figure 5.5**. We point out that the electron drift velocity appears to peak at ~2×10^7 cm/s at 1 V/μm field for $n = 10^{13}$ cm^{-2} at 300 K. The onset of negative differential velocity (NDV) at 0.3–0.5 V/μm for lower carrier densities is an intriguing prediction, and here it is attributed to electrons gaining enough energy at high fields to populate the flatter regions of the BZ (**Figure 5.2a**), causing a decrease of the drift velocity. The appearance of NDV has been presented in previous simulations as well (Akturk and Goldsman, 2008, Shishir and Ferry, 2009, Shishir et al., 2009).

Using Monte-Carlo simulations, Akturk and Goldsman (2008) predicted peak drift velocities for intrinsic graphene as high as 4.6×10^7 cm/s and slight NDV for fields above ~10 V/μm. Shishir and Ferry (2009) estimate similar peak velocities and NDV for $n \leq 2 \times 10^{12}$ cm^{-2} at fields above ~0.5 V/μm. Chauhan and Guo (2009) show carrier velocity as a function of electric field up to 1 V/μm with v_{sat} near 4×10^7 cm/s, but without NDV ostensibly due to their use of a simple linear dispersion. However, the carrier density was fixed at 5.29×10^{12} cm^{-2} in their simulations so it is uncertain if NDV would have been observed at lower densities. The calculated v_d versus electric field for these works is summarised in **Figure 5.6**. The variation among different theoretical works for transport in ideal graphene may, at least in part, be associated

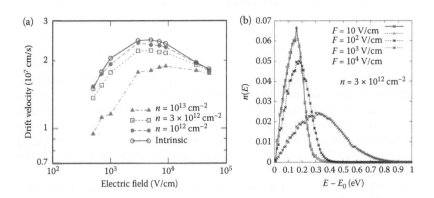

FIGURE 5.5 (a) Drift velocity versus electric field in graphene at 300 K calculated using the Monte-Carlo method (Fischetti et al., 2013), and considering only phonon-limited electron transport, that is, electron–electron scattering, substrate effects and self-heating are ignored. The various curves are distinguished by the electron density (n). (b) Corresponding electron energy distribution function calculated for various values of the electric field (F) for the case of $n = 3 \times 10^{12}$ cm^{-2}. (Adapted from Fischetti, M. V. et al., Pseudopotential-based studies of electron transport in graphene and graphene nanoribbons. *J. Phys. Condens. Matter*, 25, 473202. Copyright 2013, Institute of Physics.)

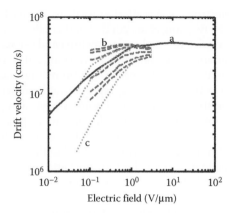

FIGURE 5.6 Calculated drift velocity versus electric field from Monte-Carlo simulations from (solid) Akturk and Goldsman (2008) (a), (dashed) Shishir and Ferry (2009) (b) and (dotted) Chauhan and Guo (2009) (c). The solid line corresponds to intrinsic graphene. The dashed lines vary in carrier density from 5×10^{11} to 10^{13} cm^{-2} from top to bottom. The multiple dotted lines have impurity densities of $n_{imp} = 0$, 10^{11} cm^{-2} and 10^{12} cm^{-2} from top to bottom.

with the different choices made for deformation potentials, band structures and phonon dispersions. Therefore, drawing qualitative, rather than quantitative, conclusions may be more appropriate from these simulation studies. For example, a clear observation in Fischetti et al. (2013) and Shishir et al. (2009) is the decrease in high-field drift velocity with increasing carrier density, which is an interesting phenomenon considering typical non-degenerate semiconductors like Si have a constant v_{sat} (e.g. ~10^7 cm/s for Si at 300 K) (Jacoboni et al., 1977). Graphene has no band gap, and thus is a degenerate semiconductor where the dependence of saturation velocity on carrier density is due to the degeneracy of carriers and the Pauli exclusion principle (Fang et al., 2011).

Continuing the discussion of the carrier-density-dependent electron transport in graphene, we note that simulations by Ferry (2012) used an impurity density that increased with carrier density, indicating that this increasing impurity density was the only way to match the experimental data of Dorgan et al. (2010) and Kim et al. (2012). It is plausible that a change in impurity density may occur during measurements due to the application of a high gate field and subsequent motion of impurities within the oxide. Another explanation is that a more dynamic form of screening, as proposed by Ong and Fischetti (2012c), is necessary to properly understand the dependence of transport on carrier density. For future analysis of experimental work, we note that a constant impurity density (independent of gate voltage) is assumed.

Returning to the previously mentioned high-field NDV in graphene shown in several theoretical studies (Akturk and Goldsman, 2008, Shishir and Ferry, 2009, Shishir et al., 2009, Fischetti et al., 2013), Fang et al. (2011) suggest that the inclusion of electron–electron (e–e) scattering actually removes the NDV effect. Because high-energy electrons exchange their momentum and energy with low-energy electrons, e–e scattering weakens the backscattering effect that would lead to NDV (Fang et al., 2011). Another effect not yet discussed is carrier multiplication due to interband tunnelling and e–e scattering (Winzer et al., 2010, Girdhar and Leburton, 2011, Pirro et al., 2012). If the Fermi level is near the Dirac point, we could expect Zener tunnelling and/or impact ionisation to generate electrons and holes, especially at high fields, and increase the carrier concentration. Experimental studies have shown that it is very difficult to obtain saturation of high-field current in graphene transistors (Meric et al., 2008, Barreiro et al., 2009, DaSilva et al., 2010, Dorgan et al., 2010, Meric et al., 2011), a characteristic credited to the lack of a band gap in graphene, which facilitates ambipolar transport and the easy transition from electron-dominated to hole-dominated transport (or vice versa) during high-field operation.

We now turn to substrate effects, a necessary discussion in light of the fact that a majority of the experimental work on graphene has been performed with graphene on solid insulating substrates (e.g. SiO_2/Si). An additional scattering process that limits electrical transport in supported graphene is scattering with charged impurities (Chen et al., 2008) and 'remote' substrate phonons (Serov et al., 2014). Substrate impurities that remain after the fabrication of graphene transistors (Martin et al., 2008, Deshpande et al., 2009, Zhang et al., 2009) play a significant role in limiting the low-field mobility of graphene transistors (Adam et al., 2007), but do not significantly affect high-field transport (Perebeinos and Avouris, 2010, Meric et al., 2011). In addition, we expect that if a graphene transistor is used in the future for nanoscale electronics, then it will most likely be a top- or multi-gated transistor with thin high-k insulator as the gate dielectric (Kim et al., 2009b, Zou et al., 2010). Detailed models have been provided for analysing the effect of charged impurity scattering and its dependence on the surrounding dielectric environment (Chen et al., 2009a, Ponomarenko et al. 2009, Ong and Fischetti, 2012a, 2013b), but as we are primarily concerned with high-field transport here, we will move on to a more relevant scattering process.

For insulators like SiO_2 and HfO_2, there are bulk dipoles associated with the ionicity of the metal–oxide bonds. These dipoles generate fringing fields on the substrate surface such that the frequencies of the dipoles are typically determined by the bulk LO phonons of the insulator. Electrons in close proximity to the substrate surface may interact with these surface optical (SO) phonons via a process commonly referred to as remote-phonon scattering. For Si inversion layers, remote-phonon scattering has been investigated (Hess and Vogl, 1979, Moore and Ferry, 1980), where it was found to have a small effect on the drift velocity in the regime just beyond Ohmic transport, but it was not strong enough to affect the saturation velocity (Leburton and Dorda, 1981), which is determined by the bulk Si optical phonons. In supported graphene, it is reasonable to consider electrons interacting strongly with SO phonons, considering electrons are essentially confined to the graphene sheet and the van der Waals gap between the graphene and substrate is small, ~0.3 nm (Ishigami et al., 2007). Furthermore, recent work shows that graphene may even emit longitudinal out-of-plane acoustic phonons into the substrate, representing an energy dissipation mechanism in the vertical direction (Chen et al., 2014).

When discussing remote-phonon scattering in graphene, Ong and Fischetti (2012b,c, 2013a) indicated the importance of considering dynamic screening and the charge density response of graphene due to the electric field created by SO phonons. **Figure 5.7** shows the remote-phonon-limited mobility (μ_{RP}) in graphene as a function of carrier density for various supporting substrates. The mobility is calculated by accounting for the hybrid IPP modes formed from the hybridisation of the SO phonons with graphene plasmons (Ong and Fischetti, 2012c). We see from the plot that μ_{RP} is saturating and weakly dependent on n at high carrier densities since most of the remote phonons are dynamically screened. The observed behaviour agrees with the findings of Zou et al. (2010), who extract μ_{RP} showing a slight increase with carrier density for a graphene transistor with HfO_2 as the top-gate dielectric.

The experimental works of Meric et al. (2008, 2011) suggest that current saturation in a graphene transistor may be observed due to remote-phonon scattering, or more specifically, scattering with the low-energy SO phonon of SiO_2 at $\hbar\omega_{OP} \approx 55$ meV (Fischetti et al., 2001). We use the term 'low-energy' here for comparison to the graphene (zone-edge) optical phonon with energy $\hbar\omega_{OP} \approx 160$ meV (see **Figure 5.2b**) (Borysenko et al., 2010). Furthermore, Barreiro et al. (2009) also suggested that the calculated current (from the analytic model shown in **Figure 5.9a**) using an average phonon energy of 149 meV overestimates the experimentally observed current in the

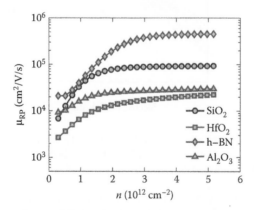

FIGURE 5.7 Remote-phonon-limited mobility (μ_{RP}) versus electron density (n) calculated using theory accounting for hybrid IPP modes (Ong and Fischetti, 2012b,c). Graphene is supported by various substrates: SiO_2 (circles), HfO_2 (squares), h-BN (diamonds) and Al_2O_3 (triangles), where degradation in mobility with increasing substrate dielectric constant is evident. (Adapted from Ong, Z.-Y. and Fischetti, M. V. 2012b. Erratum: Theory of interfacial plasmon–phonon scattering in supported graphene. *Phys. Rev. B*, 86, 199904. Copyright 2012 by the American Physical Society.)

high-field limit. Unfortunately, these works do not account for self-heating effects in their analysis, which we know is evident in graphene transistors at high currents and high fields (Chae et al., 2009, Freitag et al. 2009, Bae et al., 2010, 2011, Berciaud et al., 2010) and appears necessary to explain high-field transport in graphene (DaSilva et al., 2010, Perebeinos and Avouris, 2010, Li et al., 2011b, Islam et al., 2013).

Figure 5.8 shows the extracted drift velocity as a function of electric field from experimental *I–V* data of multiple studies for graphene-on-SiO_2 devices (Barreiro et al., 2009, DaSilva et al., 2010, Dorgan et al., 2010, Meric et al., 2011). Direct comparison of these curves is difficult as they correspond to different values of carrier density and temperature, as well as different device structures. Nevertheless, we are able to highlight some key features of high-field transport in graphene. Saturation velocity in graphene, even at high carrier densities >10^{13} cm^{-2}, appears to be larger than 10^7 cm/s (i.e. larger than v_{sat} in Si at room temperature). For suspended graphene, the extracted saturation velocity from high-field measurements at high temperature (>1000 K) is between 1 and 3×10^7 cm/s (Dorgan et al., 2013). **Figure 5.8** also shows the decrease

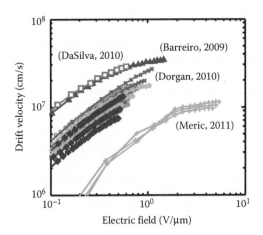

FIGURE 5.8 Experimentally extracted drift velocity versus electric field corresponding to data from Barreiro et al. (2009) for $T_0 = 300$ K at $n \approx 1.7 \times 10^{12}$ cm^{-2} (blue triangles), DaSilva et al. (2010) for $T_0 = 20$ K at $n \approx 2.1 \times 10^{12}$ cm^{-2} (red squares), Meric et al. (2011) for $T_0 = 300$ K at $n \approx 1.1 \times 10^{12}$–$1.4 \times 10^{12}$ cm^{-2} from top to bottom (cyan crosses) and Dorgan et al. (2010) for $T_0 = 80$ K at $n \approx 1.9 \times 10^{12}$–$5.9 \times 10^{12}$ cm^{-2} from top to bottom (magenta crosses), for $T_0 = 300$ K at $n \approx 2.8 \times 10^{12}$–$6.6 \times 10^{12}$ cm^{-2} from top to bottom (tan stars), for $T_0 = 80$ K at $n \approx 2.9 \times 10^{12}$–$1.23 \times 10^{13}$ cm^{-2} from top to bottom (green circles) and for $T_0 = 450$ K at $n \approx 2.7 \times 10^{12}$–$7.5 \times 10^{12}$ cm^{-2} from top to bottom (black diamonds). Data from Meric et al. (2011) correspond to a top-gated device with $L = 130$ nm and a relatively low low-field mobility due to high-impurity scattering. Consequently, velocity saturation effects are observed at relatively higher fields than the other devices shown here, which are all larger, back-gated SiO_2-supported graphene transistors.

in high-field drift velocity with increasing carrier density, as mentioned above and shown in previous theoretical works. Similarly, the high-field drift velocity decreases with increasing temperature, which is an expected trend if we assume high-field transport is limited by emission of optical phonons.

Lastly, in **Figure 5.9a**, we summarise the practical analytic models from Dorgan et al. (2010), Freitag et al. (2009) and Fang et al. (2011). These models assume that velocity saturation in graphene is caused by a single optical phonon of energy $\hbar\omega_{OP}$. The first model (Dorgan et al., 2010), as shown in **Figure 5.9a**, approximates the high-field electron distribution with two half-disks such that positive-k_x electrons are populated to an energy $\hbar\omega_{OP}$ higher than negative-k_x moving electrons. The assumption here is that the inelastic process of emitting an OP causes the electron to instantly backscatter. The second model (Freitag et al., 2009) shown in **Figure 5.9b** assumes that for a given electron density defined by $n = k_F^2/\pi$, the high-field transport regime simply consists of electrons within a $\pm\hbar\omega_{OP}/2$ window around the Fermi energy E_F, such that $E_F = \hbar v_F k_F = \hbar v_F(\pi n)^{1/2}$. The last model (Fang et al., 2011) (**Figure 5.9c**) considers a low-energy disk (k_l) and a high-energy disk (k_h) defined by $\omega_{OP}/v_F = k_h - k_l$, such that if an electron travelling along the k_x-direction reaches the high-energy circle, then it instantly emits an OP and backscatters. This 'streaming' model (Fang et al., 2011) accounts for carrier degeneracy in graphene by assuming that the occupation of the low-energy circle is so

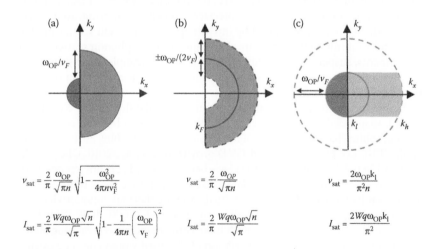

$$v_{sat} = \frac{2}{\pi}\frac{\omega_{OP}}{\sqrt{\pi n}}\sqrt{1 - \frac{\omega_{OP}^2}{4\pi n v_F^2}}$$

$$I_{sat} = \frac{2}{\pi}\frac{Wq\omega_{OP}\sqrt{n}}{\sqrt{\pi}}\sqrt{1 - \frac{1}{4\pi n}\left(\frac{\omega_{OP}}{v_F}\right)^2}$$

$$v_{sat} = \frac{2}{\pi}\frac{\omega_{OP}}{\sqrt{\pi n}}$$

$$I_{sat} = \frac{2}{\pi}\frac{Wq\omega_{OP}\sqrt{n}}{\sqrt{\pi}}$$

$$v_{sat} = \frac{2\omega_{OP}k_l}{\pi^2 n}$$

$$I_{sat} = \frac{2Wq\omega_{OP}k_l}{\pi^2}$$

FIGURE 5.9 Simplified models for high-field electron distribution and velocity saturation in graphene. The electric field is oriented along the x-axis. The shaded regions represent the portions of k-space populated by electrons, as the high-field electron distribution is distorted from equilibrium (circular disk). Analytic models from (a) Dorgan et al. (2010) and Barreiro et al. (2009); (b) Freitag et al. (2009) and Perebeinos and Avouris (2010) and (c) Fang et al. (2011). These models assume a single optical phonon with energy $\hbar\omega_{OP}$ is responsible for velocity saturation. The equations for velocity saturation v_{sat} and current saturation I_{sat} based on each model are provided as well.

full that the Pauli exclusion principle prohibits an electron with energy less than $E_h = \hbar v_F k_h$ to emit an OP, which forces the distribution function to be squeezed and elongated along the direction of the electric field.

We emphasise that these analytic models are more suited for empirical fitting and compact modelling of transistors. It is known that even small changes in the electron distribution can significantly affect charge transport, and thus, these models most likely oversimplify the distribution of carriers during high-field operation. Nevertheless, the compact models give the correct dependence of velocity saturation on carrier density and the magnitude of the phonon responsible (assuming that a single phonon energy is dominant). For example, Dorgan et al. (2010) fit against experimental data using a dominant phonon of ~81 meV, which is a value between that of the SiO_2 substrate phonon and the intrinsic graphene OP, suggesting that both maybe playing a role – a scenario examined in more depth by Serov et al. (2014).

5.3 Thermal Transport in Graphene

5.3.1 Intrinsic Thermal Conductivity of Graphene

We now turn to the discussion of thermal transport in graphene, with implications both for transistors and interconnects. First, we focus on the 'intrinsic' thermal conductivity (k) of SLG. Here, by 'intrinsic', we mean isolated graphene without impurities, defects, interfaces and edge scattering, so its thermal conductivity is only limited by intrinsic phonon–phonon scattering due to crystal anharmonicity (Balandin, 2011) and electron–phonon scattering. In experiments, suspended, micrometer scale graphene samples have properties close to intrinsic ones. We thus first summarise current experimental observations of k in suspended SLG.

Using the Raman thermometry technique (Balandin, 2011, Xu et al., 2014b), suspended micro-scale graphene flakes obtained by both exfoliation from graphite (Balandin et al., 2008, Ghosh et al., 2008, 2009, 2010, Faugeras et al., 2010; Lee et al., 2011) and CVD growth (Cai et al., 2010, Chen et al., 2011, 2012a,b, Vlassiouk et al., 2011) have been measured at room temperature and above. Some representative data versus temperature from these studies are shown in **Figure 5.10a**. The obtained in-plane thermal conductivity values of suspended SLG generally fall in the range of ~2000–4000 W/m/K at room temperature, and decrease with increasing temperature, reaching about 700–1500 W/m/K at 500 K. The variation of obtained values could be attributed to different choices of graphene optical absorbance (in the analysis of the Raman data), thermal contact resistance, different sample geometries, sizes and qualities.

For comparison, we also plot the experimental thermal conductivity of diamond (Ho et al., 1972), graphite (Ho et al., 1972) and carbon nanotubes (CNTs) (Kim et al., 2001, Pop et al., 2006a) in **Figure 5.10a**. It is clear that suspended graphene has thermal conductivity as high as these carbon allotropes near room temperature, even higher than its three-dimensional counterpart, graphite, whose in-plane thermal conductivity in highly oriented pyrolytic

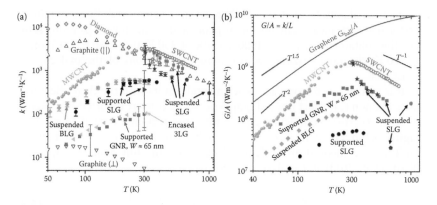

FIGURE 5.10 (a) Experimental thermal conductivity k as a function of temperature T: representative data for suspended CVD SLG by Chen et al. (2011) (solid red squares), suspended exfoliated SLG by Lee et al. (2011) (solid purple asterisk) and Faugeras et al. (2010) (solid brown pentagon), suspended SLG by Dorgan et al. (2013) (solid grey hexagon), suspended exfoliated BLG by Pettes et al. (2011) (solid orange diamond), supported exfoliated SLG by Seol et al. (2010) (solid black circle), supported CVD SLG by Cai et al. (2010) (solid blue right-triangle), encased exfoliated 3LG by Jang et al. (2010) (solid cyan left-triangle), supported exfoliated GNR of $W \approx 65$ nm by Bae et al. (2013) (solid magenta square), type IIa diamond (Ho et al., 1972) (open gold diamond), graphite in-plane (Ho et al., 1972) (open blue up-triangle), graphite cross-plane (open blue down-triangle), suspended SWCNT by Pop et al. (2006a) (open dark-green circle) and multi-walled CNT by Kim et al. (2001) (solid light-green circle). (b) Thermal conductance per cross-sectional area, $G/A = k/L$, converted from thermal conductivity data in (a), compared with the theoretical ballistic limit of graphene (solid line) (Bae et al., 2013). Data in (a) whose sample L is unknown or not applicable are not shown in (b). (Xu, Y., Li, Z. and Duan, W.: Thermal and thermoelectric properties of graphene. *Small.* 2014. 10. 2182–2199. Copyright Wiley-VCH Verlag GmbH & Co. KGaA. Reproduced with permission.)

graphite is ~2000 W/m/K at 300 K. The currently available data of graphene based on the Raman thermometry technique only cover the temperature range of ~300–600 K, except one at ~660 K reported by Faugeras et al. (2010) showing $k \approx 630$ W/m/K. For higher temperature, Dorgan et al. (2013) used the electrical breakdown method for thermal conductivity measurements and found $k \approx 310$ W/m/K at 1000 K for suspended SLG. The overall trend of the current graphene data from 300 to 1000 K shows a steeper temperature dependence than graphite (see **Figure 5.10a**), consistent with the extrapolation of thermal conductivity by Dorgan et al. (2013). This behaviour could be attributed to stronger second-order three-phonon scattering (relaxation time $\tau \sim T^{-2}$) in graphene than graphite enabled by the flexural (ZA) phonons of suspended graphene (Nika et al., 2012), similar to the observations in single-walled CNTs (SWCNTs) (Mingo and Broido, 2005, Pop et al., 2006a). For temperature below 300 K, the micro-resistance thermometry technique

(Sadeghi et al., 2012, Xu et al., 2014b) needs to be employed. Xu et al. (2014a) performed such measurements for suspended SLG, and for a 7-μm-long sample, they obtained $k \sim 1500 - 1800$ W/m/K at 300 K and $\sim 500 - 600$ W/m/K at 120 K, the range arising from different estimations of thermal contact resistance.

It is instructive to compare experimental results with the theoretical ballistic phonon transport limit of graphene, so we convert reported k in **Figure 5.10a** to thermal conductance per unit cross-sectional area, $G/A = k/L$ in **Figure 5.10b**, comparing them with the ballistic limit, G_{ball}/A. The ballistic limit can be theoretically calculated based on the full phonon dispersion (Serov et al., 2013) and analytically approximated as

$$\frac{G_{ball}}{A} \approx \left[\frac{1}{4.4 \times 10^5 T^{1.68}} + \frac{1}{1.2 \times 10^{10}} \right]^{-1}, \quad (5.8)$$

over the temperature range 1–1000 K (Bae et al., 2013). Above room temperature, measured G/A of suspended, few-micron long SLG is over one order of magnitude lower than G_{ball}/A, indicating the diffusive transport regime. The value of Faugeras et al. (2010) is much lower than others because of a much larger $L = 22$ μm (radius) of their suspended graphene. More importantly, graphene phonon mean free path (MFP), λ can be estimated based on G_{ball}/A and diffusive thermal conductivity (k_{diff}), that is, $\lambda = (2/\pi)k_{diff}/(G_{ball}/A)$ (Pop et al., 2012, Bae et al., 2013; Xu et al., 2014b). Since most k values measured on few-micron graphene samples can be approximated as k_{diff}, the estimated phonon MFP is $\sim$300–600 nm for suspended SLG at room temperature.

5.3.2 Extrinsic Thermal Conductivity of Graphene

The long phonon MFP in pristine graphene would suggest that it is possible to tune thermal conductivity more effectively by introducing extrinsic scattering mechanisms which dominate over intrinsic scattering mechanisms in graphene. For example, isotope scattering, normally unimportant with respect to other scattering processes, could become significant in graphene thermal conduction. In the following, we discuss various scattering mechanisms and their influences on thermal conduction in graphene.

5.3.2.1 Isotope Effects

The knowledge of isotope effects on thermal transport properties is valuable for tuning heat conduction in graphene. Natural abundance carbon materials are made up of two stable isotopes of ^{12}C (98.9%) and ^{13}C (1.1%). Changing isotope composition can modify the dynamic properties of crystal lattices and affect their thermal conductivity (Hu et al., 2010, Lindsay et al., 2013). For instance, at room temperature, isotopically purified diamond has a thermal conductivity of $\sim$3300 W/m/K (Anthony et al., 1990, Berman, 1992), about 50% higher than that of natural diamond, $\sim$2200 W/m/K (Ho et al., 1972). Similar effects have also been observed in one-dimensional nanostructures, boron nitride nanotubes (Chang et al., 2006). Very recently, the first experimental

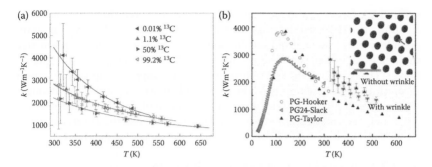

FIGURE 5.11 (a) Thermal conductivity of suspended CVD graphene as a function of temperature for different ^{13}C concentrations, showing isotope effect. (Reprinted by permission from Macmillan Publishers Ltd. *Nat. Mater.*, Chen, S. S. et al., 11, 203–207, copyright 2012.) (b) Thermal conductivity of suspended CVD graphene with (down-triangle) and without (up-triangle) wrinkles as a function of temperature. Also shown in comparison are the literature thermal conductivity data of pyrolytic graphite samples (Slack, 1962, Hooker et al., 1965, Taylor, 1966). Inset shows the SEM image of CVD graphene on the Au-coated SiN$_x$ holey membrane. The arrow indicates a wrinkle. Scale bar is 10 µm. (Adapted from Chen, S. S. et al., Thermal conductivity measurements of suspended graphene with and without wrinkles by micro-Raman mapping, *Nanotechnology*, 23, 365701, Copyright 2012, Institute of Physics.)

work to show the isotope effect on graphene thermal conduction was reported by Chen et al. (2012b), who synthesized CVD graphene with various percentages of [13]C. Their graphene flakes were suspended over 2.8-µm-diameter holes and thermal conductivity was measured by the Raman thermometry technique. As shown in **Figure 5.11a**, compared with natural abundance graphene (1.1% [13]C), the k values were enhanced in isotopically purified samples (0.01% [13]C), and reduced in isotopically mixed ones (50% [13]C).

5.3.2.2 Structural Defect Effects

Structural defects are common in fabricated graphene, especially in CVD grown graphene (Wood et al., 2011, Koepke et al., 2013). The effects of wrinkles (Chen et al., 2012a) and grain size (Vlassiouk et al., 2011) on the thermal conduction of suspended single-layer CVD graphene have been examined in experiments by using the Raman thermometry technique. Chen et al. (2012a) found that the thermal conductivity of graphene with obvious wrinkles (indicated by arrows in the inset of **Figure 5.11b**) is about 15%–30% lower than that of wrinkle-free graphene over their measured temperature range, ~330–520 K (**Figure 5.11b**). Vlassiouk et al. (2011) measured suspended graphene with different grain sizes obtained by changing the temperature of CVD growth. The grain sizes were estimated to be ℓ_G = 150 nm, 38 nm and 1.3 nm in different samples in terms of the intensity ratio of the G peak to D peak in Raman spectra (Cançado et al., 2006). Since grain boundaries in graphene serve as extended defects and scatter phonons, graphene with smaller grain sizes are expected to suffer more frequent phonon scattering. Their measured thermal conductivity shows the expected decrease for smaller grain sizes, indicating the grain boundary effect on thermal conduction. The dependence on the grain size shows a weak power law, $k \sim \ell_G^{1/3}$, for which there is no theoretical explanation yet (Vlassiouk et al., 2011). However, for SiO_2-supported graphene, recent theoretical work based on the NEGF method showed a similar but stronger dependence of k on the grain size ℓ_G in the range of $\ell_G < 1$ μm (Serov et al., 2013). Further experimental studies are required to reveal the grain size effects on the thermal transport of both suspended and supported graphene.

5.3.2.3 Substrate Effects in Supported Graphene

For practical applications, graphene is usually in contact with a substrate in electronic and optoelectronic devices, so it is important to understand substrate effects on the thermal properties of supported graphene (Ong and Pop, 2011, Guo et al., 2012, Xu and Buehler, 2012, Chen et al., 2013). Seol et al. (2010, 2011) measured exfoliated SLG on a 300-nm-thick SiO_2 membrane by using the micro-resistance thermometry technique. The observed thermal conductivity is $k \sim 600$ W/m/K near room temperature (solid black circles in **Figure 5.10a**). This value is much lower than those reported for freely suspended SLG via the Raman thermometry technique, but is still relatively high compared with those of bulk silicon (~150 W/m/K) and copper (~400 W/m/K). Another study by Cai et al. (2010) showed CVD SLG supported on Au also has a decreased thermal conductivity, ~370 W/m/K, and this lower value could be caused by grain boundary scattering (Serov et al., 2013). If using 600 W/m/K as k_{diff}, phonon MFP for SiO_2-supported graphene is estimated to be ~90 nm at room temperature (Pop et al., 2012, Bae et al., 2013, Xu et al., 2014b). The thermal conductivity reduction in supported graphene is attributed to substrate scattering, which strongly affects the out-of-plane flexural (ZA) mode of graphene (Seol et al., 2010, Ong and Pop, 2011, Qiu and Ruan, 2012). This effect becomes stronger in encased graphene, where graphene is sandwiched between bottom and top SiO_2. Jang et al. (2010) measured such SiO_2-encased graphene samples using the micro-resistance thermometry technique. The obtained k for encased SLG is reported to be below 160 W/m/K, and that for encased three-layer graphene (3LG) is shown as cyan triangles in **Figure 5.10a**. For encased graphene, besides the phonon scattering by bottom and top oxides, the deposition of the top SiO_2 layer could cause defects in graphene, which can further lower thermal conductivity. Knowing the thermal conductivity of supported and encased graphene is useful for analysing heat dissipation in graphene transistors and interconnects.

5.3.2.4 Size Effects and Boundary Scattering

In macroscopic bulk materials, thermal conductance satisfies Fourier's scaling law in the diffusive region, $G = kA/L$, where the thermal conductivity k is an intrinsic material property, independent of system size. This scaling law can break down in nanostructures through two mechanisms: (i) non-diffusive (quasi-ballistic) thermal transport and (ii) boundary effects become important. For non-diffusive transport, k becomes length dependent. It is proportional to

L as $k_{ball} = (G_{ball}/A)L$ in the ballistic limit, gets saturated in the diffusive limit, and typically grows gradually with increasing L in the intermediate region. In nanostructures, boundary effects cannot be neglected, and k becomes dependent on lateral sizes due to boundary scattering. The relatively long intrinsic phonon MFP of graphene (100–600 nm) makes the observations of such size effects on thermal transport possible.

The length-dependent k has been reported by Bae et al. (2013) and Xu et al. (2014a) in SiO_2-supported and suspended SLG, respectively. The measured k is reduced as the transport length decreases over a wide temperature range (see **Figure 5.12a**), and both studies show that the thermal transport enters the quasi-ballistic regime when L is a few hundred nanometres, comparable to phonon MFP of graphene. The observed L-dependent k can be captured by a simple ballistic-diffusive model (Pop et al., 2012, Bae et al., 2013, Xu et al., 2014b)

$$k(L) = \left[\frac{1}{(G_{ball}/A)L} + \frac{1}{k_{diff}} \right]^{-1}, \qquad (5.9)$$

with theoretically calculated G_{ball}/A and properly fitted k_{diff} (see solid lines in **Figure 5.12a**).

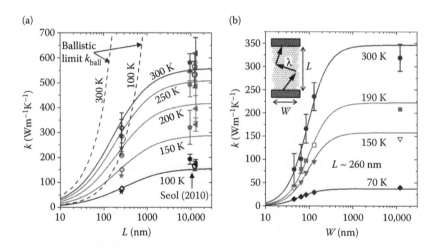

FIGURE 5.12 (a) Thermal conductivity reduction with length for 'wide' graphene samples ($W \gg \lambda$), compared to the ballistic limit ($k_{ball} = G_{ball}L/A$) at several temperatures. Symbols are data for 'short' (Bae et al., 2013) and 'long' (Seol et al., 2010) samples. Solid lines are model from **Equation 5.9**. (b) Thermal conductivity reduction with width for GNRs, all with $L \approx 260$ nm. Symbols are experimental data (Bae et al., 2013) and lines are fitted model. Inset shows a schematic of edge scattering in GNRs with dimensions comparable to the MFP λ. (Reprinted by permission from Macmillan Publishers Ltd. *Nat. Commun.*, Bae, M.-H. et al., Ballistic to diffusive crossover of heat flow in graphene ribbons, 4, 1734, copyright 2013.)

The lateral size effects, that is, width-dependence of graphene thermal transport has also been measured by Bae et al. (2013) on GNRs with $W \sim 45$–$130\,nm$, implemented by developing a substrate-supported thermometry platform (Li et al., 2014). The significant decrease of k is observed when $W < 200\,nm$ (see **Figure 5.12b**), due to increased edge scattering in narrower GNRs (see inset of **Figure 5.12b**). Through electrical breakdown measurements, Liao et al. (2011) were able to estimate k of CNT-unzipped GNRs (Jiao et al., 2009). Although the obtained values are slightly higher than those of Bae et al. (2013) for similar widths, considering that CNT-unzipped GNRs have smoother edges (Jiao et al., 2009, Kosynkin et al., 2009) (less edge roughness scattering), the two studies are essentially consistent.

5.3.2.5 Interlayer Effects in Few-Layer Graphene

Interlayer scattering as well as top and bottom boundary scattering could take place in few-layer graphene (FLG), which could be another mechanism to modulate graphene thermal conductivity. It is interesting to investigate the evolution of the thermal conductivity of FLG with increasing thickness, denoted by the number of atomic layers (N), and the critical thickness needed to recover the thermal conductivity of graphite.

Several experimental studies on this topic have been conducted for encased (Jang et al., 2010), supported (Sadeghi et al., 2013) and suspended FLG (Ghosh et al., 2010, Jang et al., 2013), and their results are summarised in **Figure 5.13**. Jang et al. (2010) measured the thermal transport of SiO_2-encased FLG by using the substrate-supported, micro-resistance thermometry platform. They found that the room-temperature thermal conductivity increases from ~ 50 to $\sim 1000\,W/m/K$ as the FLG thickness increases from 2 to 21 layers, showing a trend to recover natural graphite k. This strong thickness dependence was explained by the top and bottom boundary scattering and disorder penetration into FLG induced by the evaporated top oxide (Jang et al., 2010). More recently, another similar yet less pronounced trend was observed in SiO_2-supported FLG by Sadeghi et al. (2013) using the suspended micro-resistance thermometry platform. As shown by filled circles in **Figure 5.13**, the measured room-temperature k increases slowly with increasing thickness, and the recovery to natural graphite would occur after more than 34 layers. The difference between the results by Jang et al. (2010) and Sadeghi et al. (2013) is not unexpected, because encased FLG k could be suppressed much more in thin layers than thick ones due to the effect of the top oxide, hence showing a stronger thickness dependence.

For suspended FLG, there are two contradictory observations in the thickness dependence. At first, based on the Raman thermometry technique, Ghosh et al. (2010) showed a decrease of suspended FLG k from the SLG high value to regular graphite value as thickness increases from 2 to 8 layers (open diamonds in **Figure 5.13**). The k reduction was explained by the interlayer coupling and increased phase-space states available for phonon Umklapp scattering in thicker FLG (Ghosh et al., 2010). However, a very recent study (Jang et al., 2013) seems to show a different thickness trend for suspended FLG. They measured thermal conductivity of suspended graphene of 2–4 and 8 layers by using a

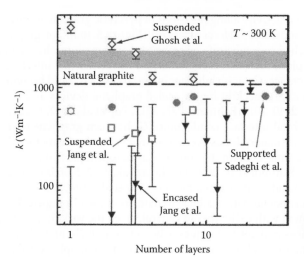

FIGURE 5.13 Experimental in-plane thermal conductivity near room temperature as a function of the number of layers N for suspended graphene by Ghosh et al. (2010) (open diamond) and Jang et al. (2013) (open square), SiO_2-supported graphene by Seol et al. (2010) (open circle) and Sadeghi et al. (2013) (solid circle) and SiO_2-encased graphene by Jang et al. (2010) (solid black triangle). The data show a trend to recover the value (dashed line) measured by Sadeghi et al. (2013) for natural graphite source used to exfoliate graphene. The grey-shaded area shows the highest reported k values of pyrolytic graphite (Slack, 1962, Hooker et al., 1965, Taylor, 1966). (Sadeghi, M. M., Jo, I. and Shi, L. Phonon-interface scattering in multilayer graphene on an amorphous support. *Proc. Natl. Acad. Sci. USA*, 110, 16321–16326, Copyright 2013 National Academy of Sciences, U.S.A.)

modified T-bridge micro-resistance thermometry technique. The obtained room-temperature k for 2–4 layers is about 300–400 W/m/K with no apparent thickness dependence, whereas k for 8-layer shows an increase to ~600 W/m/K (open squares in **Figure 5.13**). Surprisingly, this trend is qualitatively in agreement with that of Sadeghi et al. (2013) for *supported* FLG; both show similar increasing amounts of k from 2 to 8 layers (**Figure 5.13**), despite a small decrease from 2 to 4 layers in the former, which could arise from different sample qualities and measurement uncertainty. Given the opposite thickness trends of Ghosh et al. (2010) and Jang et al. (2013), further experiments are required to clarify the real thickness-dependent k in *suspended* FLG.

5.3.2.6 Cross-Plane Thermal Conduction

A remarkable feature of graphite and graphene is that their thermal properties are highly anisotropic. Despite high thermal conductivity along the in-plane direction, heat flow along the cross-plane direction (c-axis) is hundreds of times weaker, limited by weak van der Waals interactions between layers (for graphite) or with adjacent materials (for graphene). For example, the thermal conductivity along the c-axis of pyrolytic graphite is only ~6 W/m/K at room temperature (Ho et al., 1972) (**Figure 5.10a**). For graphene, as it is often attached to a substrate or embedded in a medium, heat conduction in the cross-plane direction is characterised by the thermal interface resistance (or conductance, $G_\perp$) with adjacent materials, which could become a limiting dissipation bottleneck in highly scaled graphene devices and interconnects (Bae et al., 2010, 2011, Behnam et al., 2012, Pop et al., 2012, Islam et al., 2013).

The thermal interface conductance $G_\perp$ between graphene (or thin graphite) and other materials has been measured using the 3ω method (Chen et al., 2009b), the time-domain thermoreflectance technique (Koh et al., 2010, Mak et al., 2010, Schmidt et al., 2010; Hopkins et al., 2012, Norris et al., 2012, Zhang et al., 2013) and Raman-based method (Cai et al., 2010, Ermakov et al., 2013). Most experimental data available to date are shown in **Figure 5.14**, and they are consistent with each other in general, given the variations of sample quality and measurement techniques. Chen et al. (2009b) and Mak et al. (2010) showed the thermal interface conductance of the graphene/SiO$_2$ interface is

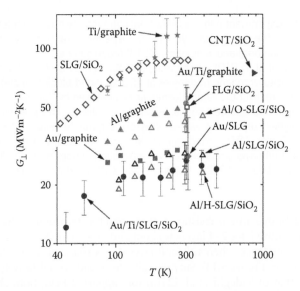

FIGURE 5.14 Experimental thermal interface conductance $G_\perp$ versus temperature for SLG/SiO$_2$ by Chen et al. (2009b) (open diamond), FLG/SiO$_2$ by Mak et al. (2010) (open square), CNT/SiO$_2$ by Pop et al. (2007) (solid right-triangle), Au/SLG by Cai et al. (2010) (solid diamond), Au/Ti/SLG/SiO$_2$ (solid circle) and Au/Ti/graphite (solid circle) by Koh et al. (2010), interfaces of graphite with Au (solid square), Al (solid up-triangle), Ti (solid asterisk) by Schmidt et al. (2010), interfaces of Al/SLG/SiO$_2$ without treatment (open black up-triangle), with oxygen treatment (Al/O-SLG/SiO$_2$, open up-triangle), and with hydrogen treatment (Al/H-SLG/SiO$_2$, open up-triangle) by Hopkins et al. (2012). (Xu, Y., Li, Z. and Duan, W.: Thermal and thermoelectric properties of graphene. *Small*, 2014, 10, 2182–2199. Copyright Wiley-VCH Verlag GmbH & Co. KGaA. Reproduced with permission.)

$G_\perp \sim 50\text{--}100 \ \text{MW/m}^2/\text{K}$ at room temperature, with no strong dependence on the FLG thickness. Their values are close to that of CNT/SiO$_2$ (Pop et al., 2007), reflecting the similarity between graphene and CNT. Schmidt et al. (2010) measured $G_\perp$ of graphite–metal interfaces, including Au, Cr, Al and Ti. Among them, the graphite–Ti interface has the highest $G_\perp \sim 120 \ \text{MW/m}^2/\text{K}$ and the graphite–Au interface has the lowest $G_\perp \sim 30 \ \text{MW/m}^2/\text{K}$ near room temperature. The $G_\perp$ of graphite–Au is consistent with the value by Norris et al. (2012) and values of SLG–Au by Cai et al. (2010) and FLG/Au by Ermakov et al. (2013). Koh et al. (2010) later measured heat flow across the Au/Ti/N-LG/ SiO$_2$ interfaces with the layer number $N = 1 - 10$. Their observed room-temperature $G_\perp$ was $\sim 25 \ \text{MW/m}^2/\text{K}$, which shows a very weak dependence on the layer number N and is equivalent to the total thermal conductance of Au–Ti–graphite and graphene–SiO$_2$ interfaces acting in series. This indicates that the thermal resistance of two interfaces between graphene and its environment dominates over that between graphene layers. Interestingly, Hopkins et al. (2012) showed the thermal conduction across the Al–SLG–SiO$_2$ interface could be manipulated by introducing chemical adsorbates between the Al and SLG. As shown in **Figure 5.14**, their measured $G_\perp$ of untreated Al/SLG/ SiO$_2$ is $\sim 30 \ \text{MW/m}^2/\text{K}$ at room temperature, in agreement with Zhang et al. (2013). The $G_\perp$ increases to $\sim 42 \ \text{MW/m}^2/\text{K}$ for oxygen-functionalised graphene (O-SLG), and decreases to $\sim 23 \ \text{MW/m}^2/\text{K}$ for hydrogen-functionalised graphene (H-SLG). These effects were attributed to changes in chemical bonding between the metal and graphene, and are consistent with the observed enhancement in $G_\perp$ from the Al/diamond (Stoner et al., 1992) to Al/O-diamond interfaces (Collins et al., 2010).

References

Adam, S., Hwang, E. H., Galitski, V. M. and Das Sarma, S. 2007. A self-consistent theory for graphene transport. *Proc. Natl. Acad. Sci. USA*, 104, 18392–18397.

Adam, S., Hwang, E. H., Rossi, E. and Sarma, S. D. 2009. Theory of charged impurity scattering in two-dimensional graphene. *Solid State Commun.*, 149, 1072–1079.

Akturk, A. and Goldsman, N. 2008. Electron transport and full-band electron–phonon interactions in graphene. *J. Appl. Phys.*, 103, 053702.

Ancona, M. G. 2010. Electron transport in graphene from a diffusion-drift perspective. *IEEE Trans. Electron Devices*, 57, 681–689.

Anthony, T. R., Banholzer, W. F., Fleischer, J. F., Wei, L. H., Kuo, P. K., Thomas, R. L. and Pryor, R. W. 1990. Thermal-diffusivity of isotopically enriched C-12 diamond. *Phys. Rev. B*, 42, 1104–1111.

Bae, M.-H., Islam, S., Dorgan, V. E. and Pop, E. 2011. Scaling of high-field transport and localized heating in graphene transistors. *ACS Nano*, 5, 7936–7944.

Bae, M.-H., Li, Z., Aksamija, Z., Martin, P. N., Xiong, F., Ong, Z.-Y., Knezevic, I. and Pop, E. 2013. Ballistic to diffusive crossover of heat flow in graphene ribbons. *Nat. Commun.*, 4, 1734.

Bae, M. H., Ong, Z. Y., Estrada, D. and Pop, E. 2010. Imaging, simulation and electrostatic control of power dissipation in graphene devices. *Nano Lett.*, 10, 4787–4793.

Balandin, A. A. 2011. Thermal properties of graphene and nanostructured carbon materials. *Nat. Mater.*, 10, 569–581.

Balandin, A. A., Ghosh, S., Bao, W. Z., Calizo, I., Teweldebrhan, D., Miao, F. and Lau, C. N. 2008. Superior thermal conductivity of single-layer graphene. *Nano Lett.*, 8, 902–907.

Barreiro, A., Lazzeri, M., Moser, J., Mauri, F. and Bachtold, A. 2009. Transport properties of graphene in the high-current limit. *Phys. Rev. Lett.*, 103, 076601.

Behnam, A., Lyons, A. S., Bae, M.-H., Chow, E. K., Islam, S., Neumann, C. M. and Pop, E. 2012. Transport in nanoribbon interconnects obtained from graphene grown by chemical vapor deposition. *Nano Lett.*, 12, 4424–4430.

Berciaud, S., Han, M. Y., Mak, K. F., Brus, L. E., Kim, P. and Heinz, T. F. 2010. Electron and optical phonon temperatures in electrically biased graphene. *Phys. Rev. Lett.*, 104, 227401.

Berman, R. 1992. Thermal-conductivity of isotopically enriched diamonds. *Phys. Rev. B*, 45, 5726–5728.

Biro, L. P., Nemes-Incze, P. and Lambin, P. 2012. Graphene: Nanoscale processing and recent applications. *Nanoscale*, 4, 1824–1839.

Bistritzer, R. and Macdonald, A. H. 2009. Hydrodynamic theory of transport in doped graphene. *Phys. Rev. B*, 80, 085109.

Bolotin, K. I., Sikes, K. J., Jiang, Z., Klima, M., Fudenberg, G., Hone, J., Kim, P. and Stormer, H. L. 2008. Ultrahigh electron mobility in suspended graphene. *Solid State Commun.*, 146, 351–355.

Borysenko, K. M., Mullen, J. T., Barry, E. A., Paul, S., Semenov, Y. G., Zavada, J. M., Nardelli, M. B. and Kim, K. W. 2010. First-principles analysis of electron–phonon interactions in graphene. *Phys. Rev. B*, 81, 121412.

Cai, W. W., Moore, A. L., Zhu, Y. W., Li, X. S., Chen, S. S., Shi, L. and Ruoff, R. S. 2010. Thermal transport in suspended and supported monolayer graphene grown by chemical vapor deposition. *Nano Lett.*, 10, 1645–1651.

Cançado, L. G., Takai, K., Enoki, T., Endo, M., Kim, Y. A., Mizusaki, H., Jorio, A., Coelho, L. N., Magalhães-Paniago, R. and Pimenta, M. A. 2006. General equation for the determination of the crystallite size L_a of nanographite by Raman spectroscopy. *Appl. Phys. Lett.*, 88, 163106.

Castro Neto, A. H., Guinea, F., Peres, N. M. R., Novoselov, K. S. and Geim, A. K. 2009. The electronic properties of graphene. *Rev. Mod. Phys.*, 81, 109–162.

Chae, D.-H., Krauss, B., von Klitzing, K. and Smet, J. H. 2009. Hot phonons in an electrically biased graphene constriction. *Nano Lett.*, 10, 466–471.

Chang, C. W., Fennimore, A. M., Afanasiev, A., Okawa, D., Ikuno, T., Garcia, H., Li, D. Y., Majumdar, A. and Zettl, A. 2006. Isotope effect on the thermal conductivity of boron nitride nanotubes. *Phys. Rev. Lett.*, 97, 085901.

Chauhan, J. and Guo, J. 2009. High-field transport and velocity saturation in graphene. *Appl. Phys. Lett.*, 95, 023120.

Chen, F., Xia, J. and Tao, N. 2009a. Ionic screening of charged-impurity scattering in graphene. *Nano Lett.*, 9, 1621–1625.

Chen, I. J., Mante, P. A., Chang, C. K., Yang, S. C., Chen, H. Y., Huang, Y. R., Chen, L. C., Chen, K. H., Gusev, V. and Sun, C. K. 2014. Graphene-to-substrate energy transfer through out-of-plane longitudinal acoustic phonons. *Nano Lett.*, 14, 1317–1323.

Chen, J., Zhang, G. and Li, B. W. 2013. Substrate coupling suppresses size dependence of thermal conductivity in supported graphene. *Nanoscale*, 5, 532–536.

Chen, J. H., Jang, C., Adam, S., Fuhrer, M. S., Williams, E. D. and Ishigami, M. 2008. Charged-impurity scattering in graphene. *Nat. Phys.*, 4, 377–381.

Chen, S. S., Li, Q. Y., Zhang, Q. M., Qu, Y., Ji, H. X., Ruoff, R. S. and Cai, W. W. 2012a. Thermal conductivity measurements of suspended graphene with and without wrinkles by micro-Raman mapping. *Nanotechnology*, 23, 365701.

Chen, S. S., Moore, A. L., Cai, W. W., Suk, J. W., An, J. H., Mishra, C., Amos, C. et al. 2011. Raman measurements of thermal transport in suspended monolayer graphene of variable sizes in vacuum and gaseous environments. *ACS Nano*, 5, 321–328.

Chen, S. S., Wu, Q. Z., Mishra, C., Kang, J. Y., Zhang, H. J., Cho, K. J., Cai, W. W., Balandin, A. A. and Ruoff, R. S. 2012b. Thermal conductivity of isotopically modified graphene. *Nat. Mater.*, 11, 203–207.

Chen, Z., Jang, W., Bao, W., Lau, C. N. and Dames, C. 2009b. Thermal contact resistance between graphene and silicon dioxide. *Appl. Phys. Lett.*, 95, 161910.

Collins, K. C., Chen, S. and Chen, G. 2010. Effects of surface chemistry on thermal conductance at aluminum–diamond interfaces. *Appl. Phys. Lett.*, 97, 083102.

Dasilva, A. M., Zou, K., Jain, J. K. and Zhu, J. 2010. Mechanism for current saturation and energy dissipation in graphene transistors. *Phys. Rev. Lett.*, 104, 236601.

Deshpande, A., Bao, W., Miao, F., Lau, C. N. and Leroy, B. J. 2009. Spatially resolved spectroscopy of monolayer graphene on SiO_2. *Phys. Rev. B*, 79, 205411.

Dorgan, V. E., Bae, M.-H. and Pop, E. 2010. Mobility and saturation velocity in graphene on SiO_2. *Appl. Phys. Lett.*, 97, 082112.

Dorgan, V. E., Behnam, A., Conley, H. J., Bolotin, K. I. and Pop, E. 2013. High-field electrical and thermal transport in suspended graphene. *Nano Lett.*, 13, 4581–4586.

Ermakov, V. A., Alaferdov, A. V., Vaz, A. R., Baranov, A. V. and Moshkalev, S. A. 2013. Nonlocal laser annealing to improve thermal contacts between multi-layer graphene and metals. *Nanotechnology*, 24, 155301.

Fang, T., Konar, A., Xing, H. and Jena, D. 2007. Carrier statistics and quantum capacitance of graphene sheets and ribbons. *Appl. Phys. Lett.*, 91, 092109.

Fang, T., Konar, A., Xing, H. and Jena, D. 2011. High-field transport in two-dimensional graphene. *Phys. Rev. B*, 84, 125450.

Faugeras, C., Faugeras, B., Orlita, M., Potemski, M., Nair, R. R. and Geim, A. K. 2010. Thermal conductivity of graphene in corbino membrane geometry. *ACS Nano*, 4, 1889–1892.

Ferry, D. K. 2012. The role of substrate for transport in graphene. *IEEE Nanotechnology Materials and Devices Conference*, 16–19 October 2012, Waikiki Beach, HI, 43–48.

Fischetti, M. V., Kim, J., Narayanan, S., Ong, Z.-Y., Sachs, C., Ferry, D. K. and Aboud, S. J. 2013. Pseudopotential-based studies of electron transport in graphene and graphene nanoribbons. *J. Phys. Condens. Matter*, 25, 473202.

Fischetti, M. V., Neumayer, D. A. and Cartier, E. A. 2001. Effective electron mobility in Si inversion layers in metal–oxide–semiconductor systems with a high-κ insulator: The role of remote phonon scattering. *J. Appl. Phys.*, 90, 4587–4608.

Freitag, M., Steiner, M., Martin, Y., Perebeinos, V., Chen, Z., Tsang, J. C. and Avouris, P. 2009. Energy dissipation in graphene field-effect transistors. *Nano Lett.*, 9, 1883–1888.

Geim, A. K. 2011. Nobel lecture: Random walk to graphene. *Rev. Mod. Phys.*, 83, 851–862.

Geim, A. K. and Novoselov, K. S. 2007. The rise of graphene. *Nat. Mater.*, 6, 183–191.

Ghosh, S., Bao, W., Nika, D. L., Subrina, S., Pokatilov, E. P., Lau, C. N. and Balandin, A. A. 2010. Dimensional crossover of thermal transport in few-layer graphene. *Nat. Mater.*, 9, 555–558.

Ghosh, S., Calizo, I., Teweldebrhan, D., Pokatilov, E. P., Nika, D. L., Balandin, A. A., Bao, W., Miao, F. and Lau, C. N. 2008. Extremely high thermal conductivity of graphene: Prospects for thermal management applications in nanoelectronic circuits. *Appl. Phys. Lett.*, 92, 151911.

Ghosh, S., Nika, D. L., Pokatilov, E. P. and Balandin, A. A. 2009. Heat conduction in graphene: Experimental study and theoretical interpretation. *New J. Phys.*, 11, 095012.

Girdhar, A. and Leburton, J. P. 2011. Soft carrier multiplication by hot electrons in graphene. *Appl. Phys. Lett.*, 99, 043107.

Guo, Z. X., Ding, J. W. and Gong, X. G. 2012. Substrate effects on the thermal conductivity of epitaxial graphene nanoribbons. *Phys. Rev. B*, 85, 235429.

Han, M. Y., Ozyilmaz, B., Zhang, Y. B. and Kim, P. 2007. Energy band-gap engineering of graphene nanoribbons. *Phys. Rev. Lett.*, 98, 206805.

Hess, K. and Vogl, P. 1979. Remote polar phonon scattering in silicon inversion layers. *Solid State Commun.*, 30, 797–799.

Ho, C. Y., Powell, R. W. and Liley, P. E. 1972. Thermal conductivity of the elements. *J. Phys. Chem. Ref. Data*, 1, 279–421.

Hooker, C. N., Ubbelohd, A. R. and Young, D. A. 1965. Anisotropy of thermal conductance in near-ideal graphite. *Proc. R. Soc. Lon. A*, 284, 17–31.

Hopkins, P. E., Baraket, M., Barnat, E. V., Beechem, T. E., Kearney, S. P., Duda, J. C., Robinson, J. T. and Walton, S. G. 2012. Manipulating thermal conductance at metal–graphene contacts via chemical functionalization. *Nano Lett.*, 12, 590–595.

Hu, J. N., Schiffli, S., Vallabhaneni, A., Ruan, X. L. and Chen, Y. P. 2010. Tuning the thermal conductivity of graphene nanoribbons by edge passivation and isotope engineering: A molecular dynamics study. *Appl. Phys. Lett.*, 97, 133107.

Huang, B., Yan, Q., Li, Z. and Duan, W. 2009. Towards graphene nanoribbon-based electronics. *Front. Phys. China*, 4, 269–279.

Huang, P. Y., Ruiz-Vargas, C. S., van der Zande, A. M., Whitney, W. S., Levendorf, M. P., Kevek, J. W., Garg, S. et al. 2011. Grains and grain boundaries in single-layer graphene atomic patchwork quilts. *Nature*, 469, 389–392.

Ishigami, M., Chen, J. H., Cullen, W. G., Fuhrer, M. S. and Williams, E. D. 2007. Atomic structure of graphene on SiO_2. *Nano Lett.*, 7, 1643–1648.

Islam, S., Li, Z., Dorgan, V. E., Bae, M.-H. and Pop, E. 2013. Role of joule heating on current saturation and transient behavior of graphene transistors. *IEEE Electron Device Lett.*, 34, 166–168.

Jacoboni, C., Canali, C., Ottaviani, G. and Alberigi Quaranta, A. 1977. A review of some charge transport properties of silicon. *Solid-State Electron.*, 20, 77–89.

Jacoboni, C. and Reggiani, L. 1983. The Monte Carlo method for the solution of charge transport in semiconductors with applications to covalent materials. *Rev. Mod. Phys.*, 55, 645–705.

Jang, W., Bao, W., Jing, L., Lau, C. N. and Dames, C. 2013. Thermal conductivity of suspended few-layer graphene by a modified T-bridge method. *Appl. Phys. Lett.*, 103, 133102.

Jang, W. Y., Chen, Z., Bao, W. Z., Lau, C. N. and Dames, C. 2010. Thickness-dependent thermal conductivity of encased graphene and ultrathin graphite. *Nano Lett.*, 10, 3909–3913.

Jiao, L., Zhang, L., Wang, X., Diankov, G. and Dai, H. 2009. Narrow graphene nanoribbons from carbon nanotubes. *Nature*, 458, 877–880.

Jimenez, D. 2011. Explicit drain current, charge and capacitance model of graphene field-effect transistors. *IEEE Trans. Electron Devices*, 58, 4377–4383.

Kedzierski, J., Hsu, P.-L., Healey, P., Wyatt, P. W., Keast, C. L., Sprinkle, M., Berger, C. and de Heer, W. A. 2008. Epitaxial graphene transistors on SiC substrates. *IEEE Trans. Electron Devices*, 55, 2078–2085.

Kim, E., Jain, N., Jacobs-Gedrim, R., Xu, Y. and Yu, B. 2012. Exploring carrier transport phenomena in a CVD-assembled graphene FET on hexagonal boron nitride. *Nanotechnology*, 23, 125706.

Kim, K. S., Zhao, Y., Jang, H., Lee, S. Y., Kim, J. M., Kim, K. S., Ahn, J. H., Kim, P., Choi, J. Y. and Hong, B. H. 2009a. Large-scale pattern growth of graphene films for stretchable transparent electrodes. *Nature*, 457, 706–710.

Kim, P., Shi, L., Majumdar, A. and Mceuen, P. L. 2001. Thermal transport measurements of individual multiwalled nanotubes. *Phys. Rev. Lett.*, 87, 215502.

Kim, S., Nah, J., Jo, I., Shahrjerdi, D., Colombo, L., Yao, Z., Tutuc, E. and Banerjee, S. K. 2009b. Realization of a high mobility dual-gated graphene field-effect transistor with Al_2O_3 dielectric. *Appl. Phys. Lett.*, 94, 062107.

Koepke, J. C., Wood, J. D., Estrada, D., Ong, Z. Y., He, K. T., Pop, E. and Lyding, J. W. 2013. Atomic-scale evidence for potential barriers and strong carrier scattering at graphene grain boundaries: A scanning tunneling microscopy study. *ACS Nano*, 7, 75–86.

Koh, Y. K., Bae, M.-H., Cahill, D. G. and Pop, E. 2010. Heat conduction across monolayer and few-layer graphenes. *Nano Lett.*, 10, 4363–4368.

Kosynkin, D. V., Higginbotham, A. L., Sinitskii, A., Lomeda, J. R., Dimiev, A., Price, B. K. and Tour, J. M. 2009. Longitudinal unzipping of carbon nanotubes to form graphene nanoribbons. *Nature*, 458, 872–875.

Leburton, J. P. and Dorda, G. 1981. Remote polar phonon scattering for hot electrons in Si-inversion layers. *Solid State Commun.*, 40, 1025–1026.

Lee, C., Wei, X., Kysar, J. W. and Hone, J. 2008. Measurement of the elastic properties and intrinsic strength of monolayer graphene. *Science*, 321, 385–388.

Lee, J. U., Yoon, D., Kim, H., Lee, S. W. and Cheong, H. 2011. Thermal conductivity of suspended pristine graphene measured by Raman spectroscopy. *Phys. Rev. B*, 83, 081419.

Li, Q., Hwang, E. H. and Das Sarma, S. 2011a. Disorder-induced temperature-dependent transport in graphene: Puddles, impurities, activation and diffusion. *Phys. Rev. B*, 84, 115442.

Li, X., Barry, E. A., Zavada, J. M., Nardelli, M. B. and Kim, K. W. 2010. Surface polar phonon dominated electron transport in graphene. *Appl. Phys. Lett.*, 97, 232105.

Li, X., Cai, W., An, J., Kim, S., Nah, J., Yang, D., Piner, R., Velamakanni, A., Jung, I. and Tutuc, E. 2009. Large-area synthesis of high-quality and uniform graphene films on copper foils. *Science*, 324, 1312–1314.

Li, X., Kong, B., Zavada, J. and Kim, K. 2011b. Strong substrate effects of Joule heating in graphene electronics. *Appl. Phys. Lett.*, 99, 233114.

Li, X. L., Wang, X. R., Zhang, L., Lee, S. W. and Dai, H. J. 2008. Chemically derived, ultrasmooth graphene nanoribbon semiconductors. *Science*, 319, 1229–1232.

Li, Z., Bae, M.-H. and Pop, E. 2014. Substrate-supported thermometry platform for nanomaterials like graphene, nanotubes and nanowires. *Appl. Phys. Lett.*, 105, 023107.

Liao, A. D., Wu, J. Z., Wang, X., Tahy, K., Jena, D., Dai, H. and Pop, E. 2011. Thermally limited current carrying ability of graphene nanoribbons. *Phys. Rev. Lett.*, 106, 256801.

Lin, Y.-M., Dimitrakopoulos, C., Jenkins, K. A., Farmer, D. B., Chiu, H.-Y., Grill, A. and Avouris, P. 2010. 100-GHz transistors from wafer-scale epitaxial graphene. *Science*, 327, 662–662.

Lindsay, L., Broido, D. A. and Mingo, N. 2010. Flexural phonons and thermal transport in graphene. *Phys. Rev. B*, 82, 115427.

Lindsay, L., Broido, D. A. and Reinecke, T. L. 2013. Phonon-isotope scattering and thermal conductivity in materials with a large isotope effect: A first-principles study. *Phys. Rev. B*, 88, 144306.

Lundstrom, M. 2000. *Fundamentals of Carrier Transport*. Cambridge University Press, Cambridge, UK.

Mak, K. F., Lui, C. H. and Heinz, T. F. 2010. Measurement of the thermal conductance of the graphene/SiO_2 interface. *Appl. Phys. Lett.*, 97, 221904.

Martin, J., Akerman, N., Ulbricht, G., Lohmann, T., Smet, J. H., Von Klitzing, K. and Yacoby, A. 2008. Observation of electron–hole puddles in graphene using a scanning single-electron transistor. *Nat. Phys.*, 4, 144–148.

Meric, I., Dean, C. R., Young, A. F., Baklitskaya, N., Tremblay, N. J., Nuckolls, C., Kim, P. and Shepard, K. L. 2011. Channel length scaling in graphene field-effect transistors studied with pulsed current – voltage measurements. *Nano Lett.*, 11, 1093–1097.

Meric, I., Han, M. Y., Young, A. F., Ozyilmaz, B., Kim, P. and Shepard, K. L. 2008. Current saturation in zero-bandgap, top-gated graphene field-effect transistors. *Nat. Nanotechnol.*, 3, 654–659.

Mingo, N. and Broido, D. A. 2005. Length dependence of carbon nanotube thermal conductivity and the 'problem of long waves'. *Nano Lett.*, 5, 1221–1225.

Mohr, M., Maultzsch, J., Dobardzic, E., Reich, S., Milosevic, I., Damnjanovic, M., Bosak, A., Krisch, M. and Thomsen, C. 2007. Phonon dispersion of graphite by inelastic x-ray scattering. *Phys. Rev. B*, 76, 035439.

Moore, B. T. and Ferry, D. K. 1980. Remote polar phonon scattering in Si inversion layers. *J. Appl. Phys.*, 51, 2603–2605.

Muller, R. S., Kamins, T. I. and Chan, M. 2003. *Device Electronics for Integrated Circuits*. John Wiley and Sons, New York.

Nika, D. L., Askerov, A. S. and Balandin, A. A. 2012. Anomalous size dependence of the thermal conductivity of graphene ribbons. *Nano Lett.*, 12, 3238–3244.

Norris, P. M., Smoyer, J. L., Duda, J. C. and Hopkins, P. E. 2012. Prediction and measurement of thermal transport across interfaces between isotropic solids and graphitic materials. *J. Heat Transf.*, 134, 020910.

Novoselov, K. S. 2011. Nobel lecture: Graphene: Materials in the flatland. *Rev. Mod. Phys.*, 83, 837–849.

Novoselov, K. S., Geim, A. K., Morozov, S. V., Jiang, D., Zhang, Y., Dubonos, S. V., Grigorieva, I. V. and Firsov, A. A. 2004. Electric field effect in atomically thin carbon films. *Science*, 306, 666–669.

Ong, Z.-Y. and Fischetti, M. V. 2012a. Charged impurity scattering in top-gated graphene nanostructures. *Phys. Rev. B*, 86, 121409.

Ong, Z.-Y. and Fischetti, M. V. 2012b. Erratum: Theory of interfacial plasmon–phonon scattering in supported graphene. *Phys. Rev. B*, 86, 199904.

Ong, Z.-Y. and Fischetti, M. V. 2012c. Theory of interfacial plasmon–phonon scattering in supported graphene. *Phys. Rev. B*, 86, 165422.

Ong, Z.-Y. and Fischetti, M. V. 2013a. Theory of remote phonon scattering in top-gated single-layer graphene. *Phys. Rev. B*, 88, 045405.

Ong, Z.-Y. and Fischetti, M. V. 2013b. Top oxide thickness dependence of remote phonon and charged impurity scattering in top-gated graphene. *Appl. Phys. Lett.*, 102, 183506.

Ong, Z.-Y. and Pop, E. 2011. Effect of substrate modes on thermal transport in supported graphene. *Phys. Rev. B*, 84, 075471.

Perebeinos, V. and Avouris, P. 2010. Inelastic scattering and current saturation in graphene. *Phys. Rev. B*, 81, 195442.

Pettes, M. T., Jo, I. S., Yao, Z. and Shi, L. 2011. Influence of polymeric residue on the thermal conductivity of suspended bilayer graphene. *Nano Lett.*, 11, 1195–1200.

Pirro, L., Girdhar, A., Leblebici, Y. and Leburton, J. P. 2012. Impact ionization and carrier multiplication in graphene. *J. Appl. Phys.*, 112, 093707.

Ponomarenko, L. A., Yang, R., Mohiuddin, T. M., Katsnelson, M. I., Novoselov, K. S., Morozov, S. V., Zhukov, A. A., Schedin, F., Hill, E. W. and Geim, A. K. 2009. Effect of a high-κ environment on charge carrier mobility in graphene. *Phys. Rev. Lett.*, 102, 206603.

Pop, E., Dutton, R. W. and Goodson, K. E. 2005. Monte Carlo simulation of Joule heating in bulk and strained silicon. *Appl. Phys. Lett.*, 86, 082101.

Pop, E. and Goodson, K. E. 2006. Thermal phenomena in nanoscale transistors. *J. Electron. Packag.*, 128, 102–108.

Pop, E., Mann, D., Wang, Q., Goodson, K. E. and Dai, H. J. 2006a. Thermal conductance of an individual single-wall carbon nanotube above room temperature. *Nano Lett.*, 6, 96–100.

Pop, E., Mann, D. A., Goodson, K. E. and Dai, H. J. 2007. Electrical and thermal transport in metallic single-wall carbon nanotubes on insulating substrates. *J. Appl. Phys.*, 101, 093710.

Pop, E., Sinha, S. and Goodson, K. E. 2006b. Heat generation and transport in nanometer-scale transistors. *Proc. IEEE*, 94, 1587–1601.

Pop, E., Varshney, V. and Roy, A. K. 2012. Thermal properties of graphene: Fundamentals and applications. *MRS Bull.*, 37, 1273–1281.

Qiu, B. and Ruan, X. 2012. Reduction of spectral phonon relaxation times from suspended to supported graphene. *Appl. Phys. Lett.*, 100, 193101.

Sadeghi, M. M., Jo, I. and Shi, L. 2013. Phonon-interface scattering in multilayer graphene on an amorphous support. *Proc. Natl. Acad. Sci. USA*, 110, 16321–16326.

Sadeghi, M. M., Pettes, M. T. and Shi, L. 2012. Thermal transport in graphene. *Solid State Commun.*, 152, 1321–1330.

Schmidt, A. J., Collins, K. C., Minnich, A. J. and Chen, G. 2010. Thermal conductance and phonon transmissivity of metal–graphite interfaces. *J. Appl. Phys.*, 107, 104907.

Sekwao, S. and Leburton, J. P. 2013. Electrical tunability of soft parametric resonance by hot electrons in graphene. *Appl. Phys. Lett.*, 103, 143108.

Seol, J. H., Jo, I., Moore, A. L., Lindsay, L., Aitken, Z. H., Pettes, M. T., Li, X. et al. 2010. Two-dimensional phonon transport in supported graphene. *Science*, 328, 213–216.

Seol, J. H., Moore, A. L., Shi, L., Jo, I. and Yao, Z. 2011. Thermal conductivity measurement of graphene exfoliated on silicon dioxide. *J. Heat Transf.*, 133, 022403.

Serov, A. Y., Ong, Z.-Y., Fischetti, M. V. and Pop, E. 2014. Theoretical analysis of high-field transport in graphene on a substrate. *J. Appl. Phys.*, 116, 034507.

Serov, A. Y., Ong, Z.-Y. and Pop, E. 2013. Effect of grain boundaries on thermal transport in graphene. *Appl. Phys. Lett.*, 102, 033104.

Shishir, R. S. and Ferry, D. K. 2009. Velocity saturation in intrinsic graphene. *J. Phys. Condens. Matter*, 21, 344201.

Shishir, R. S., Ferry, D. K. and Goodnick, S. M. 2009. Room-temperature velocity saturation in intrinsic graphene. *J. Phys. Conf. Ser.*, 193, 012118.

Slack, G. A. 1962. Anisotropic thermal conductivity of pyrolytic graphite. *Phys. Rev.*, 127, 694–701.

Son, Y.-W., Cohen, M. L. and Louie, S. G. 2006. Energy gaps in graphene nanoribbons. *Phys. Rev. Lett.*, 97, 216803.

Stoner, R. J., Maris, H. J., Anthony, T. R. and Banholzer, W. F. 1992. Measurements of the Kapitza conductance between diamond and several metals. *Phys. Rev. Lett.*, 68, 1563–1566.

Svintsov, D., Vyurkov, V., Yurchenko, S., Otsuji, T. and Ryzhii, V. 2012. Hydrodynamic model for electron–hole plasma in graphene. *J. Appl. Phys.*, 111, 083715.

Taur, Y., Hsu, C. H., Wu, B., Kiehl, R., Davari, B. and Shahidi, G. 1993. Saturation transconductance of deep-submicron-channel MOSFETs. *Solid-State Electron.*, 36, 1085–1087.

Taur, Y. and Ning, T. H. 2009. *Fundamentals of Modern VLSI Devices*. Cambridge University Press, New York.

Taylor, R. 1966. Thermal conductivity of pyrolytic graphite. *Philos. Mag.*, 13, 157–166.

Vlassiouk, I., Smirnov, S., Ivanov, I., Fulvio, P. F., Dai, S., Meyer, H., Chi, M. F., Hensley, D., Datskos, P. and Lavrik, N. V. 2011. Electrical and thermal conductivity of low temperature CVD graphene: the effect of disorder. *Nanotechnology*, 22, 275716.

Winzer, T., Knorr, A. and Malic, E. 2010. Carrier multiplication in graphene. *Nano Lett.*, 10, 4839–4843.

Wood, J. D., Doidge, G. P., Carrion, E. A., Koepke, J. C., Kaitz, J. A., Datye, I., Behnam, A. et al. 2015. Annealing free, clean graphene transfer using alternative polymer scaffolds. *Nanotechnology*, 26, 055302.

Wood, J. D., Schmucker, S. W., Lyons, A. S., Pop, E. and Lyding, J. W. 2011. Effects of polycrystalline Cu substrate on graphene growth by chemical vapor deposition. *Nano Lett.*, 11, 4547–4554.

Wu, Y., Jenkins, K. A., Valdes-Garcia, A., Farmer, D. B., Zhu, Y., Bol, A. A., Dimitrakopoulos, C., Zhu, W., Xia, F. and Avouris, P. 2012. State-of-the-art graphene high-frequency electronics. *Nano Lett.*, 12, 3062–3067.

Xu, X. F., Pereira, L. F. C., Wang, Y., Wu, J., Zhang, K. W., Zhao, X. M., Bae, S. et al. 2014a. Length-dependent thermal conductivity in suspended single-layer graphene. *Nat. Commun.*, 5, 3689.

Xu, Y., Li, Z. and Duan, W. 2014b. Thermal and thermoelectric properties of graphene. *Small*, 10, 2182–2199.

Xu, Z. P. and Buehler, M. J. 2012. Heat dissipation at a graphene-substrate interface. *J. Phys. Condens. Matter*, 24, 475305.

Yanagisawa, H., Tanaka, T., Ishida, Y., Matsue, M., Rokuta, E., Otani, S. and Oshima, C. 2005. Analysis of phonons in graphene sheets by means of HREELS measurement and *ab initio* calculation. *Surf. Interface Anal.*, 37, 133–136.

Zhang, C. W., Zhao, W. W., Bi, K. D., Ma, J., Wang, J. L., Ni, Z. H., Ni, Z. H. and Chen, Y. F. 2013. Heat conduction across metal and nonmetal interface containing imbedded graphene layers. *Carbon*, 64, 61–66.

Zhang, Y., Brar, V. W., Girit, C., Zettl, A. and Crommie, M. F. 2009. Origin of spatial charge inhomogeneity in graphene. *Nat. Phys.*, 5, 722–726.

Zhu, W., Perebeinos, V., Freitag, M. and Avouris, P. 2009. Carrier scattering, mobilities and electrostatic potential in monolayer, bilayer and trilayer graphene. *Phys. Rev. B*, 80, 235402.

Zou, K., Hong, X., Keefer, D. and Zhu, J. 2010. Deposition of high-quality HfO_2 on graphene and the effect of remote oxide phonon scattering. *Phys. Rev. Lett.*, 105, 126601.

SECTION II

Transition Metal Dichalcogenides

6

Theoretical Study of Transition Metal Dichalcogenides

Emilio Scalise

Contents

In this chapter, we will focus on a class of layered materials made up of transition metal (Mo, W, etc.) and chalcogen (S, Se, etc.) atoms, so-called transition metal dichalcogenides (TMDCs). The electronic properties of these layered materials vary with the number of layers: bulk MoS$_2$ has an indirect electronic bandgap, which increases while the number of layers decrease and finally

2D Materials for Nanoelectronics edited by Michel Houssa, Athanasios Dimoulas and Alessandro Molle © 2016 CRC Press/Taylor & Francis Group, LLC. ISBN: 978-1-4987-0417-5.

evolves to a direct gap in the monolayer MoS_2. The predicted structural and electronic properties of TMDCs are reviewed, focussing on the most notable compound of the TMDC family, MoS_2. Other possible ways to engineer the electronic band structure of TMDCs, such as the application of mechanical strain, are discussed.

6.1 Introduction

The exfoliation of graphene by K. Novoselov and A. Geim (Novoselov et al. 2004) is certainly responsible for opening a new field in condensed matter physics with thousands of scientific publications coming out every year. The unique electronic, optical and magnetic properties (Neto et al. 2009) of graphene and its two-dimensional (2D) nature, which favour the integration in the current integrated circuits technology, are the main reasons of this success. Recently, scientific interest has been extended to several other 2D materials with major scope: finding new candidates that could replace traditional materials in the next generation of nano(opto)-electronic devices. Graphene has no bandgap in its electronic structures, in its pristine form, and this limits its use in applications where the electronic bandgap is fundamental, such as field-effect transistors (FETs) or emitters and detectors in optoelectronic devices.

Unlike graphene, several layered TMDCs are semiconducting, with a gap varying from 0 (semimetals, such as TiS_2 [Bao et al. 2011] or WT_2 [Ali et al. 2014]) to more than 2 eV (2.1 eV for monolayer WS_2 [Kuc et al. 2011]). Some TMDCs also show exotic properties, such as superconductivity (Castro Neto 2001) and charge density waves (Zhu et al. 2012).

Furthermore, the bandgap of semiconducting TMDCs can be sized by varying the number of layers, and it can be changed from indirect to direct approaching the single layer. The bandgap can also be engineered by applying an external electric field perpendicular to the TMDC layers (Ramasubramaniam et al. 2011; Liu et al. 2012) or by straining the material, as discussed below. In single-layer MoS_2 and other TMDCs, the lack of inversion symmetry and the strong spin–orbit coupling are responsible for strong spin–orbit band splitting (Zhu et al. 2011) and for the coupling of spin and valley physics (Xiao et al. 2012).

All these features make TMDCs very promising for several applications, such as transistors (Radisavljevic et al. 2011a,b; Wang et al. 2012a), optoelectronic (Lee et al. 2012; Shanmugam et al. 2012) and valleytronic devices (Rycerz et al. 2007; Xiao et al. 2007). Besides, recent advances in the epitaxial growth of 2D crystals have opened up new opportunities towards novel devices based on van der Waals (vdW) heterostructures (Geim and Grigorieva 2013), in which TMDCs could play a major role.

In this chapter, we first give an overview of the structural and electronic properties of MoS_2 and other important TMDCs. Then, using a first-principles approach, we study the evolution of the band structure with the number of

MoS$_2$ layers. Finally, the effect of strain on the electronic and vibrational properties of monolayer MoS$_2$ is considered.

6.1.1 Crystal Structures and Electronic Properties of TMDCs

TMDCs have the general formula MX$_2$, where M is the transition metal element and X is the chalcogen.

The transition metal elements, which can be from group IV (Ti, Zr, etc.), V (V, Nb, etc.) or VI (Mo, W, etc.), are sandwiched in between two planes of chalcogen anions, as illustrated in **Figure 6.1**. Ionic–covalent bonds ensure the in-plane stability of the MX$_2$ layers, whereas vdW interactions hold the layers together, forming the bulk MX$_2$ crystal. The most notable compound of the TMDC family, MoS$_2$, has a hexagonal (H) or rhomboidal (R) symmetry, with trigonal prismatic coordination of the Mo atoms, which are bonded to six S atoms (see **Figure 6.2**). These two polytypes of bulk MoS$_2$, known as 2H- and the 3R-MoS$_2$ (Wilson and Yoffe 1969), are illustrated in **Figure 6.2**. The mineral molybdenite occurs in nature mainly as 2H-MoS$_2$, with two S–Mo–S

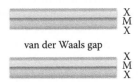

FIGURE 6.1 Schematic representation of the MX$_2$ layered structure.

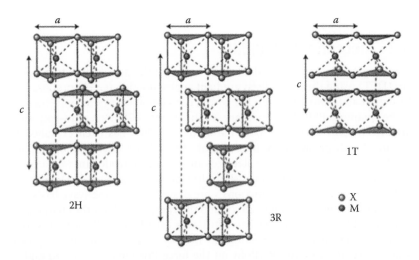

FIGURE 6.2 Schematic representation of the 2H, 3R and 1T-MoX$_2$. (Reprinted by permission from Macmillan Publishers Ltd. *Nat. Nanotechnol.*, Wang, Q.H. et al., 699, copyright 2012.)

units per unit cell, whereas the 3R-MoS$_2$ has three S–Mo–S units per unit cell. The 2H and 3R polytypes are also common for other TMDCs, such as MoSe$_2$, WS$_2$ and WSe$_2$.

A different polytype with octahedral coordination of the Mo atoms has also been reported for single-layer MoS$_2$ (Eda et al. 2012; Lin et al. 2014), the 1T-MoS$_2$. It has tetragonal symmetry with one S–Mo–S unit per unit cell, and contrary to the 2H-MoS$_2$, 1T-MoS$_2$ is metallic (Lin et al. 2014). The 1T polytype is typical for TaS$_2$, TaSe$_2$, NbS$_2$, NbSe$_2$ (Wilson and Yoffe 1969) and VSe$_2$ (Claessen et al. 1990), which are all metallic TMDCs.

The electronic properties of TMDCs can also be grouped in few classes, having very similar features: MoX$_2$, WX$_2$ and most of the compounds where the transition metal is Hf, Zr, Re, Pd and Pt are semiconducting, whereas NbX$_2$, TaX$_2$, VX$_2$ and other TMDCs (listed in **Table 6.1**) are metallic.

In this chapter, we focus on the semiconducting TMDCs having 2H polytype, which becomes 1H in the case of monolayers TMDCs, having just one X–M–X sandwich per unit cell.

The electronic structure of these compounds can be tuned by changing the number of layers forming their structure. The trend of the variations is very similar in different compounds, as reported both in theoretical (Lebègue and Eriksson 2009; Ding et al. 2011a; Kuc et al. 2011; Liu et al. 2011; Scalise et al. 2011; Ataca et al. 2012) and experimental studies (Kam and Parkinson 1982a,b; Hengge and Schlogl 1992; Mak et al. 2010). The bulk phase of these semiconducting TMDCs has an indirect bandgap, which changes to direct in the case of a single layer, as reported in **Table 6.2**. The evolution of the electronic structure of semiconducting TMDCs, particularly MoS$_2$, is detailed in the next sections.

6.2 Theoretical Methods

We present first-principles calculations based on the density-functional theory (DFT) using the generalised gradient approximation (GGA) proposed by Perdew et al. (1996) for the exchange and correlation energy, as implemented in the Quantum ESPRESSO package (Giannozzi et al. 2009). A semi-empirical dispersion term (Grimme 2006; Barone et al. 2009) has been exploited to account for dispersive vdW interactions. The valence electrons for sulphur and the valence and semi-core states ($4s$ and $4p$) for molybdenum were explicitly treated in the calculations using ultrasoft pseudopotentials (Vanderbilt 1990). The electronic wavefunctions were described by plane-wave basis sets with a kinetic energy cutoff of 28 Ry, and the energy cutoff for the charge density was set to 280 Ry. For the band structure calculations, 300 k-points were chosen, along the lines connecting the high-symmetric points in reciprocal space. Thresholds of 10^{-4} Ry/Bohr on the force (for ionic relaxation) and of 0.05 GPa on the pressure (for cell relaxation) were used.

The phonon frequency was obtained by diagonalisation of the dynamical matrix calculated by the density-functional perturbation theory (Baroni et al. 2001), as implemented in the Quantum ESPRESSO package.

Table 6.1   Electronic Bandgap and Magnetic Ordering (FM for Ferromagnetic, AFM for Antiferromagnetic) of TMDCs Obtained by *Ab Initio* Calculations

2D Chemical Formula	Gap (eV)	Magnetism	2D Chemical Formula	Gap (eV)	Magnetism
YS_2	Metal		CrS_2	Metal	AFM
TiS_2	0.02		$CrSe_2$	Metal	FM
$TiSe_2$	Metal		$CrTe_2$	Metal	FM
$TiTe_2$	Metal		MoS_2	1.6	
ZrS_2	1.1		$MoSe_2$	1.4	
$ZrSe_2$	0.4		$MoTe_2$	1.15	
$ZrTe_2$	Metal		WS_2	1.8	
HfS_2	1.3		WSe_2	1.5	
$HfSe_2$	0.6		WTe_2	Metal	
$HfTe_2$	Metal		TcS_2	1.2	
VS_2	Metal	FM	ReS_2	1.4	
VSe_2	Metal	FM	$ReSe_2$	1.3	
VTe_2	Metal	FM	$CoTe_2$	Metal	
NbS_2	Metal		$RhTe_2$	Metal	
$NbSe_2$	Metal		$IrTe_2$	Metal	
$NbTe_2$	Metal		$NiTe_2$	Metal	
TaS_2-*AB*	Metal		PdS_2	1.1	
TaS_2-*AA*	Metal		$PdSe_2$	1.3	
$TaSe_2$-*AB*	Metal		$PdTe_2$	0.2	
$TaSe_2$-*AA*	Metal		PtS_2	1.8	
$TaTe_2$	Metal		$PtSe_2$	1.4	
			$PtTe_2$	0.8	

Source: Lebègue, S. et al., *Phys. Rev. X*, 2013: 031002.
Note: The cells of the table are left blank if the material is not magnetically ordered.

Table 6.2 Summary of Semiconducting TMDC Bandgap

		$-S_2$		$-Se_2$		$-Te_2$
		Electronic Character	Referencs	Electronic Character	References	Electronic Character	References
Mo		Semiconducting		Semiconducting		Semiconducting	
		1L: 1.8 eV	Mak et al. (2010) (E), Scalise et al. (2011) (T)	1L: 1.5 eV	Liu et al. (2011) (T)	1L: 1.1 eV	Liu et al. (2011) (T)
		Bulk: 1.2 eV	Kam and Parkinson (1982a,b) (E)	Bulk: 1.1 eV	Kam and Parkinson (1982a,b) (E)	Bulk: 1.0 eV	165 (E)
W		Semiconducting		Semiconducting		Semiconducting	
		1L: 1.9–2.1 eV	Liu et al. (2011) (T), Kuc et al. (2011) (T)	1L: 1.7 eV	Ding et al. (2011a) (T)	1L: 1.1 eV	Ding et al. (2011a) (T)
		Bulk: 1.4 eV	Kam and Parkinson (1982a,b) (E)	Bulk: 1.2 eV	Kam and Parkinson (1982a,b) (E)		

Source: Wang, Q.H. et al., *Nat. Nanotechnol.*, 2012: 699. With permission.
Note: The references are indicated as experimental (E) and theoretical (T).

6.3 Structural, Electronic and Vibrational Properties of MoS$_2$: From Bulk to Monolayer

The experimental lattice parameters of 2H-MoS$_2$ are $a = b = 3$, 16 Å and $c = 12.29$ Å (Young 1968; Kam and Parkinson 1982a,b; Boker et al. 2001) (see **Figure 6.3**). Theoretical values of the in-plane lattice parameter reproduce quite well the experimental values, with few percentages of discrepancy: 3.13 Å for local density approximation (LDA) (Scalise et al. 2012) and 3.23 Å for GGA (Scalise et al. 2011; Kumar and Ahluwalia 2012). The discrepancy between experiment and theory is more evident for the out-of-plane (c) parameter (Ataca et al. 2011), because non-covalent interactions play a crucial role in bonding the MoS$_2$ layers and they are not well described both in LDA and GGA calculations. Diverse solutions are possible for the description of non-covalent interactions, giving slightly different c parameter and layer-by-layer binding energy (Bjorkman et al. 2012). By adding a correction for the weak non-covalent interactions, as detailed in the theoretical methods, we obtained reasonable values for the c parameter (11.89 Å [Scalise et al. 2012] and 13.07 Å [Scalise et al. 2011] for LDA and GGA, respectively).

Note that the multilayer MoS$_2$ structures with different number of S–Mo–S layers, from monolayer to the bulk limit, do not show significant differences with respect to the bulk structure in term of a and b lattice parameters, as well as for the Mo–S bond lengths and S–S distances.

The GGA calculated energy bandgap for the MoS$_2$ structure described above and illustrated in **Figure 6.3** is 1.18 eV (Scalise et al. 2011) (0.62 eV for LDA calculations [Scalise et al. 2012]), in excellent agreement with the experimental value (1.23 eV [Kam and Parkinson 1982a,b]).

6.3.1 Electronic Structure of MoS$_2$

The calculated band structure of bulk MoS$_2$ is illustrated in **Figure 6.4**. The minimum energy gap of about 1.2 eV is an indirect bandgap transition from the top of the valence band (VB) located at the Γ point and the bottom of the conduction band (CB) located at the midpoint, in between the Γ and the K high symmetry points of the Brillouin zone. The minimum direct-gap transition energy is predicted to be about 1.65 eV and located at the K-point.

2H-MoS$_2$

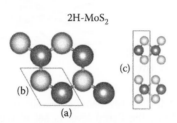

FIGURE 6.3 Top and side view of the 2H-MoS$_2$ crystal structure; (a), (b) and (c) are the in-plane (out-of-plane) lattice parameters. S and Mo are represented in light and dark color, respectively.

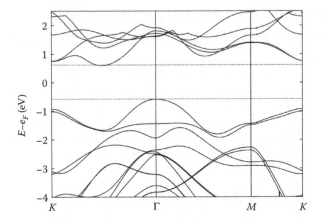

FIGURE 6.4 Electronic band structure of bulk MoS_2.

As the number of S–Mo–S layer sandwiches of multilayer MoS_2 decreases, one observes a progressive increment in the indirect energy gap up to 2 eV for monolayer MoS_2, as illustrated in **Figure 6.5a**. In contrast, the change in the direct gap is not significant, with a variation of about 0.1 eV from the bulk to the monolayer. As a result, the indirect–direct-gap crossover is reached in the limit

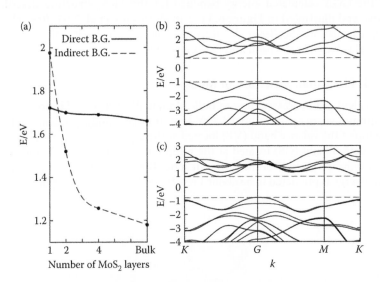

FIGURE 6.5 (a) Plot of the energy bandgap versus number of S–Mo–S sandwiches in multilayer MoS_2, the Fermi level is set at 0 eV. Energy band structures of monolayer (b) and bilayer (c) MoS_2. (With kind permission from Springer Science+Business Media: *Nano Res.*, Strain-induced semiconductor to metal transition in the two-dimensional honeycomb structure of MoS_2. 43, 2011, Scalise, E. et al.)

of the monolayer MoS_2 (see **Figure 6.5b**), with the bilayer MoS_2 showing only a small energy difference between the direct and indirect energy gap (see also **Figure 6.5c**). These changes in the bandgap with the number of layers are due to variations in the hybridisation between the p_z orbitals of S atoms and d orbitals of Mo atoms, resulting from the quantum confinement effect (Mak et al. 2010).

The computed direct (indirect) bandgap energies for the monolayer and bilayer MoS_2 are 1.72 and 1.7 eV (1.98 and 1.52 eV), respectively, in good agreement with experimental results (Mak et al. 2010; Splendiani et al. 2010; Han et al. 2011). Note that DFT typically underestimates the energy gaps of solids (up to 50%) (Seidl et al. 1996), and its success in predicting the bandgap of monolayer MoS_2 may sound surprising. Contrary, simulations based on the Green's function based screened Coulomb interaction (GW) approach, which are expected to be more accurate in calculating the energy gaps, predicted a bandgap of 2.7–2.9 eV (Ding et al. 2011b; Cheiwchanchamnangij and Lambrecht 2012; Komsa and Krasheninnikov 2012) for the monolayer MoS_2. These theoretical works based on GW calculations, attributed the difference between the experimental optical gap (1.9 eV) and the GW calculated bandgap, to a large exciton binding energy (0.8–1.0 eV) in monolayer MoS_2. Consequently, the fundamental gap of monolayer MoS_2 is expected to be about 2.8 eV, but any evident experimental confirmation is still missing, to the best of our knowledge.

6.3.2 Charge Carrier Effective Masses

Several works have been conducted to investigate the electrical conductivity of few- and single-layer MoS_2 (Novoselov et al. 2005; Ayari et al. 2007; Lee et al. 2011; Radisavljevic et al. 2011a). Novoselov et al. (2005) reported values of the carrier mobility in the range of 0.5–3 cm^2 V^{-1} s^{-1}, measured by using a device based on thin MoS_2 with back-gated configuration. By using a similar approach, Ayari et al. (2007) measured mobility values for ultrathin MoS_2 in tens of cm^2 V^{-1} s^{-1}. Much higher mobility values have been reported for a top-gated transistor based on single-layer MoS_2 by Kis et al. They exploited a high-k gate dielectric layer to enhance the field-effect mobility to values exceeding 200 cm^2 V^{-1} s^{-1} (Radisavljevic et al. 2011a). Other experiments have been conducted using an Al_2O_3 gate dielectric layer, leading to a mobility value for the bulk MoS_2 of about 500 cm^2 V^{-1} s^{-1} (Liu and Ye 2012); although these values will not compete with carrier mobility values measured for transistors based on III–V semiconductors, features such as the on/off current ratio and the feasibility for transparent and flexible devices keep MoS_2 very attractive for some electronic applications. It is also evident that further improvements in the carrier mobility can contribute to the success of TMDs in electronics.

In this section, we discuss the carrier effective masses of bulk and monolayer MoS_2. Even though it is not straightforward to directly relate the experimental carrier mobility to the effective mass calculated from the energy band structure, because of many factors influencing the former (i.e. phonon scattering, impurity scattering and effect of substrate), a qualitative analysis of the effective mass of MoS_2 is very interesting in order to corroborate solutions for the enhancement of the carrier mobility, such as the application of strain.

From the band structures plotted in **Figures 6.4** and **6.5**, one can notice that the curvature at the edges of both the VB and CB at the high symmetry points of the Brillouin zone is markedly different, comparing bulk, bilayer and monolayer MoS_2. Since the energy (ε) versus momentum (k) curve is nearly parabolic at the proximity of these high symmetry k-points, one can use its curvature to calculate the effective mass (m^*) for holes and electrons by exploiting the relation:

$$m^* = \frac{\hbar^2}{\delta^2\varepsilon/\delta k^2},$$

where $\delta^2\varepsilon/\delta k^2$ is the average curvature at the proximity of these high symmetry k-points.

The (GGA) calculated values for the carrier effective masses in multilayer MoS_2, listed in **Table 6.3**, are in agreement with other theoretical works (Cheiwchanchamnangij and Lambrecht 2012; Kadantsev and Hawrylak 2012; Yun et al. 2012). For bulk MoS_2, we found an electron effective mass of 0.52 m_0 and a hole effective mass of 0.56 m_0. These values are not much larger than the well-known carrier effective masses of Si (0.98 and 0.19 m_0 for longitudinal and transverse electron effective mass and 0.49 and 0.16 m_0 for heavy and light holes effective masses, respectively).

Table 6.3 Calculated GGA Energy Bandgap for Pristine Multilayer MoS_2, and Hole and Electron Effective Mass m^* in Unit of the Electron Mass m_0

	Bandgap (eV)		Effective Mass Holes (m_h^*/m_0)		Effective Mass Electron (m_e^*/m_0)	
	Indirect	Direct	Γ	K	K	Midpoint K–Γ
Structure						
Bulk	0.62	1.75	0.56 (VBM)	0.55	0.76	0.52 (CBM)
Trilayer	0.94	1.79	0.75 (VBM)	0.56	0.68	0.56 (CBM)
Bilayer	1.23	1.8	0.9 (VBM)	0.62	0.56	0.58 (CBM)
Monolayer	1.98	1.86	3.55	0.62 (VBM)	0.48 (CBM)	0.58

Source: Scalise, E. et al., Phys. E, 2012: 416. With permission.
Note: CBM (VBM) indicates the value of the electron (hole) effective mass at the CB minimum (VB maximum).

By decreasing the number of MoS_2 layers, both the electron and hole effective mass values raise, and particularly the hole effective mass at Γ reaches the value of 3.55 m_0 for the monolayer MoS_2. However, the monolayer MoS_2 changes the nature of the bandgap from indirect to direct, and the effective mass calculated at the CB minimum and VB maximum are 0.48 and 0.62 m_0, for electrons and holes, respectively. The relative small difference between these values and the carrier effective mass calculated for bulk MoS_2 do not explain the substantial higher experimental field-effect mobility for bulk compared to monolayer MoS_2 (Ayari et al. 2007; Liu and Ye 2012). Therefore, the higher measured current in the metal-oxide-semiconductor field-effect-transistors (MOSFETs) with multilayer MoS_2 channel compared to monolayer MoS_2 channel cannot be attributed to a difference in the intrinsic carrier effective mass between bulk and single-layer MoS_2 but likely to the lower carrier density in the latter (Scalise et al. 2012).

6.3.3 Vibrational Properties of MoS_2

Raman and infrared spectroscopy are powerful and widely used techniques for the structural characterisation of solids and molecules. Raman scattering and infrared absorption are strictly related to the vibrational properties of materials, which can be predicted by first-principles calculations.

The $2H\text{-}MoS_2$ – point group symmetry D_{6h} – has 18 vibrational modes, with 12 irreducible representations: two translation acoustic modes, two infrared modes and four Raman active modes, listed in **Table 6.3**. Monolayer MoS_2 also named $1H\text{-}MoS_2$ and having D_{3h} point group symmetry has nine vibrational modes and six irreducible representations, two translational acoustic, one infrared and three Raman modes.

Obviously, there is a correspondence between the vibrational modes of the monolayer MoS_2 and bulk or few-layer MoS_2 (see **Table 6.4**), except for vibrational modes with atomic displacements involving adjacent layers, such as the E_{2g}^2 shear mode and the E_{1u} mode. The atomic displacements of the four Raman active modes or the two infrared modes in $2H\text{-}MoS_2$ are illustrated in **Figure 6.6**. Their frequency calculated by LDA is also listed in **Table 6.4**, both for bulk and monolayer MoS_2. These results are in agreement with other theoretical (Molina-Sanchez and Wirtz 2011) and experimental works (Wieting and Verble 1971; Lee et al. 2010; Zeng et al. 2012), confirming that this functional is well suited for the correct prediction of the vibrational properties of bulk and few-layer MoS_2 with accuracy even better than GGA functional, attributable to the better estimation of the interlayer interaction of the local functional.

An interesting comparison of the phonon modes of bulk and monolayer MoS_2 is also possible through their phonon dispersions plotted in **Figure 6.7**. Both of them show three acoustic modes, longitudinal acoustic (LA) and transverse acoustic (TA) out-of-plane acoustic (ZA) mode. In the bulk MoS_2, low-frequency optical modes (E_{2g}^2 and B_{2g}) are present, corresponding to rigid shear and vertical motion of the MoS_2 layers, which are missing in the phonon dispersion of monolayer MoS_2.

Table 6.4   Phonon Frequencies for the 12 Irreducible Representations Calculated by LDA

Irreducible Representation	Bulk MoS_2 (D_{6h}) Frequency in cm^{-1}	Irreducible Representation	Monolayer MoS_2 (D_{3h}) Frequency in cm^{-1}	Character
A_{2u}	0	A''	0	Acoustic
E_{1u}	0	E'	0	Acoustic
E_{2g}^2	39.7			Raman
B_{2g}	66.4			Silent
E_{1g}	289.4	E''	289.3	Raman
E_{2u}	286.8			Silent
E_{2g}^1	387.2	E'	391.8	Raman
E_{1u}	387.9			Infrared
B_{1u}	407.6			Silent
A_{1g}	413.5	A'	410.6	Raman
B_{2g}	468.4			Silent
A_{2u}	472.7	A_2''	476.5	Infrared

Source: Scalise, E. et al., *Phys. E,* 2012: 416.

Between the low-frequency modes and the high-frequency optical modes, both dispersions have a gap of about 50 cm^{-1}. The high-frequency E_{2g}^1 and E_{1u} bulk modes collapse into the mode E$'$ in the single-layer MoS_2. This is also evident from the atomic displacements in **Figure 6.6**: approaching to the monolayer, the displacements of the atoms in the two (independent) layers are equivalent in the E_{2g}^1 and E_{1u} modes. Interestingly, one can observe an increase in frequency for the E_{2g}^1 Raman mode in bulk with respect to the E$'$ Raman mode in monolayer MoS_2. Contrary, the out-of-plane A_{1g}, being the other main Raman peak in bulk MoS_2, is found to be at lower frequency in the monolayer MoS_2 (named as A' mode). These shifts in the position of the Raman peaks have been measured experimentally (Lee et al. 2010; Korn et al. 2011) and are used to identify the layer thickness through the Raman spectrum. In fact, the plot of the frequency of the E_{2g}^1 and A_{1g} Raman peaks with respect to the number of layers (see **Figure 6.8**) evidences a clear 'anomalous' trend in the frequency of the two Raman peaks. The A_{1g} mode increases in frequency with an increasing number of layers, whereas the E_{2g}^1 mode decreases. The increase in frequency of the A_{1g} mode is due to the influence of the neighbouring layers on the restoring force on the atoms, which increases in multilayers due to the additional interlayer interactions. The origin of the decrease in frequency of the E_{2g}^1 with an increase of the number of layers has been identified in the enhancement of

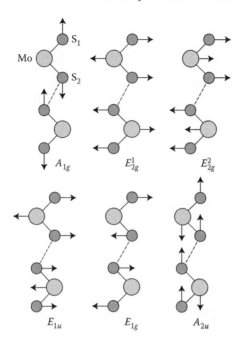

FIGURE 6.6 Atomic displacements of the four Raman active modes and the two infrared active modes of bulk MoS_2, as viewed along the [1000] direction. (Cheng, Y., Zhu, Z., Schwingenschlögl, U. Role of interlayer coupling in ultra thin MoS_2. *RSC Adv.*, 2012: 7798. Reproduced by permission of The Royal Society of Chemistry.)

the dielectric screening of the long-range Coulomb interactions that overcompensates for the interlayer interactions, resulting in a reduction of the overall restoring force on the atoms (Molina-Sanchez and Wirtz 2011).

6.4 Strain Effect on the Electronic Properties of Monolayer MoS_2

Engineering the electronic properties of a material, primarily its bandgap, is highly desirable for applications in nano- and opto-electrics, and strain is one of the possible ways. Quasi-2D materials such as mono- and few-layer TMDCs can sustain much higher strain compared to bulk materials, therefore, they are very interesting both as possible stain-engineered materials as well as semiconductors for flexible electronic and optoelectronic devices. Strain engineering is also very interesting as a possibility to increase carrier mobility of 2D TMDCs. In fact, enhancement of mobility by strain has been exploited since 2003 (Thompson et al. 2002) in commercial N and p-type metal-oxide-semiconductor (PMOS) silicon-based transistors.

Contrary to graphene, which is known as one of the strongest materials ever measured (Huang et al. 2011), MoS_2 is predicted to be much softer (Yue

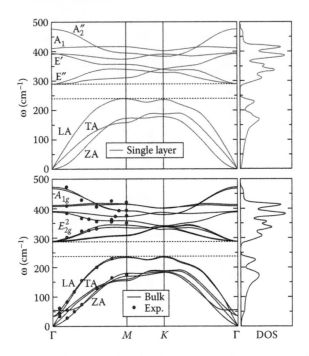

FIGURE 6.7 Phonon dispersion curves and density of states of single-layer and bulk MoS_2. Points are experimental data extracted from Wakabayashi (1975). (Reprinted with permission from Molina-Sanchez, A., Wirtz, L. *Phys. Rev. B,* 155413. Copyright 2011 by the American Physical Society.)

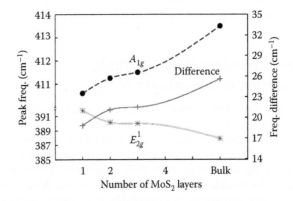

FIGURE 6.8 Calculated frequency of the E_{2g}^1, A_{1g} Raman modes versus applied strain (compressive, square and star bullets; tensile, + and × bullets). (Reprinted from *Phys. E*, Scalise, E. et al., 416. Copyright 2012, with permission from Elsevier.)

et al. 2012) and the elastic properties measured experimentally confirm these predictions (Castellanos-Gomez et al. 2012). Experiments also evidenced that single-layer MoS_2 can sustain relatively large (uniaxial) strain, greater than 11% (Bertolazzi et al. 2011). In **Figure 6.9**, we plot the calculated trend of the energy bandgap with applied biaxial strain for monolayer and bilayer MoS_2. Interestingly, a progressive reduction of the energy bandgap with respect to the applied tensile strain is observed up to about 10% (8% for the bilayer MoS_2), where a semiconductor–metal transition point is predicted. In agreement with the experimental works mentioned above, our stress–strains analysis (Scalise et al. 2011) indicates that the semiconductor–metal transition limit could be reached in realistic situations, although there are not yet any experimental works that corroborate this transition.

The reduction of the bandgap with an applied strain can be explained by the variation of the interlayer atomic distance, both in plane and out-of-plane, which leads to different superposition of their atomic orbitals and consequent shift in the energy states. Particularly, all the significant changes in the band structure of the mono- and bilayer MoS_2 occurring on application of biaxial strain can be related to shifts in the energy states mainly originating from the $3p$ orbitals of S atoms and $4d$ orbitals of Mo atoms. Upon application of the tensile strain, the energy at the top of the VB near the Γ point and at the bottom of the conduction at the K–Γ midpoint rises, as shown in **Figure 6.10**. This

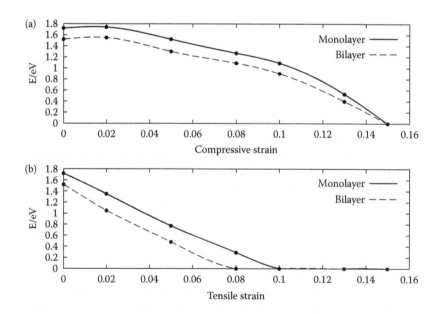

FIGURE 6.9 Plot of the energy bandgap versus applied strain, in monolayer and bilayer MoS_2. (With kind permission from Springer Science+Business Media: *Nano Res.*, Strain-induced semiconductor to metal transition in the two-dimensional honeycomb structure of MoS_2, 43, 2011, Scalise, E. et al.)

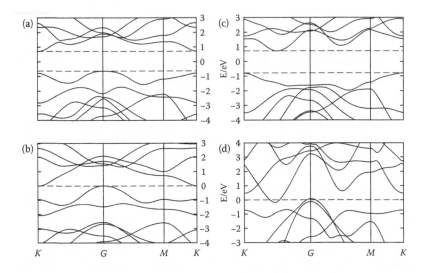

FIGURE 6.10 Energy band structures of monolayer MoS$_2$ with different applied strain: 2% (a) and 10% (b) tensile; 5% (c) and 15% (d) compressive. The Fermi level is set at 0 eV. (With kind permission from Springer Science+Business Media: *Nano Res.*, Strain-induced semiconductor to metal transition in the two-dimensional honeycomb structure of MoS$_2$, 43, 2011, Scalise, E. et al.)

is likely induced by the reduction of the distance between the S atoms and the increase of the vertical distance between the Mo and S atoms, due to the deformation of the layer in the direction perpendicular to the applied strain. The top of the VB and the bottom of the CB near the K-point undergo much less significant changes. Such a shift of the energy levels results in a change in the nature of the energy bandgap for the monolayer MoS$_2$, from direct to indirect, even for an applied tensile strain of less than 2%. The strain induced direct-to-indirect transition of the optical band has also been observed experimentally for an applied uniaxial strain of about 1% (Conley et al. 2013). The trend in the bandgap with applied uniaxial tensile strain is predicted to be not so different with respect to the biaxial case illustrated above, but the gap variation with applied uniaxial strain is less marked and the semiconductor-to-metal transition is reached under a uniaxial strain of about 20% (Kumar and Ahluwalia 2013). Different experimental studies support the theoretical predictions on the reduction of the bandgap with applied tensile strain (Conley et al. 2013; He et al. 2013; Zhu et al. 2013).

An opposite trend for the shifts of the energy bands is observed in the case of compressive biaxial strain. In this case, both the top of the VB near the Γ point and the bottom of the conduction at the K–Γ midpoint drop substantially with increasing applied strain. In contrast, the valence and conduction bands at the K-point rises. For a relatively small strain (below 5%), compressive and tensile strains show opposite effects on the band structure. But if larger compressive strain is applied, the bandgap shows a reduction with the applied compressive strain (similarly to tensile strain), due to the significant shifts of inner energy states induced by the over-lapping of the in-plane xy atomic orbitals.

Since the application of strain deeply perturb the band structure of monolayer MoS_2, the carrier effective mass is also changed. The calculated values of electron and hole effective mass are reported in **Table 6.5**. One can note a decrease of the carrier effective mass with applied tensile strain. Particularly drastic is the reduction of the hole effective mass at the Γ point. A considerable reduction (about 25%) of the electron effective mass at the K-point (minimum of the CB) is also observed. On the contrary, compressively strained MoS_2

Table 6.5 Calculated Electron and Hole Effective Mass for Strained Monolayer MoS_2				
	Effective Mass Holes (m_h^*/m_0)		Effective Mass Electron (m_e^*/m_0)	
	Γ	K	K	Midpoint K–Γ
Tensile strain				
0	3.5	0.62	0.48	0.58
0.01	2.7	0.61	0.44	0.58
0.02	2.17	0.6	0.41	0.57
0.03	1.81	0.59	0.39	0.54
0.05	1.37	0.58	0.36	0.51
0.08	0.99	0.61	0.36	
0.1	0.85	0.65	0.37	
Compressive strain				
0	3.5	0.62	0.48	0.58
0.01	5.15	0.64	0.55	0.57
0.02	9	0.66	0.65	0.56
0.03	4.28	0.69	0.85	0.53
0.05	2	0.74	3.98	0.48
0.08	3.53/0.6	0.83	0.74	0.39
0.1	1.6/0.56	0.88	0.65	0.34

Source: Scalise, E. et al., *Phys. E*, 2012: 416. With permission.

Note: Double values for m^* in the same cell are given in case of degenerate energy states at the top of the VB.

shows an increase of both holes and electrons effective masses, except in cases of high-applied strain (more than 5%).

We briefly discussed that smaller carrier effective mass can lead to larger carrier mobility. Therefore, strain is expected to enhance the performance of FETs based on MoS_2. Recent computational study on the impact of strain on monolayer MoS_2 FETs evidenced that 1.75% uniaxial strain could provide improvement in the PMOS ON currents of two to three times (Sengupta et al. 2013).

6.5 Strain Effect on the Vibrational Properties of Monolayer MoS_2

Table 6.6 shows the vibrational frequencies for monolayer MoS_2 for different magnitudes of applied tensile and compressive strain and the trend of the two Raman active modes (E_{2g}^1 and A_{1g}) is illustrated in **Figure 6.11**. Interestingly, both the out-of-plane and in-plane Raman active modes stiffen with an applied compressive strain, whereas they soften with tensile strain.

The variation of the A' mode is linear with respect to the applied strain, while for the E' mode, the linear trend is observed only below 3% of applied strain. For a larger strain, the E' vibrational mode becomes constant with increasing compressive strain while its slope increases for the tensile case. The predicted trend of the Raman active modes with respect to the strain can be a precious support to identify and quantify strain in experiments, as much as Raman spectroscopy is important for the identification of the number of layers in multilayer MoS_2. Experimental measurements support the trend of the Raman frequencies with applied tensile strain: Novololov et al.

Table 6.6 Calculated Vibrational Frequencies for Monolayer MoS_2 with Applied Tensile and Compressive Strain

Tensile Strain	0.01	0.02	0.03	0.05	0.08	0.1
A'-Raman (cm^{-1})	409.3	407.87	406.5	403.5	397.3	393.1
E'-Raman (cm^{-1})	387.2	382.1	376.9	364.6	341	323.4
E''-Raman (cm^{-1})	286.9	284	280.9	273.6	258	246.4
A_2'' (cm^{-1})	472	467.2	461.9	450.4	429.4	414
Compressive Strain	0.01	0.02	0.03	0.05	0.08	0.1
A'-Raman (cm^{-1})	412	413.2	414.4	416.6	419.3	420.3
E'-Raman (cm^{-1})	395.8	398.8	401.55	404.81	404.4	400.8
E''-Raman (cm^{-1})	291.5	293.2	294.5	295.8	295	292.2
A_2'' (cm^{-1})	480.5	484.2	487.2	491.7	495.3	494.8

Source: Reprinted from *Phys. E*, Scalise, E. et al., First-principles study of strained 2D MoS2. 416. Copyright 2012, with permission from Elsevier.

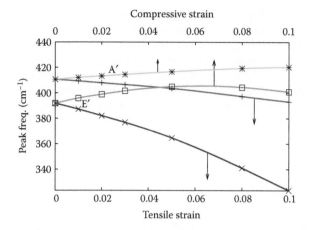

FIGURE 6.11 Calculated frequency of the E′ and A′ Raman modes versus applied strain (compressive, green and blue lines; tensile, red and magenta). (Reprinted from *Phys. E*, Scalise, E. et al., 416. Copyright 2012, with permission from Elsevier.)

(Rice et al. 2013) reported shifts at a rate of −0.4 cm⁻¹/% for the A′ mode and of −2.1 cm⁻¹/% for the E′ mode for single-layer MoS_2 under uniaxial tensile strain, although the amount of strain applied experimentally is very limited compared to the one considered in our study.

6.6 Conclusions

We introduced a very interesting class of layered materials, namely TMDCs, giving a brief overview of the properties that make these materials, especially in their 2D phase, very attractive for several electronic, photonic and energy applications. We focussed on the (theoretically) predicted electronic and vibrational properties of bulk and few-layer MoS_2. The effect of strain on the electronic and vibrational properties of monolayer MoS_2 was also discussed. A decrease of the energy bandgap of monolayer MoS_2 upon the application of strain is predicted, together with a change in its nature (from direct to indirect), even upon the application of only 1% strain. A semiconductor–metal transition is also predicted for a tensile biaxial strain of about 10%. As expected, the carrier effective mass is affected by the strain, and predicted to be strongly reduced by a tensile strain applied to monolayer MoS_2. Strain is thus an important method to engineer the electronic properties of MoS_2 and, consequently, the performances of MoS_2-based devices: the reduction of bandgap together with a lower effective electron mass would result in a decrease of the energy-delay product for an metal-oxide-semiconductor (MOS) transistor based on strained monolayer MoS_2. In addition, the tuning of the MoS_2 bandgap by strain could also be useful in other applications such as tunnel field-effect-transistor (TFET), photonic and piezoelectric nano-devices.

Finally, the vibrational properties of monolayer MoS_2, with particular emphasis on the Raman active modes, were discussed. The two main Raman active modes (A' and E') of monolayer MoS_2 are characterised by a shift of frequencies with applied strain, which is downwards for the tensile strain and upwards for the compressive one. This predicted trend of the Raman active modes with respect to the strain is markedly different from the anomalous trend of the Raman modes observed for increasing layer numbers. Measurements of Raman modes are thus potentially important for the experimental characterisation of MoS_2 and the quantification of strain in the layers.

References

Ali, M.N. et al., *Nature*, 514, 2014, 205.

Ataca, C., Sahin, H., Akturk, E., Ciraci, S. *J. Phys. Chem. C*, 115, 2011, 3934.

Ataca, C., Şahin, H., Ciraci, S. *J. Phys. Chem. C*, 116, 2012, 8983.

Ayari, A., Cobas, E., Ogundadegbe, O., Fuhrer, M.S. *J. Appl. Phys.*, 101, 2007, 014507.

Bao, L., Jie, Y., Yonghao, H., Tingjing, H., Wanbin, R., Cailong, L., Yanzhang, M., Chunxiao, G. *J. App. phys.*, http://scitation.aip.org/content/aip/journal/jap/109/5/10.1063/1.3552299. 109, 2011, 53717.

Barone, V., Casarin, M., Forrer, D., Pavone, M., Sambi, M., Vittadini, A. *J. Comput. Chem.*, 30, 2009, 934.

Baroni, S., de Gironcoli, S., Dal Corso, A., Giannozzi, P. *Rev. Mod. Phys.*, 73, 2001, 515.

Bertolazzi, S., Brivio, J., Kiss, A. *ACS Nano*, 5, 2011, 9703.

Bjorkman, T., Gulans, A., Krasheninnikov, A.V., Nieminen, R.M. *J. Phys., Condens. Matter*, 24, 2012, 424218.

Boker, T. et al. *Phys. Rev. B*, 64, 2001, 235305.

Castellanos-Gomez, A., Poot, M., Steele, G.A., van der Zant, H.S.J., Agrait, N., Rubio-Bollinger, G. *Adv. Mater.*, 24, 2012, 772.

Castro Neto, A.H. *Phys. Rev. Lett.*, 86, 2001, 4382.

Cheiwchanchamnangij, T., Lambrecht, W.R.L. *Phys. Rev. B*, 85, 2012, 205302.

Cheng, Y., Zhu, Z., Schwingenschlögl, U. *RSC Adv.*, 2, 2012, 7798.

Claessen, R., Schafert, I., Skibowski, M. *J. Phys. Condens. Matter*, 2, 1990, 10045.

Conley, H.J., Wang, B., Ziegler, J.I., Haglund, R.F., Pantelides, S.T., Bolotin, K.I. *Nano Lett.*, 13, 2013, 3626.

Ding, Y., Wang, Y.L., Ni, J., Shi, L., Shi, S.Q., Tang, W.H. *Phys. B*, 406, 2011, 2254.

Eda, G., Fujita, T., Yamaguchi, H., Voiry, D., Chen, M., Chhowalla, M. *ACS Nano*, 6, 2012, 7311.

Geim, A.K., Grigorieva, I.V. *Nature*, 499, 2013, 419.

Giannozzi, P. et al. *J. Phys. Condens. Matter*, 21, 2009, 395502.

Grimme, S. *J. Comput. Chem.* 27, 2006, 1787.

Han, S.W. et al. *Phys. Rev. B*, 84, 2011, 045409.

He, K., Poole, C., Mak, K.F., Shan, J. *Nano Lett.*, 13, 2013, 2931.

Hengge, E., Schlogl, L. *Gmelin Handbook of Inorganic and Organometallic Chemistry*, 8th edn., Vol. B7, Springer, Berlin, 1992.

Huang, X., Yin, Z., Wu, S., Qi, X., He, Q., Zhang, Q., Yan, Q., Boey, F., Zhang, H. *Small*, 7, 2011, 1876.

Kadantsev, E.S., Hawrylak, P. *Solid State Commun.*, 152, 2012, 909.

Kam, K.K., Parkinson, B.A. *J. Phys. Chem.*, 86, 1982a: 463.

Kam, K. K., Parkinson, B. *J. Chem. Phys.*, 86, 1982b: 463.

Komsa, H.P., Krasheninnikov, A.V. *Phys. Rev. B*, 86, 2012, 241201(R).

Korn, T., Heydrich, S., Hirmer, M., Schmutzler, J., Schuller, C. *Appl. Phys. Lett.*, 99, 2011, 102109

Kuc, A., Zibouche, N., Heine, T. *Phys. Rev. B*, 83, 2011, 245213.

Kumar, A., Ahluwalia, P.K. *Mater. Chem. Phys.*, 135, 2012, 755.

Kumar, A., Ahluwalia, P.K. *Model. Simul. Mater. Sci. Eng.*, 21, 2013, 065015.

Lebègue, S., Björkman, T., Klintenberg, M., Nieminen, R.M., Eriksson, O. *Phys. Rev. X*, 3, 2013, 031002.

Lebègue, S., Eriksson, O. *Phys. Rev. B*, 79, 2009, 115409.

Lee, C., Yan, H., Brus, L.E., Heinz, T.F., Hone, J., Ryu, S. *ACS Nano*, 4, 2010, 2695.

Lee, K., Kim, H.K., Lotya, M., Coleman, J.N., Kim, G.T., Duesberg, G.S. *Adv. Mater.*, 23, 2011, 4178.

Lee, S. H., Min, S.W., Chang, Y.G., Park, M.K., Nam, T., Kim, H., Kim, J.H., Ryu, S., Im, S. *Nano Lett.*, 12, 2012, 3695.

Lin, Y.C., Dumcenco, D.O., Huang, Y.S., Suenaga, K. *Nat. Nanotechnol.*, 9, 2014, 391.

Liu, H., Ye, P. D. *IEEE Electron Device Lett.*, 33, 2012, 546.

Liu, L., Kumar, S.B., Ouyang, Y., Guo, J. *IEEE Trans. Electron Devices*, 58, 2011, 3042.

Liu, Q., Li, L., Li, Y., Gao, Z., Chen, Z., Lu, J. *Phys. J. Chem. C*, 116, 2012, 21556.

Mak, K.F., Lee, C., Hone, J., Shan, J., Heinz, T.F. *Phys. Rev. Lett.*, 105, 2010, 136805.

Molina-Sanchez, A., Wirtz, L. *Phys. Rev. B*, 84, 2011, 155413.

Neto, A.H.C., Guinea, F., Peres, N.M.R., Novoselov, K.S., Geim, A.K. *Rev. Mod. Phys.*, 81, 2009, 109.

Novoselov, K., Giem, A., Morozov, S., Jiang, D., Zhang, Y., Dubonos, S., Grigorieva, I., Firsov, A. *Science*, 306, 2004, 666.

Novoselov, K.S., Jiang, D., Schedin, F., Booth, T.J., Khotkevich, V.V., Morozov, S.V., Geim, A.K. *Proc. Natl. Acad. Sci. USA*, 102, 2005, 10451.

Perdew, J.P., Burke, K., Ernzerhof, M. *Phys. Rev. Lett.*, 77, 1996, 3865.

Radisavljevic, B., Radenovic, A., Brivio, J., Giacometti, V., Kis, A. *Nat. Nanotechnol.*, 6, 2011a, 147.

Radisavljevic, B., Whitwicj, M.B., Kis, A. *ACS. Nano*, 5, 2011b, 9934.

Ramasubramaniam, A., Naveh, D., Towe, E. *Phys. Rev. B*, 84, 2011: 205325.

Rice, C., Young, R.J., Zan, R., Bangert, U., Wolverson, D., Georgiou, T., Jalil, R., Novoselov, K.S. *Phys. Rev. B*, 87, 2013: 081307.

Rycerz, A., Tworzydlo, J., Beenakker, C.W.J. *Nat. Phys.*, 3, 2007, 172.

Scalise, E., Houssa, M., Stesmans, A., Geoffrey, P., Afanas'ev, V. *Nano Res.*, 5, 2011, 43.

Scalise, E., Houssa, M., Stesmans, A., Geoffrey, P., Afanas'ev, V. *Phys. E*, 54, 2012, 416.

Seidl, A., Go ̈rling, A., Vogl, P., Majewski, J.A., Levy, M. *Phys. Rev. B*, 53, 1996, 3764.

Sengupta, A., Ghosh, R.K., Mahapatra, S. *IEEE Trans. Electron Devices*, 60, 2013, 2782.

Shanmugam, M., Bansal, T., Durcan, C.A., Yu, B. *Appl. Phys. Lett.*, 100, 2012, 153901.

Splendiani, A. et al. *Nano Lett.*, 10, 2010, 1271.

Thompson, S. et al. *Electron Devices Meeting Technical Digest International*, San Francisco, CA, 61, 2002.

Vanderbilt, D. *Phys. Rev. B*, 41, 1990, 7892.

Wakabayashi, N., Smith, H.G., Nicklow, R.M. *Phys. Rev. B*, 12, 1975, 659.

Wang, H., Yu, L., Lee, Y.H., Shi, Y., Hsu, A., Chin, M.L., Li, L.J., Dubey, M., Kong, J., Palacios, T. *Nano Lett.*, 12, 2012a, 4674.

Wang, Q.H., Kalantar-Zadeh, K., Kis, A., Coleman, J.N., Strano, M.S. *Nat. Nanotechnol.*, 7, 2012, 699.

Wieting, T.J., Verble, J.L. *Phys. Rev. B*, 3, 1971, 4286.

Wilson, J.A., Yoffe, A.D. *Adv. Phys.*, 73, 1969, 193.

Xiao, D., Liu, G.-B., Feng, W., Xu, X., Yao, W. *Phys. Rev. Lett.*, 108, 2012, 196802.

Xiao, D., Yao, W., Niu, Q. *Phys. Rev. Lett.* 99, 2007, 236809.

Young, P.A. *J. Phys. D*, 1, 1968, 936.

Yue, Q., Kang, J., Shao, Z., Zhang, X., Chang, S., Wang, G., Qin, S., Li, J. *Phys. Lett. A*, 376, 2012, 1166.

Yun, W.S., Han, S.W., Hong, S.C., Kim, I.G., Lee, J.D. *Phys. Rev. B.*, 85, 2012, 033305.

Zeng, H., Zhu, B., Liu, K., Fan, J., Cui, X., Zhang, Q.M. *Phys. Rev. B*, 86, 2012, 241301.

Zhu, C., Wang, G., Liu, B., Marie, X., Feng, Q., Wu, X., Fan, H., Tan, P., Amand, T., Urbaszek, B. *Phys. Rev. B.*, 88, 2013, 121301R.

Zhu, Z.Y., Cheng, Y.C., Schwingenschlögl, U. *Phys. Rev. B*, 84, 2011, 15342.

Zhu, Z.Y., Cheng, Y.C., Schwingenschlögl, U. *Phys. Rev. B*, 85, 2012, 245133.

7

Physico-Chemical Characterisation of MoS$_2$/ Metal and MoS$_2$/Oxide Interfaces

Stephen McDonnell, Rafik Addou, Christopher L. Hinkle and Robert M. Wallace

Contents

2D Materials for Nanoelectronics edited by Michel Houssa, Athanasios Dimoulas and Alessandro Molle © 2016 CRC Press/Taylor & Francis Group, LLC. ISBN: 978-1-4987-0417-5.

7.1 Introduction

In recent years, there has been a renewed interest in two-dimensional (2D) materials research. These materials, which had primarily been used for their tribological properties, have become exciting candidates for beyond silicon and beyond CMOS nanoelectronic device applications. This renewed focus has coincided with an increased interest in monolayer MoS$_2$ for its optoelectronic and photocatalytic properties. For this wide range of potential MoS$_2$ applications, the surface and interfaces play critical roles in device behaviour. The characterisation of these interfaces will be the focus of this chapter and particular efforts will be made to highlight non-idealities and deviations from expected behaviour.

7.1.1 Background on MoS$_2$ Research and Properties

7.1.1.1 Tribology

Of the transition metal dichalcogenide (TMD) family of materials, molybdenite (MoS$_2$) appears most in literature and industrial applications, most likely due to its natural abundance and commercial availability, relative to other

TMDs. While in the early twenty-first century, interest has revolved around the electronic properties of MoS_2, the material itself has been well studied during the twentieth century due to its application as a dry lubricant (Winer 1967), since the layered structure of MoS_2 gives it intrinsic lubrication properties similar to graphite and other graphite analogues (Brainard 1968). The structure of MoS_2 has thus been studied extensively (Dickinson and Pauling 1923). Unlike graphite, where each layer consists of a 2D sheet of sp^2-bonded carbon in a hexagonal structure, each 'monolayer' of MoS_2 is actually a three layered sheet three-dimensionally bonded sulphur and molybdenum. Each molybdenum atom is bonded to six sulphur atoms, with three above and three below in a trigonal prismatic or octahedral structure, as seen in **Figure 7.1** (Kuc 2014). The result is a 2D sheet that is three atoms thick with no dangling bonds on the basal plan. These sheets are then bonded to each through van der Waals forces, which can easily be overcome by shearing stress leading to the lubrication properties of these materials.

Dry lubricants are particularly important in applications involving high temperatures where liquid lubricants would evaporate. Thermal desorption experiments suggest that MoS_2 is stable to temperatures of greater than 1000°C and high-temperature friction tests in vacuum suggest that MoS_2 works as an effective lubricant up to 650°C in vacuum (Brainard 1968). MoS_2 has also been used as an additive to oil to enhance lubrication, particularly in the automotive industry (Black et al. 1966, 1967), with a patent for this application being

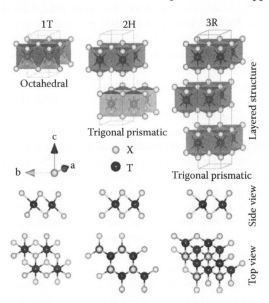

FIGURE 7.1 Structural representation of 1T, 2H and 3R TMD polytypes and their corresponding metal atoms coordination. The side and top views of layered forms are shown. (Kuc, A. Low-dimensional transition-metal dichalcogenides, *Chemical Modelling* 11;2014:1–29. Reproduced by permission of The Royal Society of Chemistry.)

filed as early as 1934 (Cooper and Damerell 1939). This early research was covered in a review of over 100 publications from 1926 to 1966 by Winer (1967).

The tribological applications of MoS_2 are, however, not limited to the automotive industry and advances are continually being made in research on MoS_2 as a lubricant (Brudnyi and Karmadonov 1975, Takahashi and Okada 1975). For example, it has been shown that doping MoS_2 with copper nanoparticles can reduce wear on the friction body (An and Irtegov 2014). MoS_2 also exhibits a very low friction coefficient in ultra-high vacuum (UHV) in the 10^{-3} range and the 'superlubricity' state of the MoS_2 has been identified in which two contacting layers show no resistance to sliding (Martin et al. 1993). Solid lubricant coatings for vacuum applications have seen considerable developments for many years, because of the use of advanced coating techniques, such as physical or chemical vapour deposition (CVD) processes (Donnet 1996, Hirvonen et al. 1996). Voumard et al. describe the applications for lubrication by MoS_2 in spacecraft, dry machining and anti-adhesion uses in extruding and moulding (Voumard et al. 2001).

7.1.1.2 Electronic

Bulk MoS_2 is known to be a semiconductor with an *indirect* bandgap of ~1.2 eV as determined by independent photoconductivity measurements (Kautek et al. 1980, Kam and Parkinson 1982, McDonnell et al. 2013). The material is often reported as being intrinsically n-type (Fivaz and Mooser 1967, Kam and Parkinson 1982, Radisavljevic et al. 2011a, Das et al. 2012, Kim et al. 2012, Lee et al. 2012, Li et al. 2012, Sik Hwang et al. 2013), particularly in recent literature. However, early reports (Wilson and Yoffe 1969) and also a recent detailed study of the variations across a geological MoS_2 crystal (McDonnell et al. 2014) suggest that geological MoS_2 can display both n-type and p-type behaviour.

There has been some debate on the extraction of mobility for monolayer MoS_2 devices (Fuhrer and Hone 2013). Initial reports claimed that the electron mobility for monolayer MoS_2 suspended on SiO_2 was 1–10 cm^2/Vs, but that the suppression of Coulomb scattering afforded by the deposition of HfO_2 on top (i.e. a HfO_2/MoS_2/SiO_2 structure) results in significant mobility enhancement, with values as high as 200 cm^2/Vs reported in that study (Radisavljevic et al. 2011a). Fuhrer and Hone (2013) have pointed out that the capacitive coupling between the top and back gates for such measurements can result in an overestimation of such mobilities and suggested that the actually mobility was likely less than 10 cm^2/Vs. The electron and hole mobilities of MoS_2 are both thickness and substrate dependent. For example, a study that investigated the correlation of device mobility with MoS_2 thickness showed that the measured mobility of MoS_2 flakes on a SiO_2 substrate peaked in a thickness range of 6–12 nm at 130–180 cm^2/Vs (Das et al. 2012). The mobilities for both thin (<5 nm) and thick (20–70 nm) fell in the range of 20–90 cm^2/Vs. A separate study measuring MoS_2 mobility in a four probe set-up reported that mobility on a SiO_2 substrate of 30–60 cm^2/Vs was essentially thickness independent in the range of 15–90 nm. They also highlight the role of the substrate in the measured mobility and reported electron (hole) mobilities of 470 (480) cm^2/Vs for 50 nm MoS_2 on polymethyl methacrylate (PMMA) substrates have been reported (Bao et al. 2013). MoS_2 has now been integrated into a number of nanoelectronic devices including a full integrated circuit based on MoS_2 (Radisavljevic et al. 2011b).

7.1.1.3 Optical

The band structure of bulk MoS_2 is known to have an indirect bandgap of ~1.2 eV at the Γ point as well as a direct transition at the K point with a gap of 1.8 eV. Mak et al. (2010) studied the band structure of MoS_2 as a function of the number of layers using a combination of photoluminescence (PL), photoconductivity and optical absorption. They showed that as MoS_2 is thinned from six to two layers, the indirect gap increases to 1.6 eV due to confinement, whereas the direct gap increases by less than 0.1 eV. For suspended monolayer MoS_2, only the direct gap of 1.9 eV was observed. Owing to this direct bandgap of monolayer MoS_2, the material has the potential for a number of optoelectronic applications such as photovoltaics, photodetectors, light-emitting diodes (LEDs) and flexible optoelectronics as discussed in a review by Wang et al. (2012). MoS_2 has been integrated with silicon and graphene in attempts to make high-efficiency solar cells (Bernardi et al. 2013, Tsai et al. 2014). A molybdenum disulphide amorphous silicon heterojunction photodetector with a photoresponse rate 10 times faster than conventional amorphous silicon-based photodetectors has also been demonstrated (Esmaeili-Rad and Salahuddin 2013).

MoS_2 also exhibits some interesting optical properties. For example, when a typical semiconductor is illuminated by light, its conductivity tends to increase. The opposite effect was seen in MoS_2 monolayers. When intense laser pulses illuminate the 2D semiconductor, the conductivity is reduced by 1/3 of its initial value. This finding is additional evidence of strong Coulomb interaction in MoS_2, which could lead to next-generation excitonic devices (Lui et al. 2014). In separate work, a laser was used to excite plasmons at the surface of a silver wire and a MoS_2 flake at the far end of the wire generated strong light emission. Going in the other direction, as the excited electrons relaxed, they were collected by the wire and converted back into plasmons, which emitted light of the same wavelength. Combining electronics and photonics on the same integrated circuits could drastically improve the performance and efficiency of mobile technology (Goodfellow et al. 2014).

A hybrid nanomaterial of MoS_2–MoO_3 was prepared for LEDs (Yin et al. 2014). This material exhibiting p-type behaviour was used as an active layer in the n-SiC/p-MoS_2–MoO_3 heterojunction applied in LEDs. The hybrid material was prepared in air by heat-assisted partial oxidation of MoS_2 nanoflakes followed by the subsequent thermal-annealing-driven crystallisation. The radiative recombination processes due to band edges and defect energy levels dictate the measured electroluminescence. The LED device was also realised using a vertical heterojunction between monolayer n-MoS_2 and p-Si (Lopez-Sanchez et al. 2014). The electroluminescence measurement shows optical transitions between the conduction and valence bands. The electroluminescence in this device is active on the entire device surface rather than being localised only at the heterojunction edge as had been reported for a device with a similar geometry (Ye et al. 2014). This opens a potential pathway for increasing the emitted light intensity by increasing the device geometry. The same p–n diode can operate as solar cells with 4% external quantum efficiency.

7.1.1.4 Other Applications

MoS_2 also has other interesting applications including catalysis and biosensing. For example, dichalcogenides like MoS_2 and WS_2 have been employed for decades in desulphurisation applications in the petroleum industry (Furimsky 1980, Prins et al. 1989). The utilisation of the reaction at defects and step edges of MoS_2 (and WS_2) crystalline particles on alumina in catalysis beds has proved effective in large-scale industrial desulphurisation applications of petroleum and petroleum products.

Another application in catalysis is in hydrogen production. Precious metals are not ideal for low-cost applications and it has been suggested that nanoparticle MoS_2–TiO_2 heterostructures may be a suitable alternative for a range of photocatalytic applications including hydrogen production (Ataca and Ciraci 2012, Xiang et al. 2012) and water purification (Tacchini et al. 2011). The rate of hydrogen evolution on CdS is also increased by more than 30 times when loaded with only 0.2 wt.% of MoS_2. The activity of MoS_2/CdS is even higher than that of the CdS photocatalysts loaded with different noble metals, such as Pt, Ru, Rh, Pd and Au (Zong et al. 2008). For MoS_2 nanoparticles, control of the particle size and edge sites has been shown to be critical for achieving the highest efficiency in H_2 production (Jaramillo et al. 2007, Karunadasa et al. 2012). However, first-principles calculations suggest that even pristine p-doped monolayers may act as good photocatalysts (Li et al. 2013). Also, the combination of MoS_2 and graphene as cocatalysts with TiO_2 has been found to further enhance the H_2 production, presumably because the graphene enhances electron transport while the edge sites of the MoS_2 promote H_2 production, giving the combined effect of enhancing H production while suppressing charge recombination (Xiang et al. 2012).

In the biosensing field, MoS_2 biosensors (for protein and pH-sensing) with 74 times greater sensitivity than graphene-based field-effect transistors (FETs) have been reported (Sarkar et al. 2014). Current graphene-based biosensors do not allow the sequencing of DNA, because the molecules stick to the nanopores. The DNA interacting with the graphene introduces a lot of noise that makes it hard to read the current. MoS_2 nanopores benefit from controllable pore architecture with combinations of Mo and S atoms at the edge, which can be engineered to obtain the optimum sequencing signals (Farimani et al. 2014). MoS_2 nanoflakes can also be used for DNA hybridisation detection where the mechanism of detection is based on the differential affinity of MoS_2 nanoflakes towards single-stranded DNA and double-stranded DNA (Loo et al. 2014).

7.1.2 Phase Diagram/Structures

7.1.2.1 Common Polytypes

The three most commonly reported polytypes of MoS_2 are the 1T, 2H and 3R. In this notation, the integer is representative of the number of layers in a unit cell, while the letters T, H and R represent the trigonal (octahedral), hexagonal (trigonal prismatic) and rhombohedral symmetry of the systems, respectively (Mattheiss 1973). While both 2H-MoS$_2$ and 3R-MoS$_2$ can be found in nature (Yang et al. 1991, Wypych et al. 1998, Tiong and Shou 2000), the 2H polytype is the most common and many studies (Shafer 1973, Schonfeld et al. 1983) on 3R-MoS$_2$ utilise synthesised crystals. In both of these polytypes, the molybdenum atom is a trigonal prismatic coordination as shown in **Figure 7.1** (Kuc 2014). Therefore, monolayer 1H-MoS$_2$ (Yang et al. 1991, Ataca et al. 2011, Ataca and Ciraci 2012, Kumar and Ahluwalia 2012) can be described as a monolayer from the parent 2H-MoS$_2$ or 3R-MoS$_2$. The trigonal prismatic structure is stable for MoS$_2$ (and most other group 6B transition metal TMDs) because there are enough d-electrons to fully fill the d_z^2 sub-band that resides at lower energies due to crystal field splitting.

The 2H-MoS$_2$ structure is the focus of most studies on tribological or nanoelectronic MoS$_2$ applications primarily because of its relative natural abundance. In geological MoS$_2$, only the hexagonal coordination of the Mo atoms is observed, although the metastable octahedral (1T) phase can be found under certain conditions. For example, during alkali intercalation of MoS$_2$, the resultant electron transfer induces a phase transition which shifts the Mo atoms from 2H to the reported 1T coordination (Py and Haering 1983, Chrissafis et al. 1989, Sandoval et al. 1991, Wypych and Schöllhorn 1992, Papageorgopoulos and Jaegermann 1995, Wypych et al. 1998, Eda et al. 2011, Kappera et al. 2014). The 1T structure looks somewhat similar to the 2H polytype with the exception that the molybdenum atoms are now in the octahedral coordination and the unit cell consists of only a single layer of the MoS$_2$. One can simplistically think of the 1T as being a monolayer of 2H where the top and bottom sulphur atoms are rotated by 60° rather than being directly on top of each other as they are in the 2H. In actuality, the structure of alkali intercalated MoS$_2$, is a *distorted* 1T. Such a structure can be thought of as 1T-MoS$_2$ where the molybdenum atoms trimerise, resulting in a √3 × √3 superstructure which has been observed experimentally by a number of characterisation techniques (Py and Haering 1983, Chrissafis et al. 1989, Sandoval et al. 1991, Wypych and Schöllhorn 1992, Papageorgopoulos and Jaegermann 1995, Wypych et al. 1998, Eda et al. 2011, Wang et al. 2013).

Alkali intercalated distorted 1T-MoS$_2$ is known to be metallic and 1T-MoS$_2$ is predicted to be metallic even in the absence of alkali metals. It is because of this that phase-stable, spatially localised, alkali free 1T-MoS$_2$ has been proposed as a potential route to achieving low resistivity contacts with 2H-MoS$_2$ (Kappera et al. 2014, Lin et al. 2014). It has been reported that Li-intercalated 1T-MoS$_2$ produced by organolithium solution exposure can remain in the 1T phase after Li removal by 'thoroughly washing' the samples (Kappera et al. 2014), while spatially controlled electron-beam-induced phase transitions from 2H- to 1T-MoS$_2$ (0.6% Re doped) are recently reported without introducing any alkali intercalation step, and is likely a more practical approach in the context of device fabrication (Lin et al. 2014).

7.1.2.2 Growth Methods

The synthesis of MoS_2, as well as other TMDs, has been the focus of a number of recent reviews and as such, we will simply provide a brief summary and focus on recent advances in the field (Wang et al. 2012, Huang et al. 2013, Lv et al. 2015). The top-down exfoliation method can be used to generate high-quality single- or few-layer flakes that can be used for device fabrication (Frindt 1966, Novoselov et al. 2005, Radisavljevic et al. 2011). While this method is extremely useful for researchers seeking to examine the electronic and optical properties of these materials, it is clearly not scalable to wafer-scale processing. The liquid exfoliation method is one that can provide large-area thin-film TMD nanosheets; however, the process can induce structural changes such as the phase transition to the 1T-MoS_2 that were discussed in Section 7.1.2.1 (Dines 1975, Joensen et al. 1986, Yang et al. 1991).

A number of bottom-up approaches exist for TMD synthesis. For example, atomic layer deposition (ALD) has been used to grow other TMDs such as WS_2 (Scharf et al. 2004) and has recently been used to grow MoS_2 (Tan et al. 2014) on sapphire using $MoCl_5$ and H_2S precursors. Van der Waals epitaxy (VDWE) is a physical vapour deposition (PVD) technique that was pioneered by the Koma group in the 1980s and demonstrated the growth of several Se-based TMDs (Koma et al. 1985). Recently, it has been employed to fabricate TMDs such as $HfSe_2$ and $MoSe_2$ (Vishwanath et al. 2015, Yue et al. 2014). For $HfSe_2$, it was shown that the TMDs could be grown on materials with large lattice mismatches while still achieving rotational alignment to the substrate. Rotational alignment will likely be important for device applications such as the bilayer pseudospin FET (Banerjee et al. 2009). CVD is currently being investigated by many groups for the synthesis of MoS_2 (Lee et al. 2012, Shi et al. 2012, Zhan et al. 2012, Amani et al. 2013, van der Zande et al. 2013, Ling et al. 2014, McCreary et al. 2014, Tarasov et al. 2014). These processes generally take the form of either the sulphurisation of molybdenum metal or the thermal decomposition of precursor compounds. The Cao group reported that self-limiting monolayer growth of MoS_2 on SiO_2 was achieved by controlling the partial pressure of MoS_2 to be between the vapour pressure of MoS_2 for monolayer and bilayer MoS_2 (Yu et al. 2013). Since a condition for precipitation is that the partial pressure of MoS_2 be greater than the vapour pressure, this process ensures that this condition for condensation is only met during the growth of the first monolayer. If sufficient differences in the vapour pressure of MoS_2 exist with increasing layers, one could envision controlling the exact number of layers until the bulk properties are reached.

As will be discussed in the later sections, geological and synthetic MoS_2 suffer from both a high concentration of defects as well as impurities which further motivates the need for high-quality TMD synthesis. Reports of defects in CVD-grown MoS_2 would suggest that the current processes are still under optimisation, but also that defects such as sulphur vacancies will most likely always exist at some concentration and will require passivation for optimising nanoelectronic devices (Zhou et al. 2013). Similar to early work on CVD graphene, a major area of optimisation will be maximising the grain size obtained and currently grain sizes for CVD MoS_2 are limited to ~120 µm (van der Zande et al. 2013).

7.1.3 Nanoelectronics and Device Interfaces

All practical nanoelectronic devices contain heterostructures which, by definition, result in interfaces between different materials. Typically, these interfaces play a critical role in device performance, and become dominant in the case of incorporating atomically thin channels. Defects close to the channel of a transistor affect carrier mobility and can inhibit Fermi-level modulation. Reactions at a metal–semiconductor interface influence the contact resistance which in turn affects the charge injection into devices. Since a full discussion of TMD nanoelectronic devices will be covered in a later chapter, this brief overview will serve purely to highlight specific device relevant interfaces whose chemistry will need to be understood if the industrial-level integration of MoS_2 into nanoelectronic devices is to be realised (**Figure 7.2**).

7.1.3.1 FETs and Interfaces

In any FET, the semiconductor–insulator and semiconductor–metal interfaces are of critical importance to device performance. The insulator is required to reduce the gate leakage current in a field-effect device. In MoS_2-based devices with atomically thin channels, the use of a high-κ insulator is expected to enhance the channel mobility due to the screening of Coulomb scattering centres (Jena and Konar 2007, Ma and Jena 2014). In theory, a major advantage of MoS_2 (or any similar 2D material) over three-dimensional (3D) crystalline semiconductors is the lack of surface dangling bonds which contribute to the interface trap density (D_{it}) if not passivated. An Al_2O_3–MoS_2-based

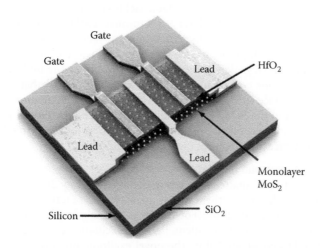

FIGURE 7.2 Integrated circuit based on single-layer MoS_2 (not to scale). (Reprinted with permission from Radisavljevic, B., M.B. Whitwick and A. Kis. Integrated circuits and logic operations based on single-layer MoS_2. *ACS Nano* 5;2011:9934–9938. Copyright 2011 American Chemical Society.)

metal-oxide-semiconductor field effect transistor (MOSFET) with interface trap density of 2.4×10^{12} cm^{-2} was reported by Liu and Ye (2012) and suggests two things. Firstly, as the authors point out, the interface trap density is indeed relatively low compared to other alternative channel materials ($In_{0.53}Ga_{0.47}As$ for example) even without the years of optimisation afforded to silicon-based technology. Secondly, the interface trap density is detectable, which suggests that the MoS$_2$ surface is *not* free of dangling bonds as is expected for the ideal case. It is also noteworthy that even the best interface employed by the semiconductor industry, viz. SiO$_2$/Si, has a similar D_{it} magnitude unless passivated with hydrogen. Hydrogen passivation, typically accomplished by post-deposition anneals of the transistor gate stack structure in an ambient such as forming gas (N_2:H_2) reduces the interfacial D_{it} from ~10^{13}/cm^2 to ~5×10^{10}/cm^2 when optimised (Poindexter and Caplan 1983; Thoan et al. 2011). It is expected that the propensity for interface defects to undergo a suitable passivation reaction will continue to be important in the case of TMDs such as MoS$_2$ as well.

This chapter will discuss in detail the difficulties in obtaining such an ideal surface as well as highlight the impurities present in geological samples. Similarly, the semiconductor–metal interface plays a critical role in determining device performance. In this chapter, we discuss metal Fermi-level pinning in MoS$_2$-based devices as well as the impact of surface defects in the electronic characteristics of such devices.

7.1.3.2 TFETs and Interfaces

For tunnelling-based FETs (TFET), the current flows through band-to-band tunnelling in the 'on state'. Such devices are proposed for low-power applications since they are predicted to allow less than 60 mV/decade subthreshold swing (Seabaugh and Zhang 2010, Jena 2013). Some device architectures put these semiconductors in direct contact, yielding a new interface of importance. Others utilise a buffer layer (to minimise TMD waverfunction overlap) which provides two more interfaces of importance. In the case of TMDs, such a buffer layer might be a wide bandgap 2D material such as hexagonal boron nitride which makes this interface distinct from the high-κ interface with MoS$_2$ used for conventional FETs. It should be noted that for TFETs, the high-κ interface with the semiconductor is still important for minimising gate leakage, and the metal–semiconductor interface is extremely important since low-resistance contacts are thought to be critical for achieving the promise of these device architectures (Seabaugh and Zhang 2010).

7.1.3.3 Photonics and Interfaces

The significant light–matter interaction demonstrated by 2D crystals (MoS$_2$, WS$_2$ and WSe$_2$) has made them highly attractive for practical optoelectronic applications (Xia et al. 2014). For example, a single layer of MoS$_2$ absorbs around 10% of vertically incident light at excitonic resonances (615 and 660 nm) (Eda and Maier 2013). The novel excitonic properties of 2D MoS$_2$ that make it very interesting for both physics research and device fabrication include: (i) the improved direct-bandgap photoluminescence quantum yield of the monolayer compared with multilayers (Mak et al. 2010, Splendiani et al. 2010), (ii) first-principles calculations show that monolayer MoS$_2$ possesses a small effective exciton Bohr radius (~1 nm) and associated large exciton with a binding energy of 0.96 eV, providing the opportunity to realise excitonic devices that operate at room temperature (Qiu et al. 2013) and (iii) the quantum confinement in layered d-electron materials provides new opportunities for engineering the electronic structure and novel optical properties which are not detected in sp-bonded materials and holds promise for new nanophotonic applications (Splendiani et al. 2010).

The integration of 2D materials with external photonic structures provides a route to obtaining the desired strong interaction. In particular, integration with optical cavities allows for the significant manipulation of the local optical density of states surrounding the 2D materials, leading to greatly modified emission/absorption properties (Liu et al. 2015). A new type of quasi-particle: plasmon–phonon polaritons are created by coupling of plasmons in graphene and phonons in a polar dielectric, which can be utilised to enhance and tune the interaction between light and 2D materials. Although graphene is not an optimal material for light emission, many single-layer TMDs including MoS$_2$ are direct-bandgap semiconductors and exhibit strong excitonic emission properties that are gate-tunable, which makes them promising candidates for light emission in the near-infrared. Single-layer MoS$_2$ also displays valley polarisation due to the breaking of inversion symmetry and the strong spin–orbit coupling. This unique phenomenon has motivated the emerging field of valleytronics (Xia et al. 2014). It is also possible to build photonic devices using heterostructures containing both narrow bandgap of black phosphorus and larger bandgap MoS$_2$ and WSe$_2$. The injection of electrons (holes) can be realised by using n-MoS$_2$ (p-WSe$_2$). The injected carriers are trapped with black phosphorus for light emission due to the band offset at the interface of black phosphorous with TMDs (Eda and Maier 2013). To minimise the contact resistance, graphene can be used. Interfaces will be critically important in optoelectronic devices since many defects act as non-radiative recombination sites or as radiative recombination sites with unwanted optical properties (Tongay et al. 2013, van der Zande et al. 2013).

7.2 MoS$_2$ Surfaces

7.2.1 Surface Chemistry

7.2.1.1 Contaminated versus Clean MoS$_2$

When compared to materials with a 3D bonding structure, layered crystals are often thought of as being 'inert', that is, having diminished reactivity due to the expected dearth of dangling bonds. The absence of dangling bonds at the (ideal) surface forces any oxidation reaction to overcome the kinetic barriers involved in breaking the covalent bonds within the layer. In reality, all TMDs are less energetically favourable than their corresponding metal oxides and for many, there is sufficient thermal energy, even at room temperature, to oxidise the top surface of actual crystals where step edges and defects, such as S vacancies, are also present. Graphite and MoS$_2$, however, are good examples of relatively 'inert' 2D crystals and neither form detectable native oxides even after long exposures to air at room temperature. This does not mean that the surface is free of contaminants, however, and physisorbed organic adsorbates are routinely observed on both surfaces from both atmospheric exposure and chemical exposure during device fabrication (Pirkle et al. 2010, McDonnell et al. 2013).

Figure 7.3 shows an example of typical x-ray photoelectron spectroscopy (XPS) spectra acquired on natural MoS_2 after mechanical exfoliation. It is clear that adventitious carbon is always detected, whereas oxygen is often below the limit of detection. The Mo $3d$ core level positioned at ~229.5 eV is typical of molybdenum in MoS_2 and can readily be identified by the spin–orbit split $3d_{3/2}$ and $3d_{5/2}$ features separated by 3.1 eV and with an intensity ratio of 0.66 corresponding to the electron occupancies in those levels (Moulder et al. 1992, McDonnell et al. 2013, 2014). Note also that atomic resolution scanning tunnelling microscopy (STM) is routinely possible without any *in situ* UHV treatment requirement (such as annealing) which is evidence of the relatively low reactivity of this surface enabling stable tunnelling currents to be detected. A more thorough review of impurities that exist below the XPS detection limit of XPS (<0.1%) will be given in Section 7.2.4.

7.2.1.2 Spatial Variations of XPS Spectra

One interesting phenomena observed in natural MoS_2 is the local (spatial) variations in the Fermi-level position. Since core-level positions in XPS are measured with respect to the Fermi level, such local variations result in the observation of multiple components in the XPS spectra. **Figure 7.4** shows an example of the Mo $3d$ core level obtained from five different locations on the same 1×1 cm geological MoS_2 sample after exfoliation. Comparison of the spectra in **Figure 7.4** to that in **Figure 7.3c** demonstrates that a convolution of two states are clearly observed with different binding energy positions. These features are labelled S 1 and S 2 in **Figure 7.4**. Their relative intensities also vary with spatial location on the sample. It is noted that this result is for a MoS_2 crystal that has no impurities within the limits of detection by XPS.

Using only the spectra obtained from a single element, it can be difficult to confidently distinguish between normal core-level chemical shifts and variations in Fermi-level position. Core-level chemical shifts would be caused by the electron distribution around the Mo atom being altered by a change in the local chemical bonding environment, while variations in the Fermi-level position could be induced by even low concentrations of dopants but with no significant change in the local Mo bonding environment. However, the difference can be seen when we examine other elements such as sulphur. When peak shifts in the Mo features are induced by changes in the Mo chemical bonding environment, concurrent but not identical changes in the S features should be observed. For example, with the hypothetical formation of metal Mo and elemental S, the Mo features would shift to *lower* binding energy with respect to MoS_2, whereas the S feature would shift to *higher* binding energy (Moulder et al. 1992). An example of this can be seen in **Figure 7.5** where chemical changes are induced by sputtering with inert He^+ ions. It can be seen that the Mo $3d$ and S $2p$ core levels show evidence of new chemical states on their low and high binding energy sides, respectively.

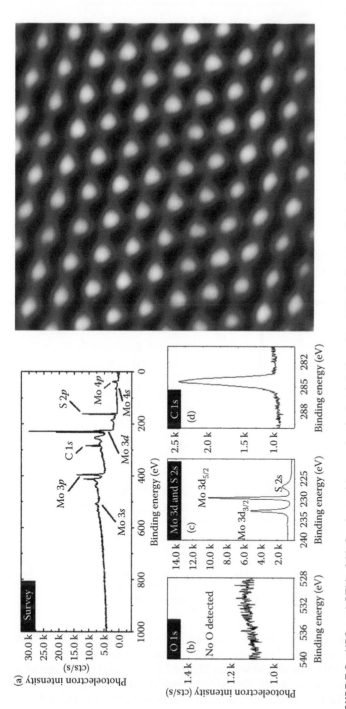

FIGURE 7.3 XPS and STM of MoS$_2$ cleaned by mechanical exfoliation. The XPS (Left: a,b,c,d) shows that carbon is the only repeatedly observed contaminant and the STM image (Right: 3 nm × 3 nm acquired at 0.5 V and 3 nA) shows that atomic resolution can be obtained without any UHV outgassing or high-temperature annealing treatments. (Reprinted with permission from McDonnell, S. et al. HfO$_2$ on MoS$_2$ by atomic layer deposition: Adsorption mechanisms and thickness scalability. *ACS Nano* 7;2013:10354–10361. Copyright 2013 American Chemical Society.)

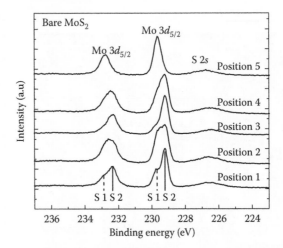

FIGURE 7.4 Mo 3*d* and S 2*s* core levels acquired from five different locations on the same 1×1 cm MoS_2 crystal after exfoliation.

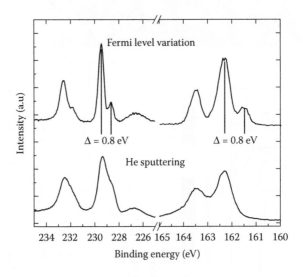

FIGURE 7.5 Comparison of the Mo 3*d* and S 2*p* core-level variations observed in the case of chemical changes (induced by ion sputtering) or local Fermi-level variations (identified by a similar shift in all core levels).

In contrast, when core-level features are shifted by a change in the Fermi-level position, all peaks should move in the same direction and by the same amount. That shift can also be seen in **Figure 7.5** where certain regions of 'as exfoliated' MoS_2 show identically shifted shoulders in both the Mo 3*d* and S 2*p* spectra. When a variation in the Fermi level occurs, *all* of the molybdenum and sulphur core levels will be seen to shift. In **Figure 7.4**, these identical shifts are actually observed in the S $2s$ core level; however, since the S $2s$ core level is significantly broader than the Mo $3d$, the corresponding features are more difficult to observe. The most obvious example of this can be seen by applying a voltage bias to a sample during analysis and observing all features move by an amount equal to the bias. This phenomenon observed in **Figures 7.4** and **7.5** is similar to that of lateral differential charging that is often observed on insulating substrates (Tielsch and Fulghum 1997). However, the shifts described here were on non-insulating substrates and, as will be discussed in Section 7.3.2.3, complementary scanning tunnelling spectroscopy (STS) and current–voltage measurements reveal that the cause appears to be local variations (stoichiometry, defects, etc.) in geological MoS₂ samples.

With this knowledge, four important conclusions can be obtained from the observed variations in the two core-level features observed on MoS₂. (1) The spatial variations clearly indicate that the properties of MoS₂ are not uniform across the sample, and such variations are expected to manifest themselves in typical device demonstrations to date. (2) The identical shifts in both Mo and S core levels suggest a Fermi-level shift rather than a chemical change. (3) The observation of two features simultaneously suggests that the variations in the Fermi level occur over areas smaller than the XPS spot size of 500 μm^2. (4) The observation of two distinct peaks, rather than simply a broadening, suggests two well-defined Fermi-level positions rather than gradual variations across the sample surface.

As discussed previously, the intercalation of alkali metals in 2H-MoS₂, such as Li, results in a phase transition to the distorted 1T-MoS₂ polytype (Py and Haering 1983, Chrissafis et al. 1989, Sandoval et al. 1991, Wypych and Schöllhorn 1992, Papageorgopoulos and Jaegermann 1995, Wypych et al. 1998, Eda et al. 2011, Kappera et al. 2014). This polytype is also predicted to be metallic and since binding energies in XPS are actually a measure of the energy difference between the initial and final (ionised) state of the atom with respect to the Fermi level, such a change in a material's electronic structure could likely result in detectable shifts in the core-level features. Therefore, the photoelectron spectra from any area exhibiting a mixture of 2H-MoS₂ and 1T-MoS₂ might be expected to yield two unique spectral features. This intercalation-induced effect has been demonstrated independently by two groups (Papageorgopoulos and Jaegermann 1995, Eda et al. 2011). *In situ* deposition of Li on MoS₂ surfaces under UHV conditions resulted in clearly distinguishable shoulders in the XPS spectra, the intensity of which can be modified by annealing, and indicated a combination of Li diffusion and desorption upon thermal activation (Papageorgopoulos and Jaegermann 1995). Similar XPS spectral behaviour has been recently reported for MoS₂ immersed in a butyl-lithium solution (Eda et al. 2011).

Importantly, the example shown in **Figures 7.4** and **7.5**, where *no* intercalation has taken place, indicates that this spectra observation in XPS spectra is *not* direct, unique evidence of either doping variations *or* phase transitions (2H and 1T). Rather, it is evidence of the local variation of the Fermi level that is being detected by XPS. Supporting electrical and structural characterisation techniques (e.g. conductivity, Raman and electron microscopy) are required in each case to unambiguously distinguish between the two phases.

7.2.2 Scanning Tunnelling Microscopy

7.2.2.1 Ideal MoS$_2$ Surface

The hexagonal structure (0001) surface of MoS$_2$ was reported a few years after the invention of the scanning tunnelling microscope. For example, G.W. Stupian et al. appear to have reported the first study of STM imaging of an MoS$_2$(0001) surface (Stupian and Leung 1987). The inert nature of the basal surface facilitates the imaging even in air (Stupian and Leung 1987, Permana et al. 1992). The hexagonal pattern observed by STM corresponds to the tunnelling from the sulphur p orbitals (Altibelli et al. 1996, Magonov and Whangbo 2007a). The underlying Mo-layer could be also imaged under certain conditions (Altibelli et al. 1996, Perrot et al. 2000, Addou et al. 2015a, 2015b). The lattice constant measured from the STM images (Stupian and Leung 1987, Permana et al. 1992, Ha et al. 1994, Perrot et al. 2000, Magonov and Whangbo 2007a, Addou et al. 2015a, 2015b) is in agreement with the value of 3.12 Å from the model of a bulk MoS$_2$ crystal (Dickinson and Pauling 1923). Simulated STM images from a defect-free surface reveal that the sulphur atoms forming the top-most surface layer generate the hexagonal structure observed on experimental STM images (Fuhr et al. 2004, Santosh et al. 2014). The MoS$_2$ bandgap of ~1.2–1.3 eV was estimated from the dI/dV spectra (Abe et al. 1995, Schlaf et al. 1997, Park et al. 2005, Inoue et al. 2013, Addou et al. 2015a, 2015b). The STS shows that the Fermi level is located near the conduction band minimum (CBM) characteristic of n-type MoS$_2$ induced possibly by the presence of impurities (Dolui et al. 2013).

7.2.2.2 Defects/Features Observed

The surface of geological and the synthetic MoS$_2$ that has been cleaned by mechanical exfoliation routinely shows a range of different types of defects (Magonov and Whangbo 2007b). STM images such as those in **Figure 7.6a** show the presence of large pits with a lateral size of ~3–5 nm and depths spanning the range of 0.3–1.8 nm. Large structural defects are also observed with a depth of 3 nm. By examining bias dependent measurements, it can be seen that the nature of the dark holes has two distinctly different root causes: (1) structural defects caused by a missing fragment of the S–Mo–S structure (**Figure 7.6b**) or (2) metal-like defects caused by sulphur desorption (**Figure 7.6c**). Nanometre ring-shaped structures are reported to be a purely electronic effect in the STM images since those features are not present in corresponding atomic force microscopy (AFM) images (Schlaf et al. 1997). The presence of impurities such as Mn, V and Ti was also suggested as the cause of such ring structures (Heckl et al. 1991, Ha et al. 1994). The presence of impurities in the vicinity of the surface is manifested by local changes: a topographic depression induced by acceptor impurities (**Figure 7.6d**) and a bright (higher conductivity) spot induced by donor atoms.

On other TMD surfaces such as WS$_2$ and WSe$_2$, the depressions are caused by acceptors located in the topmost surface layer or the first or second subsurface layer (Matthes et al. 1998, Sommerhalter et al. 1999). The force curves measurements using AFM and varying the voltage between the cantilever and

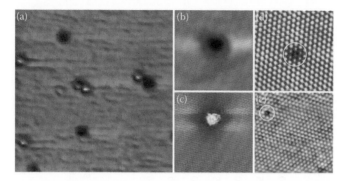

FIGURE 7.6 (a) STM image 50 nm × 50 nm shows two different type of defects (bright and dark), (b) STM image 12 nm × 12 nm of a dark defect, (c) STM image 10 nm × 10 nm of bright defect, (d) STM image 3 nm × 3 nm of local depression (highlighted by a white circle) caused by an impurity and (e) STM image 7 nm × 7 nm of single S-vacancy (white circle).

the surface demonstrate the presence of localised dopants (Schlaf et al. 1997). STM images (Addou et al. 2015a, 2015b) recorded on an as-exfoliated MoS₂ surface without any thermal treatment or metal deposition show defects similar to those obtained after Na and Sn deposition on MoS₂ and Re on MoSe₂ (Abe et al. 1995, Murata et al. 2001). Another frequently observed defect is the vacancies (**Figure 7.6e**), which are likely S-vacancies due to the relative formation energies as elucidated in density functional theory (DFT) calculations (Liu et al. 2013, Santosh et al. 2014). Experimental STM images show the formation of single vacancy (Addou et al. 2015a, 2015b) and the images were found to be consistent with simulated STM images. It is important to note that an S-vacancy induces a defect state into the MoS₂ bandgap and therefore needs to be prevented or passivated (Fuhr et al. 2004, Santosh et al. 2014).

For optoelectronic applications, defects may in some instances actually be beneficial. Recent studies on MoS₂, WSe₂ and MoSe₂ showed that defects could be used as adsorption sites at cryogenic temperatures for molecules such as N₂ that could in turn enhance the observed photoluminescence and induce a new defect-related feature in the spectra (Tongay et al. 2013, Nan et al. 2014). This suggests that defect engineering may be a route to tuning the properties in optoelectronic devices based on TMDs.

7.2.2.3 STS Results

Spectroscopy measurements are employed to depict the electronic properties for the MoS₂ surfaces. The first derivative dI/dV, calculated from the current–voltage I–V curve, is proportional to the local electron density of states of the surface and allows determining the conduction and valence band edges that define the bandgap. Several different characteristics of the STS obtained from MoS₂ cleaned by mechanical exfoliation are identified in **Figure 7.7**: (1) regions with the expected n-type conductivity (**Figure 7.7a**), (2) regions with p-type conductivity (**Figure 7.7b**) and (3) with zero current in negative bias (**Figure 7.7c**).

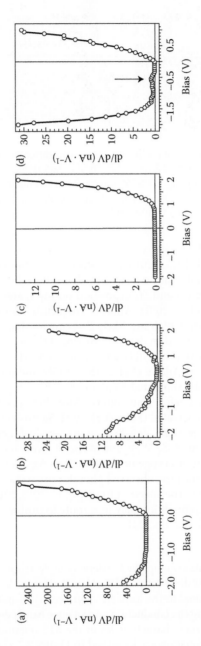

FIGURE 7.7 (a) STS of a n-type MoS$_2$ with expected bandgap (1.27 eV), (b) STS of p-type MoS$_2$ with a small bandgap, (c) STS with flat valence band (zero conductivity at negative bias), the Fermi level is shifted by 1 V below the conduction band and (d) STS showing a band-gap feature (indicated by an arrow) at −0.65 V.

The Fermi level located near the CBM as in **Figure 7.7a** is representative of n-type MoS_2. The Fermi level located near the valence band maximum (VBM) as in **Figure 7.7b** is representative of p-type conduction. The n-type spectrum is routinely obtained from surface areas with a low defect density (McDonnell et al. 2014, Addou et al. 2015a, 2015b). The p-type behaviour is observed from areas with apparent structural defects exhibiting a high defect density. The bandgap is quite variable on the p-type/defective regions. Local stress, S-rich/Mo-deficiencies, as well as impurities may explain such observations.

One of the unusual electronic signatures measured on a rough region is where dI/dV is zero over negative bias (**Figure 7.7c**). This behaviour has previously been attributed to tip-induced band-bending effect on WSe_2 (Yoshida et al. 2013). However, since the STS variation seen in **Figure 7.7** can be observed using the same tip across the single sample and the same imaging conditions (applied bias and tunnelling current) and the observed zero conductance occurs on several spots independent of the tip-surface distance or the tip shape, this strongly suggests that the observed phenomenon is not tip induced. Instead, the behaviour of zero current while probing filled states is likely due to a defect-induced local depletion of electrons. Similar effects in TiO_2 have been described by other groups as defect-induced band bending (Batzill et al. 2002, Tao et al. 2011). This is not entirely surprising since there are many local changes in the stoichiometry, defect concentration and the impurities level across a sample. The spectroscopy measurements show another electronic signature that is illustrated in **Figure 7.7d**, where there is the clear presence of a feature in the bandgap region located at 0.65 eV below the Fermi level (Addou et al. 2015a, 2015b). A comparable feature was observed above the VBM in calculated density of states on MoS_2 layers with induced defects (Fuhr et al. 2004).

7.2.3 Raman

7.2.3.1 Modes

Raman spectroscopy is considered a powerful non-destructive characterisation instrument. It has been used to study different crystalline structures of MoS_2 (Sandoval et al. 1991, Lee et al. 2010b, Molina-Sanchez and Wirtz 2011, Li et al. 2012b). Four first-order Raman active modes in bulk MoS_2 can be detected in off resonance, E^2_{2g}, E_{1g}, E^1_{2g} and A_{1g} which are located at 32, 286, 383 and 408 cm⁻¹, respectively (Sandoval et al. 1991, Lee et al. 2010b, Molina-Sanchez and Wirtz 2011, Li et al. 2012b). On resonance, several Raman peaks can be observed caused by the strong electron–phonon couplings. The E^2_{2g} mode arises from the vibration of an S–Mo–S layer against adjacent layers. The E_{1g} mode is forbidden in backscattering experiments on a basal plane. The in-plane E^1_{2g} mode is produced by the opposite vibration of two S atoms with respect to the Mo atom, whereas the A_{1g} mode is associated with the out-of-plane vibration of only S atoms in opposite directions. The most prominent mode on resonance, measured at around 460 cm⁻¹, arises from a second-order process involving the longitudinal acoustic phonons at the M point, LA(M). Most of the new peaks are typically assigned to multiphonon bands involving

LA(M) and other phonons at the M point (Molina-Sanchez and Wirtz 2011, Li et al. 2012b). Unlike XPS, the Raman spectra of alkali intercalated MoS_2 provides clear evidence of a 2H to 1T phase transition (Eda et al. 2011, Kappera et al. 2014).

7.2.3.2 Variations with Thickness, Phase and Doping

The intensity and the width of the Raman E^1_{2g} and A_{1g} peaks vary monotonically with the thickness of ultrathin MoS_2 and can be used as reliable features to identify the number of layers, as well as to determine the changes in material properties with thickness variation. As MoS_2 becomes single layer, the in-plane mode upshifts to 386 cm^{-1} and the out-of-plane downshifts to 404 cm^{-1}. The difference of these two modes (~18 cm^{-1}) can be used as a reliable identification for monolayer MoS_2 (Lee et al. 2010b, Molina-Sanchez and Wirtz 2011, Li et al. 2012b). An example of this is shown in **Figure 7.8**. Owing to the differences in the crystallography symmetry, 2H and 1T phases can be differentiated by Raman spectra. A shift in both E and A mode occurs when the phase changes. Moreover, new frequencies appear in the distorted 1T phase at 156, 226 and 333 cm^{-1}, which are absent from the $2H-MoS_2$ spectra. Similar to graphene, the Raman modes can also be used to monitor doping in the MoS_2 (Das et al. 2008, Chakraborty et al. 2012). It was shown by doping a MoS_2 crystal with Au nanoparticles and carrying out Raman analysis over multiple areas that a significant increase the A_{1g}/E_{2g} ratio occurs with p-type doping. To a lesser extent, there is also a small upshift detected in the A_{1g} peak position (Shi et al. 2013).

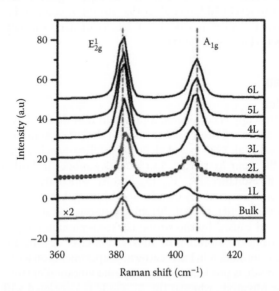

FIGURE 7.8 Sample Raman spectra for n-layers of MoS_2 and bulk MoS_2. (Reprinted with permission from Lee, C., H. Yan, L.E. Brus et al. Anomalous lattice vibrations of single- and few-layer MoS_2. *ACS Nano* 4;2010:2695–2700. Copyright 2011 American Chemical Society.)

7.2.4 Stoichiometry and Impurities

7.2.4.1 Stoichiometry Measurements

High-resolution Rutherford backscattering spectrometry (HR-RBS) is widely used for near-surface layer analysis of materials, elemental composition and depth profiling of individual elements. The analysis of the depth profiles obtained from two MoS_2 crystals that came from different geographical locations (Canada, labelled as 'c-MoS_2', and Australia, labelled as 'a-MoS_2') is shown in **Figure 7.9** showed that the atomic concentration varies significantly throughout the depth investigated. Also, the S atomic % indicates that the c-MoS_2 is more sulphur-deficient and more Mo-rich in comparison to a-MoS_2. The composition averaged over 80 nm in depth shows that the c-MoS_2 is sulphur-deficient with $(S/Mo)_{c-MoS2} = 1.78$, and the a-MoS_2 sample is slightly S-rich with $(S/Mo)_{a-MoS2} = 2.03$. This is consistent with similar stoichiometry variations observed in previous work (McDonnell et al. 2014).

7.2.4.2 Impurity Measurements

A recent report on the elemental analysis of MoS_2 minerals by inductively coupled plasma mass spectrometry (ICPMS) collected from 135 localities over the earth with known geological ages shows the detection of notable amounts of transition elements of Fe (~1000 ppm), W (~4000 ppm) and Re (5000 ppm), as well as high concentrations of Ni, Mn, Cu, Ru, Ca and Co (>20 ppm and <1000 ppm) (Golden et al. 2013). Rhenium content varies significantly depending on geological age and location on the earth. *Ab initio* calculations also reveal that Re results in donor levels and alkali metals shift the Fermi level close to the conduction band edge making the MoS_2 n-type (Dolui et al. 2013).

Recent ICPMS analysis, performed by the authors and collaborators, from acid-digested geological and synthetic (vapour phase transport (VPT) growth) MoS_2 reveals many impurities with concentrations greater than 1000

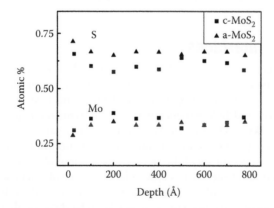

FIGURE 7.9 Comparison of the stoichiometry between two different geological MoS₂ samples using HR-RBS. This highlights the large sample–sample variation in stoichiometry. (After Addou, R., S. McDonnell, D. Barrera et al. *ACS Nano* 9;2015b:9124–9133.)

Table 7.1 Comparison of Selected Impurity Concentrations from ICPMS Analysis of Geological MoS$_2$ and Synthetic MoS$_2$ Grown by VPT

Source	Impurity (ppbw)						
	Al	Ba	Cd	K	Pb	Sb	W
Geological	8041	68	1876	40,008	9798	43,908	2378
VPT	223	1329	3119	337	568	66	798

Note: The abundance unit is parts per billion by weight (pbbw).

parts per billion by weight (ppbw) in bulk MoS$_2$. A selection of these is shown in **Table 7.1**, including Al, Ba, Cd, K, Pb, Sb and W, all of which have energy levels in the bandgap of silicon (Sivasubramani et al. 2006). Conversion of these levels to equivalent impurity concentrations (shown on **Table 7.1**) leads to the conclusion that concentrations can easily exceed 10^{10}/cm^2, and would be expected to impact transport measurements if present as ionised impurities (Sze and Ng 2006, Ma and Jena 2014). In comparison, XPS has a detection limit of ~0.05% – well above the ICPMS concentrations (Addou et al. 2015b).

Positive secondary-ion mass spectrometry (SIMS) was performed to estimate the concentration of elements in MoS$_2$. The SIMS spectra reveal, in addition to Mo and S, the presence of several impurities: C (0.5%), Mn (0.5%), Al (0.4%) and Ca (1.5%) (Heckl et al. 1991). Glow discharge mass spectrometry shows the presence of different types of impurities considered as part of the MoS$_2$ sample. In addition to the major impurities of Ti (46 ppm) and V (19 ppm), small quantities (<1 ppm) were detected for F, Na, Al, Ca, K, Mn and Fe (Ha et al. 1994). The detection of such metals in the bulk or near the surface of MoS$_2$ may explain the large variability found in the topography and the electronic structure of molybdenite samples.

7.3 MoS$_2$ Interfaces

In all nanoelectronic and optoelectronic devices there exist interfaces which critically influence the operation and performance of the device. Contact resistance is one such limiting factor in devices and is strongly influenced by the electron and hole barriers at the metal–semiconductor interface. In an ideal Schottky diode, this electron (hole) barrier can be predicted by the difference between the contact metals work function and electron affinity (photoionisation potential) of the semiconductor. In real interfaces, one must consider reactions at the interface. Another interface of importance for field-effect devices is the insulator–semiconductor interface, which is required to reduce gate leakage currents. For 2D materials, it has been predicted (Jena and Konar 2007) and experimentally demonstrated (Jang et al. 2008, Radisavljevic et al. 2011a) that high-κ dielectrics can enhance channel mobility by reducing the enhanced Coulomb impurity scattering that is present in free-standing 2D materials.

7.3.1 Oxide/MoS_2 Interfaces

7.3.1.1 Oxide ALD

7.3.1.1.1 Cleaning

MoS_2, like graphite, is often thought of as an inert material which is a reputation earned due to its resistance to oxidation in air and its usefulness as a solid lubricant. However, both still accumulate physisorbed organic surface adsorbates, possibly at surface defect sites. Spurious contamination from materials and chemicals used in device fabrication is also a potential source. To obtain a nominally clean surface, layered materials are typically exfoliated mechanically by adhering scotch tape to the top surface and removing the top layers from the bulk crystal, leaving behind a fresh surface that had not previously been exposed to the air (Frindt and Yoffe 1963, Frindt 1966). While this method is used routinely to obtain 'clean' 2D materials, organic contaminants adsorb to the surface if the process is not carried out in vacuum despite the inert nature of the materials. This is shown in **Figure 7.10** for MoS_2 and was shown previously for graphite (McDonnell et al. 2012).

For graphite crystals, annealing to 430°C in UHV was found to be sufficient to remove surface contaminants to below the limits of detection of XPS (Pirkle et al. 2010). Similar experiments on MoS_2 (**Figure 7.10**) have shown that annealing to 400°C is sufficient to remove oxygen and adventitious carbon to below the limit of detection. However, high-temperature annealing is typically not a preferred method for cleaning binary systems such as MoS_2 since there is a risk of generating sulphur vacancies.

Ultraviolet (UV) ozone is routinely used to remove organic residues from a range of inorganic substrates (Vig and LeBus 1976, McClintock et al. 1982, Vig 1985, Baunack and Zehe 1989, Hansen et al. 1993, Zhang et al. 1993). A study using room-temperature UV–ozone, generated by the UV excitation of

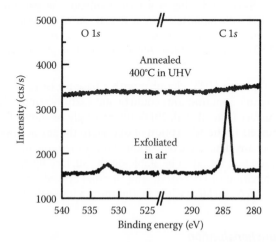

FIGURE 7.10 XPS spectra of the O 1s and C 1s core level after exfoliation in air and after annealing to 400°C in UHV. Carbon is always detected on MoS_2 after exfoliation in air; however, oxygen is often below the limit of detection.

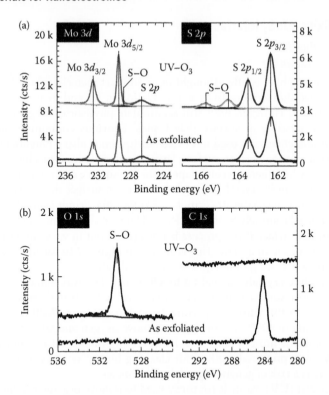

FIGURE 7.11 XPS spectra of (a) Mo $3d$ and S $2p$, (b) O $1s$ and C $1s$ core levels before and after room-temperature UV–ozone treatment. The treatment clearly removes carbon contamination and also adds oxygen to the surface through the formation of an S–O bond. This is achieved without the generation of Mo–O bonds. (Reprinted with permission from Azcatl, A. et al. MoS$_2$ functionalization for ultra-thin atomic layer deposited dielectrics. *Applied Physics Letters* 104;2014:111601. Copyright 2014, American Institute of Physics.)

1000 mbar of O$_2$ in an UHV chamber with a base pressure of 1×10^{-9} mbar, showed that organic contaminants could be successfully removed from a MoS$_2$ bulk crystal surface (Azcatl et al. 2014). Interestingly, as shown in **Figure 7.11**, this process results in the adsorption of oxygen to the surface (S–O bond formation) without any detectable oxidation of the underlying molybdenum. The adsorbed oxygen was found to be stable on the surface up to ~250°C and as such the combination of UV–ozone and thermal annealing provides a route to cleaning the MoS$_2$ surfaces while maintaining a lower thermal budget than annealing alone.

7.3.1.1.2 Functionalisation

ALD requires chemical reactions between precursor species and the substrate in order to form a self-terminating layer that can be further reacted by a second precursor exposure. Since the ideal basal plane of 2D crystals has no dangling bonds, this reaction process is expected to be inhibited and early work on graphite substrates highlighted this point by demonstrating that deposition took place only on the more reactive step edges (Lee et al. 2008).

The earliest work on single-layer MoS_2 transistors utilised a modified ALD process to obtain a uniform film of HfO_2 on MoS_2 (Radisavljevic et al. 2011a). Given that the reported growth rate was 1.9 Å/cycle rather than 0.93 Å/cycle which is the known rate of self-limiting growth (Hausmann et al. 2002), this process cannot be considered true ALD. This work did, however, prompt a further study by Liu et al. on the comparative susceptibility of 2D crystals to nucleating ALD and it was found that MoS_2 does in fact seem to be more reactive than graphite (Liu et al. 2012b). Further work demonstrated that when self-limiting ALD conditions are employed, HfO_2 does not grow uniformly on MoS_2 (McDonnell et al. 2013). It was also shown that organic contaminants and solvent residues, often present in device fabrication experiments incorporating 'lift-off' methods, enhance nucleation. In studies on graphene surfaces, it was shown that such residues persist even after deliberate attempts to remove them (Lin et al. 2011, Pirkle et al. 2011, Chan et al. 2012, Gong, Floresca et al. 2013). The presence of such residues may explain why MoS_2 devices fabricated on exfoliated flakes and patterned by electron-beam lithography did not require any intentional functionalisation to grow very thick, uniform, high-κ dielectrics. The impact of such residual contamination on device results remains to be systematically studied, but variations in performance appear to be widespread in the literature.

In order to obtain controlled deposition of high-κ dielectrics using true ALD processes, a number of approaches have been taken to functionalise the MoS_2 surface. Among the first of these was the work (Yang et al. 2013) that utilised an oxygen plasma which had previously been shown (Cui et al. 1999) to oxidise (and thus etch) the top surface of the MoS_2. XPS was used to verify that the oxygen plasma reacted with the surface and chemical states assigned to MoO_3 were identified. Following this treatment, both Al_2O_3 and HfO_2 were deposited on MoS_2 and transmission electron microscopy (TEM) showed uniform layers as well as the presence of an interlayer. While this process showed a clear enhancement in ALD nucleation, the formation of MoO_3 implies etching of the MoS_2 and therefore makes it impractical for monolayer-based devices.

Low-temperature ozone-based ALD has been used as a seed layer for subsequent optimal-temperature ALD growth on graphene (Lee et al. 2008, 2010a). A similar process does improve the nucleation of ALD on MoS_2; however, uniform (pin-hole free) films were not formed under the conditions employed in that study (Cheng et al. 2014).

Another oxidation process used a room-temperature UV–ozone exposure to adsorb oxygen on the MoS_2 surface which was discussed earlier from the point of view of MoS_2 cleaning (Azcatl et al. 2014). Somewhat surprisingly, this process resulted in a clear S–O feature in XPS, but there was no indication of any disturbance to the underlying molybdenum bonding (viz. Mo–O formation). It was hypothesised that the top sulphur atoms can re-hybridise and accommodate a covalent bond to the adsorbed oxygen without breaking the Mo–S back bonds. Such a structure was also shown to be thermodynamically stable and with an electronic structure that yielded no states within the MoS_2 bandgap. This process also showed enhanced ALD with both Al_2O_3 and HfO_2 layers being deposited uniformly using a self-limiting water-based ALD process. **Figure 7.12** shows an example of Al_2O_3 on MoS_2 with and without functionalisation. The ALD process temperature was limited to 200°C due to the low thermal stability of the adsorbed oxygen. Also shown in **Figure 7.12** is a separate experiment comparing MoS_2 exposure to atomic oxygen and remotely generated ozone (no UV light in either case). These resultant XPS spectra confirmed that atomic oxygen plays a major role in the generation of the S–O functionalisation layer. This process has subsequently been investigated on other TMD surfaces (Azcatl et al. 2015).

PVD has also been used to seed ALD growth on 2D materials. Thin aluminium layers deposited by PVD subsequently oxidised in air were previously employed to seed the growth of HfO_2 on graphene flakes (Fallahazad et al. 2010). A similar process has been used to facilitate the growth of HfO_2 on MoS_2 (Zhu et al. 2014). Zou et al. carried out a comparative study of MgO, Al_2O_3 and Y_2O_3 seeding layers for subsequent HfO_2 deposition and found the $HfO_2/Y_2O_3/MoS_2$ structure resulted in an electron mobility of 63.7 cm²/Vs, an on/off ratio of 10^8, and a subthreshold swing of 65 mV/dec (Zou et al. 2014). The benefit of PVD methods is that nucleation is not as dependent on surface reactivity. One potential limitation is that PVD is 'line-of-sight' and thus nonconformal in all geometries, and, should future device architectures employ 3D structures (e.g. 2D crystals wrapped around 3D fins), then seeding by PVD will not be practical.

7.3.1.1.3 Current Status of Top-Gated Devices

In theory, one of the main advantages of MoS_2 and other layered 2D materials is the lack of surface dangling bonds. However, as seen in Section 7.2.2.2, that ideal surface is far from reality in exfoliated, geological MoS_2 on which most reports of device characteristics are fabricated. The non-ideal surfaces in MoS_2 would of course degrade device performance and a measurable D_{it} should be expected for metal-oxide-semiconductor (MOS) devices. It should again be noted, however, that the surface of MoS_2 is much more inert than other materials, particularly 3D semiconductors, as evidenced by the difficulties in nucleating ALD dielectrics on MoS_2 as seen in the preceding section. To date, very few experimental studies have been undertaken to elucidate the interface state density of MoS_2-based devices. Liu reported D_{it} of 2.4×10^{12}/cm² eV in 2012 for top-gated devices fabricated with ~16 nm ALD Al_2O_3 on MoS_2 without intentional functionalisation (Liu and Ye 2012). The subthreshold slope (SS) measured in that report was 140 mV/dec. A more recent investigation by Zou et al. utilised thermally evaporated thin metals (Y, Mg), subsequently oxidised in an oven, as a seed layer prior to 9 nm ALD HfO_2 deposition. The resultant top-gated devices on 3–5-layer MoS_2 exhibited promising SS of 65 mV/dec for 3 μm channel length devices and an extracted D_{it} of 2.3×10^{12}/cm² eV (Zou et al. 2014). Whether this impressive device performance is due to an improved interface from dielectric processing, from a higher-quality MoS_2

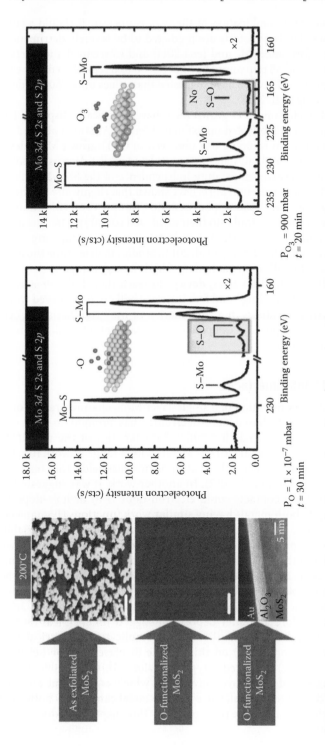

FIGURE 7.12 AFM images showing the enhanced ALD nucleation after UV–ozone treatment resulting in a conformal Al₂O₃ layer as seen in the TEM image. The XPS spectra of the Mo 3*d* and S 2*p* core levels for separate atomic oxygen and remote ozone MoS₂ treatments (no UV light) show that atomic oxygen clearly plays a role in S–O bond formation. (After Azcatl, A. et al. *Applied Physics Letters* 104;2014:111601.)

starting crystal, or a combination of the two, is unclear. It should again be noted that these levels of defects are predicted to be severely detrimental to tunnel FET performance (Ma and Jena 2014) and so reconciliation between the extracted D_{it} and SS remains to be understood and reproduced. For example, recent reports of TFET SS have suggested that values under 60 mV/decade are possible (Kang et al. 2015).

Devices fabricated on CVD-grown MoS_2 have historically displayed lower mobilities than those fabricated on exfoliated flakes (Amani et al. 2013, Zhu et al. 2014). Significant trapping was observed in CVD-grown MoS_2 devices and the presence of band tail states was postulated for the decreased mobility (~10 cm²/Vs) and relative temperature independence of the SS (~200 mV/dec) (Zhu et al. 2014). Some recent reports have shown that chemically synthesised films (Hwang et al. 2013) and sulphurised Mo (Tarasov et al. 2014) films can exhibit device properties similar to their exfoliated counterparts. This has led to speculation that traps in the high-k dielectric (e.g. border traps) are responsible for the degraded SS (Hwang et al. 2013). Studies to determine the impact of dielectric functionalisation methods and high-k border traps on device performance will be critical for MoS_2 devices to reach their full potential. It is likely that the ideal dangling bond free interface will not be achieved without significant further advances in a combination of MoS_2 synthesis and defect passivation. Recent reports on wet passivation treatments of MoS_2 are promising (Amani et al. 2015).

7.3.2 Metal/MoS$_2$ Interfaces

Since the first demonstration of a single-layer MoS_2 channel transistor, the investigation of MoS_2-based nanoelectronics has brought to light some interesting and unexpected electronic characteristics of the metal/MoS_2 interface (Radisavljevic et al. 2011a). Metals with a range of work functions appear to have a pinned Fermi level at a position just below the conduction band of the MoS_2 (Das et al. 2012). In another work, it was suggested that gold and palladium contacts could electron-dope (Au) or hole-dope (Pd) the MoS_2 despite both metals having similar work functions (Fontana et al. 2013). Later work, which focussed on demonstrating the large variability in the electronic properties of MoS_2, highlighted the dangers in drawing conclusions from the electrical properties determined from a single sample (McDonnell et al. 2014). An example of this is shown in **Figure 7.13** where analysis of multiple 500 × 500 μm gold contacts is deposited on a single piece of MoS_2. The overlapping XPS core-level spectra for the Mo 3d and S 2s core level before and after deposition suggest no chemical reactions are detected with the substrate that would complicate the electrical analysis. However one contact (b) on this sample displayed n-type behaviour, whereas the other (c) displayed p-type behaviour. This incredible variability casts a shadow of doubt over any results which suggest that MoS_2 may be doped by the contact metal, since a single metal can display both n- and p-type behaviour on the same piece of MoS_2. This chapter will discuss these issues in detail.

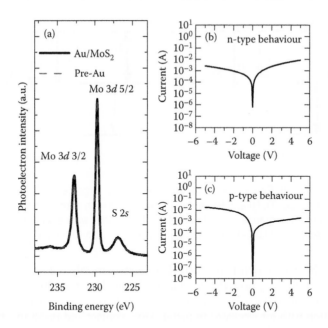

FIGURE 7.13 (a) The Mo 3d and S 2s core-level spectra before and after the deposition of Au showing no evidence of chemical reaction, (b) a gold contact showing n-type behaviour, (c) a gold contact on the same sample as (b) showing p-type behaviour. (Reprinted with permission from McDonnell, S. et al. Defect dominated doping and contact resistance in MoS_2. *ACS Nano* 8;2014:2880–2888. Copyright 2011 American Chemical Society.)

7.3.2.1 Contacts and Ideal Schottky Behaviour Based on Work Functions

One of the most prominent anomalies of the metal–MoS_2 contact is that high work function metals appear to result in unexpectedly low electron Schottky barriers with n-type MoS_2 with some authors even reporting ohmic contacts (Radisavljevic et al. 2011a, Liu et al. 2012a, Neal et al. 2012). This was highlighted in a deliberate study carried out by Das et al. using metal contacts with work functions ranging from 3.5 to 5.9 eV, that the MoS_2–metal interface was strongly impacted by Fermi-level pinning close to the conduction band of MoS_2 which is illustrated in **Figure 7.14** (Das et al. 2012). Multiple theories have been presented to explain this phenomenon.

Such Fermi-level pinning has been explained by first-principles calculations (Gong et al. 2014) which suggest that two mechanisms play a role in this predicted Fermi-level pinning. The first is the metal work function modification due to an interface dipole formation caused by the charge redistribution, whereas the other process is the gap states produced primarily in the Mo d-orbital by the weakening of the S–Mo interlayer bonding due to the metal–S interaction. Other theoretical work using the screen exchange hybrid functional showed that S vacancies introduce transition states in the upper half of the bandgap whose formation energy becomes small or negative for

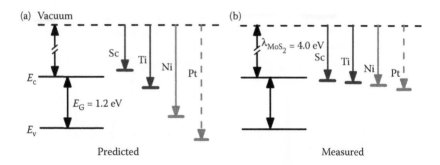

FIGURE 7.14 (a) Predict and (b) observed band alignments in MoS_2–metal contacts for a Sc, Ti, Ni and Pt contacts. (Reprinted with permission from Das, S. et al. High performance multi-layer MoS_2 transistors with scandium contacts. *Nano Letters* 2012:100–105. Copyright 2011 American Chemical Society.)

E_F near the conduction band such that reactive metal contacts will lead to Fermi-level pinning near the conduction band (Liu et al. 2013). It has also been reported that experimentally observed defects can provide a parallel conduction path for electrons in metal–MoS_2 contact and result in a reduced electron Schottky barrier. Even areal defect densities of less than 1% can have a profound effect on the extracted Schottky barrier if the defect–MoS_2 contact has a low Schottky barrier since the current through each path has an exponential dependence on its local Schottky barrier. As such, the combination of low work function defect and high work function metal will actually result in the majority of the current flowing through the defect even when it makes up ~1% of the contact area (McDonnell et al. 2014). In reality, it is quite likely that the MoS_2–metal interface in devices fabricated from geological MoS_2 suffers from both the defect-mediated parallel conduction path as well as metal Fermi-level pinning.

7.3.2.2 Metal Interactions

While the MoS_2 surface is typically described as being inert, the possibility of chemical reactions between deposited metals and the MoS_2 surface should not be overlooked. In a comparative study of palladium, gold and silver thin films deposited on MoS_2, it was shown that the resultant films had noticeably varied topology which was attributed to differing wettability of the surface for each metal (Gong et al. 2013b). For these metals, no direct evidence of chemical reactions with the substrate was observed but the variations in the MoS_2 Raman modes suggested that palladium more strongly interacted with the surface. An XPS investigation of the core-level shifts induced in the Mo $3d$ and S $2p$ core level after the deposition of a wide range of metals (Ag, Al, Au, Co, Fe, In, Mn, Pd, Rh, Ti and V) onto MoS_2 suggested that of the metals studied, only Mn was found to react with the surface (Lince et al. 1987). A similar expansive study using soft x-ray photoemission examined the reactions between metals (Al, Cu, In, Mg, Ni and Ti) and found that only Mg and Ti showed clear signs of chemical reactions (McGovern et al. 1985). The apparent contradiction between the results for Ti is because the former work actually deposited TiO_2 which they verified with XPS and therefore discounted the result. In studies focussing on the lubricating properties of MoS_2 in contact with stainless steel, it was found that the Cr also reacts with the MoS_2 surface (Durbin et al. 1992, Durbin et al. 1994). Shown in **Figure 7.15** is our recent observance of the Mo 3*d* and S 2*s* core levels for a range of metals deposited by electron-beam deposition in UHV where it can be seen that the lower work function metals clearly react with the MoS_2. Since the low work function metals readily react with the surface, they would result in metal–MoS_2 Schottky diodes with behaviours that would not have been predicted by simply comparing the metal work function with the semiconductor electron affinity.

We should also consider the unintentional deposition of TiO_2 instead of Ti in the work by Lince et al. (1987). In that work, the authors correctly identified that pure metal had not been deposited and speculated that there must have been a sufficient oxygen partial pressure in their chamber to result in Ti oxidation. However, it is important to expand upon the implications of this result. While researchers focussed on interface chemistry studies to typically carry out their work in UHV (base pressure $<1 \times 10^{-9}$ mbar), those focussed

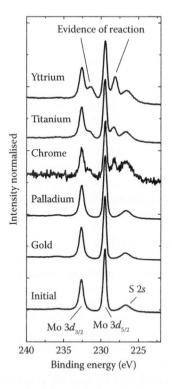

FIGURE 7.15 Mo 3*d* and S 2*s* core levels for a range of metals deposited by electron-beam deposition in UHV (base pressure $<1 \times 10^{-9}$ mbar) on MoS_2. Chrome, titanium and yttrium all clearly react with the substrate.

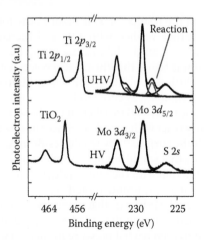

FIGURE 7.16 Ti $2p$, Mo $3d$ and S $2s$ core-level spectra for UHV and HV electron-beam deposition of Ti on MoS_2. It is clear that Ti deposits as TiO_2 in HV and does not react with the substrate. In UHV, the pure Ti metal reacts with the MoS_2.

on device fabrication will often use commercial high-vacuum (HV) electron-beam deposition tools with base pressures on the order of 1×10^{-7} mbar and would, therefore, be subject to much higher oxygen partial pressures. This begs the question, what is really deposited when we deposit a Ti contact metal in such a tool? (McDonnell et al. 2015). **Figure 7.16** shows a comparison of the Mo $3d$ and Ti $2p$ core-level features after the deposition of titanium in both UHV and in a commercial HV tool. It is clear that, in the case of the HV deposition, it is TiO_2 that is deposited and there are no reactions with the substrate; however, in the case of UHV deposition, the unoxidised titanium is free to aggressively react with the MoS_2 substrate. From a device performance perspective, each of these contacts would be expected to behave very differently and so knowing what has actually been deposited should obviously be of the utmost important to researchers.

7.3.2.3 Defect-Dominated Contacts

In our efforts to understand the metal/MoS_2 interface, it is critical to understand the role of the variability in the electronic properties of MoS_2. It has been clear since an early review of TMDs (Wilson and Yoffe 1969) that the properties of MoS_2 varied widely from one report to another. While recent reports tend to imply a consensus has been reached on the properties of this material, it would be unwise to assume that the previous, widely differing reports can all be ignored. In truth, the properties of MoS_2 vary not only from sample to sample, but also on the macroscopic, microscopic and nanoscopic scales which has been confirmed by spatially resolved studies that were carried out across a single sample (McDonnell et al. 2014). An example of this was discussed earlier with respect to the variation in the Fermi level that can be inferred from the XPS spectra. The high density of defects that is observed in

STM in conjunction with the impurities detected by ICPMS also suggest that we should expect spatial variations in the properties as defect and impurities concentrations vary locally. **Figure 7.17** shows an example of a controlled experiment when a sample was probed by multiple techniques using a shadow mask as a mapping tool (McDonnell et al. 2014). In this way, it could be insured that electrical characterisation, XPS, STM and STS were all carried out in the same 500×500 μm region. It is clear from these results that areas from which the I–V characteristic displayed n-type behaviour also showed a low concentration of structural defects (STM), a Fermi-level position just below the condition band (STS) and symmetric Mo $3d$ and S $2p$ core-level features with relative intensities suggesting sulphur deficiency (or excess Mo). Areas from which the I–V characteristic displayed p-type behaviour exhibited a high concentration of structural defects (STM), a Fermi-level position further from the condition band (STS) and asymmetric Mo $3d$ and S $2p$ core-level features shifted to low binding energy with relative intensities suggesting excess sulphur or molybdenum deficiency (XPS). To ensure that the observed effects were not simply damaged induced by the probing of the sample, multiple samples were investigated and the sequence of the measurements was varied. The correlations between the techniques were consistently observed regardless of the sequence of measurements. The defect concentration detected in some p-type (**Figure 7.17h**) regions approaches 5% of the surface, whereas those is n-type regions is less than 1%. In the n-type regions, a simple model based on thermionic emission shows that such high conductivity defects can dominate the contact properties.

7.4 Summary

7.4.1 State-of-Art Understanding of Surface and Interface

It should now be clear that while MoS₂ and other 2D materials are predicted to have perfect dangling bond free surfaces, the reality is very different. Geological and synthetic MoS₂ crystals are both highly impure and defective relative to the specifications currently employed in the semiconductor industry. Similar issues are anticipated for synthetic MoS₂ films as well. The impact of such impurities and defects can cause local variations in the Fermi level resulting in p-type and n-type MoS₂ being observed on the same sample. The defects can play a role in determining the contact resistance in Schottky diodes by creating a parallel conduction path. Similar defects can also enhance PL and generate unique optical properties which provide a route for controlled defect engineering of TMDs, but also emphasise that uncontrolled defects can result in unexpected optoelectronic properties that would strongly impact photonic devices. Even when defect-free MoS₂ has been achieved, overcoming the metal Fermi-level pinning issues will likely be a considerable challenge. Significant advances have been made in functionalising the MoS₂ surface and a number of techniques including electron-beam seeding and oxygen adsorption have successfully allowed for conformal ALD thin films to be realised.

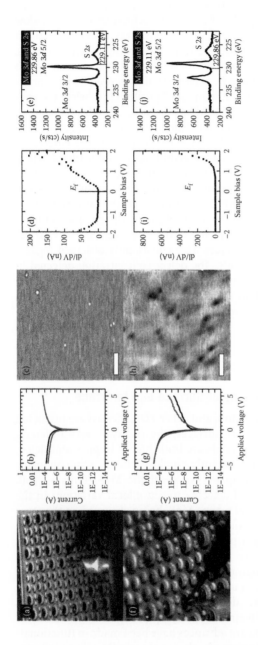

FIGURE 7.17 Spatially resolved analysis showing how n-type behaviour can be correlated with low defect density and sulphur deficiency, whereas p-type behaviour is correlated with sulphur excess and a highly defective surface. (a) MoS₂ sample covered with a shadow mask to allow positional mapping of the sample. (b) I–V characteristics acquired from within a specific mask aperture showing n-type behaviour. (c) STM acquired in the same mask aperture as 7.17b with both bright and dark defects observed. (d) STS acquired in the same mask aperture as 7.17b showing the Fermi level close to the conduction band. (e) XPS of the Mo 3d and S 2s core-levels observed. (e) XPS of the Mo 3d and S 2s core-levels acquired from the same mask aperture as 7.17b consistent with n-type MoS₂. (f) STM tip scanning in a different shadow mask aperture. (g) I–V characteristics acquired from within a different mask aperture to 7.17b showing p-type behaviour. (h) STM acquired in the same mask aperture as 7.17g with a comparatively high concentration of dark defects observed. (i) STS acquired in the same mask aperture as 7.17g showing the Fermi level position more than 1 eV away from the conduction band. (j) XPS of the Mo 3d and S 2s core-levels acquired from the same mask aperture as 7.17g showing lower binding energy than 7.17e. STM scale bars are 20 nm, and the scanning conditions are (c) −1.5 V, 0.1 nA, (h) 1.5 V, 1 nA. (Reprinted with permission from McDonnell, S. et al. Defect dominated doping and contact resistance in MoS₂. *ACS Nano* 8;2014:2880–2888. Copyright 2011 American Chemical Society.)

7.4.2 Future Directions

The need for high-quality MoS$_2$ synthesis is clear and the current challenges appear to be minimising defects (sulphur/molybdenum vacancies, anti-sites and interstitials), minimising impurities, maximising the grain size and achieving uniform large-area growth with controllable thickness. It remains to be seen whether the purity and large grain size criteria can be met by CVD-based processes, but it is clear that significant strides have been made to demonstrate that this technique does offer large-area growth capability with accurate thickness control. The PVD method, specifically VDWE, offers the potential for higher purity and has already been shown to deliver rotational alignment with the substrate. However, this technique is primarily employed for Se- and Te-based TMDs rather than MoS$_2$ where the associated vapour pressures are suitable.

Finally, it is interesting to note that, as studies expand from MoS$_2$ to include other TMDs, it has become clear that many of these materials offer their own unique challenges with respect to surface and interface quality. The stability of these materials varies widely and as such the surface engineering advances that are being made on MoS$_2$ will not necessarily be immediately transferable to the whole family of TMDs. This represents numerous opportunities for improved materials and further discovery.

Acknowledgements

The authors acknowledge the extensive contributions by their colleagues and research assistants at the University of Texas at Dallas who performed many experiments presented in this work and cited in the references. The authors also thank Professor M. Quevedo (UT Dallas) and Professor H. Alshareef (KAUST) for the HR-RBS analysis. This work is supported in part by the Center for Low Energy Systems Technology (LEAST), one of six centres supported by the STARnet phase of the Focus Center Research Program (FCRP), a Semiconductor Research Corporation programme sponsored by MARCO and DARPA. It is also supported by the SWAN Center, an SRC centre sponsored by the Nanoelectronics Research Initiative and NIST. The work was also supported in part by the National Science Foundation (NSF) award no. 1407765 under the US–Ireland R&D Partnership.

References

Abe, H., K. Kataoka, K. Ueno and A. Koma. Scanning tunneling microscope observation of the metal-adsorbed layered semiconductor surfaces. *Japanese Journal of Applied Physics, Part 1: Regular Papers and Short Notes and Review Papers* 34;1995:3342–3345.

Addou, R., L. Colombo and R.M. Wallace. Surface defects on natural MoS$_2$. *ACS Applied Materials and Interfaces* 2015a;7:11921–11929.

Addou, R., S. McDonnell, D. Barrera et al. Impurities and electronic property variations of natural MoS$_2$ crystal surfaces. *ACS Nano* 2015b;9: 9124–9133.

Altibelli, A., C. Joachim and P. Sautet. Interpretation of STM images: The MoS_2 surface. *Surface Science* 367;1996:209–220.

Amani, M., M.L. Chin, A. Glen Birdwell et al. Electrical performance of monolayer MoS_2 field-effect transistors prepared by chemical vapor deposition. *Applied Physics Letters* 102;2013:193107.

Amani, M., D.-H. Lien, D. Kiriya et al. Near-unity photoluminescence quantum yield in MoS_2. *Science*, 2015; in press.

An, V. and Y. Irtegov. Tribological properties of nanolamellar MoS_2 doped with copper nanoparticles. *Journal of Nanomaterials* 2014;2014:731073.

Ataca, C. and S. Ciraci. Dissociation of H_2O at the vacancies of single-layer MoS_2. *Physical Review B* 85;2012:195410.

Ataca, C., M. Topsakal, E. Akturk and S. Ciraci. A comparative study of lattice dynamics of three- and two-dimensional MoS_2. *The Journal of Physical Chemistry C* 115;2011:16354–16361.

Azcatl, A., S. McDonnell, K.C. Santosh et al. MoS_2 functionalization for ultra-thin atomic layer deposited dielectrics. *Applied Physics Letters* 104;2014:111601.

Azcatl, A., K.C. Santosh, X. Peng et al. HfO_2 on $UV-O_3$ exposed transition metal dichalcogenides: Interfacial reactions study. *2D Materials* 2;2015:014004.

Banerjee, S.K., L.F. Register, E. Tutuc, D. Reddy and A. MacDonald. Bilayer pseudo-spin field-effect transistor (BiSFET): A proposed new logic device. *IEEE Electron Device Letters* 30;2009:158–160.

Bao, W.Z., X.H. Cai, D. Kim, K. Sridhara and M.S. Fuhrer. High mobility ambipolar MoS_2 field-effect transistors: Substrate and dielectric effects. *Applied Physics Letters* 102;2013:042104.

Batzill, M., K. Katsiev, D.J. Gaspar and U. Diebold. Variations of the local electronic surface properties of $TiO_2(110)$ induced by intrinsic and extrinsic defects. *Physical Review B* 66;2002:235401.

Baunack, S. and A. Zehe. A study of UV/ozone cleaning procedure for silicon surfaces. *Physica Status Solidi (a)* 115;1989:223–227.

Bernardi, M., M. Palummo and J.C. Grossman. Extraordinary sunlight absorption and one nanometer thick photovoltaics using two-dimensional monolayer materials. *Nano Letters* 13;2013:3664–3670.

Black, A.L., R.W. Dunster and J.V. Sanders. Molybdenum disulphide deposits – Their formation and characteristics on steel balls. *Wear* 9;1966:462–476.

Black, A.L., R.W. Dunster, J.V. Sanders and F.K. McTaggart. Molybdenum bisulphide deposits – Their formation and characteristics on automotive engine parts. *Wear* 10;1967:17–32.

Brainard, W.A. *The Thermal Stability and Friction of the Disulfides, Diselenides, and Ditellurides of Molybdenum and Tungsten in Vacuum (10^{-9} to 10^{-6} Torr)*. National Aeronautics and Space Administration, Technical Note TN D-1541, Washington, DC, USA, 1968.

Brudnyi, A.I. and A.F. Karmadonov. Structure of molybdenum disulphide lubricant film. *Wear* 33;1975:243–249.

Chakraborty, B., A. Bera, D.V.S. Muthu et al. Symmetry-dependent phonon renormalization in monolayer MoS_2 transistor. *Physical Review B* 85;2012:161403.

Chan, J., A. Venugopal, A. Pirkle et al. Reducing extrinsic performance-limiting factors in graphene grown by chemical vapor deposition. *ACS Nano* 6;2012:3224–3229.

Cheng, L., X. Qin, A.T. Lucero et al. Atomic layer deposition of a high-k dielectric on MoS_2 using trimethylaluminum and ozone. *ACS Applied Materials and Interfaces* 6;2014:11834–11838.

Chrissafis, K., M. Zamani, K. Kambas et al. Structural studies of MoS$_2$ intercalated by lithium. *Materials Science and Engineering: B* 3;1989:145–151.

Cooper, H.S. and V.R. Damerell. Lubricant. US Patent 2,156,803, 1939.

Cui, N.-Y., N.M.D. Brown and A. McKinley. An AFM study of the topography of natural MoS$_2$ following treatment in an RF–oxygen plasma. *Applied Surface Science* 151;1999:17–28.

Das, A., S. Pisana, B. Chakraborty et al. Monitoring dopants by Raman scattering in an electrochemically top-gated graphene transistor. *Nature Nanotechnology* 3;2008:210–215.

Das, S., H.-Y. Chen, A.V. Penumatcha and J. Appenzeller. High performance multilayer MoS$_2$ transistors with scandium contacts. *Nano Letters* 13;2012:100–105.

Dickinson, R.G. and L. Pauling. The crystal structure of molybdenite. *Journal of the American Chemical Society* 45;1923:1466–1471.

Dines, M.B. Lithium intercalation via n-butyllithium of the layered transition metal dichalcogenides. *Materials Research Bulletin* 10;1975:287–291.

Dolui, K., I. Rungger, C. Das Pemmaraju and S. Sanvito. Possible doping strategies for MoS$_2$ monolayers: An *ab initio* study. *Physical Review B* 88;2013:075420.

Donnet, C. Advanced solid lubricant coatings for high vacuum environments. *Surface and Coatings Technology* 80;1996:151–156.

Durbin, T.D., J.R. Lince, S.V. Didziulis, D.K. Shuh and J.A. Yarmoff. Soft-x-ray photoelectron-spectroscopy study of the interaction of Cr with MoS$_2$(0001). *Surface Science* 302;1994:314–328.

Durbin, T.D., J.R. Lince and J.A. Yarmoff. Chemical interaction of thin Cr films with the MoS$_2$(0001) surface studied by x-ray photoelectron spectroscopy and scanning Auger microscopy. *Journal of Vacuum Science and Technology A: Vacuum, Surfaces, and Films* 10;1992:2529–2534.

Eda, G. and S.A. Maier. Two-dimensional crystals: Managing light for optoelectronics. *ACS Nano* 7;2013:5660–5665.

Eda, G., H. Yamaguchi, D. Voiry et al. Photoluminescence from chemically exfoliated MoS$_2$. *Nano Letters* 11;2011:5111–5116.

Esmaeili-Rad, M.R. and S. Salahuddin. High performance molybdenum disulfide amorphous silicon heterojunction photodetector. *Scientific Reports* 3;2013:2345.

Fallahazad, B., S. Kim, L. Colombo and E. Tutuc. Dielectric thickness dependence of carrier mobility in graphene with HfO$_2$ top dielectric. *Applied Physics Letters* 97;2010:123105.

Farimani, A.B., K. Min and N.R. Aluru. DNA base detection using a single-layer MoS$_2$. *ACS Nano* 8;2014:7914–7922.

Fivaz, R. and E. Mooser. Mobility of charge carriers in semiconducting layer structures. *Physical Review* 163;1967:743.

Fontana, M., T. Deppe, A.K. Boyd et al. Electron–hole transport and photovoltaic effect in gated MoS$_2$ Schottky junctions. *Scientific Reports* 3;2013:1634.

Frindt, R.F. Single crystals of MoS$_2$ several molecular layers thick. *Journal of Applied Physics* 37;1966:1928–1929.

Frindt, R.F. and A.D. Yoffe. Physical properties of layer structures: Optical properties and photoconductivity of thin crystals of molybdenum disulphide. *Proceedings of the Royal Society of London. Series A. Mathematical and Physical Sciences* 273;1963:69–83.

Fuhr, J.D., A. Saúl and J.O. Sofo. Scanning tunneling microscopy chemical signature of point defects on the MoS$_2$ (0001) surface. *Physical Review Letters* 92;2004:026802.

Fuhrer, M.S. and J. Hone. Measurement of mobility in dual-gated MoS_2 transistors. *Nature Nanotechnology* 8;2013:146–147.

Furimsky, E. Role of MoS_2 and WS_2 in hydrodesulfuritation. *Catalysis Reviews Science and Engineering* 22;1980:371–400.

Golden, J., M. McMillan, R.T. Downs et al. Rhenium variations in molybdenite (MoS_2): Evidence for progressive subsurface oxidation. *Earth and Planetary Science Letters* 366;2013:1–5.

Gong, C., L. Colombo, R.M. Wallace and K. Cho. The unusual Fermi level pinning mechanism at metal-MoS_2 interfaces. *Nano Letters* 14;2014:1714.

Gong, C., H.C. Floresca, D. Hinojos et al. Rapid selective etching of PMMA residues from transferred graphene by carbon dioxide. *The Journal of Physical Chemistry C* 117;2013a:23000–23008.

Gong, C., C. Huang, J. Miller et al. Metal contacts on physical vapor deposited monolayer MoS_2. *ACS Nano* 7;2013b:11350–11357.

Goodfellow, K.M., R. Beams, C. Chakraborty, L. Novotny and A.N. Vamivakas. Integrated nanophotonics based on nanowire plasmons and atomically thin material. *Optica* 1;2014:149–152.

Ha, J.S., H.-S. Roh, S.-J. Park, J.-Y. Yi and E.-H. Lee. Scanning tunneling microscopy investigation of the surface structures of natural MoS_2. *Surface Science* 315;1994:62–68.

Hansen, R.W.C., M. Bissen, D. Wallace, J. Wolske and T. Miller. Ultraviolet/ozone cleaning of carbon-contaminated optics. *Applied Optics* 32;1993:4114–4116.

Hausmann, D.M., E. Kim, J. Becker and R.G. Gordon. Atomic layer deposition of hafnium and zirconium oxides using metal amide precursors. *Chemistry of Materials* 14;2002:4350–4358.

Heckl, W.M., F. Ohnesorge, G. Binnig, M. Specht and M. Hashmi. Ring structures on natural molybdenum disulfide investigated by scanning tunneling and scanning force microscopy. *Journal of Vacuum Science and Technology B* 9;1991:1072–1078.

Hirvonen, J.-P., J. Koskinen, J.R. Jervis and M. Nastasi. Present progress in the development of low friction coatings. *Surface and Coatings Technology* 80;1996:139–150.

Huang, X., Z. Zeng and H. Zhang. Metal dichalcogenide nanosheets: Preparation, properties and applications. *Chemical Society Reviews* 42;2013:1934–1946.

Hwang, W.S., M. Remskar, R. Yan et al. Comparative study of chemically synthesized and exfoliated multilayer MoS_2 field-effect transistors. *Applied Physics Letters* 102;2013:043116.

Inoue, A., T. Komori and K.-I. Shudo. Atomic-scale structures and electronic states of defects on Ar^+-ion irradiated MoS_2. *Journal of Electron Spectroscopy and Related Phenomena* 189;2013:11–18.

Jang, C., S. Adam, J.-H. Chen et al. Tuning the effective fine structure constant in graphene: Opposing effects of dielectric screening on short- and long-range potential scattering. *Physical Review Letters* 101;2008:146805.

Jaramillo, T.F., K.P. Jørgensen, J. Bonde et al. Identification of active edge sites for electrochemical H_2 evolution from MoS_2 nanocatalysts. *Science* 317;2007:100–102.

Jena, D. Tunneling transistors based on graphene and 2-D crystals. *Proceedings of the IEEE* 101;2013:1585–1602.

Jena, D. and A. Konar. Enhancement of carrier mobility in semiconductor nanostructures by dielectric engineering. *Physical Review Letters* 98;2007:136805.

Joensen, P., R.F. Frindt and S.R. Morrison. Single-layer MoS_2. *Materials Research Bulletin* 21;1986:457–461.

Kam, K.K. and B.A. Parkinson. Detailed photocurrent spectroscopy of the semiconducting group VIB transition metal dichalcogenides. *The Journal of Physical Chemistry* 86;1982:463–467.

Kang, K., S. Xie, L. HUang et al., High-mobility three-atom-thick semiconducting films with wafer-scale homogeneity. *Nature* 520;2015:656–660.

Kappera, R., D. Voiry, S.E. Yalcin et al. Phase-engineered low-resistance contacts for ultrathin MoS_2 transistors. *Nature Materials* 13;2014:1128–1134.

Karunadasa, H.I., E. Montalvo, Y.J. Sun et al. A molecular MoS_2 edge site mimic for catalytic hydrogen generation. *Science* 335;2012:698–702.

Kautek, W., H. Gerischer and H. Tributsch. The role of carrier diffusion and indirect optical transitions in the photoelectrochemical behavior of layer type d-band semiconductors. *Journal of the Electrochemical Society* 127;1980:2471–2478.

Kim, S., A. Konar, W.S. Hwang et al. High-mobility and low-power thin-film transistors based on multilayer MoS_2 crystals. *Nature Communications* 3;2012:1011.

Koma, A., K. Sunouchi and T. Miyajima. Fabrication of ultrathin heterostructures with van der Waals epitaxy. *Journal of Vacuum Science and Technology B: Microelectronics and Nanometer Structures* 3;1985:724.

Kuc, A. Low-dimensional transition-metal dichalcogenides. *Chemical Modelling* 11;2014:1–29.

Kumar, A. and P.K. Ahluwalia. Electronic structure of transition metal dichalcogenides monolayers $1H$-MX_2 (M = Mo, W; X = S, Se, Te) from *ab-initio* theory: New direct band gap semiconductors. *The European Physical Journal B – Condensed Matter and Complex Systems* 85;2012:1–7.

Lee, B., G. Mordi, M.J. Kim et al. Characteristics of high-k Al_2O_3 dielectric using ozone-based atomic layer deposition for dual-gated graphene devices. *Applied Physics Letters* 97;2010a:043107.

Lee, B., S.-Y. Park, H.-C. Kim et al. Conformal Al_2O_3 dielectric layer deposited by atomic layer deposition for graphene-based nanoelectronics. *Applied Physics Letters* 92;2008:203102.

Lee, C., H. Yan, L.E. Brus et al. Anomalous lattice vibrations of single- and few-layer MoS_2. *ACS Nano* 4;2010b:2695–2700.

Lee, Y.H., X.Q. Zhang, W.J. Zhang et al. Synthesis of large-area MoS_2 atomic layers with chemical vapor deposition. *Advanced Materials* 24;2012:2320–2325.

Li, H., Z. Yin, Q. He et al. Fabrication of single- and multilayer MoS_2 film-based field-effect transistors for sensing NO at room temperature. *Small* 8;2012a:63–67.

Li, H., Q. Zhang, C.C.R. Yap et al. From bulk to monolayer MoS_2: Evolution of Raman scattering. *Advanced Functional Materials* 22;2012b:1385–1390.

Li, Y.G., Y.L. Li, C.M. Araujo, W. Luo and R. Ahuja. Single-layer MoS_2 as an efficient photocatalyst. *Catalysis Science and Technology* 3;2013:2214–2220.

Lin, Y.-C., D.O. Dumcenco, Y.-S. Huang and K. Suenaga. Atomic mechanism of the semiconducting-to-metallic phase transition in single-layered MoS_2. *Nature Nanotechnology* 9;2014:391–396.

Lin, Y.-C., C.-C. Lu, C.-H. Yeh et al. Graphene annealing: How clean can it be? *Nano Letters* 12;2011:414–419.

Lince, J.R., D.J. Carré and P.D. Fleischauer. Schottky-barrier formation on a covalent semiconductor without Fermi-level pinning: The metal–MoS_2 (0001) interface. *Physical Review B* 36;1987:1647.

Ling, X., Y.-H. Lee, Y. Lin et al. Role of the seeding promoter in MoS_2 growth by chemical vapor deposition. *Nano Letters* 14;2014:464–472.

Liu, D., Y. Guo, L. Fang and J. Robertson. Sulfur vacancies in monolayer MoS_2 and its electrical contacts. *Applied Physics Letters* 103;2013:183113.

Liu, H., A.T. Neal and P.D. Ye. Channel length scaling of MoS$_2$ MOSFETs. *ACS Nano* 6;2012a:8563–8569.

Liu, H., K. Xu, X. Zhang and P.D. Ye. The integration of high-k dielectric on two-dimensional crystals by atomic layer deposition. *Applied Physics Letters* 100;2012b:152115.

Liu, H. and P.D. Ye. MoS$_2$ dual-gate MOSFET with atomic-layer-deposited Al$_2$O$_3$ as top-gate dielectric. *IEEE Electron Device Letters* 33;2012:546–548.

Liu, X., T. Galfsky, Z. Sun et al. Strong light–matter coupling in two-dimensional atomic crystals. *Nature Photonics* 9;2015:30–34.

Loo, A.H., A. Bonanni, A. Ambrosi and M. Pumera. Molybdenum disulfide (MoS$_2$) nanoflakes as inherently electroactive labels for DNA hybridization detection. *Nanoscale* 6;2014:11971–11975.

Lopez-Sanchez, O., E.A. Llado, V. Koman et al. Light generation and harvesting in a van der Waals heterostructure. *ACS Nano* 8;2014:3042–3048.

Lui, C.H., A.J. Frenzel, D.V. Pilon et al. Trion induced negative photoconductivity in monolayer MoS$_2$. *Physical Review Letters* 113;2014:166801.

Lv, R., J.A. Robinson, R.E. Schaak et al. Transition metal dichalcogenides and beyond: Synthesis, properties and applications of single- and few-layer nanosheets. *Accounts of Chemical Research* 48;2015:56.

Ma, N. and D. Jena. Charge scattering and mobility in atomically thin semiconductors. *Physical Review X* 4;2014:011043.

Magonov, S.N. and M.-H. Whangbo. STM and AFM images of layered inorganic compounds. In *Surface Analysis with STM and AFM*, pp. 83–111. Wiley-VCH Verlag GmbH, Weinheim, Germany, 2007a.

Magonov, S.N. and M.-H. Whangbo. STM images associated with point defects of layered inorganic compounds. In *Surface Analysis with STM and AFM*, pp. 113–134. Wiley-VCH Verlag GmbH, Weinheim, Germany, 2007b.

Mak, K.F., C. Lee, J. Hone, J. Shan and T.F. Heinz. Atomically thin MoS$_2$: A new direct-gap semiconductor. *Physical Review Letters* 105;2010:136805.

Martin, J.-M., C. Donnet, T.L. Mogne and T. Epicier. Superlubricity of molybdenum disulphide. *Physical Review B* 48;1993:10583.

Mattheiss, L.F. Band structures of transition-metal-dichalcogenide layer compounds. *Physical Review B* 8;1973:3719.

Matthes, T.W., C. Sommerhalter, A. Rettenberger et al. Imaging of dopants in surface and sub-surface layers of the transition metal dichalcogenides WS$_2$ and WSe$_2$ by scanning tunneling microscopy. *Applied Physics A* 66;1998:1007–1011.

McClintock, J.A., R.A. Wilson and N.E. Byer. UV–ozone cleaning of GaAs for MBE. *Journal of Vacuum Science and Technology* 20;1982:241–242.

McCreary, K.M., A.T. Hanbicki, J.T. Robinson et al. Large-area synthesis of continuous and uniform MoS$_2$ monolayer films on graphene. *Advanced Functional Materials* 24;2014:6449–6454.

McDonnell, S., R. Addou, C. Buie, R.M. Wallace and C.L. Hinkle. Defect dominated doping and contact resistance in MoS$_2$. *ACS Nano* 8;2014:2880–2888.

McDonnell, S., A. Pirkle, J. Kim, L. Colombo and R.M. Wallace. Trimethyl-aluminum and ozone interactions with graphite in atomic layer deposition of Al$_2$O$_3$. *Journal of Applied Physics* 112;2012:104110.

McDonnell, S., B. Brennan, A. Azcatl et al. HfO$_2$ on MoS$_2$ by atomic layer deposition: Adsorption mechanisms and thickness scalability. *ACS Nano* 7;2013:10354–10361.

McDonnell, S., C. Smyth, C.L.Hinkle and R.M.Wallace, MoS$_2$-titanium contact interface reactions, 2015; submitted.

McGovern, I.T., E. Dietz, H.H. Rotermund et al. Soft x-ray photoemission spectroscopy of metal–molybdenum bisulphide interfaces. *Surface Science* 152;1985:1203–1212.

Molina-Sanchez, A. and L. Wirtz. Phonons in single-layer and few-layer MoS$_2$ and WS$_2$. *Physical Review B* 84;2011:155413.

Moulder, J.F., W.F. Stickle, P.E. Sobol and K.D. Bomben. *Handbook of X-Ray Photoelectron Spectroscopy: A Reference Book of Standard Data for Use in X-Ray Photoelectron Spectroscopy.* Physical Electronics Division, Perkin-Elmer Corp., Eden Praire, Minnesota, USA, 1992.

Murata, H., K. Kataoka and A. Koma. Scanning tunneling microscope images of locally modulated structures in layered materials, MoS$_2$ (0001) and MoSe$_2$ (0001), induced by impurity atoms. *Surface Science* 478;2001:131–144.

Nan, H., Z. Wang, W. Wang et al. Strong photoluminescence enhancement of MoS$_2$ through defect engineering and oxygen bonding. *ACS Nano* 8;2014: 5738–5745.

Neal, A.T., H. Liu, J.J. Gu and P.D. Ye. Metal contacts to MoS$_2$: A two-dimensional semiconductor. *2012 70th Annual IEEE Device Research Conference (DRC),* Digest, University Park, PA, USA, 2012. DOI: 10.1109/DRC.2012.6256928.

Novoselov, K.S., D. Jiang, F. Schedin et al. Two-dimensional atomic crystals. *Proceedings of the National Academy of Sciences of the United States of America* 102;2005:10451–10453.

Papageorgopoulos, C.A. and W. Jaegermann. Li intercalation across and along the van der Waals surfaces of MoS$_2$(0001). *Surface Science* 338;1995:83–93.

Park, J.B., C. Brian France and B.A. Parkinson. Scanning tunneling microscopy investigation of nanostructures produced by Ar$^+$ and He$^+$ bombardment of MoS$_2$ surfaces. *Journal of Vacuum Science and Technology B: Microelectronics and Nanometer Structures* 23;2005:1532–1542.

Permana, H., S. Lee and K.Y. Simon Ng. Observation of protrusions and ring structures on MoS$_2$ by scanning tunneling microscopy. *Journal of Vacuum Science and Technology B* 10;1992:2297–2301.

Perrot, E., A. Humbert, A. Piednoir, C. Chapon and C.R. Henry. STM and TEM studies of a model catalyst: Pd/MoS$_2$ (0001). *Surface Science* 445;2000:407–419.

Pirkle, A., J. Chan, A. Venugopal et al. The effect of chemical residues on the physical and electrical properties of chemical vapor deposited graphene transferred to SiO$_2$. *Applied Physics Letters* 99;2011:122108.

Pirkle, A., S. McDonnell, B. Lee et al. The effect of graphite surface condition on the composition of Al$_2$O$_3$ by atomic layer deposition. *Applied Physics Letters* 97;2010:082901.

Poindexter, E.H. and P.J. Caplan. Characterization of Si/SiO$_2$ interface defects by electron spin resonance. *Progress in Surface Science* 14;1983:201–294.

Prins, R., V.H.J. De Beer and G.A. Somorjai. Structure and function of the catalyst and the promoter in Co–Mo hydrodesulfurization catalysts. *Catalysis Reviews – Science and Engineering* 31;1989:1–41.

Py, M.A. and R.R. Haering. Structural destabilization induced by lithium intercalation in MoS$_2$ and related compounds. *Canadian Journal of Physics* 61;1983:76–84.

Qiu, D.Y., H. Felipe and S.G. Louie. Optical spectrum of MoS$_2$: Many-body effects and diversity of exciton states. *Physical Review Letters* 111;2013:216805.

Radisavljevic, B., A. Radenovic, J. Brivio, V. Giacometti and A. Kis. Single-layer MoS$_2$ transistors. *Nature Nanotechnology* 6;2011a:147–150.

Radisavljevic, B., M.B. Whitwick and A. Kis. Integrated circuits and logic operations based on single-layer MoS$_2$. *ACS Nano* 5;2011b:9934–9938.

Sandoval, S.J., D. Yang, R.F. Frindt and J.C. Irwin. Raman study and lattice dynamics of single molecular layers of MoS$_2$. *Physical Review B* 44;1991:3955.

Santosh, K.C., R.C. Longo, R. Addou, R.M. Wallace and K. Cho. Impact of intrinsic atomic defects on the electronic structure of MoS_2 monolayers. *Nanotechnology* 25;2014:375703.

Sarkar, D., W. Liu, X.J. Xie et al. MoS_2 field-effect transistor for next-generation label-free biosensors. *ACS Nano* 8;2014:3992–4003.

Scharf, T.W., S.V. Prasad, T.M. Mayer, R.S. Goeke and M.T. Dugger. Atomic layer deposition of tungsten disulphide solid lubricant thin films. *Journal of Materials Research* 19;2004:3443–3446.

Schlaf, R., D. Louder, M.W. Nelson and B.A. Parkinson. Influence of electrostatic forces on the investigation of dopant atoms in layered semiconductors by scanning tunneling microscopy/spectroscopy and atomic force microscopy. *Journal of Vacuum Science and Technology A* 15;1997:1466–1472.

Schonfeld, B., J.J. Huang and S.C. Moss. Anisotropic mean-square displacements (MSD) in single-crystals of 2H- and 3R-MoS_2. *Acta Crystallographica Section B: Structural Science* 39;1983:404–407.

Seabaugh, A.C. and Q. Zhang. Low-voltage tunnel transistors for beyond CMOS logic. *Proceedings of the IEEE* 98;2010:2095–2110.

Shafer, M.W. Electron-paramagnetic-resonance studies on arsenic acceptors in natural (2H) and synthetic (3R) MoS_2 crystals. *Physical Review B* 8;1973:615.

Shi, Y.M., W. Zhou, A.Y. Lu et al. van der Waals epitaxy of MoS_2 layers using graphene as growth templates. *Nano Letters* 12;2012:2784–2791.

Shi, Y., J.-K. Huang, L. Jin et al. Selective decoration of Au nanoparticles on monolayer MoS_2 single crystals. *Scientific Reports* 3;2013:1839.

Sik Hwang, W., M. Remskar, R. Yan et al. Comparative study of chemically synthesized and exfoliated multilayer MoS_2 field-effect transistors. *Applied Physics Letters* 102;2013:043116-1–043116-3.

Sivasubramaniani, P., M.A. Quevedo-Lopez, T.H. Lee et al. Interdiffusion studies of high-k gate dielectric stack constituents. In *Defects in High-k Gate Dielectric Stacks*, E. Gusev, Ed., pp. 135–146. Springer, Dodrecht, Netherlands, 2006.

Sommerhalter, C., T.W. Matthes, J. Boneberg, M.C. Lux-Steiner and P. Leiderer. Investigation of acceptors in p-type WS_2 by standard and photo-assisted scanning tunneling microscopy/spectroscopy. *Applied Surface Science* 144;1999:564–569.

Splendiani, A., L. Sun, Y. Zhang et al. Emerging photoluminescence in monolayer MoS_2. *Nano Letters* 10;2010:1271–1275.

Stupian, G.W. and M.S. Leung. Imaging of MoS_2 by scanning tunneling microscopy. *Applied Physics Letters* 51;1987:1560–1562.

Sze, S.M. and K.K. Ng. *Physics of Semiconductor Devices*. Wiley, Hoboken, New Jersey, USA, 2006.

Tacchini, I., E. Terrado, A. Anson and M.T. Martinez. Preparation of a TiO_2–MoS_2 nanoparticle-based composite by solvothermal method with enhanced photoactivity for the degradation of organic molecules in water under UV light. *Micro and Nano Letters* 6;2011:932–936.

Takahashi, N. and K. Okada. Microscopical study of frictional properties of molybdenum disulphide. *Wear* 33;1975:153–167.

Tan, L.K., B. Liu, J.H. Teng et al. Atomic layer deposition of a MoS_2 film. *Nanoscale* 6;2014:10584–10588.

Tao, J., T. Luttrell and M. Batzill. A two-dimensional phase of TiO_2 with a reduced bandgap. *Nature Chemistry* 3;2011:296–300.

Tarasov, A., P.M. Campbell, M.-Y. Tsai et al. Highly uniform trilayer molybdenum disulfide for wafer-scale device fabrication. *Advanced Functional Materials* 24;2014:6389–6400.

Tielsch, B.J. and J.E. Fulghum. Differential charging in XPS. Part III. A comparison of charging in thin polymer overlayers on conducting and non-conducting substrates. *Surface and Interface Analysis* 25;1997:904–912.

Thoan, N.H., K. Keunen, V.V.Afanas'ev and A. Stesmans. Interface state energy distributions and Pb defects at Si(110)/SiO$_2$ interfaces: Comparison to (111) and (100) silicon orientations. *Journal of Applied Physics* 109;2011:013710.

Tiong, K.K. and T.S. Shou. Anisotropic electrolyte electroreflectance study of rhenium-doped MoS$_2$. *Journal of Physics: Condensed Matter* 12;2000:5043.

Tongay, S., J. Suh, C. Ataca et al. Defects activated photoluminescence in two-dimensional semiconductors: Interplay between bound, charged, and free excitons. *Scientific Reports* 3;2013:2657.

Tsai, M.-L., S.-H. Su, J.-K. Chang et al. Monolayer MoS$_2$ heterojunction solar cells. *ACS Nano* 8;2014:8317–8322.

van der Zande, A.M., P.Y. Huang, D.A. Chenet et al. Grains and grain boundaries in highly crystalline monolayer molybdenum disulphide. *Nature Materials* 12;2013:554–561.

Vig, J.R. UV/ozone cleaning of surfaces. *Journal of Vacuum Science and Technology A* 3;1985:1027–1034.

Vig, J.R. and J. LeBus. UV/ozone cleaning of surfaces. *IEEE Transactions on Parts, Hybrids, and Packaging* 12;1976:365–370.

Vishwanath, S., X. Liu, S. Rouvimov et al. Comprehensive structural and optical characterization of MBE grown MoSe$_2$ on graphite, CaF$_2$ and graphene. *2D Materials* 2;2015:024007.

Voumard, P., A. Savan and E. Pflüger. Advances in solid lubrication with MoS$_2$ multilayered coatings. *Lubrication Science* 13;2001:135–145.

Wang, H., Z. Lu, S. Xu et al. Electrochemical tuning of vertically aligned MoS$_2$ nanofilms and its application in improving hydrogen evolution reaction. *Proceedings of the National Academy of Sciences* 110;2013:19701–19706.

Wang, Q.H., K. Kalantar-Zadeh, A. Kis, J.N. Coleman and M.S. Strano. Electronics and optoelectronics of two-dimensional transition metal dichalcogenides. *Nature Nanotechnology* 7;2012:699–712.

Wilson, J.A. and A.D. Yoffe. Transition metal dichalcogenides discussion and interpretation of observed optical, electrical and structural properties. *Advances in Physics* 18;1969:193.

Winer, W.O. Molybdenum disulfide as a lubricant: A review of the fundamental knowledge. *Wear* 10;1967:422–452.

Wypych, F., T. Weber and R. Prins. Scanning tunneling microscopic investigation of 1T-MoS$_2$. *Chemistry of Materials* 10;1998:723–727.

Wypych, F. and R. Schöllhorn. 1T-MoS$_2$, a new metallic modification of molybdenum disulfide. *Journal of the Chemical Society, Chemical Communications* 19;1992:1386–1388.

Xia, F., H. Wang, D. Xiao, M. Dubey and A. Ramasubramaniam. Two-dimensional material nanophotonics. *Nature Photonics* 8;2014:899–907.

Xiang, Q., J. Yu and M. Jaroniec. Synergetic effect of MoS$_2$ and graphene as cocatalysts for enhanced photocatalytic H$_2$ production activity of TiO$_2$ nanoparticles. *Journal of the American Chemical Society* 134;2012:6575–6578.

Yang, D., S. Jiménez Sandoval, W.M.R. Divigalpitiya, J.C. Irwin and R.F. Frindt. Structure of single-molecular-layer MoS$_2$. *Physical Review B* 43;1991:12053–12056.

Yang, J., S. Kim, W. Choi et al. Improved growth behavior of atomic-layer-deposited high-k dielectrics on multilayer MoS$_2$ by oxygen plasma pretreatment. *ACS Applied Materials and Interfaces* 5;2013:4739–4744.

Ye, Y., Z. Ye, M. Gharghi et al. Exciton-related electroluminescence from monolayer MoS_2. *CLEO: Science and Innovations*, paper STh4B.4;2014.

Yin, Z., X. Zhang, Y. Cai et al. Preparation of MoS_2–MoO_3 hybrid nanomaterials for light-emitting diodes. *Angewandte Chemie* 126;2014:12768–12773.

Yoshida, S., Y. Terada, M. Yokota et al. Direct probing of transient photocurrent dynamics in p-WSe_2 by time-resolved scanning tunneling microscopy. *Applied Physics Express* 6;2013:6601.

Yu, Y., C. Li, Y. Liu et al. Controlled scalable synthesis of uniform, high-quality monolayer and few-layer MoS_2 films. *Scientific Reports* 3;2013:66.

Yue, R., A.T. Barton, H. Zhu et al. $HfSe_2$ thin films: 2D transition metal dichalcogenides grown by molecular beam epitaxy. *ACS Nano* 9;2014:474–480.

Zhan, Y., Z. Liu, S. Najmaei, P.M. Ajayan and J. Lou. Large-area vapor-phase growth and characterization of MoS_2 atomic layers on a SiO_2 substrate. *Small* 8;2012:966–971.

Zhang, X.-J., G. Xue, A. Agarwal et al. Thermal desorption of ultraviolet–ozone oxidized Ge (001) for substrate cleaning. *Journal of Vacuum Science and Technology A* 11;1993:2553–2561.

Zhou, W., X.L. Zou, S. Najmaei et al. Intrinsic structural defects in monolayer molybdenum disulfide. *Nano Letters* 13;2013:2615–2622.

Zhu, W., T. Low, Y.-H. Lee et al. Electronic transport and device prospects of monolayer molybdenum disulphide grown by chemical vapour deposition. *Nature Communications* 5;2014:3087.

Zong, X., H. Yan, G. Wu et al. Enhancement of photocatalytic H_2 evolution on CdS by loading MoS_2 as cocatalyst under visible light irradiation. *Journal of the American Chemical Society* 130;2008:7176–7177.

Zou, X., J. Wang, C.-H. Chiu et al. Interface engineering for high-performance top-gated MoS_2 field-effect transistors. *Advanced Materials* 26;2014:6255–6261.

8

Transition Metal Dichalcogenide Schottky Barrier Transistors
A Device Analysis and Material Comparison

*Joerg Appenzeller, Feng Zhang,
Saptarshi Das and Joachim Knoch*

Contents

2D Materials for Nanoelectronics edited by Michel Houssa, Athanasios Dimoulas and Alessandro Molle © 2016 CRC Press/Taylor & Francis Group, LLC. ISBN: 978-1-4987-0417-5.

8.1 Introduction

Conventional three-terminal devices like metal–oxide–semiconductor field-effect transistors (MOSFETs) are often characterised by an n/p/n or p/n/p doping profile [1]. **Figure 8.1a** displays how degenerately doped source and drain segments may be attached to the gated p-doped channel region of an n-type field effect transistor (n-FET) marked by the two vertical dashed lines. E_c and E_v are the conduction and valence band edge of the semiconductor, respectively. E_{Fs} and E_{Fd} denote the position of the Fermi level in source and in drain. A positive drain voltage moves E_{Fd} relative to E_{Fs} downwards. An 'ohmic' contact with a constant resistance is created in case (a) if the doped source and drain regions are (i) not under gate control, (ii) display a linear dependence of current as a function of drain voltage and (iii) are attached to the channel region such that ideally no carriers are reflected back into the contacts. In contrast, **Figures 8.1b** and **8.2** illustrate the case of metal source/drain contacts with a much larger spacing between the Fermi level and the conduction band edge if compared to **Figure 8.1a**, indicating a large mismatch of the carrier concentrations in the contact and the channel. Most importantly, the absence of a band gap in case of a metal contact allows easy injection of *both* carrier types, that is, electrons and holes, while the band gap in case of **Figure 8.1a** prohibits hole injection to a large extent.

To mimic the scenario of **Figure 8.1a** in terms of electron injection from source with metal contacts, a metal-to-channel interface must exhibit a negative Schottky barrier (SB) as shown in **Figures 8.1b** and **8.2**. While the graphic representation in **Figure 8.1** makes this point rather obvious, it has frequently confused scientists dealing with metal contacts to semiconducting nano-structures. For example, in the context of carbon nanotubes (CNs), only

(a)

(b)

(c)

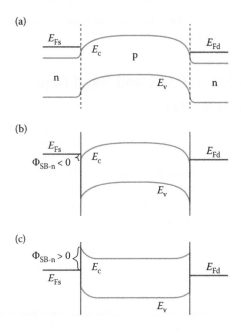

FIGURE 8.1   Top to bottom shows bands for (a) a conventional n/p/n FET, (b) a SB FET with negative barrier and (c) a SB FET with positive barrier.

in 2002, 4 years after the first CN FET was built, it was noted that a negative SB of $\Phi_{SB-n} \approx -0.25$ eV is required to achieve on-current levels in ballistic SB-CNFETs on par with ballistic CN MOSFETs [2]. In reality, negative SBs are rarely obtained and the scenario in most SB-FETs is described by **Figure 8.1c**.

Another point of confusion is related to the impact of doping in SB-FETs. Often it is argued that transparent enough SBs can be created by means of high doping levels. This statement, however, ignores the fact that a doped channel is still under gate control and thus exhibits unavoidably a gate voltage-dependent contact resistance in this case. In fact, doping changes $\log(I_d)$ versus V_{gs} (subthreshold) – characteristics mainly in terms of the threshold voltage when ignoring the impact of doping on scattering in the channel. The injection properties at the contacts are frequently not impacted by doping since the doping-dependent maximum depletion width W_{DM} is typically larger than

FIGURE 8.2   'Zoom-out' of the contact region in **Figure 8.1b**, showing the conduction band edge of the metal contact.

the body thickness t_{body} of the gated channel region – a point that will be discussed in greater detail below.

8.2 How to Extract SBs and Band Gaps

8.2.1 SBs in Nano-Devices

Next, we turn our attention to the device characteristics that are obtained for nano SB-FETs, where 'nano' refers in particular to the body thickness, that is, the thickness of the electrically active channel region. For any applied gate voltage V_{gs}, the current through the device can be determined by evaluating the transmission probability through the conduction band in the channel whose position is determined by the gate voltage (we currently focus on electron conduction only). If we ignore scattering inside the channel as appropriate for a device operating below threshold voltage V_{th},* the transmission from source to drain is determined by the SBs at the contacts. Let us assume that the drain voltage V_{ds} had been sufficiently positive to move the drain barrier out of the electron path. In this case, once carriers are injected from the source into the channel, they count towards the current flow. We start our discussion by evaluating the SB device in its deep off-state. The *yellow* line and star in **Figure 8.3a** and **b**, respectively, capture the situation in this case. Carriers are injected through thermionic emission from source over the barrier posed by the highest conduction band edge position inside the channel depending on the actual temperature defining the 'tail' of the Fermi distribution in source. When a more positive gate voltage is applied, the current increases since the conduction band moves downwards allowing for more carriers injected into the channel (*brown* line and star). The change in band position is ideally (assuming the absence of interface traps and a fully depleted device with a vanishing depletion capacitance C_d [3] as appropriate for most ultrathin body devices) one-to-one dependent on the change in gate voltage. This implies that the current changes exponentially with gate voltage and that the so-called inverse subthreshold slope is given by $d[\log(I_d)]/dV_{gs} = k_B T/q \cdot \ln 10 \approx 60$ mV/dec at room temperature in the thermal branch of the device characteristics [4]. This trend continuous until flat band conditions are reached at $V_{gs} = V_{FB}$ (*green* line and star).

Up to this point, the gate voltage response was no different from what could have been found in a conventional device. However, for gate voltages beyond flat band, the situation in an ultrathin body SB-FET becomes drastically different from its conventional counterpart. The key to understand this difference lies in the characteristic length scale λ (see **Figure 8.3c**) that determines the decay of the potential at the source/drain interface into the semiconducting nano-channel. As discussed by Frank et al. [5], the doping-dependent screening length in a bulk device is given by

---

* V_{th} is the gate voltage that determines the transition between the FET off-state, where the current exponentially depends on gate voltage and the on-state that typically exhibits a power law dependence of current on V_{gs}.

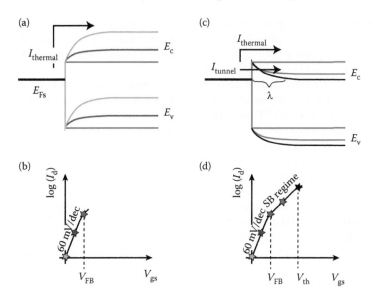

FIGURE 8.3 (a) and (c) A hypothetical band situation at the source contact for various gate voltages. (b) and (d) The corresponding currents and inverse subthreshold slopes as a function of gate voltage, with the star colour being correlated to the respective colour of the bands.

$$\lambda_{\text{bulk}} = \sqrt{\frac{\varepsilon_{\text{body-x}}}{\varepsilon_{\text{ox}}} W_{\text{DM}} t_{\text{ox}}} \qquad (8.1)$$

In **Equation 8.1**, t_{ox} denotes the gate dielectric film thickness, ε_{ox} is the value of the dielectric constant of the gate oxide and $\varepsilon_{\text{body-x}}$ captures the dielectric constant of the channel material in the direction of transport. W_{DM} is the maximum depletion width given by

$$W_{\text{DM}} = \sqrt{\frac{4\varepsilon_{\text{body-y}} \Psi_{\text{B}}}{q N_{\text{A,D}}}} \qquad (8.2)$$

with $\Psi_{\text{B}} = kT/q \ln(N_{\text{A}}/n_{\text{i}})$. $\varepsilon_{\text{body-y}}$ is the dielectric constant of the channel material in the gate direction, n_{i} is the intrinsic carrier concentration of the channel material used and $N_{\text{A,D}}$ is the acceptor or donor doping level. For a doping level of $N_{\text{A,D}} = 10^{18}$ cm^{-3}, $t_{\text{ox}} = 5$ nm, $\varepsilon_{\text{body-x}} = \varepsilon_{\text{body-y}} = 12$ and $\varepsilon_{\text{ox}} = 4$, a typical value of $\lambda_{\text{bulk}} \approx 20$ nm and $W_{\text{DM}} \approx 25$ nm is found for silicon.

On the other hand, if the channel material is thinner than the maximum depletion width, W_{DM} in **Equation 8.1** has to be replaced by the channel (also called body) thickness t_{body} and **Equation 8.1** becomes

$$\lambda_{\text{geo}} = \sqrt{\frac{\varepsilon_{\text{body-x}}}{\varepsilon_{\text{ox}}} t_{\text{body}} t_{\text{ox}}} \qquad (8.3)$$

In this case, the geometrical, doping-independent screening length λ_{geo} rather than λ_{bulk} describes the SB width in **Figure 8.3c**. Using the same parameters as above and a value of $t_{body} = 0.5$ nm for a single-layer transition metal dichalcogenide (TMD) film, $\lambda_{geo} \approx 3$ nm is found.

It is the apparent difference between λ_{bulk} and λ_{geo} for thin body devices in particular at low doping levels that make reevaluating the transport across the SB for a nano SB-FET necessary. While it is justified to ignore tunnelling through the source and drain SBs for most 'bulk-type' devices because of the large λ_{bulk}-value, it is crucially important in order to understand the performance of nano-FETs in the on- *and* off-state to consider the gate voltage-dependent tunnelling through the thin SBs. Thermal assisted tunnelling plays a major role even when the doping level in these SB-FETs is very low and conventional bulk devices would exhibit a large depletion width.

Figure 8.3c and **d** graphically illustrates how the 'below-threshold' region of SB-FETs is impacted by SB-tunnelling. Once the gate voltage is sufficient to move the conduction band edge below the flat band position, in addition to the thermal excitation of electrons 'over' the SB, thermal assisted tunnelling 'through' the SB needs to be considered. In fact, if I_{tunnel} does not exist because λ is too large, the current beyond V_{FB} will not increase. As mentioned above, in case of sufficiently small body thicknesses, λ-values are of the order of nanometres and a gate voltage-dependent tunnelling component adds to $I_{thermal}$. If we first assume a λ-value close to zero, it is clear that the transmission probability for tunnelling T_{tunnel} becomes unity. In this case, the conduction band edge determines – like in the case below flat band – the amount of carriers injected into the channel and the inverse subthreshold slope remains 60 mV/dec at room temperature even above V_{FB}. An increasing λ-value decreases T_{tunnel} and the inverse subthreshold slope (S) now exhibits a value larger than 60 mV/dec, that is, the amount of gate voltage required to change the drain current by one order of magnitude is increased ($S \propto \lambda$ [6]). The general trend of changing λ is captured by **Figure 8.4a**.

On the other hand, when asking the question, how an increase in SB-height Φ_{SB-n} manifests itself in the device characteristics, the answer lies in the fact that it is the current in the gate voltage range between V_{FB} and V_{th} that is impacted by the SB tunnelling current. The smaller Φ_{SB-n}, the smaller this gate voltage range. Since the threshold voltage remains almost* unaffected by the actual SB-height – the threshold voltage is approximately reached in SB-devices when the conduction band edge coincides with the source Fermi level (see **Figure 8.3c** and **d**) – the above means that the flat band voltage occurs the closer to V_{th} the smaller Φ_{SB-n} as illustrated in **Figure 8.4b**.

Note that while we started the above argument assuming a vanishing SB on the drain side by applying a sufficiently large drain voltage V_{ds}, the trends described in **Figures 8.3** and **8.4** hold true also for small V_{ds}. While the exact current level I_d depends on the applied V_{ds}-value, the device off-state below threshold

---

* Note that for large SB-heights and/or large λ-values, the bands in the channel have to be moved 'below' the Fermi level in the source to reach threshold conditions, that is, to enable large enough tunnelling currents.

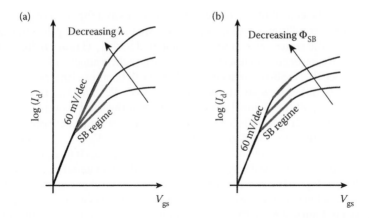

FIGURE 8.4 (a) The impact of decreasing λ on the device characteristics below threshold. (b) The impact of a change in SB height for a given λ-value.

with its current exponentially depending on V_{gs} does not occur substantially different for different V_{ds} when $\log(I_d) - V_{gs}$ characteristics are considered (until carrier injection from the drain becomes relevant – see the following sections).

8.2.2 Ambipolar Device Characteristics

So far, we have focussed on electron injection and the so-called electron branch of the subthreshold characteristics. Correspondingly, V_{th} was in fact V_{th-n}. However, the absence of a band gap in the source and drain regions of an SB-FET allows for both, electron and hole injection, resulting in so-called ambipolar device characteristics. The situation is graphically illustrated in **Figure 8.5**. Using the arguments from **Figure 8.3** now also for hole injection from the drain contact into the valence band of the gated channel, **Figure 8.5a**

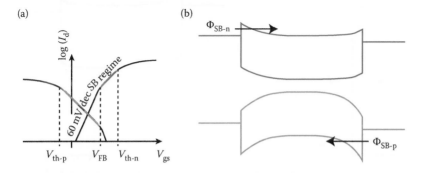

FIGURE 8.5 (a) Qualitative illustration of the contributions of electrons (curve with positive slope) and holes (curve with negative slope) for an SB-FET resulting in ambipolar device characteristics. (b) Electron injection through the source SB and hole injection through Φ_{SB-p} for positive V_{ds}.

summarises the contributions of electrons and holes as a function of gate voltage. Flat band conditions occur for sufficiently small V_{ds} at exactly the same gate voltage V_{FB}. The larger value of Φ_{SB-p} if compared to Φ_{SB-n} (**Figure 8.5b**) manifests itself in a lower current level in the p-branch (light orange) and a larger gate voltage range that is SB-dominated (length of the light orange line) as explained in the context of **Figure 8.4b**. The difference between V_{th-n} and V_{th-p} is a measure of the band gap of the channel material. Indeed, if the band movement would be one-to-one with gate voltage over the entire voltage range between the two threshold voltages, the band gap E_g would be $E_g = q \cdot (V_{th-n} - V_{th-p})$. **Figure 8.6** illustrates the difference in ambipolar device characteristics if Φ_{SB-n} is kept constant and Φ_{SB-p} is increased implying a larger band gap material in the channel ($E_g = \Phi_{SB-n} + \Phi_{SB-p}$). The curve in light blue (1) is a copy of the scenario in **Figure 8.5**. The dark blue characteristics (2) belong to a scenario of a larger band gap as displayed in **Figure 8.6b**. If Φ_{SB-n} is kept constant, V_{th-n} remains the same, while the larger band gap and higher SB to the valence band edge results in

1. A more negative threshold voltage V_{th-p2} (compared to V_{th-p1})

2. A lower minimum current level I_{MIN}

3. A larger gate voltage range over which Φ_{SB-p} determines the inverse subthreshold slope (i.e. from V_{th-p2} to V_{FB} line in **Figure 8.6a**)

4. A lower current level I_p in the p-branch on-state

Note that the discussion above is only valid for a finite value of λ that is not too small. Once λ decreases, both branches of the ambipolar characteristics will show a steeper inverse subthreshold slope value S that eventually becomes 60 mV/dec as discussed in the context of **Figure 8.4a**. This implies I_{MIN}, I_p and I_n increase for decreasing λ and the V-shaped ambipolar characteristics becomes more symmetric, that is, $I_n = I_p$ for $\lambda = 0$.

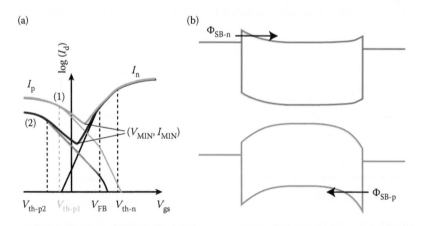

FIGURE 8.6 (a) Ambipolar characteristics for two devices with (1) band gap as in **Figure 8.5** (light blue) and (2) enlarged band gap with larger Φ_{SB-p} if compared with **Figure 8.5** (dark blue). (b) Electron and hole injection.

All of the above-described dependences can be readily expressed quantitatively as discussed in Sections 8.2.3 to 8.2.5. The exact method that is most suitable for the extraction of $\Phi_{SB\text{-}n}$, $\Phi_{SB\text{-}p}$ and E_g depends on a number of aspects that will be considered in the following subsections.

8.2.3 Utilising Temperature and Gate Voltage Dependence to Identify the Flat Band Voltage

After the discussion above, it is clear that determining the flat band voltage V_{FB} is one method to identifying *one* SB of a metal contacted nano-FET. As we will describe now, determining the gate voltage at which the current injected into the channel no longer contains any thermionic tunnelling component allows extracting *the smaller* of the two SBs.

8.2.3.1 Method of SB Extraction

From the discussion in Sections 8.2.1 and 8.2.2, one might conclude that finding V_{FB} is straightforward since it only requires searching for the 'knee' in the $\log(I_d)$ versus V_{gs} – characteristics. In reality, however, the entire inverse subthreshold slope – including the thermal branch – is impacted by finite temperature effects and gradual transitions between the various transport regimes as well as interface trap contributions, making the identification of V_{FB} difficult. A viable approach to extract V_{FB} and thus Φ_{SB} in this case makes use of the distinctly different temperature dependences of the thermal and the SB regime. The technique to extract the flat band voltage from a temperature-dependent measurement of the inverse subthreshold characteristics is characterised by four steps:

1. Measure $\log(I_d) - V_{gs}$ as a function of temperature (**Figure 8.7a**)

2. Create gate voltage-dependent Arrhenius plots (**Figure 8.7b**)

3. Extract 'apparent' barrier versus gate voltage assuming thermal emission theory (**Figure 8.7c**)

4. Determine the 'real' SB-height at V_{FB} (**Figure 8.7d**)

Figure 8.7a shows qualitatively the variation of $\log(I_d) - V_{gs}$ with temperature. It is not important in this context whether or not a 'knee' is visible in the device characteristics. The deeper the device operates in its off-state, the larger the change in current as a function of temperature. In other words, the thermal branch of the device characteristics shows a stronger dependence on temperature than the tunnelling dominated SB regime. This situation is qualitatively illustrated in **Figure 8.7b**. Next, slopes of the Arrhenius plots are analysed using a simple thermal emission model that assumes only injection 'over' a barrier of heights Φ_B. As discussed in the context of CN devices [7], from $I_d \propto \exp(-q\Phi_B/k_B T)$, Φ_B can be extracted as a function of gate voltage as displayed in **Figure 8.7c**. It is obvious from the above discussion that an effective barrier Φ_B extracted in the SB regime – the gate voltage range in which thermal assisted tunnelling impacts the current flow through the device – will always be smaller than the actual SB-height. This is the case since the above I_d-expression assumes thermally activated injection 'over' (not through) a barrier only.

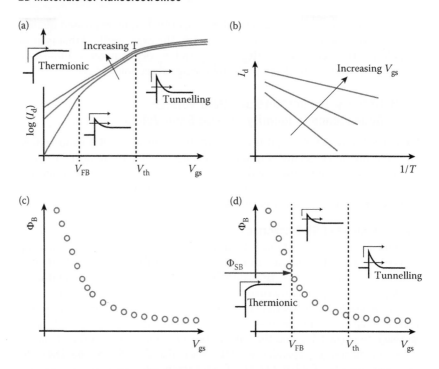

FIGURE 8.7 (a) The qualitative temperature dependence of $\log(I_d)$ versus V_{gs}. (b) Arrhenius plots schematically at various gate voltages. (c) and (d) The trend of 'apparent' barrier height Φ_B as a function of V_{gs}, the different transport regimes and the extraction of the 'real' SB height Φ_{SB}.

For gate voltage values below V_{FB}, as long as the conduction band's response to the external gate voltage is constant, Φ_B changes linearly with V_{gs}, see **Figure 8.3b**. If interface traps are present, the response of the bands to the gate in the device off-state may not be one-to-one, but remains proportional to the gate voltage as long as the impact of interface traps can be captured by a constant interface trap capacitance C_{it}. Once the thermal assisted tunnelling regime is reached for $V_{gs} > V_{FB}$, the linear dependence between Φ_B and V_{gs} no longer prevails as has been shown numerically by us before [7]. As indicated in **Figure 8.7d**, it is this gate voltage point that defines V_{FB} and the actual SB height Φ_{SB}.

An important comment that indicates one limitation of the above approach is in order at this point: The above discussion assumed that at no point ambipolar currents impact the Φ_B-extraction. In order to ensure that the flat band voltage can be properly determined, V_{FB} must not lie close to I_{MIN} (see **Figure 8.6a**). If this condition is not fulfilled, the thermal branch may be impacted by hole transport through the valence band if an electron SB Φ_{SB-n} is supposed to be extracted. If $\Phi_{SB-p} < \Phi_{SB-n}$, the above argument is just reversed in terms of carrier type. We will see in Section 8.2.4 in particular that highly symmetric ambipolar device characteristics require using a different approach to extract SB-height information in SB-devices.

Next, we will apply the above-described method to extract the SB for electron injection Φ_{SB-n} in case of MoS_2.

8.2.3.2 Device Fabrication

All devices characterised in this chapter are fabricated as follows: TMD flakes are exfoliated onto a silicon/silicon dioxide substrate. The thicknesses t_{ox} of the various SiO_2 layers used are 20, 90 and 100 nm, respectively. Next, flakes are mapped out relative to predefined alignment marks on the substrate. An electronic file is created that defines source and drain contacts relative to the TMD positions identified earlier. Electron beam lithography and lift-off are utilised to create source/drain contacts from various metals with different work functions. Channel lengths vary between 150 nm and 4 μm and channel widths between 500 nm and 20 μm and are recorded for the analysis presented in the following sections. The flakes under investigation exhibit thicknesses between 3 and 12 nm as characterised by atomic force microscopy (AFM). **Figure 8.8** shows a typical device layout (top right), optical and scanning electron microscope (SEM) image of a readily fabricated back-gated TMDFET (left) as well an AFM scan (lower right) that is used to determine the respective flake thickness. Devices are characterised in terms of so-called output ($I_d - V_{ds}$), subthreshold ($\log(I_d) - V_{gs}$) and transfer ($I_d - V_{gs}$) characteristics.

8.2.3.3 Determining Φ_{SB} for MoS_2

Next, we employ the above-described method to extract the SB heights for MoS_2 devices with electrodes from scandium, titanium, nickel and platinum. As discussed by us before [8], transistor characteristics of MoS_2 FETs had been

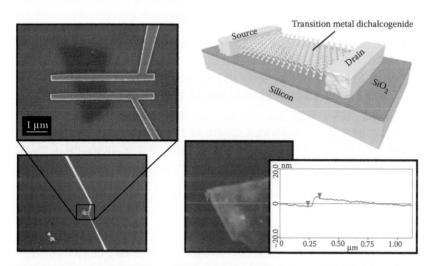

FIGURE 8.8 Left side: An optical and a false coloured SEM image of a back-gated TMDFET – here from exfoliated $MoTe_2$. Top right: The layout of a TMD device schematically. Bottom right: An AFM image and corresponding height measurement of a representative multi-layer TMD flake.

interpreted by other groups as 'ohmic' despite the use of large work function source/drain metals [9–11]. Frequently, it is the apparently linear response of the current I_d to the drain voltage V_{ds} in the device on-state and the clear V_{gs}-dependence of the output characteristics that is taken as evidence for an 'ohmic' contact. From the above discussion, it is clear that this conclusion is not justified since linear $I_d - V_{ds}$ characteristics are readily obtained for devices with small λ_{geo}, that is, for devices with a small body thickness t_{body}, and a gated SB *always* gives rise to a non-constant, meaning gate voltage-dependent contact resistance. In particular, the continuous change of Φ_B with V_{gs} towards very small values as illustrated in **Figure 8.7d** due to substantial tunnelling currents through the thin SB gives rise to a device behaviour that is uncommon when considering bulk type SB devices. Consequently, when analysing subthreshold characteristics for Ni-contacted MoS$_2$ devices exactly the types of results as discussed in the context of **Figure 8.7** are found. For the above device, $t_{body} = 6$ nm, $t_{ox} = 100$ nm and $L = 5$ μm. **Figure 8.9** shows in (a) to (c) the extraction of Φ_{SB-n} with the characteristic change of Φ_B with applied gate voltage in **Figure 8.9c**. **Figure 8.9d** summarises the SB-values found for Sc-, Ti-, Ni- and Pt-contacted MoS$_2$ flakes. As discussed in greater detail in Reference [8], interestingly the response of the SB height to the contact metal work function is even weaker than in the case of SBs formed to silicon, indicating a previously unnoticed strong Fermi-level pinning at the metal-to-MoS$_2$ interface. A comment about the absence of the p-branch in **Figure 8.9a** is in order at this point. From **Figure 8.6**, one may conclude that ambipolar device characteristics must be always observable when source/drain metal contacts are used as source and drain. However, the minimum current level that can be measured is limited by leakage and the capabilities of the measurement set-up. The noise floor (~1 pA) observed in **Figure 8.9a** prevents observing the much smaller expected hole currents in case of MoS$_2$ due to its large band gap and Fermi-level line-up close to the conduction band for all metals used here.

8.2.4 Utilising the Asymmetry of Ambipolar Device Characteristics to Extract the SB Height

The technique described above for the extraction of the SB height works perfectly for unipolar devices which show both the thermal and the SB regimes in the $\log(I_d) - V_{gs}$ characteristics. However, in case of ambipolar devices, the thermal regime of the subthreshold characteristics and in particular the flat band voltage may be hard to identify due to the vicinity of V_{FB} to V_{MIN} (see **Figure 8.6**) masking the linear region of the $\Phi_B - V_{gs}$ characteristics (see **Figure 8.9c**).

8.2.4.1 Method of SB Extraction

In this section, we will discuss how to use the asymmetry of such ambipolar devices in order to determine the SB heights. **Figure 8.10a** shows three different positions of the metal Fermi level with respect to the band gap of a hypothetical semiconducting channel material. Note that a smaller SB height for the electrons is associated with a larger SB height for the holes

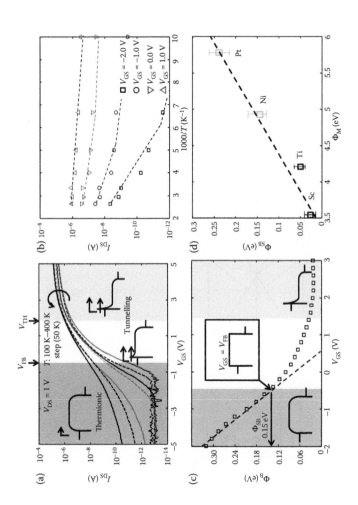

FIGURE 8.9 (a) Experimental subthreshold characteristics of a back-gated MoS$_2$ transistor with Ni contacts at different operating temperatures. (b) An Arrhenius-type plot for various gate voltages. (c) The extracted effective barrier height (Φ_B) as a function of applied gate voltage. (d) Summary of the results in terms of actual SB height Φ_{SB-n} for different work function metals. The dotted line is a guide to the eyes. (Reprinted with permission from S. Das et al., *Nano Letters* **13**, 100. Copyright 2013 American Chemical Society.)

and vice versa. The asymmetry seen in the device characteristics in **Figure 8.10b** is a direct consequence of this simple fact. While the mid-gap alignment ($\Phi_{SB-n-2} = \Phi_{SB-p-2} = E_g/2$) gives perfectly symmetric device characteristics, a slight shift towards either the conduction band (Φ_{SB-n-1}) or the valence bands (Φ_{SB-n-3}) results in asymmetric device characteristics. **Figure 8.10c** shows the ratio of the electron current I_n and the hole current I_p extracted at their respective threshold voltages V_{th-n} or V_{th-p} for these three cases qualitatively.

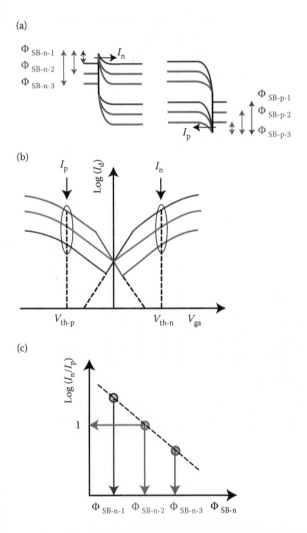

FIGURE 8.10 (a) The band diagram and (b) the qualitative log(I_d) versus V_{gs} corresponding to three different alignments of metal Fermi level with the band gap of the semiconductor. (c) Qualitative depiction of the ratio of electron current to hole current at their respective threshold voltages on a logarithmic scale as a function of the SB height.

As apparent, for the symmetric SB case, the ratio of I_n/I_p is unity. A quantitative version of **Figure 8.10c** as discussed below can therefore be used to determine the SB height for an ambipolar device.

8.2.4.2 Determining Φ_{SB-n} and Φ_{SB-p} for WSe$_2$

Next, we will apply this technique to extract information from the ambipolar WSe$_2$ devices shown in **Figure 8.11**. The black and grey curves correspond to WSe$_2$ devices with Pd and Ni contacts, respectively [12]. **Figure 8.11a** and **b** shows the device characteristics in linear and logarithmic scale. Note that the curves have been shifted relative to the V_{gs}-axis such that the minimum current point (V_{MIN}) coincides with $V_{gs} = 0$ V. Whenever the minimum current is limited by leakage and extends over a certain V_{gs}-range, the symmetry point was identified as V_{MIN}. As seen in **Figure 8.11a** and **b**, the Pd contacted device shows higher hole conduction and the Ni contacted device shows higher electron conduction. This is a direct consequence of the fact that the Fermi level of Pd lines up close to the valence band of WSe$_2$, whereas the Ni Fermi-level line-up is closer to the conduction band of WSe$_2$. As apparent from the flat minimum current region in **Figure 8.11b**, for these devices the thermal branch is beyond the detection limit defined by the gate leakage current ($\sim 10^{-6}$ µA/µm) and hence both the on- and off-state of the FET are dominated by SB-tunnelling. This prevents using the method described in Section 8.2.1 to extract information about V_{FB}. **Figure 8.11c** is the numerical representation of **Figure 8.10c** calculated using the following equations:

$$I_{\text{thermal}} = q \int_{\Phi_{SB-p,n}}^{\infty} M(E)f(E)dE \tag{8.4}$$

$$I_{\text{tunnel}} = q \int_{0}^{\Phi_{SB-p,n}} M(E)T_{\text{WKB}}(E)f(E)dE \tag{8.5}$$

with $M(E) = (2/h^2)\sqrt{2m_{\text{eff}}E}$, $f(E) = 1/(1 + \exp(E/k_BT))$ and $E_g = \Phi_{SB-n} + \Phi_{SB-p}$. The total current is the sum of the thermal and the tunnelling component at threshold as depicted in **Figure 8.3c**:

$$I_d = I_{\text{thermal}} + I_{\text{tunnel}} \tag{8.6}$$

$$T_{\text{WKB}}(E) = \exp\left(-\frac{8\pi}{3h}\sqrt{2m_{\text{eff}}(\Phi_{SB-p,n} - E)^3}\frac{\lambda}{\Phi_{SB-p,n}} \right) \tag{8.7}$$

Considering the fact that field lines from source to drain are not just confined to the TMD channel, **Equation 8.3** becomes

$$\lambda_{\text{geo}} \approx \sqrt{t_{\text{ox}}t_{\text{body}}} \tag{8.8}$$

Here, we replaced $\varepsilon_{\text{body}}$ in **Equation 8.3** with the average of the dielectric constants of air, $\varepsilon_{\text{body}}$ and ε_{ox}. To describe the transmission through a

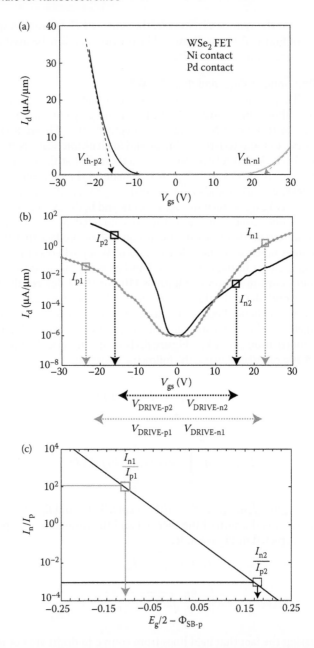

FIGURE 8.11 (a) and (b) show the transfer characteristics of WSe$_2$ devices with Ni (grey) and Pd (black) contacts in linear and logarithmic scale, respectively. (b) shows the quantitative dependence of the ratio of electron current to hole current at their respective threshold voltages in logarithmic scale as a function of the SB height. (Reprinted with permission from S. Das and J. Appenzeller, *Applied Physics Letters* **103**, 103501-1. Copyright 2013, American Institute of Physics.)

triangular SB to calculate the different current components, Landauer formalism and Wentzel-Kramers-Brillouin (WKB) approximation has been used. q is the electronic charge, h is the Planck constant, k_B is the Boltzmann constant and T is the temperature. $M(E)$ (in units of (eV m s)$^{-1}$) is the number of conducting modes calculated as the product of the density of states (DOS) and the average velocity corresponding to a two-dimensional (2D) parabolic band structure, m_{eff} is the carrier effective mass and $f(E)$ is the Fermi distribution. Note that scattering in the channel is not included in this ballistic injection model since the injection of carriers across the SB into the channel is limiting current flow rather than scattering in the channel. The electron current (I_n) and hole current (I_p) can be calculated by inserting the respective SB heights and effective masses in the above equations. Both I_n and I_p have two components as discussed before in the context of **Figure 8.3c** – the thermionic emission current through the top of the SB ($I_{thermal}$) and the tunnelling current through the triangular SB (I_{tunnel}). As mentioned earlier, when the metal Fermi level coincides with the middle of the band gap, the ratio of the electron current to the hole current is unity. On the other hand, if the device characteristics show an asymmetry, the SB heights can be extracted from **Figure 8.11c** by employing the following approach:

1. Determine the threshold voltage for the dominant conduction branch of the device characteristics from the linear $I_d - V_{gs}$ characteristics (V_{th-p2} for the Pd contacted device and V_{th-n1} for the Ni contact device from **Figure 8.11a**).

2. Extract the dominant current corresponding to the threshold voltage from the logarithmic $I_d - V_{gs}$ characteristics (I_{p2} for the Pd contact device and I_{n1} for the Ni contact device from **Figure 8.11b**).

3. Identify the overdrive voltage $V_{DRIVE-p,n} = |V_{th-p,n} - V_{MIN}|$, where V_{MIN} is the gate voltage corresponding to the minimum current point ($V_{DRIVE-p2}$ for the Pd contact device and $V_{DRIVE-n1}$ for the Ni contact device from **Figure 8.11b**).

4. Extract the current corresponding to the non-dominant branch using the same overdrive voltage (I_{n2} for the Pd contact device and I_{p1} for the Ni contact device from **Figure 8.11b**).

5. Finally, calculate the ratio of the electron current to the hole current and compare it to the simulated ratio in **Figure 8.11c** to extract the SB heights.

Note that the determination of the current at threshold for the non-dominant branch is not accurate. However, the fact that we are using the same overdrive voltage is expected to result in similar band bending situations for both the electrons and the holes. Moreover, the exponential dependence of the ratio of electron to hole current (I_n/I_p) on the actual SB heights (**Figure 8.11c**) makes this technique very robust and not strongly dependent on an error in the extraction of the threshold voltages. As apparent from **Figure 8.11c**, even an order of magnitude error in the determination of I_n/I_p results only in about an error of 50 meV in the extraction of the SB height which is well within the limit of our extraction method and the typically observed device-to-device variations.

For Ni contacts to WSe_2, the SB heights for electron and hole injection were found to be 0.54 and 0.76 eV, respectively, and for Pd contacts to WSe_2, 0.82 and 0.48 eV, respectively [12]. This technique requires prior knowledge of the carrier effective masses and the band gap of the material which in the case of WSe_2 were assumed to be $m_{eff} = 0.4\ m_0$ (for both, electrons and holes), where m_0 is free electron mass and $E_g = 1.3$ eV [13]. Also note that this technique relies on the asymmetry of the device characteristics which tends to disappear when the tunnelling distance λ becomes very small. In other words, for $\lambda = 0$, the device characteristics will always be symmetric. A comparison of the various extracted SB heights for different materials can be found in Section 8.3.2.

8.2.5 Utilising the Flat Band and Minimum Conduction Points to Extract the SB Height and the Band Gap

8.2.5.1 Method of SB Extraction

In case of ambipolar device characteristics for which V_{FB} and V_{MIN} (see **Figure 8.6a**) can be identified with reasonable accuracy, a third technique can be employed to determine both the height of the SBs as well as the band gap of the semiconductor. **Figure 8.12** qualitatively illustrates this technique. Let us consider an ambipolar device where the metal Fermi level is aligned closer to the conduction band of the semiconducting channel. This will result in an asymmetric device characteristic with dominating electron branch as shown in **Figure 8.6a**. At V_{FB}, electrons are injected into the channel via thermionic emission over the top of the SB (Φ_{SB-n}) as discussed in Sections 8.2.1 and 8.2.2. We will refer in the following to the current I_d at flat band voltage V_{FB} as flat band current (I_{FB}). I_{FB} is a pure thermal current of electrons by definition. **Figure 8.12b** shows calculated thermal currents ($I_{thermal}$) in logarithmic scale as a function of hypothetical thermal barriers with height Φ_B. By comparing I_{FB} with $I_{thermal}$, the SB height for electrons (Φ_{SB-n}) can be extracted from **Figure 8.12b**. Reducing the gate voltage ($V_{gs} < V_{FB}$), the electron current decreases as discussed before since the thermal barrier for electron injection is increased as shown in the bottom left diagram of **Figure 8.12a**. On the other hand, the hole current increases simultaneously as discussed in the context of **Figure 8.5a** and illustrated in the bottom right of **Figure 8.12a**. Finally, at the minimum current point (V_{MIN}, I_{MIN}) of the device characteristics, the electron current is exactly identical to the hole current ($I_{p-MIN} = I_{n-MIN} = 0.5I_{MIN}$). The effective barrier height ($\Phi_{SB-n} + \Delta\Phi$) for electron injection at this point can be extracted by comparing I_{n-MIN} with the calculated $I_{thermal}$-value using **Figure 8.12b** as before. In fact, for an ideal ultrathin body transistor for which the band movement in the channel of the transistor follows the applied gate bias V_{gs} one-to-one, one would find $\Delta\Phi = V_{FB} - V_{MIN}$. The knowledge of the magnitude of $\Delta\Phi$ is important in order to determine the SB height for hole injection in the same device through the comparison of I_{n-MIN} and I_{p-MIN}. Note that the hole current

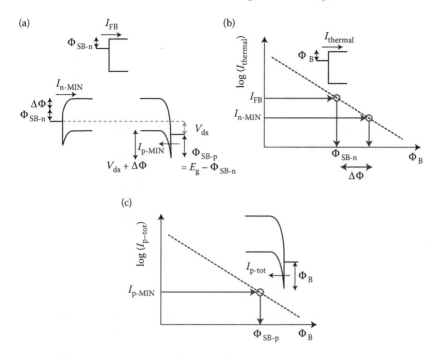

FIGURE 8.12 (a) Qualitative energy band diagrams for electron and hole injection at flat band and minimum current point. (b) The dependence of the thermionic emission current on the thermal barrier height and (c) the qualitative trend of the SB current on Φ_B.

($I_{p\text{-MIN}}$) at the minimum current point includes contributions from various tunnelling current components of as well as the thermionic emission current, as will be discussed below. **Figure 8.12c** shows such hole current ($I_{p\text{-tot}}$) as a function of the hole SB height. By comparing $I_{p\text{-MIN}}$ with $I_{p\text{-tot}}$, the SB height for holes ($\Phi_{SB\text{-p}}$) can be extracted. Finally, $E_g = \Phi_{SB\text{-n}} + \Phi_{SB\text{-p}}$.

8.2.5.2 Determining $\Phi_{SB\text{-n}}$ and $\Phi_{SB\text{-p}}$ for MoSe$_2$

Next, we will apply this technique to analyse ambipolar MoSe$_2$ devices of the type shown in **Figure 8.13a**. First, we extract V_{FB} by identifying the point in the subthreshold regime of the device characteristics where the slope deviates from the exponential trend as marked by the black dotted line. Note that the slope of this line should ideally be 60 mV/decade according to the thermal injection model, but due to non-idealities, the slope is given by 60γ mV/decade (where $\gamma = 1 + C_{IT}/C_{OX}$, where C_{IT} and C_{OX} are the interface trap capacitance and oxide capacitance, respectively. The function γ as defined above does not include any depletion capacitance (C_d) contribution as frequently relevant in bulk type MOSFET devices since for ultrathin body devices as TMDFETs ($C_d \approx 0$). **Figure 8.13b** which is the numerical analogue of **Figure 8.12b** shows the thermionic emission current calculated using **Equation 8.9** as a function

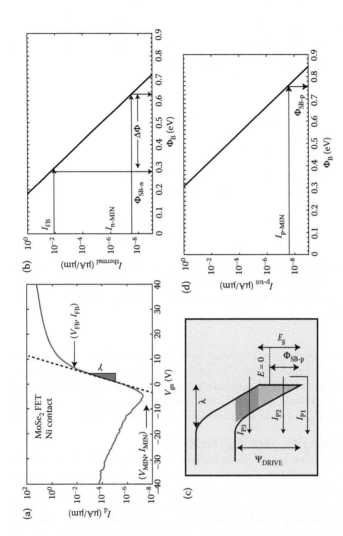

FIGURE 8.13 (a) The transfer characteristics of a MoSe$_2$ device with Ni contacts. (b) The quantitative dependence of the thermionic emission current as a function of barrier height. (c) The band diagram for hole injection from the drain. (d) The quantitative dependence of the SB current from drain as a function of barrier height.

of the barrier height. Note that **Equation 8.9** is a modified version of **Equation 8.4** taking into account the thermal barrier height Φ_B at flat band as shown in **Figure 8.12a**. Using the experimental data for the flat band current (I_{FB}), the height of the SB can be determined from **Figure 8.13b**. For the MoSe$_2$ FET from a 4.5 nm thick flake with Ni contacts, the flat band current was found to be 1.5×10^{-2} μA/μm which corresponds to a SB height of $\Phi_{SB\text{-}n} \approx 0.28$ eV assuming $m_e = 0.55$ m_0. While this approach is less precise than what was discussed in Section 8.2.3, we found good agreement between the two methods. In fact, the SB height extracted for the present case using the Arrhenius-type approach described in Section 8.2.3 was found to be 0.25 eV in good agreement with the above $\Phi_{SB\text{-}n}$-value. Also note that even for an order of magnitude variation in I_{FB}, the barrier height extraction changes by only ~50 meV, which is comparable to typical device-to-device variations as well as the error bar when employing the Arrhenius-type technique.

$$I_{\text{thermal}} = q \int_{\Phi_B}^{\infty} M(E - \Phi_B) f(E) dE \qquad (8.9)$$

Next, we focus on the minimum current point in the $\log(I_d) - V_{gs}$ characteristics. As mentioned earlier, the minimum current point marks the transition from electron to hole current and, therefore, it is the superposition of two equal components:

1. The thermionic electron emission current ($I_{n\text{-MIN}} = 0.5 I_{MIN}$) injected from the source over the top of the effective thermal barrier $\Phi_{SB\text{-}n} + \Delta\Phi$, where $\Delta\Phi$ is the band movement due to an applied gate voltage of $V_{FB} - V_{MIN}$ shown in the bottom diagram of **Figure 8.12a**).

2. The hole current ($I_{p\text{-MIN}} = 0.5 I_{MIN}$) injected from the drain which comprises both the thermionic emission current and thermally assisted tunnelling current as shown in the right diagram of **Figure 8.12a**.

The effective barrier height ($\Phi_{SB\text{-}n} + \Delta\Phi$) for electron injection can be determined from **Figure 8.13b** using the same technique as described above in the context of the $\Phi_{SB\text{-}n}$-extraction. In essence, one has to compare the experimentally determined minimum current level $0.5 I_{MIN}$ from **Figure 8.13a** with the calculated thermal emission current over a barrier with height $\Phi_{SB\text{-}n} + \Delta\Phi$ in **Figure 8.13b**. For example, $0.5 I_{MIN} = 5 \cdot 10^{-8}$ μA/μm implies $\Phi_{SB\text{-}n} + \Delta\Phi = 0.62$ eV. Since $\Phi_{SB\text{-}n} = 0.28$ eV, $\Delta\Phi = 0.34$ eV.

An alternative way to determine $\Delta\Phi$ is from the band movement factor γ in the sub-flat band ($V_{gs} < V_{FB}$) region of the device operation using the expression $\Delta\Phi' = (V_{FB} - V_{MIN})/\gamma$. For the device data shown in **Figure 8.13a**, one finds that $\Delta\Phi' = 0.35$ eV, in close agreement with the above value of $\Delta\Phi = 0.34$ eV. Note that for an ideal band movement $\gamma = 1$ and $\Delta\Phi' = (V_{FB} - V_{MIN})$ as mentioned earlier, while we extract $\gamma = 30$ from **Figure 8.13a**.

Next, we analyse the hole current ($I_{\text{p-MIN}}$) to determine the SB-height ($\Phi_{\text{SB-p}}$) for hole injection. The band bending situation at the drain end is depicted in **Figure 8.13c** with Ψ_{DRIVE} being the bias across the tunnelling barrier. Comparing **Figures 8.12a** and **8.13c**, one can see that $\Psi_{\text{DRIVE}} = V_{\text{ds}} + \Delta\Phi$. **Figure 8.13d** shows the SB hole current $I_{\text{p-tot}}$ (which is the sum of both the thermionic emission current I_{P1} and the thermally assisted tunnelling currents I_{P2} and I_{P3}) as a function of the SB height calculated using **Equations 8.10** through **8.15**:

$$I_{\text{P1}} = q \int_{\Phi_B}^{\infty} M(E + \Psi_{\text{DRIVE}} - \phi)f(E)dE, \quad \Psi_{\text{DRIVE}} = V_{\text{ds}} + \Delta\phi \qquad (8.10)$$

$$I_{\text{P2}} = q \int_{\Phi_B - E_G}^{\Phi_B} M(E + \Psi_{\text{DRIVE}} - \phi)T_{\text{WKB}-2}(E)f(E)dE \qquad (8.11)$$

$$I_{\text{P3}} = q \int_{\Phi_B - \Psi_{\text{DRIVE}}}^{\Phi_B - E_G} M(E + \Psi_{\text{DRIVE}} - \phi)T_{\text{WKB}-3}(E)f(E)dE \qquad (8.12)$$

$$I_{\text{p-tot}} = I_{\text{P1}} + I_{\text{P2}} + I_{\text{p3}} \qquad (8.13)$$

$$T_{\text{WKB}-2}(E) = \exp\left(-\frac{8\pi}{3h}\sqrt{2m_p(\Phi_B - E)^3}\,\frac{\lambda}{\Psi_{\text{DRIVE}}}\right) \qquad (8.14)$$

$$T_{\text{WKB}-3}(E) = \exp\left(-\frac{8\pi}{3h}\sqrt{2m_p E_G^3}\,\frac{\lambda}{\Psi_{\text{DRIVE}}}\right) \qquad (8.15)$$

Using the same value of $0.5I_{\text{MIN}} = 5 \cdot 10^{-8}$ µA/µm as above, for MoSe$_2$ FETs with Ni contacts, the SB height for hole injection can be extracted from **Figure 8.13d** to be $\Phi_{\text{SB-p}} = 0.76$ eV when assuming a hole effective mass of $m_p = 0.64\,m_0$ [14]. The band gap of the material is then determined as the sum of the SB heights for hole and electron injection ($E_g = \Phi_{\text{SB-n}} + \Phi_{\text{SB-p}}$). For MoSe$_2$, our approach thus results in an experimentally determined $E_g = 1.04$ eV.

8.3 TMD(SB)FETs

8.3.1 On the Experimental On-State Performance of TMDFETs

So far, we have exclusively focussed on the subthreshold characteristics of TMDFETs. The main emphasis had been on extracting information about the formation of SBs. Now, we want to evaluate the device on-state. From the above arguments about λ_{geo}, one might conclude that contact effects play a small role for the on-state of ultrathin body devices since a short tunnelling distance through a SB implies a small contact resistance R_c. However, the actual magnitude of R_c strongly depends on the details of the SB contact, that is, the barrier height and λ_{geo}. **Figure 8.14** illustrates how the contact resistance for a TMDFET with a body thickness of 6 nm depends on the gate oxide thickness t_{ox} when a moderate SB height of $\Phi_{SB} = 0.2$ eV is assumed. To illustrate how the situation changes for different TMDs, R_c is calculated for 0.3 $m_0 < m_{eff} < 0.5$ m_0. **Equations 8.4** through **8.8** were employed for small drain voltages $qV_{ds} = \Phi_{SB}$ to determine the impact of the SB on the contact resistance at threshold. While the effective mass variation that is chosen to cover the typical range of TMD materials results in a rather small change of R_c, the strong dependence of the contact resistance on gate oxide thickness is apparent. The trends observed in **Figure 8.14** are readily understood in the context of our discussion on thermal assisted tunnelling through a contact SB and thermal emission over the same. For small gate oxide thicknesses, tunnelling dominates and the contact resistance at threshold changes substantially. For larger t_{ox}-values, tunnelling becomes less relevant and the total resistance as a function of gate oxide thickness approaches a constant R_c-value that is determined by thermal emission over the SB. Note, that increasing the gate voltage beyond threshold would change the set of curves in **Figure 8.14** in such that the slope of R_c versus gate oxide thickness t_{ox} would decrease and extend over a larger t_{ox}-range. When comparing these calculated values with silicon MOSFETs, where R_c-values in the 150 Ω μm range are achievable (see, e.g., Reference [18]), it is obvious that only very small t_{ox}-values in combination with not too high SBs allow ignoring the contact resistance contributions in case of SB TMDFETs. Ultimately, it is the frequent absence of a negative SB that makes contact resistances more relevant in SB-FETs if compared with conventional MOSFETs as discussed in Section 8.1.

Accordingly, output characteristics of TMDFETs frequently show large R_c-values in the low V_{ds}-region in particular for prototype devices with a

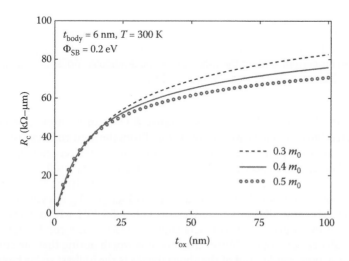

FIGURE 8.14 Contact resistance versus gate oxide thickness for various effective mass values and a SB height of 0.2 eV.

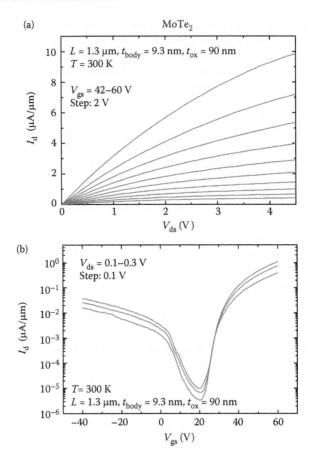

FIGURE 8.15 (a) Output and (b) subthreshold characteristics of a nickel contacted back-gated MoTe$_2$ FET with $t_{ox} = 90$ nm, showing the highest so far reported on-current level in the n-branch of the ambipolar MoTe$_2$ device characteristics (compare References 15 through 17).

thick gate dielectric film thickness. **Figure 8.15** is an example of a nickel contacted MoTe$_2$ device with a t_{ox} of 90 nm. From the high contact resistance and the analysis performed on the subthreshold characteristics using the method described in Section 8.2.5, we extract a SB of $\Phi_{SB-n} = 0.35$ eV as further discussed in Section 8.3.2. While output characteristics occur similar to those of conventional FETs, the ambipolar $\log(I_d) - V_{gs}$ curves in **Figure 8.15b** clearly show all the expected features of an SB-FET with a Fermi-level line-up of the source/drain contacts slightly closer to the conduction than to the valence band of the MoTe$_2$ channel. It is worth noting that the current level for n-type conduction of the above device is the highest so far reported for MoTe$_2$, however, as apparent from our discussion, clearly not limited by scattering inside the device channel but rather the SBs at the source and drain ends of the FET, leaving ample space for significant performance improvements of TMDFETs.

8.3.2 Experimental Extraction of Materials Parameters

In this section, we will summarise our findings on the various TMD materials – MoS_2, $MoSe_2$, $MoTe_2$ and WSe_2 – and discuss differences, commonalities and trends in the device characteristics of different SB TMDFETs using the techniques described in Section 8.2. In particular, we will discuss how the band gaps and the alignments of the band gaps with the metal Fermi levels at the contact interfaces determine the transport properties of these SB TMDFETs. All devices described in this section were fabricated using the procedure mentioned in Section 8.2.3.2 and **Figure 8.8**.

8.3.2.1 Φ_{SB-n}, Φ_{SB-p} and E_g for MoS_2

Figure 8.16a shows transfer characteristics of a 12 nm thick MoS_2 flake with 100 nm SiO_2 as the gate dielectric and Ni as the source and drain contact electrode. This device exhibits characteristics similar to what was discussed in Section 8.2.3 and displayed in **Figure 8.9a**. As argued in Section 8.2.3, the complete absence of the hole branch is a result of the fact that for MoS_2 devices, the Ni Fermi level is aligned close to the conduction band, which implies a large SB height for hole injection (Φ_{SB-p}) into the valence band and the fact that λ_{geo} is rather large for $t_{ox} = 100$ nm. The SB height for electron injection (Φ_{SB-n}) at the Ni to MoS_2 interface was determined to be 0.15 eV following the Arrhenius technique described in Section 8.2.3, whereas Φ_{SB-n} was found to be 0.16 eV using the technique described in Section 8.2.5. The above values are in good agreement considering device-to-device variations impacting the exact Fermi-level line-up. The device characteristics shown in **Figure 8.16a**, however, do not allow to determine Φ_{SB-p} and hence the band gap ($E_g = \Phi_{SB-n} + \Phi_{SB-p}$) of MoS_2 since the minimum current point (V_{MIN}, I_{MIN}) is screened by the leakage floor.

In order to remove this limitation, the minimum current point and thus the hole current needs to be increased. This could be achieved in three ways – by (i) reducing the SB height for the hole injection (Φ_{SB-p}) and/or by (ii) reducing the SB tunnelling distance (λ_{geo}) and/or (iii) increasing the drain voltage. **Figure 8.16b** shows the transfer characteristics of a 3 nm thick MoS_2 flake with 20 nm SiO_2 as the gate dielectric with Pd as the source and drain contact electrode. Φ_{SB-n} at the Pd to MoS_2 interface was determined to be 0.23 eV following the technique described in Section 8.2.5. This number is consistent with $\Phi_{SB-n} = 0.23$ eV for Pt contacted MoS_2 FETs [8], where the Arrhenius technique described in Section 8.2.3 was employed. Note that the use of a higher work function metal as Pd for the source/drain contact electrodes increases Φ_{SB-n} and hence reduces Φ_{SB-p}. However, this change is only about 70 meV in spite of a work function difference $\Delta\Phi_M$ of 500–600 meV between Ni and Pd (all work function references are from Reference 19) similar to our findings for Ni and Pt as presented in **Figure 8.9d**. Strong Fermi-level pinning at the metal-to-TMD interface is responsible for a ratio of $d\Phi_{SB}/d\Phi_M$ around 0.1 which is even

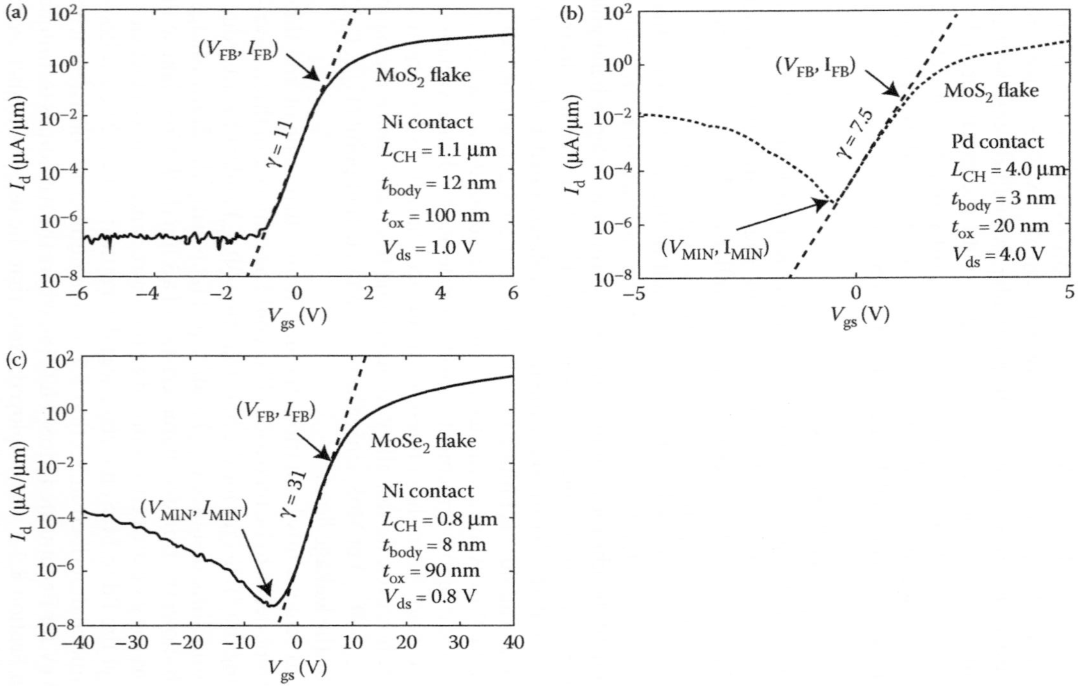

FIGURE 8.16 Transfer characteristics of back-gated TMDFETs with Ni (left column) and Pd (right column) source/drain contacts. (a), (b) MoS$_2$ FETs, (c) MoSe$_2$ FETs, (d), (e) MoTe$_2$ FETs and (f), (g) WSe$_2$ FET. ((f) and (g) Reprinted with permission from S. Das and J. Appenzeller, *Applied Physics Letters* **103**, 103501-1. Copyright 2013, American Institute of Physics.) *(Continued)*

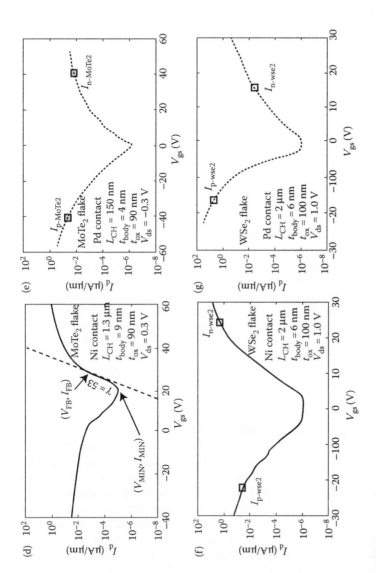

FIGURE 8.16 (CONTINUED) Transfer characteristics of back-gated TMDFETs with Ni (left column) and Pd (right column) source/drain contacts. (a), (b) MoS$_2$ FETs, (c) MoSe$_2$ FETs, (d), (e) MoTe$_2$ FETs and (f), (g) WSe$_2$ FET. ((f) and (g) Reprinted with permission from S. Das and J. Appenzeller, *Applied Physics Letters* **103**, 103501-1. Copyright 2013, American Institute of Physics.)

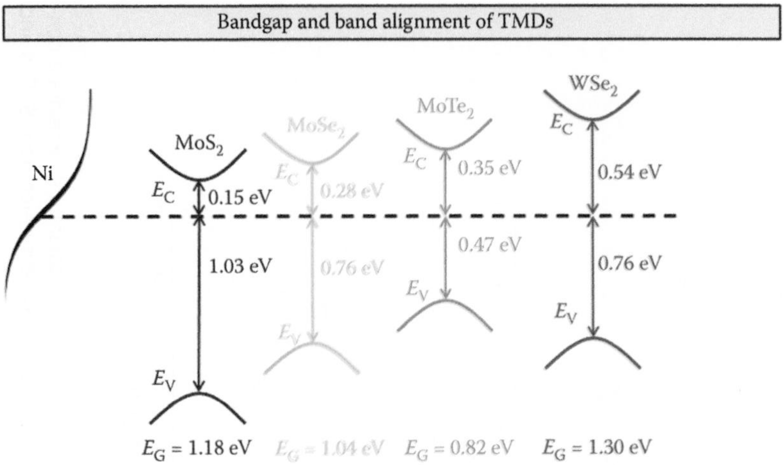

FIGURE 8.17 Band gaps and band alignments of Ni and Pd Fermi levels for different TMDs.

smaller than what is typically observed in case of metal-to-silicon interfaces [8]. The enhancement in the hole conduction in **Figure 8.16b** is, therefore, a result of the fact that the SB tunnelling distance (λ_{geo}) was scaled down by a factor of 2.7 by the use of a thinner flake and scaled gate oxide and not the choice of metal contact. Also note that the large drain bias of $V_{ds} = 4.0$ V facilitates hole tunnelling by increasing the bias (Ψ_{DRIVE}) across the tunnelling barrier as evident from **Figure 8.13c** and **Equations 8.10** through **8.15**. **Figure 8.16b** allows us to extract Φ_{SB-p} and E_g of MoS_2 using the technique discussed in Section 8.2.5. We find $\Phi_{SB-p} = 0.95$ eV for Pd contacts and $E_g = 1.18$ eV. The knowledge of E_g can then be used to obtain $\Phi_{SB-p} = 1.03$ eV for Ni contacts. The results for MoS_2 FETs are summarised in **Figure 8.17**.

8.3.2.2 Φ_{SB-n}, Φ_{SB-p} and E_g for $MoSe_2$

Figure 8.16c shows the transfer characteristics of an 8 nm thick $MoSe_2$ FET with 90 nm SiO_2 as the gate dielectric and Ni as source/drain contact electrodes. The presence of both electron and hole conduction can be qualitatively understood as the result of the Ni Fermi level being aligned closer to the valence band of $MoSe_2$ than to the valence band of MoS_2. Note that while the actual hole current not only depends on Φ_{SB-p}, but also the hole effective mass m_h and the λ_{geo}-value according to **Equation 8.7** as well as the drain voltage, for the data presented in **Figure 8.16a** and c, the main difference arises due to the change in SB height when considering a hole effective mass of 0.64 m_0 for $MoSe_2$ and 0.56 m_0 for MoS_2 [20]. However, the dominance of the electron branch is an indication of the fact that the Ni Fermi level is still closer to the conduction band of $MoSe_2$ than to the valence band. Using the technique described in Section 8.2.5, we find $\Phi_{SB-n} = 0.28$ eV, $\Phi_{SB-p} = 0.76$ V for Ni contacts and hence $E_g = 1.04$ eV.

8.3.2.3 Φ_{SB-n}, Φ_{SB-p} and E_g for MoTe$_2$

Figure 8.16d shows the transfer characteristics of a 9 nm thick MoTe$_2$ FET with 90 nm SiO$_2$ as the gate dielectric and Ni contacts. This device clearly shows a higher degree of symmetry in terms of ambipolar conduction. Using the technique described in Section 8.2.5 and assuming an electron effective mass of 0.55 m_0 and a hole effective mass of 0.67 m_0 for MoTe$_2$ [20], we extract Φ_{SB-n} = 0.35 eV, Φ_{SB-p} = 0.47 V for Ni contacts and hence E_g = 0.82 eV. Replacing the Ni electrodes with higher work function Pd electrodes results in even more symmetric characteristics as shown in Figure 8.16e. Using the technique described in Section 8.2.4, and from the knowledge of the band gap of MoTe$_2$ extracted above, we conclude that Φ_{SB-n} = 0.42 eV, Φ_{SB-p} = 0.40 eV for Pd contacts. Note that the change in the SB height of ~70 meV between Ni and Pd electrodes is consistent with our earlier statement regarding the difference in SB heights for Ni and Pd contacted MoS$_2$ FETs.

8.3.2.4 Φ_{SB-n} and Φ_{SB-p} for WSe$_2$

Finally, Figure 8.16f and g shows the transfer characteristics of 6 nm thick WSe$_2$ FETs with 100 nm SiO$_2$ as the gate dielectric plus Ni and Pd source/drain contacts, respectively. As described in details in Section 8.2.4, we find that Φ_{SB-n} = 0.54 eV, Φ_{SB-p} = 0.76 eV for Ni contacts and Φ_{SB-n} = 0.82 eV, Φ_{SB-p} = 0.48 eV for Pd contacts when a band gap of E_g = 1.3 eV is assumed. Here, the difference between Ni and Pd contacted devices is substantially larger than in case of MoS$_2$ or MoTe$_2$.

Finally, we combine the findings on the band gap and the band alignment of Ni and Pd Fermi levels with the band structure of different TMDs to draw the relative position of the conduction and valence bands as shown in Figure 8.17. These results are consistent with theoretical predictions from density functional theory (DFT) calculations [20,21] and are first time experimental evidence that can provide critical feedback for the simulation effort on TMDs.

8.3.3 Comparison with NEGF Results for TMDFETs

The last section of this chapter compares the above experimental findings and analytical arguments with fully self-consistent non-equilibrium Green's functions (NEGF) formalism simulation results. The emphasis is on the impact of SBs on the device performance and the associated asymmetry of $\log(I_d) - V_{gs}$ characteristics when comparing the on-state of the n- and p-branch as well as the on/off-current ratio.

It has been shown that since the 2D electrostatics can be described well by an effective one-dimensional (1D) Poisson equation employing a quadratic approximation of the potential profile perpendicular to the direction of current transport, a 1D modified Poisson equation can be obtained which is given by [22]

$$\frac{\partial^2 \Phi(x)}{\partial x^2} - \frac{\Phi(x) - \Phi_g}{\lambda^2} = \frac{e(\rho(x) \pm N_{D,A})}{\varepsilon_0 \varepsilon_{ch}} \qquad (8.16)$$

where $\Phi(x)$ is the electrostatic potential in the direction of current transport at the interface between the channel and the gate dielectric; Φ_g is the gate potential (including a possible work function difference between the gate and the channel), $\rho(x)$ is the free charge density (note that only a dependence on x is considered) and $N_{D,A}$ is a constant dopant density of donors or acceptors, respectively. In the case of a constant charge density, the equation can be solved analytically showing that potential variations are exponentially screened on the length scale λ_{geo}. This is exactly the length scale λ_{geo} that was introduced earlier in the chapter (cf. **Equation 8.3**) to describe an ultrathin-body device.

As mentioned above, in our calculation, the charge in and current through the device are calculated using the NEGF formalism on a finite difference grid [23]. An effective mass approximation for the conduction and valence bands is used and the complex band structure in the band gap is accounted for using Flietner's dispersion relation [24,25]. To simplify the 2D quantum computation, independent 1D modes are calculated and their contribution to the charge and current is summed up. This approximation, that is, describing a 2D system with independent 1D modes is possible since ballistic transport is considered and because in the limit of a very wide (in fact infinitely wide) 2D system, there is no electric field along the device's width, implying that the momentum is conserved along this direction. Therefore, carriers will remain within a certain 1D mode even when scattered at the channel-to-contact interfaces; this situation is schematically depicted in **Figure 8.18**.

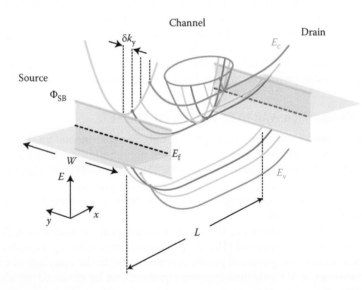

FIGURE 8.18 Different vertical momenta k_y that lead to the various 1D modes that yield increasing effective band gaps and as a result, only a limited number of these modes have to be computed to obtain a solution sufficiently close to a true 2D computation.

Along the y-direction periodic boundary conditions are assumed leading to discrete $k_y^n = 2\pi n/W$ values including $k_y = 0$. Since with increasing k_y, the effective SB for a 1D mode increases according to $\Phi_{SB}^{eff} = \Phi_{SB} + (\hbar^2(k_y^n)^2/2m^*)$ only a limited number of transverse k_y modes will contribute substantially to the current and hence needs to be taken into consideration.

After self-consistency concerning the conduction and valence band structure is obtained, the current is computed using the Fisher–Lee relation. To be specific, the current is given by

$$I_d \approx \frac{2e}{h}\frac{W}{2\pi}\sum_{k_y}\delta k_y \int_{-\infty}^{\infty} dE\, T(k_y, E)(f_s - f_d) \qquad (8.17)$$

where $T(k_y, E)$ includes the increase of the effective SB height Φ_{SB}^{eff} for each 1D mode. **Figure 8.19** shows a logarithmic plot of the local DOS in a TMDFET with a SB of 0.6 eV.

Since there is always a substantial SB present at the metal-to-TMD interface, a small λ_{geo} is mandatory in order to increase the tunnelling probability through this SB to obtain high on-state performance. We therefore performed simulations based on the formalism outlined above assuming that the channel consists of a monolayer ($t_{body} = 0.6$ nm) and the gate oxide thickness is scaled down to 10 nm. In order to study the impact of various SB heights, WSe$_2$ is considered as a representative for a TMDFET with appropriate effective masses and band gap.

Figure 8.20 shows simulated transfer characteristics for SB heights (for electron injection) ranging from 0.1 to 1 eV. Inspecting the curves for $\Phi_{SB} = 0.1$ up to 0.3 eV (black, red and green curves, respectively) one observes two different inverse subthreshold slopes in the electron branch of the off-state

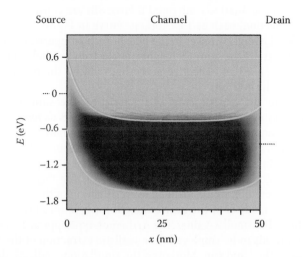

FIGURE 8.19 Logarithmic plot of the local DOS of a TMDFET with a SB of 0.6 eV.

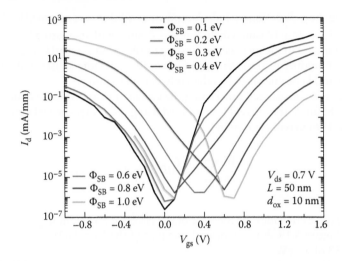

FIGURE 8.20 Transfer characteristics of a monolayer TMDFET with varying SB heights; the electron and hole effective masses are 0.34 and 0.44, respectively, and $E_g = 1.2$ eV.

characteristics. As expected from the discussion above (see Section 8.2), the 'knee' at V_{FB} in the $\log(I_d)$ versus V_{gs} characteristics, where the thermal branch changes into the SB regime, is shifted closer to the minimum current level the larger the SB height is. In the case of $\Phi_{SB} = 0.6$ eV, the characteristic becomes almost symmetric (apart from a slight difference due to the different effective electron and hole masses). Furthermore, for low SBs, the ambipolar characteristics are asymmetric with a substantially smaller current in the hole branch (i.e. for negative V_{gs}) as expected from the discussion in Section 8.2. The behaviour is qualitatively mirrored if large SBs are considered (i.e. small SBs for hole injection) such as 1 eV (orange curve in **Figure 8.20**). However, the current level is slightly lower in the hole branch compared to the electron branch (in the case of the same SB heights for electron and hole injection) due to the larger valence band effective mass that reduced the tunnelling probability through the hole SB.

Compared to the experimental characteristics, the simulated data are significantly less asymmetric even in the case of an SB as low as 0.1 eV for electron injection. The main reason for this is the small screening length $\lambda_{geo} = 4.15$ nm in our simulations that results in a high tunnelling probability (cf. **Equation 8.7**) and hence in a substantially decreased 'apparent' SB height [26] for hole injection (see also **Figure 8.7**). Extracting Φ_{SB} as described in Section 8.2.4, therefore, would yield an overestimation of the SB for electrons and an underestimation of the barrier height for hole injection for the case simulated here. For small λ-values, the Arrhenius-type approach described in Section 8.2.3 needs to be employed for a realistic extraction of the SB height and the size of the band gap. Moreover, the simulations indicate that realising TMDFETs based on a monolayer is desirable to scale λ: the relatively high

SBs present at metal-to-TMD interfaces in combination with the rather large effective masses can lead for thicker TMD flakes to a substantial deterioration of the on-state performance of TMDFETs. This argument, however, has to be weighed against a potentially reduced mobility for too thin flakes [8] due to the impact of the substrate on transport in thin TMDFETs. The details of the device layout and application will determine in this case the optimum TMD layer thickness.

8.4 Conclusion

In this chapter, we discussed in detail the impact of SBs on the performance of TMDFETs. In particular, we elucidated how the ultrathin body of TMD devices is linked with the tunnelling probability through SBs at the source and drain contacts of a TMDFET. Various approaches to extract information about SB heights and band gaps have been discussed and a comparison of experimentally determined Φ_{SB}- and E_g-values of MoS_2, $MoSe_2$, $MoTe_2$ and WSe_2 FETs has been presented. Lastly, we have compared our experimental findings and qualitative expectations with NEGF results on ballistic SB TMDFETs further highlighting the importance of the geometrical screening length in the context of SB-FETs.

Acknowledgements

This work was supported in part by the Center for Low Energy Systems Technology (LEAST) and the Center for Function Accelerated nanoMaterial Engineering (FAME), two of six centres of STARnet, a Semiconductor Research Corporation programme sponsored by MARCO and DARPA. J.K. acknowledges Deutsche Forschungsgemeinschaft for partial financial support under grant number KN545/3-1.

References

1. Y. Taur and T.H. Ning (eds), *Fundamentals of Modern VLSI Devices*, Second Edition, Cambridge, United Kingdom: Cambridge University Press, 148–201, 2009.
2. J. Guo and M.S. Lundstrom, *IEEE Transactions on Electron Devices* **49**, 1897, 2002.
3. H.-K. Lim and J.G. Fossum, *IEEE Transactions on Electron Devices* **30**, 1244–1251, 1983.
4. S.M. Sze (ed.), *Physics of Semiconductor Devices*, Third Edition, Hoboken, New Jersey: Wiley Inter Science, 314–316, 2007.
5. D.J. Frank, Y. Taur and H.S.P. Wong, *IEEE Electron Device Letters* **19**, 385, 1998.
6. J. Knoch, M. Zhang, S. Mantl and J. Appenzeller, *IEEE Transactions on Electron Devices* **53**, 1669, 2006.
7. J. Appenzeller, M. Radosavljevic, J. Knoch and Ph. Avouris, *Physical Review Letters* **92**, 048301-1, 2004.

8. S. Das, H.-Y. Chen, A.V. Penumatcha and J. Appenzeller, *Nano Letters* **13**, 100, 2013.
9. B. Radisavljevic, A. Radenovic, J. Brivio, V. Giacometti and A. Kis, *Nature Nanotechnology* **6**, 147, 2011.
10. C. Chang, W. Zhang, Y.-H. Lee, Y.-C. Lin, M.-T. Chang, C.-Y. Su, C.-S. Chang et al., *Nano Letters* **12**, 1538, 2012.
11. A.T. Neal, H. Liu, J.J. Gu and P.D. Ye, *70th Device Research Conference*, IEEE, University Park, p. 65, 2012, ISBN: 978-1-4673-1163-2.
12. S. Das and J. Appenzeller, *Applied Physics Letters* **103**, 103501-1, 2013.
13. A. Kumar and P. K. Ahluwalia, *The European Physical Journal B* **85**, 2, 2012.
14. G.B. Liu, W.Y. Shan, Y. Yao, W. Yao and D. Xiao, *Physical Review B* **88**, 085433, 2013.
15. S. Fathipour, W.S. Hwang, Th. Kosel, H. Xing, W. Haensch, D. Jena and A. Seabaugh, *71st Device Research Conference*, IEEE, Notre Dame, p. 115, 2013, ISBN: 978-1-4799-0814-1.
16. N. Haratipour and S.J. Koester, *72nd Device Research Conference*, IEEE, Santa Barbara, p. 171, 2014, ISBN: 978-1-4799-5406-1.
17. Y.-F. Lin, Y. Xu, S.-T. Wang, S.-L. Li, M. Yamamoto, A. Aparecido-Ferreira, W. Li et al., *Advanced Materials* **26**, 3263, 2014.
18. International Technology Roadmap for Semiconductors ITRS, 2012, http://www.itrs.net/links/2012itrs/home2012.htm.
19. D.R. Lide (ed.), *CRC Handbook of Chemistry and Physics*, Boca Raton, Florida: CRC Press, 2008.
20. G.-B. Liu, W.-Y. Shan, Y. Yao, W. Yao and D. Xiao, *Physical Review B* **88**, 085433, 2013.
21. C. Gong, H. Zhang, W. Wang, L. Colombo, R.M. Wallace and K. Cho, *Applied Physics Letters* **103**, 053513, 2013.
22. R.-H. Yan, A. Ourmazd and K.F. Lee, Scaling the Si MOSFET: From bulk to SOI to bulk, *IEEE Transactions on Electron Devices* **39**, 1704–1710, 1992.
23. S. Datta, *Electronic Transport in Mesoscopic Systems*, Cambridge University Press, 1995.
24. H. Flietner, *Physica Status Solidi* **54**, 201–208, 1972.
25. J. Knoch and J. Appenzeller, Carbon nanotube field-effect transistors-the importance of being small, *AmIware Hardware Technology Drivers of Ambient Intelligence*, S. Mukherjee (ed.), Dordrecht, Netherlands: Springer Science & Business Media, 371–402, 2006.
26. J. Knoch, M. Zhang, S. Mantl and J. Appenzeller, *IEEE Transactions on Electron Devices* **53**, 1669–1674, 2006.

9

TMD-Based Photodetectors, Light Emitters and Photovoltaics

Thomas Mueller

Contents

2D Materials for Nanoelectronics edited by Michel Houssa, Athanasios Dimoulas and Alessandro Molle © 2016 CRC Press/Taylor & Francis Group, LLC. ISBN: 978-1-4987-0417-5.

THE (RE-)DISCOVERY OF GRAPHENE by A.K. Geim and K.S. Novoselov[1] in 2004 has led to renewed interest in two-dimensional (2D) materials. Beyond graphene, hundreds of other 2D materials (also called van der Waals materials) exist in bulk form that can be thinned down to atomically thin layers, in a similar manner as graphene can be obtained from graphite.[2] One important class of 2D materials is layered transition metal dichalcogenides (TMDs) that show a wide range of physical properties, depending on material and thickness. TMDs have been studied for decades, and, already back in 1963, the pioneering work by Frindt and Yoffe[3] demonstrated that these materials can be exfoliated into ultrathin crystals by mechanical cleavage.

TMDs offer properties that complement those of graphene. Graphene is a semi-metal with extremely high carrier mobility and peculiar electrical transport behaviour. However, the lack of a band gap in graphene hampers its use in many practical applications. In contrast, semiconducting TMDs possess a sizable band gap of the order of 1–2 eV. The existence of a band gap is a strict necessity for the realisation of energy-efficient transistors,[4] but even more so, for the realisation of efficient optoelectronic devices, especially light emitters and photovoltaic solar cells. Moreover, properties of monolayer materials differ significantly from their bulk characteristics. For example, a layer-dependent indirect–direct band gap transition was observed in TMDs, which resulted in bright photoluminescence (PL) from monolayers.[5] PL from bulk crystals, on the other hand, was found to be practically absent. Although optical and other properties of monolayer TMDs have been investigated for decades,[3,6–8] their role as active material in optoelectronics is new. Intensive research efforts on TMD optoelectronics have started merely 4 years ago and have already led to some exciting and promising new developments.

This chapter presents an overview on TMD-based optoelectronics, with a strong focus on devices. We will start by reviewing some basic optical properties in Section 9.1. The subsequent Sections 9.2 through 9.4 will address experimental realisations of various kinds of devices – photodetectors, light emitters and photovoltaic cells. Finally, in Section 9.5, we will conclude with a discussion of future challenges and opportunities, and a comparison of the performance of state-of-the-art TMD-based optoelectronic devices with currently available technologies.

9.1 Basic Optical Properties

9.1.1 Band Structure

TMDs are materials with the chemical formula MX_2, where M is a transition metal atom (e.g. Mo, W, Ti, Zr, Ta, Hf, etc.) and X is a chalcogen atom (S, Se and Te).[9–12] Some dozen different combinations of these elements are possible, and many of them are layered and stable 2D crystals. The atoms in 2D TMDs are arranged in sheets so that M-atoms are sandwiched between two layers of X-atoms, thus forming an X–M–X structure, as depicted in **Figure 9.1a**. As the intra-layer covalent bonds are much stronger than the interlayer van der Waals forces, bulk TMDs can be mechanically exfoliated down to films with monolayer thickness. The structure of these materials is similar to that of graphite with a monolayer thickness of approximately 0.6–0.7 nm.

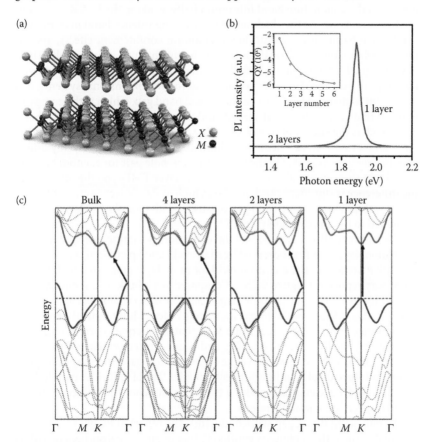

FIGURE 9.1 (a) Crystal structure of two layers of a TMD. M and X are transition metal chalcogen atoms, respectively. (Reprinted by permission from Macmillan Publishers Ltd. *Nat. Nanotechnol.*, Radisavljevic, B. et al., Single-layer MoS_2 transistors, 6:147–150, copyright 2011.) (b) PL spectra for mono- and bilayer MoS_2 flakes. Inset: PL quantum yield as a function of MoS_2 thickness (1–6 layers) on a logarithmic scale. (Reprinted with permission from Mak, K.F. et al. *Phys. Rev. Lett.* 105:136805. Copyright 2010 by the American Physical Society.) (c) Band structures of bulk, 4-layer, bilayer and monolayer MoS_2 as calculated from density-functional theory. Arrow indicates the lowest-energy transitions. (Reprinted with permission from Splendiani, A. et al. Emerging photoluminescence in monolayer MoS_2. *Nano Lett.* 10:1271–1275. Copyright 2010 American Chemical Society.)

In optoelectronics, mainly molybdenum- and tungsten-based dichalcogenides have so far been studied. These materials are semiconductors, and their properties depend critically on the number of layers. They possess an indirect band gap in bulk form, which becomes a direct one in monolayers.[5,13] An example is shown in **Figure 9.1c** that depicts density-functional theory band structure calculations for MoS_2 with various thicknesses.[13] Bulk MoS_2 has a band gap of approximately 1.3 eV, with the valence band maximum at the Γ point and the conduction band minimum halfway along the Γ–K direction in the Brillouin zone. As the thickness decreases, the valence band maxima and conduction band minima shift due to quantum confinement effects, and, in the monolayer limit, a direct semiconductor is obtained. Early publications have assigned a band gap of 1.85 eV to monolayer MoS_2, as estimated from the PL peak position. However, we note that in low-dimensional systems, it is important to distinguish between the optical band gap, which is the energy of photons that are emitted (or absorbed) and the electrical (single-particle) band gap, which is the energy of free carriers. In traditional three-dimensional (3D) semiconductors such as silicon or gallium arsenide, there is barely any difference between the two, due to the small electron–hole pair (or exciton) binding energies in these materials. In mono- and few-layer TMDs, on the other hand, the difference is significant and may be as high as 0.6 eV (see Section 9.1.2). In recent work, the electronic band gap of (suspended) monolayer MoS_2 has been estimated to be approximately 2.5 eV.[14]

The indirect–direct semiconductor transition is manifested as enhanced PL in monolayers, as first observed by Mak et al.[5] and Splendiani et al.[13] A 100-fold enhancement of the quantum yield for monolayer compared with bilayer MoS_2 was reported (**Figure 9.2b**).[5] Compared with bulk, the monolayer PL is even four orders of magnitude larger. Similar observations were made for other TMDs.[15] Mo- and W-based dichalcogenides exhibit optical band gaps in the range $E_X = 1$–2 eV (see **Table 9.1**), that is, energies corresponding to the technologically important visible and near-infrared wavelength regimes. Among them, MoS_2 and WS_2 have the largest optical gaps and may find application in future display or lighting technologies. Metal tellurides emit light at near-infrared wavelengths,[16] making them attractive for optical communications or spectroscopic applications.

Despite being direct gap semiconductors, the reported PL quantum yields (defined as the ratio of photons emitted to that of photons absorbed) of TMDs are still poor. For example, the room-temperature quantum yield of monolayer MoS_2 was found to be 4×10^{-3} only.[5] In metal selenides (WSe_2 and $MoSe_2$), more than an order of magnitude better yields were observed.[15,17] The origin of the low quantum yield is not yet fully understood. Likely, unintentional doping and crystal defects play a major role in suppressing the luminescence, but further studies will be necessary to elucidate the exact mechanism.

9.1.2 Excitons

The optical properties of 2D TMDs are dominated by excitonic rather than band-to-band transitions. Excitons are quasi-particles that form when

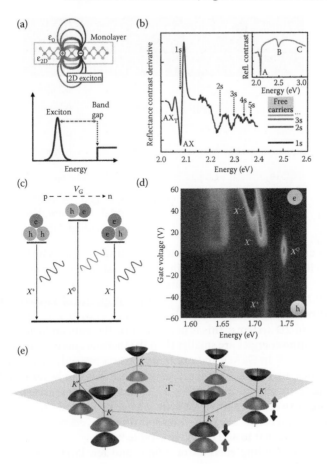

FIGURE 9.2 (a) Schematic drawing of an exciton in a 2D semiconductor (top image). ε_{2D} is the dielectric constant of the 2D material and ε_0 denotes the vacuum permittivity. The bottom image schematically illustrates exciton and band-to-band transitions. (b) Derivative of the reflectance contrast spectrum, shown in the inset, of a WS_2 monolayer. The exciton ground state is labelled 1s. The excited states are 2s, 3s and so on. Inset: reflectance contrast, showing A, B and C exciton transitions. (Reprinted with permission from Chernikov, A. et al. Exciton binding energy and nonhydrogenic Rydberg series in monolayer WS_2. *Phys. Rev. Lett.* 113:076802. Copyright 2014 by the American Physical Society.) (c) Evolution from positively charged, to neutral, and to negatively charged excitons as a function of gate voltage. (d) Gate voltage-dependent PL from monolayer WSe_2. (Reprinted by permission from Macmillan Publishers Ltd. *Nat. Nanotechnol.*, Jones, A.M. et al. Optical generation of excitonic valley coherence in monolayer WSe2. 8:634–638, copyright 2013.) (e) Band structure of monolayer MoS_2 showing opposite spin–orbit splitting of the valence bands at the K and K' points of the first Brillouin zone. (Reprinted with permission from Xiao, D. et al. Coupled spin and valley physics in monolayers of MoS_2 and other group-VI dichalcogenides. *Phys. Rev. Lett.* 108:196802. Copyright 2012 by the American Physical Society.)

Table 9.1 Basic Properties of Selected TMD Monolayers			
	E_X [eV]	ΔE_{so} [eV]	ΔE_b [eV]
MoS$_2$	1.85[7,13]	0.16[7,13]	0.57[14]
MoSe$_2$	1.55[15,18]	0.2[19]	0.55[20]
WS$_2$	2.0[17]	0.4[17]	0.32[21]
WSe$_2$	1.65[15,17]	0.4[17]	0.37[22]
MoTe$_2$	1.1[16]	0.25[16]	

Note: E_X denotes the optical band gap (A-exciton energy), ΔE_{so} is the spin–orbit (A–B exciton) splitting and ΔE_b is the exciton binding energy.

electrons and holes in a semiconductor are bound into pairs by the Coulomb force. Excitons are strongly bound in 2D materials due to the enhanced electron–hole interaction in 2D systems and the reduced dielectric screening (**Figure 9.2a**). As a result, the optical absorption spectra of TMDs exhibit pronounced peaks, rather than steps that would be expected for single-particle transitions in 2D systems. This is in contrast to 3D semiconductors, in which exciton binding energies are smaller than the thermal energy at room temperature (e.g. 4 meV in GaAs), and excitonic features are only observable at low temperatures. In TMDs, several excitonic transitions can be observed, where the two lowest-energy peaks, known as A- and B-excitons, arise from optical transitions between the spin–orbit split valence bands and the doubly degenerate conduction band.[5] The spin–orbit splitting ΔE_{SO} depends on the material and varies between 0.16 and 0.4 eV (see **Table 9.1**).

The exciton binding energies of mono- and few-layer TMDs have been determined by many groups, using various experimental and theoretical techniques.[14,20–25] As expected, it was found that the binding energy decreases with thickness due to stronger screening. It also depends on strain and the dielectric environment, which can be detected as shifts of the peak positions.[26] In **Figure 9.2b**, we show optical reflection measurements of monolayer WS$_2$ reported by Chernikov et al.[21] The A- and B-excitons are clearly visible (inset). A third peak, labelled C, stems from excitonic transitions associated with the van Hove singularity of WS$_2$. Multiple peaks were resolved in the derivative of the signal (main panel; labelled 1s, 2s, 3s and so on) that can be associated with the excitonic Rydberg series. By fitting the peaks to a hydrogenic Rydberg spectrum, an exciton binding energy of $\Delta E_b = 0.32$ eV was determined. For other monolayer TMDs, even higher values were found (see **Table 9.1**). It can be concluded that, due to their large binding energy, excitons in 2D semiconductors are thermally stable at room temperature and dominate the optical emission and absorption.

In addition to neutral excitons (X^0), negatively or positively charged excitons – so-called trions – have been studied in TMDs. This was accomplished by electrostatic control of the doping level in a TMD field-effect transistor

(**Figure 9.2c**).[19,24,27] Trions are quasi-particles that are composed of two electrons and a hole (X^-) or two holes and an electron (X^+). PL measurements by Jones et al.,[29] shown in **Figure 9.2d**, demonstrated the evolution from positively charged, to neutral, and then to negatively charged excitons as a function of gate voltage. The measurements directly revealed the trion binding energy (the energy difference between the trion $X^{+(-)}$ and neutral exciton X^0) in the range 20–40 meV, which far exceeds that of 3D semiconductors. Energies for positively and negatively charged trions were found to be similar, due to similar effective electron and hole masses. The trion energies shift as a function of gate voltage, likely as a result of the quantum confined Stark effect.[29]

Figure 9.2e shows the simplified band structure of a TMD monolayer. It exhibits two inequivalent valleys, labelled K and K', located at the corners of the Brillouin zone. Because of broken inversion symmetry, spin–orbit interactions split the valence bands by an energy ΔE_{SO}. This splitting can be assessed by spectroscopy, and gives rise to the aforementioned A- and B-exciton resonances in the optical spectra. Time-reversal symmetry requires that the spin splitting must be opposite at the K and K' valleys, as schematically illustrated by the arrows in **Figure 9.2e**. Right (left)-handed circularly polarised light only couples to the K (K') valley, as dictated by optical selection rules (conservation of energy, linear and angular momentum). Consequently, excitons are only excited into one of the two valleys, depending on the helicity of the light – spin and valley degrees of freedom are coupled.[28] Recently, several groups have demonstrated selective valley population in TMD monolayers using circularly polarised light excitation.[29,30] By measuring the polarisation of the emitted light, it was proved that valley polarisation was achieved and largely preserved during the exciton lifetime. The ability to control the valley polarisation, not only optically but also electrically, could provide a basis for constructing a new generation of 'valleytronic' devices for information processing.

9.2 Photodetectors

Photodetectors convert light into electrical signals and are among the most ubiquitous types of optoelectronic devices. The diversity of applications, ranging from optical communications to imaging and sensing, require a wide range of different materials and technologies. Most commercial photodetectors are based on elemental or compound semiconductors, but with growing demand for devices that offer flexibility, semi-transparency and better integrability at lower cost, the need for alternative photodetection materials is becoming more eminent.

Photodetection in 2D atomic crystals[31] was first studied in graphene,[32,33] which offers several distinct advantages compared to conventionally used 3D semiconductor materials. Graphene's gapless character enables photodetection over an extremely wide spectral range, from terahertz to ultraviolet wavelengths. Furthermore, graphene exhibits ultrafast carrier dynamics, enabling conversion of photons into electrical currents with multi-gigahertz electrical bandwidth.[34–37] Although graphene photodetectors are being envisaged for applications in optical communications[35-37] and long-wavelength (terahertz and mid-infrared) detection,[38,39] TMDs offer properties that complement those of graphene. In particular, TMDs are advantageous for applications that require high sensitivity and low dark currents, though they work mainly in the visible and (part of the) near-infrared spectral regions. Also, the combination of graphene with TMDs for the realisation of hybrid photodetection devices and van der Waals heterostructures is being actively explored. In the latter, graphene sheets are utilised as transparent electrodes, while TMDs are employed as photoactive material.

Broadly speaking, three groups of TMD-based photodetectors have been investigated: (i) lateral metal-TMD-metal detectors, whose device structure resembles that of field-effect transistors. These are mostly operated as phototransistors, which allows them to overcome the major drawback of photodiodes: unity gain; (ii) hybrid photodetectors, that use sensitising centres such as quantum dots, that absorb light, followed by transfer of charge carriers to the channel of a TMD transistor and (iii) vertically stacked heterostructure devices, in which a few-layer TMD semiconductor is sandwiched between graphene or transparent metal electrodes.

9.2.1 Phototransistors

Figure 9.3a shows a schematic drawing of a back-gated TMD field-effect transistor. This device structure has been employed by a number of groups for photodetection applications, and devices based on various 2D semiconductors, including MoS_2, WSe_2, WS_2, GaS, GaSe, GaTe, In_2Se_3, black phosphorus and others, have been fabricated and characterised.[40-53] In particular, we should mention the work by Yin et al.,[40] which constitutes, to our knowledge, the first demonstration of TMD-based light detection. Lopez-Sanchez et al.[41] and Zhang et al.[42] were the first who reported very high photogain and photoresponsivities of up to 2200 A/W. The photoresponsivity is one of the most important characteristics of a photodetector and measures the electrical output current I_{ph} per optical input power P_{opt}. It is defined as $R = I_{ph}/P_{opt}$ and is usually expressed in units of A/W. The term photogain or photoconductive gain refers to the fact that the number of circuit electrons generated per photogenerated electron–hole pair can be larger than one. Wavelength-dependent studies suggested that the photocurrent follows the TMD's absorption spectrum. Further, it was shown that the spectral responsivity of TMD photodetectors can be adjusted by using 2D layers of different thicknesses. Lee et al.[43] demonstrated that mono- and bi-layer MoS_2 photodetectors are effective for detecting green visible light, whereas triple-layer MoS_2 is better suited for red light detection (**Figure 9.3c**). Studies have also extended beyond mechanically exfoliated samples to chemical vapour deposition grown TMD layers.[42]

In general, a number of physical mechanisms can give rise to photocurrent generation in a TMD field-effect transistor. Most traditional semiconductor photodetectors are photodiodes, and rely on the separation of photoexcited electron–hole pairs in the built-in electric field of a p–n junction or Schottky (metal–semiconductor) junction due to the internal photoelectric effect.

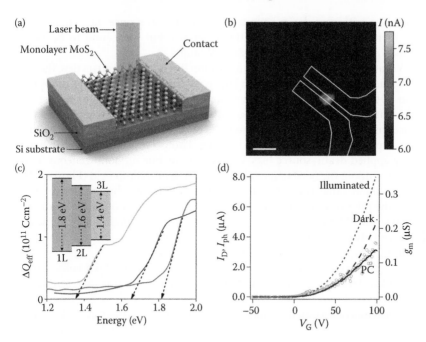

FIGURE 9.3 (a) 3D schematic of a MoS$_2$ phototransistor. (Reprinted by permission from Macmillan Publishers Ltd. *Nat. Nanotechnol.*, Lopez-Sanchez, O. et al., 8:497–501, copyright 2013.) (b) Spatial photocurrent map for a MoS$_2$ monolayer phototransistor, recorded by raster-scanning of a focused laser beam over the surface of the device. Scale bar, 5 μm. (Reprinted by permission from Macmillan Publishers Ltd. *Nat. Nanotechnol.*, Lopez-Sanchez, O. et al., 8:497–501, copyright 2013.) (c) Spectral response of triple-, bi- and monolayer MoS$_2$ photodetectors. The optical energy gaps, as indicated by arrows, are 1.35, 1.65 and 1.82 eV, respectively. Inset: schematic band diagrams for triple-, bi- and monolayer MoS$_2$. (Reprinted with permission from Lee, H.S. et al. MoS$_2$ nanosheet phototransistors with thickness-modulated optical energy gap. *Nano Lett.* 12:3695–3700. Copyright 2012 American Chemical Society.) (d) Electrical characteristics of a MoS$_2$ phototransistor in the dark (dashed line) and under illumination with P_{opt} = 5 nW (dotted line). The photocurrent is shown as solid line. The transconductance is plotted in the same graph on the right axis (symbols). (Reprinted with permission from Furchi, M.M. et al. Mechanisms of photoconductivity in atomically thin MoS$_2$. *Nano Lett.* 14:6165–6170. Copyright 2014 American Chemical Society.)

The electric field may also be produced or enhanced by means of an external bias voltage. In addition, photo-excitation can heat the electron gas and produce a photovoltage in a non-uniform device structure due to the photo-thermoelectric (Seebeck) effect. In graphene, for example, the Seebeck effect is now believed to be the dominant photocurrent generation mechanism.[54,55] Both photoelectric[46] and photo-thermoelectric[56] effects were also identified in TMD transistors using scanning photocurrent microscopy and device operation with zero drain–source bias. The photoresponse then comes from the metal–semiconductor (Schottky) interface, but was found to be rather weak with responsivities of a few mA/W only.

What makes the transistor configuration in **Figure 9.3a** attractive, though, is the fact that it can achieve much larger responsivities via internal amplification of the current. However, the trade-off between responsivity and speed, which is associated with the time required for the gain mechanism to take place, makes these devices only suitable for applications that do not require fast response times such as video imaging. Phototransistors require an externally applied drain–source bias V_D and can operate on homogeneous TMD sheets, without the need to introduce a junction or built-in field (**Figure 9.3b**). It was found[47] that two mechanisms contribute to the photoresponse under bias: the photovoltaic and the photoconductive effects. Both were studied in organic thin-film transistors before.[57] The photocurrent I_{ph}, defined as the difference between drain–source current I_D under illumination and dark current, $I_{ph} = I_{D,illum} - I_{D,dark}$, is the sum of both contributions.

The photovoltaic effect is the change in transistor threshold voltage from V_{th} in the dark to $V_{th} - \Delta V_{th}$ under optical illumination. It is therefore sometimes referred to as photogating effect. Its physical origin is the trapping of photoexcited holes into trap states that are present at the interface between the TMD and the underlying SiO_2 layer. This is exacerbated by the high surface-to-volume ratio of the 2D transistor channel, and SiO_2 surface treatment was shown to modify this effect.[41] As a result of threshold voltage shift, the photocurrent becomes $I_{ph} = I_D(V_G - V_{th} + \Delta V_{th}) - I_D(V_G - V_{th}) \approx g_m \Delta V_{th}$, where g_m is the transistor's transconductance and V_G is the gate voltage. **Figure 9.3d** shows electrical characteristics of a biased MoS_2 monolayer transistor in the dark (dash line) and under optical illumination (dotted line). The photocurrent I_{ph}, obtained by subtracting one curve from the other, is depicted as a solid line. A photoresponsivity of 10^3 A/W can be extracted, indicating strong photogain. In the same graph, we show the transconductance g_m of the transistor on the right axis (symbols). The similarity between I_{ph} and g_m is striking and suggests that the photovoltaic effect indeed dominates the photoresponse. Under high illumination intensities, trap states get filled, resulting in saturation of the photoresponse. Response times were found to be extremely slow, extending from 1 s to even tens of seconds depending on surface treatments.

In addition to the slow photovoltaic effect described above, a faster (but also weaker) response was identified and attributed to the photoconductive component of the photoresponse.[47] It is the increase in conductivity $\Delta \sigma$ due to photo-induced excess carriers, $I_{ph} = (W/L)V_D \Delta \sigma$, where W and L are the device dimensions. **Figure 9.4c** shows the responsivity for drain–source bias voltages of $V_D = \pm 5$ V. The measurements were performed at kHz-frequencies to suppress the photovoltaic response. At low incident optical power, a responsivity of 6 A/W was achieved, corresponding to a photoconductive gain of $\approx$100. The gain was attributed to valence band tail states, induced by disorder or structural defects (**Figure 9.4a**), into which photoexcited holes are trapped. Charge neutrality then requires the external circuit to provide additional electrons until recombination with the trapped holes occurs. The photoresponse was calculated by solving the

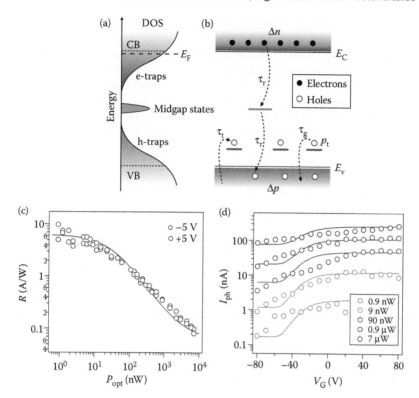

FIGURE 9.4 (a) Density-of-states in MoS$_2$. Band tail states underneath (above) the conduction (valence) band edge act as electron (hole) traps. (b) Simplified energy band diagram. The valence band tail is approximated by a discrete distribution of hole trap states. τ_r, τ_t and τ_g denote time constants with which carriers are trapped or released. (c) Power dependence of photoresponsivity (symbols, measurement; solid line, theoretical results). (d) Gate voltage dependence of photocurrent for different optical powers (symbols, measurement; solid line, theoretical results). (Reprinted with permission from Furchi, M.M. et al. Mechanisms of photoconductivity in atomically thin MoS$_2$. *Nano Lett.* 14:6165–6170. Copyright 2014 American Chemical Society.)

rate equations for the model in **Figure 9.4b** and good agreement with the measurement was obtained. The reduction of responsivity under strong illumination is due to filling of the hole traps. Electron trap states can be emptied by negative gate voltages that shift the Fermi level towards mid-gap thus, reducing the effective mobility and consequently the photocurrent (**Figure 9.4d**).

9.2.2 Hybrid Phototransistors

Photogain can also be accomplished by integration of 2D crystals with other materials to produce a hybrid phototransistor device. This was first demonstrated by Konstantatos et al. by sensitising the surface of a graphene transistor with lead sulphide (PbS) quantum dots.[58] Light absorption in the PbS quantum dots leads to charge transfer to the graphene sheet, which effectively modifies graphene's conductance. A similar concept was demonstrated Roy et al., who employed MoS_2 instead of quantum dots as photoactive material.[59] Although ultrahigh photoconductive gain and responsivities of 10^7–10^{10} A/W were achieved, these devices also suffered from high dark currents and large power consumption because of the semi-metallic character of the graphene channel.

In a second generation of hybrid phototransistors, Kufer et al. thus replaced the graphene channel with MoS_2, whose semiconducting behaviour provided the possibility to turn off the transistor channel by applying an appropriate back-gate voltage, leading to orders of magnitude lower dark current and high signal-to-noise ratio.[60] Light is absorbed in the PbS quantum dots and photoexcited carriers are separated at the PbS–MoS_2 interface. It was found that, while holes remain trapped in the dots, electrons circulate through the MoS_2-channel driven by the applied bias voltage. The high mobility in the MoS_2-channel yields an electron transit time that is much shorter than the hole lifetime in the dots, resulting in a photogain as explained above. The hybrid detector showed a responsivity of up to 10^6 A/W, and, although this value is smaller than what had been achieved with graphene before,[58] the low dark current resulted in a high detectivity of almost 10^{15} Jones, comparable to the best compound semiconductor photodetectors for the investigated wavelength regime. Response times were found to be of the order of 10^{-1} s.

9.2.3 Heterostructure Photodetectors

2D materials can also be assembled layer by layer to form so-called van der Waals heterostructures.[61] The surfaces of 2D materials display no dangling bonds and only weak van der Waals forces act between individual 2D layers. In contrast to traditional semiconductor heterostructures grown by epitaxy, the absence of lattice matching conditions in van der Waals heterostructures permits the combination of any arbitrary sequence of layered materials, and since the electronic properties of 2D materials vary considerably (conducting, semiconducting and insulating), a numerous amount of novel electronic and optoelectronic devices can be produced. For example, novel transistor concepts based on vertical electrical transport between two graphene sheets, separated by hexagonal boron nitride (hBN) layers, were demonstrated.[62-64]

The first heterostructure photodetectors, comprising vertical stacks of graphene–TMD–graphene layers, have been reported recently.[65,66] In these devices (**Figure 9.5a** and **b**), graphene was employed as work function tunable electrode, whereas a TMD semiconductor was utilised as the photoactive material, displaying strong light–matter interactions and photon absorption. The Fermi levels in graphene can be positioned by asymmetric chemical or electrostatic doping of the two graphene sheets, so that a built-in field is produced even without applying an external voltage between both electrodes. Upon illumination, photogenerated carriers are separated in the biased TMD

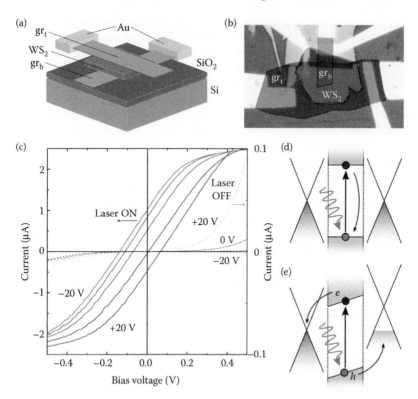

FIGURE 9.5 (a) Schematic and (b) optical micrograph of a TMD heterostructure photodetector. gr_t and gr_b are the top and bottom-graphene sheets. (c) Gate-dependent electrical characteristics taken under illumination (left axis) and in the dark (right axis). Schematic band diagrams (d) without and (e) with built-in electric field to separate the photogenerated carriers. (From Britnell, L. et al. 2013. Strong light–matter interactions in heterostructures of atomically thin films. *Science* 340:1311–1314. Reprinted with permission of AAAS.)

layer and collected within a short distance by the graphene electrodes (**Figure 9.5c** through **e**). This approach enables fast response times, primarily due to the short channel lengths. Response times down to 50 µs were achieved, limited by the parasitic capacitance of the sandwich structure. The photocurrent was mapped by scanning photocurrent microscopy, and it was verified that the photocurrent was generated across the whole graphene-TMD junction. The measured responsivities were of the order of 0.1 A/W.

9.3 Light Emitters

Electroluminescence (EL) is the result of radiative recombination of electrons and holes (or excitons) in a semiconductor in response to the passage of an electric current, and it forms the basis of light-emitting devices such as light-emitting diodes (LEDs) and semiconductor lasers. In semiconductors with a direct band gap, the radiative recombination occurs much more efficiently than in indirect gap materials. Being direct gap materials, monolayer TMDs are thus promising candidates for future low cost, large-area and flexible light emitters. Such as the PL, the EL from TMDs stems from excitonic recombination. In order to electrically induce excitons and achieve EL from TMDs, several device concepts have been realised.

9.3.1 Hot Carrier EL

Electrically driven light emission from TMDs was first obtained by exploiting hot carrier processes in monolayer MoS_2 by Sundaram et al.[67] Under high drain–source bias, strong band bending at the metal contact occurs in MoS_2 field-effect transistors. Electrons, injected into the conduction band, are accelerated by the high electric field in the vicinity of the contact. If these electrons acquire an energy that is larger than the MoS_2 optical band gap before being scattered by optical phonons, excitons can be generated via impact excitation – a process that is efficient in low-dimensional systems, as demonstrated before in carbon nanotubes,[68] nanowires[69] and other systems. The threshold electric field ε_{th} for impact excitation can roughly be estimated from[68] $\varepsilon_{th} \sim 1.5E_g/(q\lambda_{ph}) = 2$ MV/cm, where E_g is the MoS_2 band gap, λ_{ph} is the longitudinal optical (LO) photon scattering length and q denotes the electron charge. The strong electron–hole interactions and large exciton binding energies in 2D materials prevent excitons from dissociating in such strong fields.

Experimentally, light emission was only measureable at threshold power densities above 15 kW/cm^2, as electrons need to acquire sufficient kinetic energy for exciton generation. Besides the threshold behaviour, another characteristic feature of impact excitation is the exponential increase of the emission above threshold, being proportional to the impact excitation rate, $\exp(\varepsilon/\varepsilon_{th})$, where ε is the local electric field and ε_{th} is the threshold field.[68] As the electric field is the highest at the contacts, the EL is expected to be spatially localised near the MoS_2/metal interface. Mapping of the spatial distribution of the light emission confirmed that this is the case. The EL emission had a broad spectral width of 40 nm and peaked at the MoS_2 A-exciton wavelength, in accordance with PL measurements. The EL efficiency (ratio of emitted optical power and electrical input power) was estimated to be only 10^{-5} because of the impact excitation being an inherently rather inefficient process.

9.3.2 Light-Emitting Diodes

LEDs are today's most widely used solid-state light sources. A LED is a p–n junction diode, in which electrons and holes recombine radiatively in the depletion region, releasing energy in the form of photons. This process is thresholdless and more efficient than impact excitation. In traditional LEDs, semiconducting materials are doped with impurities to create the junction. As stable chemical doping is currently difficult to achieve in 2D semiconductors, electrostatic doping was used in three independent studies by Pospischil et al.,[70] Baugher et al.[71] and Ross et al.[72] to form a TMD monolayer p–n junction diode. In these devices, split-gate electrodes couple to two different regions of a mechanically exfoliated TMD flake, as illustrated in **Figure 9.6a**. By biasing one gate electrode with a positive voltage and the other one with a negative, electrons and holes, respectively, are drawn into the semiconducting channel and a lateral p–n junction is realised. In order to achieve both n- and p-type doping, ambipolar electrical transport behaviour of the 2D material is required. Ambipolar conduction is difficult to achieve in MoS$_2$ monolayers (because of its large band gap, high-electron affinity and strong Fermi-level pinning at the contacts[73]) and has, as yet, only been observed in devices with ionic liquid gating.[74] Motivated by electrical transport studies of bulk WSe$_2$ crystals that had previously shown ambipolar behaviour,[75] WSe$_2$ monolayer flakes were thus

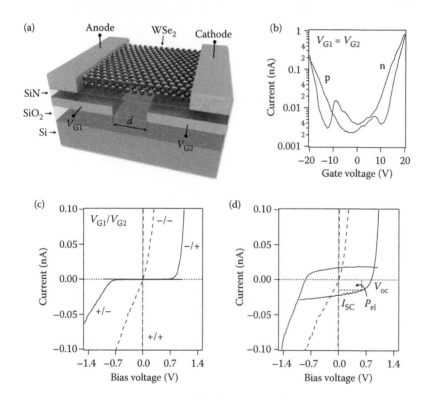

FIGURE 9.6 (a) 3D schematic of a WSe$_2$ monolayer device with split-gate electrodes. (b) Gate-dependent electrical characteristic of the device for $V_{G1} = V_{G2}$. Electrical characteristics (c) in the dark and (d) under illumination for different biasing conditions ($V_{G1} = \pm40$ V/$V_{G2} = \pm40$ V). V_{OC} is the open-circuit voltage, I_{SC} is the short-circuit current and P_{el} is the extracted electrical power. (Reprinted by permission from Macmillan Publishers Ltd. *Nat. Nanotechnol.*, Pospischil, A. et al., Solar-energy conversion and light emission in an atomic monolayer p–n diode, 9:257–261, copyright 2014.)

used in References 70, 71 and 72 for the realisation of TMD LEDs. The gating characteristics (**Figure 9.6b**), acquired by interconnecting the two gate electrodes, exhibit indeed ambipolar behaviour, demonstrating that both electrons and holes can be drawn into the WSe$_2$-channel. Applying voltages of opposite polarities to the gate electrodes leads to current rectification, whereas by applying voltages of same polarity, the device operates as a resistor (**Figure 9.6c**). **Figure 9.6d** shows that the 2D p–n junction exhibits a photovoltaic effect under illumination – we will come back to that in Section 9.4.1.

By driving a forward current through the electrostatically defined diode, EL emission was obtained. If, on the other hand, a unipolar current was driven through the device, no light emission was observed, demonstrating that the EL arises from the ambipolar injection of both n- and p-type carriers and not impact excitation. The emission spectrum, depicted in **Figure 9.7**, occurs at the same wavelength as the PL, indicating that the EL arises

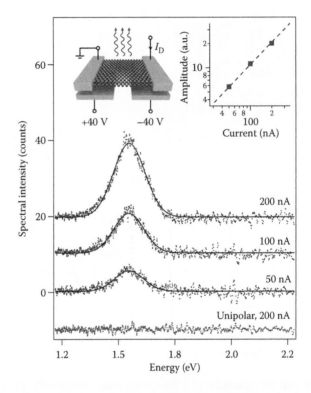

FIGURE 9.7 EL emission spectra for gate voltages as shown in the left inset and constant currents of 50, 100 and 200 nA. The bottom curve demonstrates that no light emission is obtained under unipolar conduction. Right inset: the emission amplitude versus current shows linear dependence. (Reprinted by permission from Macmillan Publishers Ltd. *Nat. Nanotechnol.*, Pospischil, A. et al., Solar-energy conversion and light emission in an atomic monolayer p–n diode. 9:257–261, copyright 2014.)

from excitonic transitions. The large exciton binding energy in TMDs may thus offer an opportunity for tailoring the emission wavelength by engineering of the dielectric environment. Band gap emission from free carriers was not observed. EL in such devices was measureable at injection currents down to[72] 200 pA – several orders of magnitude lower than in unipolar devices described in Section 9.3.1. Also, the linear dependence of emission intensity as a function of current (inset in **Figure 9.7**) is in contrast to the exponential behaviour of the impact excitation process.[67] EL efficiencies of up to 0.1% were obtained, limited by resistive losses, non-radiative recombination in WSe$_2$ and, possibly, the occupation of dark exciton states.

Low-temperature measurements performed on 2D LEDs revealed that the broad room-temperature EL peaks, shown in **Figure 9.7**, are composed of multiple emission lines, resulting in complex EL spectra.[72] The dominant exciton species are tunable with gate voltage and their origins could be deduced by comparison with PL spectra recorded on the same device. It was found that the negatively charged trion X^- dominates the EL, in accordance with the observation that X^- also dominates the PL in WSe$_2$ (see Section 9.1.2). In addition, EL peaks from positively charged trions X^+, neutral excitons X^0 and impurity-bound excitons X^I were identified. Emission from trions was found to be spectrally broader than that from X^0. As aforementioned, the trion binding energy depends on the perpendicular electric field. The spectrally broad trion EL emission thus likely stems from a field variation across the junction of the device. The X^0 peak, on the other hand, barely shifts with gate voltage and the X^0 EL thus was found to be narrow with a linewidth of 5 meV only. Similar studies, later performed on WS$_2$, yielded similar results.[76]

9.3.3 Circularly Polarised Light Emission

As outlined in Section 9.1.2, due to the strong spin–orbit coupling in (monolayer) TMDs and their peculiar band structure, excitonic interband transitions in these materials are associated not only with the spin degree of freedom but also the valley degree of freedom, which could be exploited in the future for a new valleytronics technology. All-optical generation and detection of valley polarisation in TMDs has been demonstrated in several works,[29,30] but reports on the realisation of active valley-dependent optoelectronic devices are still rare. In the following, we will review one such implementation – a circularly polarised light source in which the EL polarisation is electrically switchable.[77] These devices generate the circularly polarised light in lateral p–n junctions similar to those described above.

Zhang et al. fabricated WSe$_2$ p–n junctions with the p and n-regions being defined using an electric double-layer gating technique.[77] Interestingly, this work demonstrated efficient light emission from few-layer flakes, although WSe$_2$ few-layers are indirect band gap semiconductors, in which the valence band maximum at the Γ point and the conduction band minimum located along the Γ–K direction in the Brillouin zone. Electric double layers, formed at solid-electrolyte interfaces, are capable of inducing extremely large gate fields, resulting in carrier accumulations in the 2D channel that can be an order of magnitude or more higher than that achieved with solid dielectrics.[74] As a result, the induced carrier density is sufficiently high, such that both electrons and holes can populate the (direct gap) K and K' valleys. The strong gate field also breaks the inversion symmetry of few-layer flakes and recovers the valley circular dichroism that is otherwise only present in monolayers. We note that EL from multilayers of MoS_2 with efficiencies comparable to that in mono-layers was also obtained in $GaN–Al_2O_3–MoS_2$ and $GaN–Al_2O_3–MoS_2–Al_2O_3–$graphene vertical heterojunctions, owing to the high electric field induced redistribution of carriers from the indirect to the direct band gap.[78]

EL spectra of the WSe_2 p–n junction[77] for opposite current flow directions are shown in **Figure 9.8a**. Clearly, the circular EL polarisation can be controlled with current flow direction. If the device is operated as a p–n diode and the current is driven from left to right, as shown in the inset in the right panel of **Figure 9.8a**, the right-handed circularly polarised light emission (σ^+) dominates over the left-handed emission (σ^-). On the other hand, if the current flow direction is reversed (n–p diode; left panel in **Figure 9.8a**), the σ^- polarisation dominates over σ^+. The degree of circular polarisation reached values as high as 45%, comparable to that of PL from monolayers. To understand this behaviour, we recall that circularly polarised luminescence in TMDs occurs when the K and K' valleys are differently populated with electron–hole pairs. It was suggested, that the electron–hole overlap can be controlled by the in-plane electric field of the 2D diode. Under an in-plane (bias) field, the hole

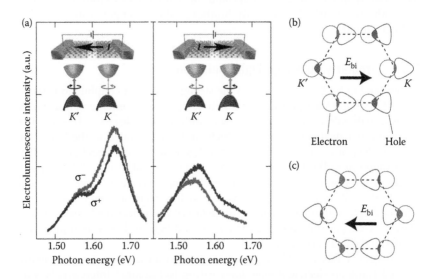

FIGURE 9.8 (a) Circularly polarised EL from a TMD lateral p–n diode for two opposite current flow directions, as shown in the insets. Schematic illustrations of electron and hole distributions for both cases are shown in (b) and (c). (From Zhang, Y.J. et al. 2014. Electrically switchable chiral light-emitting transistor. *Science* 344:725–728. Reprinted with permission of AAAS.)

and electron distributions in momentum space are shifted from their equilibrium positions in direction of the field (holes) or opposite to it (electrons). We depict these shifts in **Figure 9.8b** and **c** by plotting the contours of the electron and hole distributions in momentum space. The shaded areas represent the electron–hole overlaps, which differ in the K and K' valleys. This is because of the trigonal distortions of the band structure around the K-point – especially that of the valence band. The trigonal warping effect provides a luminescence intensity difference between both valleys, leading to circularly polarised emission. Similar effects were not only observed in WSe_2, but also other TMDs such as $MoSe_2$ and MoS_2.[77]

9.3.4 Heterostructure Light Emitters

All light-emitting devices presented so far use lateral arrangements of electrodes to achieve in-plane current flow in a TMD (mono-)layer. More recently, light emitters based on vertical structures have been demonstrated in which the current flow is perpendicular to the 2D layers. The vertical heterostructure design allows the improvement of the performance of 2D light emitters in several respects: reduced contact resistance due to the larger contact area, higher achievable current densities which allows for brighter light emission, luminescence from the whole device area, and easier scalability. Vertical structures can be fabricated by stacking of 2D materials in a layered configuration on top of each other, or on top of a traditional 3D semiconductor such as silicon. The van der Waals interaction between the 2D (or 2D and 3D) materials keeps the stack together and a van der Waals heterostructure is formed.

Electrically driven light emission from vertical p–n heterojunctions was demonstrated by Ye et al.[79] and Lopez-Sanchez et al.[80] Their devices were composed of a monolayer of MoS_2 – which is intrinsically strongly n-doped – and a silicon wafer. The silicon was heavily p-doped and served as a hole injection layer into MoS_2. Light emission was localised at the heterojunction edge[79] or – in hydrogen passivated silicon substrates – occurred across the entire area of the heterojunction.[80] EL came from direct excitons related to the optical transitions between the conduction and valence bands in MoS_2. Both A- and B-exciton emission were observed, with the B-exciton likely being excited by impact excitation. Radiative emission from silicon was not observed due to its indirect band gap. In addition, bound-exciton-related emission features were identified and, at a high injection rates, exciton–exciton annihilation[81] of the bound excitons was studied.[79]

Figure 9.9a schematically depicts the architecture of a more complex van der Waals heterojunction, demonstrated by Withers et al.[82] It was made by stacking of metallic (graphene, Gr), insulating (hBN) and semiconducting (TMD monolayers; WS_2 or MoS_2) 2D materials, to realise a vertical LED. In this device, electrons and holes are injected into a TMD monolayer from two graphene electrodes. Boron nitride layers to both sides of the TMD were used to reduce the direct tunnelling between the graphene sheets, which led to electron and hole accumulation in the semiconductor, as schematically shown in **Figure 9.9b**. EL emission (**Figure 9.9c** and **e**) was obtained above a certain

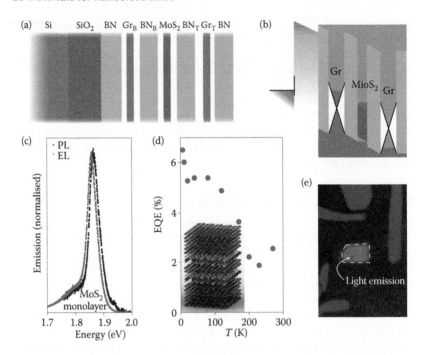

FIGURE 9.9 (a) Schematic drawing of a Si/SiO$_2$/hBN/Gr/hBN/MoS$_2$/hBN/Gr/hBN heterostructure LED. Gr, graphene. (b) Band diagram for biased heterostructure. (c) Comparison of PL and EL spectra (normalised). (d) Temperature dependence of external quantum efficiency. Inset: schematic representation of a device with two active TMD layers. (e) Optical image of EL from a heterostructure LED. The heterostructure area is highlighted by dashed lines. The other structures are metal contacts. (Reprinted by permission from Macmillan Publishers Ltd. *Nat. Mater.*, Withers, F. et al. Light-emitting diodes by band-structure engineering in van der Waals heterostructures, 14:301–306, copyright 2015.)

threshold, associated with the alignment of the top and bottom-graphene Fermi levels with the TMD conduction and valence band edges, respectively. The EL emission wavelength was found to be similar to that of the PL emission from negatively charged excitons, suggesting more efficient electron than hole injection into the semiconductor. An external quantum efficiency of more than 1% was obtained – 10 times larger than that of the planar p–n diodes presented in Section 9.3.2. This improvement is likely due to the lower contact resistance. The quantum efficiency was further improved by the stacking of multiples of those devices in series, as depicted in **Figure 9.9d**. At low temperatures (7 K), efficiencies of up to 8.4% were reported, comparable to that of state-of-the-art organic LEDs. The efficiencies at room temperature are two to three times lower than those at 7 K. Also samples on elastic and transparent substrates were prepared, demonstrating the capability of such devices for flexible and semi-transparent optoelectronics.

9.4 Photovoltaics

Photovoltaic cells allow the conversion of solar energy into electricity. Typically, inorganic or organic semiconducting materials are used for that purpose, whose band gap, according to the Shockley–Queisser theory,[83] ideally should be near 1.34 eV. Few- and monolayer TMDs are well positioned in terms of optical band gap and also mobility, making them promising candidates for photovoltaics. With an optical gap of 1.65 eV, monolayer WSe_2, for example, allows energy conversion with a theoretical maximum efficiency of approximately 30%. It is important to realise, though, that the electronic band gap in TMDs is higher than the optical one due to large exciton binding energies in TMDs (see Section 9.1.2). Excitons that are excited below the electronic gap must thus dissociate to generate a current, like in organic solar cells. From a technological point of view, abundance of source materials and low cost of production are essential requirements in photovoltaics. TMDs can potentially fulfill these requirements, making them interesting candidates as light-absorbing material in thin-film solar cells, including flexible photovoltaics.

Theoretical studies of graphene/MoS_2[84] and WS_2/MoS_2[85] (all monolayers) heterojunction solar cells suggested achievable power conversion efficiencies of 1% and 1.5%, respectively, limited by the low optical absorption of the atomically thin materials. Absorption enhancement using plasmonics[86] or vertical stacking of ultrathin cells will be necessary for these devices to become competitive.

9.4.1 Photovoltaics in Lateral p–n Junctions

The suitability of TMDs for photovoltaics was first studied in lateral p–n and Schottky junctions. **Figure 9.6d** shows electrical characteristics of the WSe_2 monolayer diode, presented in Section 9.3.2, under optical illumination with $P_{opt} = 1400$ W/m². When biased in diode configuration, it was observed that the device produces both a current and a voltage, and that electrical power $P_{el} = FFV_{OC}I_{SC}$ was delivered to an external load.[70] The monolayer p–n junction can thus be used for photovoltaic energy conversion. *FF* is the fill factor which is a key parameter in evaluating the performance of a solar cell. It is defined as the ratio of maximum obtainable power to the product of the open-circuit voltage V_{OC} and short-circuit current I_{SC}. A fill factor of $FF = 50\%$ was achieved, which is a promising result, although lower than that of commercial solar cells ($FF > 70\%$). As expected, the electrical device characteristics were not affected by light when the device was operated as resistor (dashed lines). The power conversion efficiency (the percentage of the incident light energy that is converted into electrical energy) of the device was $\eta = P_{el}/P_{opt} = 0.5\%$. This value is comparable to efficiencies reported for bulk TMD solar cells,[87,88] however, still much lower than what would be required for practical applications. The reason for the low efficiency is mainly the greater than 90% optical transparency of the ultrathin TMD monolayer. For similar devices, which employed multilayer $MoSe_2$ instead of a monolayer, power conversion efficiencies surpassing 5% and fill factors of 70% were reported.[89] Resistive dissipation losses were found to be negligible. Losses associated with carrier recombination via trap states in the WSe_2 monolayer were found to reduce the efficiency by approximately 40%, suggesting further room for improvement by improving the material quality.[70] Moreover, by choosing an appropriate TMD layer thickness, the trade-off between optical transparency and efficiency may be adjusted according to the application requirements. Similar results were obtained using other 2D semiconductors such as black phosphorus.[90] Multi-junction solar cells could hence potentially be constructed by stacking of 2D materials with different band gaps, which would allow for more efficient absorption of the solar spectrum without losses due to carrier thermalisation.

Lateral built-in fields in 2D semiconductors also occur at Schottky junctions. Fontana et al.[91] fabricated MoS_2 field-effect transistors with different drain and source contact metals, namely Pd and Au. When the device was irradiated with light, photogenerated electron–hole pairs were separated by the built-in potential from the space charge at the contacts. Electrons accumulate on the Au contacts and holes accumulate on the Pd side, giving rise to an open-circuit voltage and a sizable photovoltaic effect.

9.4.2 Photovoltaics in van der Waals Heterojunctions

Lateral p–n and Schottky junctions, as discussed above, have demonstrated the capability of TMDs for possible future applications in photovoltaics. However, the lateral arrangement of the junctions in these devices does not allow for easy scalability for which a vertical geometry would be desirable. Such vertical heterostructures can be obtained by stacking of 2D materials in a layered configuration to realise a van der Waals heterojunction. Ultrathin photovoltaic cells could potentially be manufactured on flexible substrates using low-cost manufacturing techniques such as roll-to-roll deposition or printing processes.

Photovoltaic effects in van der Waals p–n heterojunctions were first demonstrated by Furchi et al.,[92] and independently, by Lee et al.[93] **Figure 9.10b** shows the energy band diagram of one of those devices; it consists of monolayers of MoS_2 and WSe_2. The device structure is depicted in **Figure 9.10a**. The electron affinity (the energy required to excite an electron from the bottom of the conduction band to vacuum) of MoS_2 is larger than that of WSe_2, owing to the larger electronegativities of Mo and S as compared to W and Se. In the overlap region of the two semiconductors, a heterojunction is formed where the heterostructure bands can be considered as a superposition of the monolayer bands.[94] The lowest-energy electron states are spatially located in the MoS_2 layer and the highest-energy hole states lie in the WSe_2, thus forming a type-II heterojunction. An important requirement for the efficient charge transfer between neighbouring layers, and high photovoltaic efficiencies, is a good quality of junction interface. It can be improved by heating the junction in vacuum to drive out water and other molecules that are trapped between the layers. PL studies[92,93,95] and optical pump/probe measurements[96,97] confirmed efficient and ultrafast charge transfer between TMDs with almost 100% efficiency and less than 100 fs transfer time.

Figure 9.10c shows the current map as a function of gate (V_G) and bias (V) voltages. The electrical characteristics are different from those obtained in ordinary TMD field-effect transistors, but can readily be understood by the following considerations. While the MoS_2-channel remains n-type over most of the observed gate voltage range and gets fully depleted only at large negative gate voltages, the WSe_2-channel switches from n- to p-type due to its ambipolar behaviour. The electrical characteristics of the device can thus be controlled by electrostatic doping, and in the range $-71\ V < V_G < -47\ V$, a vertical p–n junction is formed. The device current as a function of bias voltage V displays diode-like rectification behaviour, consistent with the type-II band alignment. At $V_G > -11\ V$, on the other hand, MoS_2 and WSe_2 are both n-type and current can flow under positive and negative V-polarities.

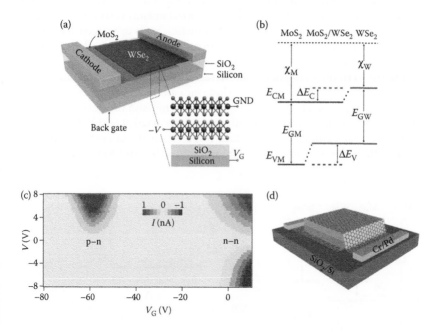

FIGURE 9.10 (a) Schematic drawing of a type-II heterojunction solar cell. (b) Schematic energy band diagram. The MoS_2/WSe_2 heterostructure bands are a superposition of the monolayer bands. The lowest-energy electron states are spatially located in the MoS_2 layer and the highest-energy hole states are located in the WSe_2 sheet. Excited states are shown as dashed lines. (c) Current map, recorded by scanning gate and bias voltages. (Adapted from Furchi, M.M. et al. 2014. *Nano Lett.* 14:4785–4791.) (d) Schematic of a van der Waals p–n heterojunction embedded between two graphene layers for vertical carrier extraction. (Reprinted by permission from Macmillan Publishers Ltd. *Nat. Nanotechnol.*, Lee, C.-H. et al. Atomically thin p–n junctions with van der Waals heterointerfaces. 9:676–681, copyright 2014.)

It is important to recognise that the rectification behaviour of a van der Waals heterojunction diode is of different origin than that of a conventional 3D semiconductor diode. Two mechanisms can contribute to the current: forward biasing the diode drives electrons (holes) in MoS_2 (WSe_2) to the heterojunction, where they may either overcome the conduction and valence band offsets to be injected into the neighbouring layer and diffuse to the opposite contact electrode, or recombine so that a continuous current is maintained. Whereas in an ideal diode the first process dominates, it was found that in atomically thin p–n junctions, the latter is more important.[93] Owing to the absence of a depletion region in an ultrathin heterojunction, majority carriers (electrons in MoS_2 and holes in WSe_2) are in close proximity to each other, which gives rise to strong interlayer carrier recombination. The recombination can either be of Langevin or Shockley–Read–Hall type and the recombination rate can be expressed as $R \sim n_{MoS2}p_{WSe2}$,[93] where n_{MoS2} and p_{WSe2} are the majority electron and hole densities in MoS_2 and WSe_2, respectively. The relative importance of both recombination mechanisms is still under discussion.[92,93] In any case, forward biasing the junction leads to an accumulation of majority carriers and enhances recombination (and current), whereas depletion under reverse bias reduces recombination (current).

Figure 9.11b shows the electrical characteristics of the junction under optical illumination with white light. In these measurements, the incident optical power P_{opt} was varied between 180 and 6400 W/m^2, and the device was operated in the p–n junction regime with gate voltage fixed at −50 V. It is obvious that the device exhibits a photovoltaic response, as the curves pass through the fourth quadrant. Photons are absorbed in WSe_2 and MoS_2, resulting in

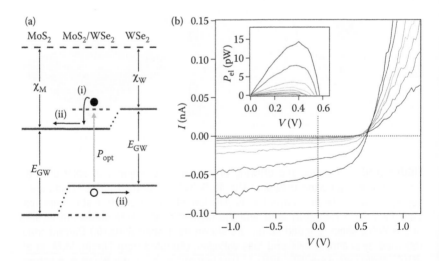

FIGURE 9.11 (a) Schematic illustration of the photovoltaic effect in a van der Waals heterostructure. (b) Electrical characteristics of the device under optical illumination with 180–6400 W/m^2. Inset: electrical power that is extracted from the device. (Adapted from Furchi, M.M. et al. 2014. *Nano Lett.* 14:4785–4791.)

electron–hole pairs in both layers. Relaxation of the photogenerated carriers then occurs, driven by the type-II band offsets. As the lowest-energy electron and hole states are spatially separated, charge transfer occurs across the 2D heterojunction. The relaxed carriers diffuse to the contacts, resulting in a photocurrent. Interlayer recombination occurs during diffusion, which reduces the efficiency of the solar cell. This process is schematically illustrated in **Figure 9.11a** for a photon that is absorbed in the WSe_2 layer (a similar diagram can be drawn for photons absorbed in MoS_2).

Power conversion efficiency and fill factor were estimated to be 0.2% and 50%, respectively. The device was also operated as a photodiode by biasing it in the third quadrant, where a responsivity of 11 mA/W was obtained. We note that these numbers need again be judged in light of the weak optical absorption of the 2D monolayers. It could be increased by stacking several junctions on top of each other or by plasmonic absorption enhancement.[86] Moreover, it was shown that better efficiency can be achieved by sandwiching the semiconductor junction between graphene electrodes for vertical carrier extraction (**Figure 9.10d**).[93] Photovoltaic effects in van der Waals heterojunctions were also demonstrated with other material combinations such as MoS_2/black phosphorus.[98]

The data presented above were obtained using a spectrally broad white-light source, which not only excites free carriers above the electronic band gap, but also below-gap excitons. These can spontaneously dissociate into free carriers and produce a photocurrent if the conduction and valence band offsets are larger than the exciton binding energy. By comparison of results obtained under resonant WSe_2 exciton excitation with those obtained by above-band gap excitation, it was verified that this condition is indeed fulfilled.[92]

Vertical p–n junctions have also been fabricated from single TMD materials using plasma-assisted doping[99] and from MoS_2/Si type-II heterojunctions. Wi et al.[99] treated a multilayer MoS_2 flake, which is intrinsically n-doped, with a plasma species to induce p-doping in the top MoS_2 layers. A vertically stacked Au/p-MoS_2/n-MoS_2/indium tin oxide (ITO) photovoltaic cell was then processed and power conversion efficiencies of up to 2.8% were achieved. Tsai et al.[100] utilised the built-in field generated at the interface between p-doped Si and a n-MoS_2 monolayer, and reported an efficiency of 5.2%.

9.5 Conclusions and Outlook

Generally speaking, TMDs offer two potential advantages over classical materials currently employed in optoelectronics. The first advantage is technological. Today, the rigidity, heavy weight and high costs of production of 3D semiconductors hinder their seamless integration into everyday objects. This especially concerns applications that require large-area dimensions such as display panels or photovoltaic solar cells. Therefore, alternative technologies are currently employed, with organic semiconductors and thin-film technologies being the most prominent ones. But these are notoriously known for their low material quality and degradation issues. In contrast, 2D semiconductors are crystalline and of high material quality and stability. They are also light-weight and bendable, thus providing opportunities for novel applications. However, the prospects for commercialisation will not only depend on device performance that can be achieved under laboratory conditions, but also on device reliability and their ability to be manufactured in large volumes. A major outstanding challenge remains the growth of high-quality and uniform TMD monolayer sheets with large-area dimensions. Almost all devices presented in the preceding section have been fabricated from mechanically exfoliated flakes with device dimensions of typically less than 100 μm^2. In recent years, considerable progress in TMD monolayer growth has been made, but it has been found to be more difficult than the growth of graphene. Also routes to control the chemical doping of TMDs, for example, for the construction of lateral and vertical p–n junctions, still need to be devised. Chemical vapour deposition is currently considered the most promising method for growing TMD monolayers, but as this process requires high temperatures, direct growth on plastic substrates seems to be not feasible. Therefore, large-area monolayer transfer techniques have to be developed. Similarly, solution-based methods for TMD deposition require further improvement.

Another advantage of TMDs is the rich physics that these materials offer. Their atomically thin nature presents unique opportunities that could eventually lead to novel device concepts. In particular, the large exciton binding energy in 2D semiconductors could be exploited for devices that allow the controlled optical or electrical generation and manipulation of excitons at room temperature. This could eventually lead to ultrafast optical switches and modulators. The ability to generate and control valley polarisations in monolayer TMDs, may provide the basis for valleytronic devices, for example, for quantum computation with valley-based qubits. Moreover, the vast number of possibilities of stacking different 2D materials into heterostructures could allow for the construction of designer materials with tailored properties.

We conclude by comparing the performance of state-of-the-art TMD optoelectronic devices with existing technologies. Given the fact that TMD optoelectronics research started merely 4 years ago, the field has made remarkable progress. It is clear now, that 2D TMDs hold great promise for optoelectronics and some device concepts have already reached some level of maturity.

TMD-based photodetectors exhibit photoresponsivities that are comparable to that of state-of-the-art compound semiconductor detectors and are beginning to achieve background-limited performance. However, their response time currently is too slow, even for video imaging applications. Other remaining challenges are the large-scale production and integration of multi-pixel arrays, but all these problems appear to be solvable. The lack of a killer application is a more serious issue. Owing to their large band gap, TMD-based photodetectors work mainly in the visible – a spectral regime that is already well covered by highly mature silicon technology. Infrared cameras (e.g. for night vision), on the other hand, are currently mainly based on compound semiconductors or micro-bolometers and are expensive. In this context, hybrid phototransistors, that combine TMDs with other materials such as quantum dots, allow the extension of the spectral response towards the infrared and could be of potential interest for infrared imaging.

Research on *TMD-based light emitters* has made large progress recently. Efficiencies comparable to that of state-of-the-art organic LEDs can now be achieved. The remaining challenge here is the extension to shorter wavelengths. Among all TMDs studied so far, WS_2 has the largest band gap and emits red light (620 nm wavelength). For full-colour displays, in addition green and blue light emitters with sufficiently high colour purity will be required. Given the large exciton binding energies of TMDs, it appears conceivable to shift the emission to somewhat shorter wavelengths, for example, by engineering of the dielectric environment, but it will not be possible to accomplish blue light emission. Other 2D semiconductors that possess larger band gaps may allow the overcoming of this limitation. At the time of writing of this chapter, lasing in 2D semiconductors had not yet been demonstrated. However, no fundamental difficulties are foreseen, and given the considerable research activity in this field, we anticipate that that it can soon be achieved.

Finally, we comment on the prospects of *TMD-based photovoltaics*. First experiments are encouraging – in mechanically exfoliated flakes power conversion efficiencies surpassing 5% and fill factors of 70% were achieved. Higher efficiencies seem feasible if the material quality can be improved. However, a major challenge will be the scaling of microscopically small proof-of-principle devices to macroscopic dimensions. This will require large-area van der Waals growth of heterostructures to realise vertical p–n junctions. Further, stacking of multiple junctions will be necessary to achieve complete light absorption.

Acknowledgement

Support by the Austrian Science Fund FWF (START Y 539-N16) is acknowledged.

References

1. Novoselov, K.S. et al. 2004. Electric field effect in atomically thin carbon films. *Science* 306:666–669.
2. Novoselov, K.S., Jiang, D., Schedin, F., Booth, T.J., Khotkevich, V.V., Morozov, S.V. and Geim, A.K. 2005. Two-dimensional atomic crystals. *Proc. Natl Acad. Sci. USA* 102:10451–10453.
3. Frindt, R.F. and Yoffe, A.D. 1963. Physical properties of layer structures: Optical properties and photoconductivity of thin crystals of molybdenum disulphide. *Proc. R. Soc. A* 273:69–83.
4. Radisavljevic, B., Radenovic, A., Brivio, J., Giacometti, V. and Kis, A. 2011. Single-layer MoS_2 transistors. *Nat. Nanotechnol.* 6:147–150.
5. Mak, K.F., Lee, C., Hone, J., Shan, J. and Heinz, T.F. 2010. Atomically thin MoS_2: A new direct-gap semiconductor. *Phys. Rev. Lett.* 105:136805.

6. Wilson, J.A. and Yoffe, A.D. 1969. Transition metal dichalcogenides: Discussion and interpretation of observed optical, electrical and structural properties. *Adv. Phys.* 18:193–335.

7. Joensen, P., Frindt, R.F. and Morrison, S.R. 1986. Single-layer MoS_2. *Mater. Res. Bull.* 21:457–461.

8. Yoffe, A.D. 1993. Layer compounds. *Annu. Rev. Mater. Sci.* 3:147–170.

9. Wang, Q.H., Kalantar-Zadeh, K., Kis, A., Coleman, J.N. and Strano, M.S. 2012. Electronics and optoelectronics of two-dimensional transition metal dichalcogenides. *Nat. Nanotechnol.* 7:699–712.

10. Chhowalla, M., Shin, H.S., Eda, G., Li, L.-J., Loh, K.P. and Zhang, H. 2013. The chemistry of two-dimensional layered transition metal dichalcogenide nanosheets. *Nat. Chem.* 5:263–275.

11. Butler, S.Z. et al. 2013. Progress, challenges and opportunities in two-dimensional materials beyond graphene. *ACS Nano* 7:2898–2926.

12. Jariwala, D., Sangwan, V.K., Lauhon, L.J., Marks, T.J. and Hersam, M.C. 2014. Emerging device applications for semiconducting two-dimensional transition metal dichalcogenides. *ACS Nano* 8:1102–1120.

13. Splendiani, A. et al. 2010. Emerging photoluminescence in monolayer MoS_2. *Nano Lett.* 10:1271–1275.

14. Klots, A.R. et al. 2014. Probing excitonic states in suspended two-dimensional semiconductors by photocurrent spectroscopy. *Sci. Rep.* 4:6608.

15. Tonndorf, P. et al. 2013. Photoluminescence emission and Raman response of monolayer MoS_2, $MoSe_2$ and WSe_2. *Opt. Express* 21:4908–4916.

16. Ruppert, C., Aslan, O.B. and Heinz, T.F. 2014. Optical properties and band gap of single- and few-layer $MoTe_2$ crystals. *Nano Lett.* 14:6231–6236.

17. Zhao, W., Ghorannevis, Z., Chu, L., Toh, M., Kloc, C., Tan, P.-H. and Eda, G. 2013. Evolution of electronic structure in atomically thin sheets of WS_2 and WSe_2. *ACS Nano* 7:791–797.

18. Tongay, S. et al. 2012. Thermally driven crossover from indirect toward direct bandgap in 2D semiconductors: $MoSe_2$ versus MoS_2. *Nano Lett.* 12:5576–5580.

19. Ross, J.S. et al. 2013. Electrical control of neutral and charged excitons in a monolayer semiconductor. *Nat. Commun.* 4:1474.

20. Ugeda, M.M. et al. 2014. Giant bandgap renormalization and excitonic effects in a monolayer transition metal dichalcogenide semiconductor. *Nat. Mater.* 13:1091–1095.

21. Chernikov, A. et al. 2014. Exciton binding energy and nonhydrogenic Rydberg series in monolayer WS_2. *Phys. Rev. Lett.* 113:076802.

22. He, K., Kumar, N., Zhao, L., Wang, Z., Mak, K.F., Zhao, H. and Shan, J. 2014. Tightly bound excitons in monolayer WSe_2. *Phys. Rev. Lett.* 113:026803.

23. Ramasubramaniam, A. 2012. Large excitonic effects in mono-layers of molybdenum and tungsten dichalcogenides. *Phys. Rev. B* 86:115409.

24. Berkelbach, T.C., Hybertsen, M.S. and Reichman, D.R. 2013. Theory of neutral and charged excitons in monolayer transition metal dichalcogenides. *Phys. Rev. B* 88:045318.

25. Qiu, D.Y., da Jornada, F.H. and Louie, S.G. Optical spectrum of MoS_2: Many-body effects and diversity of exciton states. *Phys. Rev. Lett.* 111:216805.

26. Castellanos-Gomez, A., Roldán, R., Cappelluti, E., Buscema, M., Guinea, F., van der Zant, H.S.J. and Steele, G.A. 2013. Local strain engineering in atomically thin MoS_2. *Nano Lett.* 13:5361–5366.

27. Mak, K.F., He, K., Lee, C., Lee, G.H., Hone, J., Heinz, T.F. and Shan, J.S. 2013. Tightly bound trions in monolayer MoS_2. *Nat. Mater.* 12:207–211.

28. Xiao, D., Liu, G.-B., Feng, W., Xu, X. and Yao, W. 2012. Coupled spin and valley physics in monolayers of MoS_2 and other group-VI dichalcogenides. *Phys. Rev. Lett.* 108:196802.

29. Jones, A.M. et al. 2013. Optical generation of excitonic valley coherence in monolayer WSe_2. *Nat. Nanotechnol.* 8:634–638.

30. Mak, K.F., He, K., Shan, J. and Heinz, T.F. 2012. Control of valley polarization in monolayer MoS_2 by optical helicity. *Nat. Nanotechnol.* 7:494–498.

31. Koppens, F.H.L., Mueller, T., Avouris, Ph., Ferrari, A.C., Vitiello, M.S. and Polini, M. 2014. Photodetectors based on graphene, other two-dimensional materials and hybrid systems. *Nat. Nanotechnol.* 9:780–793.

32. Lee, E.J.H., Balasubramanian, K., Weitz, R.T., Burghard, M. and Kern, K. 2008. Contact and edge effects in graphene devices. *Nat. Nanotechnol.* 3:486–490.

33. Xia, F., Mueller, T., Golizadeh-Mojarad, R., Freitag, M., Lin, Y., Tsang, J., Perebeinos, V. and Avouris, Ph. 2009. Photocurrent imaging and efficient photon detection in a graphene transistor. *Nano Lett.* 9:1039–1044.

34. Xia, F., Mueller, T., Lin, Y., Valdes-Garcia, A. and Avouris, Ph. 2009. Ultrafast graphene photodetector. *Nat. Nanotechnol.* 4:839–843.

35. Mueller, T., Xia, F. and Avouris, Ph. 2010. Graphene photodetectors for high-speed optical communications. *Nat. Photon.* 4:297–301.

36. Pospischil, A., Humer, M., Furchi, M.M., Bachmann, D., Guider, R., Fromherz, T. and Mueller, T. 2013. CMOS-compatible graphene photodetector covering all optical communication bands. *Nat. Photon.* 7:892–896.

37. Gan, X. et al. 2013. Chip-integrated ultrafast graphene photodetector with high responsivity. *Nat. Photon.* 7:883–887.

38. Vicarelli, L. et al. 2012. Graphene field-effect transistors as room-temperature terahertz detectors. *Nat. Mater.* 11:865–871.

39. Yan, J. et al. 2012. Dual-gated bilayer graphene hot-electron bolometer. *Nat. Nanotechnol.* 7:472–478.

40. Yin, Z. et al. 2012. Single-layer MoS_2 phototransistors. *ACS Nano* 6:74–80.

41. Lopez-Sanchez, O., Lembke, D., Kayci, M., Radenovic, A. and Kis, A. 2013. Ultrasensitive photodetectors based on monolayer MoS2. *Nat. Nanotechnol.* 8:497–501.

42. Zhang, W., Huang, J.-K., Chen, C.-H., Chang, Y.-H., Cheng, Y.-J. and Li, L.-J. 2013. High-gain phototransistors based on a CVD MoS_2 monolayer. *Adv. Mater.* 25:3456–3461.

43. Lee, H.S. et al. 2012. MoS_2 nanosheet phototransistors with thickness-modulated optical energy gap. *Nano Lett.* 12:3695–3700.

44. Choi, W. et al. 2012. High-detectivity multilayer MoS_2 phototransistors with spectral response from ultraviolet to infrared. *Adv. Mater.* 24:5832–5836.

45. Tsai, D.-S. et al. 2013. Few-layer MoS_2 with high broadband photogain and fast optical switching for use in harsh environments. *ACS Nano* 7:3905–3911.

46. Wu, C.-C., Jariwala, D., Sangwan, V.K., Marks, T.J., Hersam, M.C. and Lauhon, L.J. 2013. Elucidating the photoresponse of ultrathin MoS_2 field-effect transistors by scanning photocurrent microscopy. *J. Phys. Chem. Lett.* 4:2508–2513.

47. Furchi, M.M., Polyushkin, D.K., Pospischil, A. and Mueller, T. 2014. Mechanisms of photoconductivity in atomically thin MoS_2. *Nano Lett.* 14:6165–6170.

48. Perea-López, N. et al. 2013. Photosensor device based on few-layered WS_2 films. *Adv. Funct. Mater.* 23:5511–5517.

49. Buscema, M., Groenendijk, D.J., Blanter, S.I., Steele, G.A., van der Zant, H.S.J. and Castellanos-Gomez, A. 2014. Fast and broadband photoresponse of few-layer black phosphorus field-effect transistors. *Nano Lett.* 14:3347–3352.

50. Liu, F. et al. 2014. High-sensitivity photodetectors based on multilayer GaTe flakes. *ACS Nano* 8:752–760.
51. Hu, P., Wen, Z., Wang, L., Tan, P. and Xiao, K. 2012. Synthesis of few-layer GaSe nanosheets for high performance photodetectors. *ACS Nano* 6:5988–5994.
52. Hu, P. et al. 2013. Highly responsive ultrathin GaS nanosheet photodetectors on rigid and flexible substrates. *Nano Lett.* 13:1649–1654.
53. Jacobs-Gedrim, R.B. et al. 2014. Extraordinary photoresponse in two-dimensional In$_2$Se$_3$ nanosheets. *ACS Nano* 8:514–521.
54. Xu, X., Gabor, N.M., Alden, J.S., van der Zande, A.M. and McEuen, P.L. 2010. Photo-thermoelectric effect at a graphene interface junction. *Nano Lett.* 10:562–566.
55. Gabor, N.M. et al. 2011. Hot carrier-assisted intrinsic photoresponse in graphene. *Science* 334:648–652.
56. Buscema, M., Barkelid, M., Zwiller, V., van der Zant, H.S.J., Steele, G.A. and Castellanos-Gomez, A. 2013. Large and tunable photothermoelectric effect in single-layer MoS$_2$. *Nano Lett.* 13:358–363.
57. Kang, H.-S., Choi, C.-S., Choi, W.-Y., Kim, D.-H. and Seo, K.-S. 2004. Characterization of phototransistor internal gain in metamorphic high-electron-mobility transistors. *Appl. Phys. Lett.* 84:3780.
58. Konstantatos, G. et al. 2012. Hybrid graphene-quantum dot phototransistors with ultrahigh gain. *Nat. Nanotechnol.* 7:363–368.
59. Roy, K. et al. 2013. Graphene–MoS$_2$ hybrid structures for multifunctional photoresponsive memory devices. *Nat. Nanotechnol.* 8:826–830.
60. Kufer, D., Nikitskiy, I., Lasanta, T., Navickaite, G., Koppens, F.H.L. and Konstantatos, G. 2015. Hybrid 2D–0D MoS$_2$–PbS quantum dot photodetectors. *Adv. Mater.* 27:176–180.
61. Geim, A.K. and Grigorieva, I.V. 2013. Van der Waals heterostructures. *Nature* 499:419–425.
62. Britnell, L. et al. 2013. Field-effect tunneling transistor based on vertical graphene heterostructures. *Science* 335:947–950.
63. Georgiou, T. et al. 2013. Vertical field-effect transistor based on graphene–WS$_2$ heterostructures for flexible and transparent electronics. *Nat. Nanotechnol.* 8:100–103.
64. Britnell, L. et al. 2013. Resonant tunnelling and negative differential conductance in graphene transistors. *Nat. Commun.* 4:1794.
65. Britnell, L. et al. 2013. Strong light–matter interactions in heterostructures of atomically thin films. *Science* 340:1311–1314.
66. Yu, W.J., Liu, Y., Zhou, H., Yin, A., Li, Z., Huang, Y. and Duan, X. 2013. Highly efficient gate-tunable photocurrent generation in vertical heterostructures of layered materials. *Nat. Nanotechnol.* 8:952–958.
67. Sundaram, R.S., Engel, M., Lombardo, A., Krupke, R., Ferrari, A.C., Avouris, Ph. and Steiner, M. 2013. Electroluminescence in single layer MoS$_2$. *Nano Lett.* 13:1416–1421.
68. Chen, J., Perebeinos, V., Freitag, M., Tsang, J., Fu, Q., Liu, J. and Avouris, Ph. 2005. Bright infrared emission from electrically induced excitons in carbon nanotubes. *Science* 310:1171–1174.
69. Doh, Y.-J., Maher, K.N., Ouyang, L., Yu, C.L., Park, H. and Park, J. 2008. Electrically driven light emission from individual CdSe nanowires. *Nano Lett.* 12:4552–4556.

70. Pospischil, A., Furchi, M.M. and Mueller, T. 2014. Solar-energy conversion and light emission in an atomic monolayer p–n diode. *Nat. Nanotechnol.* 9:257–261.

71. Baugher, B.W.H., Churchill, H.O.H., Yang, Y. and Jarillo-Herrero, P. 2014. Optoelectronic devices based on electrically tunable p–n diodes in a monolayer dichalcogenide. *Nat. Nanotechnol.* 9:262–267.

72. Ross, J.S. et al. 2014. Electrically tunable excitonic light-emitting diodes based on monolayer WSe_2 p–n junctions. *Nat. Nanotechnol.* 9:268–272.

73. Das, S., Chen, H.-Y., Penumatcha, A.V. and Appenzeller, J. 2013. High performance multilayer MoS_2 transistors with scandium contacts. *Nano Lett.* 13:100–105.

74. Zhang, Y., Ye, J., Matsuhashi, Y. and Iwasa, Y. 2012. Ambipolar MoS_2 thin flake transistors. *Nano Lett.* 12:1136–1140.

75. Podzorov, V., Gershenson, M.E., Kloc, Ch., Zeis, R. and Bucher, E. 2004. High-mobility field-effect transistors based on transition metal dichalcogenides. *Appl. Phys. Lett.* 84:3301.

76. Jo, S., Ubrig, N., Berger, H., Kuzmenko, A.B. and Morpurgo, A.F. 2014. Mono- and bilayer WS_2 light-emitting transistors. *Nano Lett.* 14:2019–2025.

77. Zhang, Y.J., Oka, T., Suzuki, R., Ye, J.T. and Iwasa, Y. 2014. Electrically switchable chiral light-emitting transistor. *Science* 344:725–728.

78. Li, D. et al. 2015. Electric field induced strong enhancement of electroluminescence in multilayer molybdenum disulfide. *Nat. Commun.* 6:7509.

79. Ye, Y. et al. 2014. Exciton-dominant electroluminescence from a diode of monolayer MoS_2. *Appl. Phys. Lett.* 104:193508.

80. Lopez-Sanchez, O., Llado, E.A., Koman, V., Fontcuberta i Morral, A., Radenovic, A. and Kis, A. 2014. Light generation and harvesting in a van der Waals heterostructure. *ACS Nano* 8:3042–3048.

81. Sun, D. et al. 2014. Observation of rapid exciton–exciton annihilation in monolayer molybdenum disulfide. *Nano Lett.* 14:5625–5629.

82. Withers, F. et al. 2015. Light-emitting diodes by band-structure engineering in van der Waals heterostructures. *Nat. Mater.* 14:301–306.

83. Shockley, W. and Queisser, H.J. 1961. Detailed balance limit of efficiency of p–n junction solar cells. *J. Appl. Phys.* 32:510.

84. Bernardi, M., Palummo, M. and Grossman, J. C. 2013. Extraordinary sunlight absorption and one nanometer thick photovoltaics using two-dimensional monolayer materials. *Nano Lett.* 13:3664–3670.

85. Gan, L.Y., Zhang, Q., Cheng, Y. and Schwingenschlögl, U. 2014. Photovoltaic heterojunctions of fullerenes with MoS_2 and WS_2 monolayers. *J. Phys. Chem. Lett.* 5:1445–1449.

86. Echtermeyer, T.J. et al. 2011. Strong plasmonic enhancement of photovoltage in graphene. *Nat. Commun.* 2:458.

87. Späh, R., Elrod, U., LuxSteiner, M., Bucher, E. and Wagner, S. 1983. p–n junctions in tungsten diselenide. *Appl. Phys. Lett.* 43:79–81.

88. Fortin, E. and Sears, W.M. 1982. Photovoltaic effect and optical absorption in MoS_2. *J. Phys. Chem. Solids* 43:881–884.

89. Memaran, S. et al. 2015. Pronounced photovoltaic response from multi-layered transition-metal dichalcogenides PN-junctions. *Nano Lett.*, doi: 10.1021/acs.nanolett.5b03265.

90. Buscema, M., Groenendijk, D.J., Steele, G.A., van der Zant, H.S.J. and Castellanos-Gomez, A. 2014. Photovoltaic effect in few-layer black phosphorus PN junctions defined by local electrostatic gating. *Nat. Commun.* 5:4651.

91. Fontana, M., Deppe, T., Boyd, A.K., Rinzan, M., Liu, A.Y., Paranjape, M. and Barbara, P. 2012. Electron–hole transport and photovoltaic effect in gated MoS_2 Schottky junctions. *Sci. Rep.* 3:1634.
92. Furchi, M.M., Pospischil, A., Libisch, F., Burgdörfer, J. and Mueller, T. 2014. Photovoltaic effect in an electrically tunable van der Waals heterojunction. *Nano Lett.* 14:4785–4791.
93. Lee, C.-H. et al. 2014. Atomically thin p–n junctions with van der Waals hetero-interfaces. *Nat. Nanotechnol.* 9:676–681.
94. Kosmider, K. and Fernández-Rossier, J. 2013. Electronic properties of the MoS_2–WS_2 heterojunction. *Phys. Rev. B* 87:075451.
95. Rivera, P. et al. 2015. Observation of long-lived interlayer excitons in monolayer $MoSe_2$–WSe_2 heterostructures. *Nat. Commun.* 6:6242.
96. Hong, X. et al. 2014. Ultrafast charge transfer in atomically thin MoS_2/WS_2 heterostructures. *Nat. Nanotechnol.* 9:682–686.
97. Yu, Y. et al. 2015. Equally efficient interlayer exciton relaxation and improved absorption in epitaxial and nonepitaxial MoS_2/WS_2 heterostructures. *Nano Lett.* 15:486–491.
98. Deng, Y. et al. 2014. Black phosphorus-monolayer MoS_2 van der Waals heterojunction p–n diode. *ACS Nano* 8:8292–8299.
99. Wi, S., Kim, H., Chen, M., Nam, H., Guo, L.J., Meyhofer, E. and Liang, X. 2014. Enhancement of photovoltaic response in multilayer MoS_2 induced by plasma doping. *ACS Nano* 8:5270–5281.
100. Tsai, M.-L. et al. 2014. Monolayer MoS_2 heterojunction solar cells. *ACS Nano* 8:8317–8322.

Optoelectronics, Mechanical Properties and Strain Engineering in MoS$_2$

Andres Castellanos-Gomez, Michele Buscema, Herre S.J. van der Zant and Gary A. Steele

Contents

10.1 Introduction

The isolation of graphene by mechanical exfoliation in 2004[1] opened the door to study a broad family of two-dimensional (2D) materials,[2-5] almost

2D Materials for Nanoelectronics edited by Michel Houssa, Athanasios Dimoulas and Alessandro Molle © 2016 CRC Press/Taylor & Francis Group, LLC. ISBN: 978-1-4987-0417-5.

unexplored at that time. These materials are characterised by strong covalent in-plane bonds and weak interlayer van der Waals interactions which give them a layered structure. This rather weak interlayer interaction can be exploited to extract atomically thin layers by mechanical[2] or chemical exfoliation methods.[6] In the last 5 years, a wide variety of materials with very different electronic properties (ranging from wide-band gap insulators to superconductors) have been explored.[7-13] These materials are expected to complement graphene in applications for which graphene does not possess the optimal properties. 2D semiconductors with an intrinsic large band gap are a good example of materials that could complement graphene.[3,14] While the outstanding carrier mobility of graphene makes it very attractive for certain electronic applications (e.g. high-frequency electronics), the lack of a band gap dramatically hampers its applicability in digital electronics. Graphene field-effect transistors (FETs) suffer from large off-state currents due to the low resistance values at the charge neutrality point. The large off-state current also translates to a large dark current in graphene-based photodetectors which limits their performance.

The handicaps of gapless graphene-based electronic devices have motivated experimental efforts towards opening a band gap in graphene in order to combine its excellent carrier mobility with a sizeable band gap. Several methods have been employed: lateral confinement,[15-17] application of a perpendicular electric field in bilayer graphene[18,19] or hydrogenation[20-23] among others. However, in all cases, the magnitude of the band gap is not enough to ensure high performance at room temperature and/or the mobility of the treated graphene severely decreases.

The difficulty of achieving high-mobility gapped graphene has triggered the study of semiconducting analogues to graphene in materials with an intrinsic band gap.[3,24-28] For instance single-layer MoS_2, the most studied 2D semiconductor so far, presents a direct band gap (1.8 eV) and a large in-plane mobility (up to 200 cm^2 V^{-1} s^{-1}).[14,29-35] The intrinsic band gap of MoS_2 is essential for many applications, including transistors for digital electronics or certain optoelectronic applications. For instance, FETs based on single-layer MoS_2 showed room-temperature current on/off ratios of 10^7–10^8 and ultralow standby power dissipation.[14] Logic circuits, amplifiers and photodetectors based on monolayer MoS_2 have also been demonstrated recently.[33,36,37]

In this chapter, the recent progress in the study of optoelectronic and subsequently the mechanical properties of MoS_2 will be reviewed. In the last section, the use of mechanical deformation to modify the optical properties of MoS_2 will be discussed.

10.2 Optoelectronic Properties of MoS_2

Recent works on the optoelectronic properties of MoS_2 have shown that the photoresponse of externally biased MoS_2-based phototransistors is driven by a change in conductivity upon illumination.[36,38-41] The photovoltaic effect in

MoS$_2$ devices has also been reported with metal electrodes that generate large Schottky barriers (SBs).[42,43] These previous works make use of either an externally applied or a built-in electric field to separate the photogenerated carriers. In this section, we will discuss the intrinsic photocurrent generation mechanism in single-layer MoS$_2$ phototransistors with negligible SBs and under no external bias by scanning photocurrent. In contrast to previous works on biased MoS$_2$ devices, the photo-thermoelectric effect dominates the photoresponse in unbiased MoS$_2$.[44]

Phototransistor devices consist of a single-layer MoS$_2$ flake, deposited onto a Si/SiO$_2$ (285 nm) substrate by mechanical exfoliation. Electrical contacts are fabricated by standard electron beam lithography and subsequent deposition of a Ti(5 nm)/Au(50 nm) layer. **Figure 10.1a** shows an optical microscopy image of a FET fabricated with a single-layer MoS$_2$ flake. A combination of atomic force microscopy, Raman spectroscopy and optical microscopy is used to characterise the MoS$_2$ samples.[45] As reported by Lee et al.[46] the frequency difference between the two most prominent Raman peaks depends monotonically on the thickness and it can, therefore, be used to accurately determine the number of MoS$_2$ layers. **Figure 10.1b** shows an example of Raman spectra

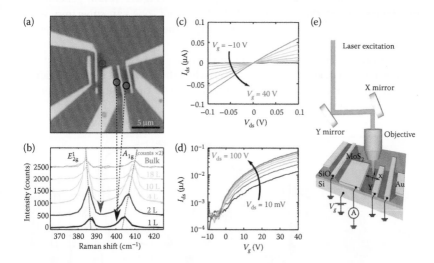

FIGURE 10.1 (a) Optical image of a fabricated single-layer MoS$_2$ FET. (b) Raman spectra of various MoS$_2$ flakes deposited on the chip (traces in lighter colour have been obtained on thicker MoS$_2$ flakes, not used to fabricate the devices studied in this work). The arrows connect the location where the Raman spectra were collected and the actual Raman data. (c) Source–drain current versus source–drain bias characteristics measured at different gate voltages. (d) Electrical transport characteristic of the single-layer MoS$_2$-based FET device (source–drain current vs. gate voltage) measured at different source–drain bias. (e) Schematic of the SPCM setup showing the excitation path and electric circuit used to perform SPCM measurements. (Adapted from M. Buscema et al., *Nano Lett.* **13**, 358, 2013.)

acquired in two different regions of the MoS_2 device. The frequency difference between the Raman peaks confirms that the device shown in **Figure 10.1a** is composed of a single-layer and a bilayer region.

Figure 10.1c shows the source–drain characteristics at different gate voltages of the device shown in **Figure 10.1a**. The current versus voltage relationship remains almost linear for a broad bias range, as expected for titanium-based contacts that are predicted to produce small SBs.[47] **Figure 10.1d** shows various gate traces measured at different source–drain voltages. The device shows a pronounced n-type behaviour, with a current on/off ratio exceeding 10^3 and a mobility of ~0.85 cm^2 V^{-1} s^{-1}, that is characteristic of single-layer MoS_2 FETs fabricated on SiO_2 surfaces.[30,31,36] Note that in order to increase the quality of the electrical contacts, the samples were annealed at 300°C for 2 hours in a Ar/H_2 flow (500 sccm/100 sccm).[14]

Scanning photocurrent microscopy (SPCM) measurements enable one to spatially resolve the local photoresponse of the MoS_2 FET device (see **Figure 10.1e**). In the experiments discussed in this section, the excitation was provided by a continuous wave green laser ($\lambda = 532$ nm) and a supercontinuum tuneable source.[48] By simultaneously recording the intensity of laser light reflection (**Figure 10.2a**) and the photocurrent (**Figure 10.2b**) generated in the device during the scanning of the laser spot, one can superimpose the two images to accurately determine where the photocurrent is generated (**Figure 10.2c**).

As it can be seen (e.g. by looking at electrode 3 in **Figure 10.2c**), there is a photogenerated current at zero bias even when the laser spot is placed microns away from the electrode edges (distances up to 10 times larger than the full

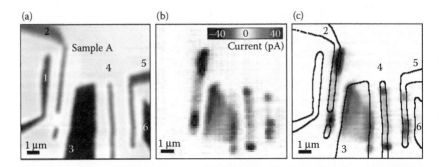

FIGURE 10.2 (a) Spatial map of the intensity of the reflected light from the device (white corresponds to low reflection). Electrodes have been numbered for clarity. (b) Photocurrent image of the MoS_2 FET. The colour scale in the inset gives the photocurrent value. (c) Superposition of the photocurrent map (from (b), same colour scale) and contours of the electrodes as obtained from the light reflection map. Reflection and photocurrent measurements are performed simultaneously with electrode number 3 connected to a current-to-voltage amplifier, whereas the other electrodes are grounded. Excitation: $\lambda = 532$ nm, $P = 1$ μW, spot radius ~400 nm. (Adapted from M. Buscema et al., *Nano Lett.* **13**, 358, 2013.)

width at half maximum of the laser spot intensity). This is in striking contrast with several earlier findings on photocurrent generation in graphene,[49,50] which is localised right at the interface between the graphene flake and the metal electrodes. In these previous works, the zero-bias photocurrent generation mechanism was attributed to the electron–hole separation by the electric field at the SBs. This mechanism, however, cannot explain the presence of a photogenerated current when the laser is illuminating far from the electrode edges where SBs are located.

The observation of photocurrent with the laser positioned deep inside the metal electrode suggests that the principal photocurrent generation mechanism in unbiased single-layer MoS$_2$ devices is different from photocurrent generation by the separation of photoexcited electron–hole pairs due to the localised electric field at the metal/semiconductor interface.

In order to gain a deeper insight into the photoresponse mechanism of single-layer MoS$_2$ FETs, one can perform scanning photocurrent measurements at different illumination wavelengths. **Figure 10.3a** shows a photoresponse map acquired with green ($\lambda = 532$ nm, $h\nu = 2.33$ eV) illumination. The photovoltage is obtained by dividing the measured photocurrent by the device resistance, measured under the same illumination conditions. A line profile of the photocurrent measured along the dashed green line in **Figure 10.3a** is presented in **Figure 10.3b** as open blue circles. The solid red line is a Gaussian fit to the data corresponding to a diffraction limited laser spot. Notice there is a significant photocurrent tail generated when the laser is scanned over the electrode (arrow in **Figure 10.3b**). For a photovoltaic response aroused from SBs, one would expect all the photocurrent to be generated right at the edge of the metal/semiconductor junction and not in a region covered by the electrode.

Figure 10.3c shows a photoresponse map acquired with red ($\lambda = 750$ nm, $h\nu = 1.65$ eV) illumination. The photovoltage under green illumination is larger because of the higher absorption of the gold electrodes at this wavelength. The observation of photoresponse even for photon energies lower than the band gap (**Figure 10.3c**) cannot be explained by the separation of photoexcited electron–hole pairs (see **Figure 10.3d**). Moreover, previous photoconductivity measurements in MoS$_2$ transistors under large source–drain bias have not shown any significant photoresponse for excitation wavelength above 700 nm (or below 1.77 eV).[25,36] These observations indicate that sub-band gap impurity states are either not present or do not contribute to photocurrents, even in the presence of a large extraction bias. One thus comes to the conclusion that sub-band gap states cannot be the source of the photocurrent observed in **Figure 10.3c**.

Furthermore, in case of a photovoltaic response from SBs, the position of the maximum photocurrent in the photovoltaic effect is expected to shift with the gate voltage. This effect, however, has not been observed in the unbiased single-layer MoS$_2$ devices. **Figure 10.4a** shows an SPCM map (at zero source–drain bias) of one device acquired with the electrical configuration sketched in **Figure 10.4b**. Since electrode 4 is connected to ground via the current amplifier the strong photocurrent inside electrode 3 is now negative. By performing SPCM measurements with gate voltages ranging from −30 to +30 V, displayed in

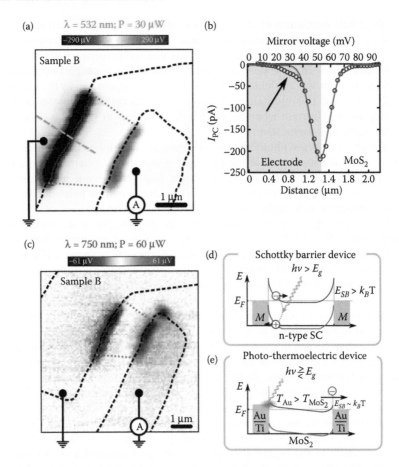

FIGURE 10.3 Photovoltage map of a single-layer MoS$_2$ FET using an excitation wavelength of 532 nm (a) and 750 nm (c). (b) Photocurrent profile across the linecut in panel (a) (open blue circles). The solid red line is a Gaussian fit to the data and the arrow points at the photocurrent tail generated when the laser spot is scanned over the electrode. The shaded blue area represents the electrode area as determined by the reflection signal. (d) Schematic of photoresponse mechanism in a typical metal–semiconductor–metal device. (e) Schematic of the photoresponse mechanism in a device dominated by the photo-thermoelectric effect. (Adapted from M. Buscema et al., *Nano Lett.* **13**, 358, 2013.)

Figure 10.4c, one can observe that within experimental resolution the position of the photocurrent peaks does not change with respect to gate voltage. This is in contrast with earlier findings on devices in which the photocurrent is generated by the separation of photoexcited electrons and holes due to the electrostatic potential in p–n junctions or SBs whose extension depends on the charge carrier concentration.[49] As the charge carrier density is reduced, the potential gradient of a p–n junction or the SB would extend deeper into the material yielding a maximum photocurrent further and further away from the electrode edges, as sketched in **Figure 10.4d**. Therefore, the absence of a shift in the position of the measured photocurrent maximum indicates that the dominant mechanism of photocurrent generation is not the separation of carriers by the built-in electrical potential along the SB near the electrodes.

Thus, the observation of a photoresponse with sub-band gap illumination, the strong measured photocurrent inside the area of the electrodes whose position is gate independent and the negligible role of the SBs in the

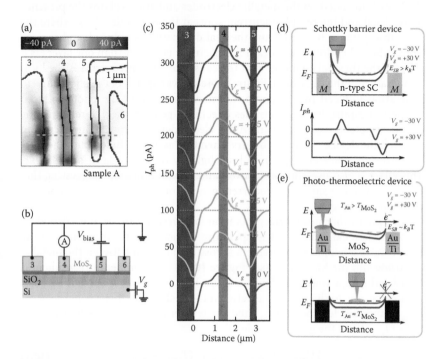

FIGURE 10.4 (a) Scanning photocurrent image of a single-layer MoS$_2$ FET. Note, that the current-to-voltage amplifier is connected to electrode 4, whereas electrodes 3, 5 and 6 are connected to ground. (b) Schematic cross-section of the device, showing the electrical circuit used to perform the measurements. (c) Photocurrent line profiles (along the green dashed line in panel (a)) for gate voltages of +30, +15, +7.5, 0, −7.5, −15 and −30 V from top to bottom. Consecutive lines are shifted by 50 pA for clarity. Shaded areas indicate the position and size of electrode numbers 3, 4 and 5 as determined from the reflection signal; the colours of the shading indicate the sign of the photocurrent. (d) Schematic of the energy diagram of a typical metal–semiconductor–metal device (top) and of the magnitude of a photocurrent line profile over such a device showing the expected gate-induced spatial shift in the photocurrent peak (bottom). (e) Schematic of the energy diagram of the single-layer MoS$_2$ device when the laser spot is incident on the electrode (top) and on the MoS$_2$ flake (bottom). In both panels (d) and (e), the conduction band is drawn at different gate voltages and the valence band is not shown for clarity. (Adapted from M. Buscema et al., *Nano Lett.* **13**, 358, 2013.)

conductance of the device all suggest that photovoltaic effects cannot be responsible for the photoresponse observed here. The photoresponse in the unbiased MoS_2 devices arises instead from a strong photo-thermoelectric effect.[44,51,52] In the photo-thermoelectric effect (**Figures 10.3e** and **10.4e**), a temperature gradient arising from light absorption generates a photo-thermal voltage across a junction between two materials with different Seebeck coefficients. This photo-thermal voltage can drive current through the device. This mechanism is consistent with the observation of strong photocurrents when the laser is focussed on the metallic electrodes and also explains the presence of localised and intense photocurrent spots at the electrode edges, where the laser absorption is increased and the heat dissipation is reduced, and at positions, where the MoS_2 underneath the electrode is thicker than one layer (as it reduces the thermal coupling with the substrate and thus increases the local temperature of the electrode). **Figure 10.4e** shows a sketch of the photocurrent generation induced by the photo-thermoelectric effect. The photo-induced temperature raise in the electrodes effectively generates a difference in the chemical potential of source and drain electrodes, thereby driving a current through the device. This only occurs when the laser is focussed onto the electrode surface since absorption of laser light by the MoS_2 flake is negligible. This effect is also present with above-band gap illumination, suggesting that the main mechanism for photocurrent generation even in that case is not the photovoltaic effect.

The photo-thermoelectric generation of current can be understood as follows: the local increase of temperature induced by the absorption of the laser creates a temperature gradient across the junction between the metallic electrode and the MoS_2 flake (see **Figure 10.4e**). This temperature gradient is translated into a voltage difference (ΔV_{PTE}), which drives a current through the device, due to the difference between the Seebeck coefficients of the MoS_2 (S_{MoS_2}) and the electrodes (S_{TiAu}) and the temperature difference between the flake (T_{MoS2}) and the gold electrode (T_{TiAu}):

$$\Delta V_{PTE} = (S_{MoS_2} - S_{TiAu}) \cdot (T_{MoS_2} - T_{TiAu}) \tag{10.1}$$

which indicates that one can estimate the Seebeck coefficient of MoS_2 if the temperature gradient is known.

The gate dependence of the photo-thermoelectric effect in single-layer MoS_2 is further studied by measuring their electrical transport characteristics while the laser spot is placed at a location with a high photoresponse. For illumination wavelengths with a photon energy higher than the band gap, the threshold voltage is shifted towards very negative values (see **Figure 10.5a**), not reachable without leading to gate leakage. It is, therefore, preferable to use illumination wavelengths with photon energy below the band gap to minimise this photogating effect. **Figure 10.5b** shows the photo-thermoelectric voltage (ΔV_{PTE}) measured for two single-layer MoS_2 devices at different gate voltages. By decreasing the gate voltage below the threshold voltage, the photo-thermoelectric voltage shows a substantial increase. As the ΔV_{PTE} is proportional to the difference in the Seebeck coefficients and the temperature gradient across

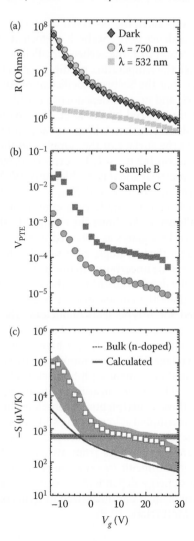

FIGURE 10.5 (a) Resistance of a single-layer MoS$_2$ device as a function of gate voltage in dark state and with the laser spot placed on the MoS$_2$/electrode interface. Two different illumination wavelengths have been used (532 and 750 nm). (b) Photo-thermoelectric voltage for two single-layer MoS$_2$ devices measured with the laser spot ($\lambda = 750$ nm) placed on the MoS$_2$/electrode interface. (c) Estimated Seebeck coefficient versus gate voltage. The values are calculated from **Equation 10.1** using the measured photovoltage (symbols). The dashed line corresponds to the Seebeck coefficient value of bulk MoS$_2$ with experimental uncertainty (shaded area). The solid line is the Seebeck coefficient calculated with the Mott formula, Expression **10.2**. (Adapted from M. Buscema et al., *Nano Lett.* **13**, 358, 2013.)

the AuTi/MoS$_2$ junction (**10.1**), the observed behaviour can be attributed to the expected gate dependence of the Seebeck coefficient of MoS$_2$.

According to Expression **10.1**, the Seebeck coefficient of MoS$_2$ can be determined from ΔV_{PTE} once the temperature gradient across the AuTi/MoS$_2$ junction is known. Notice that the Seebeck coefficient of AuTi is negligible with respect to that of MoS$_2$ and no gate dependence is expected. To estimate the increase of temperature induced by the laser illumination, we performed a finite-elements analysis calculation, taking into account the reflections losses in the objective and at the surface of the sample and the absorbed intensity through the material according to $I = I_0(1 - \exp(-\alpha d))$, where α is the absorption coefficient of the material and d is its thickness. With the employed laser excitation ($\lambda = 750$ nm, $I_0 = 60$ μW and $R_{spot} \approx 500$ nm) and if we assume that all the energy delivered by the laser is converted into heat, we obtain $\Delta T_{AuTi/MoS2} \approx 0.13$ K. **Figure 10.5c** shows the estimated values: the experimental data are represented by squares as obtained from the average of the thermovoltages measured for the two single-layer samples shown in **Figure 10.5b**. The shaded region shows the uncertainty in the Seebeck coefficient estimation. The Seebeck coefficient value for bulk MoS$_2$[53] (between −500 and −700 μV K^{-1}) is also plotted to facilitate the comparison. A negative value of the Seebeck coefficient is expected for n-type semiconductors. The obtained value for the Seebeck coefficient is remarkably large (between −10^4 to −10^5 μV K^{-1} for low doping level and −10^2 to −10^3 μV K^{-1} for high doping level) and depends strongly on the gate voltage, varying by more than two orders of magnitude. These values are in good agreement with Seebeck coefficient values recently determined by electrical transport measurements, using on-chip heaters and thermometers.[54,55] In those experiments, a Seebeck coefficient of −0.5 × 10^4 to −3 × 10^4 μV K^{-1} was determined at low doping and −10^2 μV K^{-1} at high doping.

At a microscopic level, the Seebeck effect is due to a change in the Fermi level with temperature which arises from an energy-dependent electronic density of states. As it is difficult to experimentally determine microscopic properties such as this energy-dependent density of states in a given sample, the Seebeck coefficient is often parameterised in terms of the conductivity of the sample using the Mott relation[56–58]:

$$S = \frac{\pi^2 k_B^2 T}{3e} \frac{d\ln(\sigma(E))}{dE}\bigg|_{E=E_F} \tag{10.2}$$

where k_B is the Boltzmann constant, T is the temperature, e is the electron charge, $\sigma(E)$ is the conductivity as a function of energy and the derivative is evaluated at the Fermi energy E_F.

Using Expression **10.2**, one can estimate the maximum Seebeck coefficient in the single-layer MoS$_2$ devices studied in **Figure 10.5** (see **Figure 10.5c**). At the low doping level, the Seebeck coefficient is on the order of −4 × 10^3 μV K^{-1}, roughly an order of magnitude lower than the value estimated from the photo-thermoelectric effect measurements. This discrepancy could arise from the fact that Expression 10.2 is based on the assumption that the conductor is a metal or a degenerate semiconductor, which is not the case for a single flake of MoS$_2$ near depletion.

The estimated Seebeck coefficient for single-layer MoS$_2$ is orders of magnitude larger than that of graphene[51,52,59] (±4 to 100 μV K^{-1}), semiconducting carbon nanotubes[60] (~−300 μV K^{-1}), organic semiconductors (~1 × 10^3 μV K^{-1})[61] and even larger than that of materials regularly employed for thermopower generation such as Bi$_2$Te$_3$[62,63] (±150 μV K^{-1}). In addition, the wide gate tunability of the Seebeck coefficient can be useful for applications such as on-chip power generation and thermoelectric nanodevices.

10.3 Mechanical Properties of MoS$_2$

Atomically thin MoS$_2$ has shown very interesting mechanical properties, unmatched by conventional three-dimensional (3D) semiconductors.[64,65] For instance, single-layer MoS$_2$ has shown very large breaking stress, approaching the theoretical limit predicted by Griffith for ideal brittle materials in which the fracture point is dominated by the intrinsic strength of the atomic bonds and not by the presence of defects.[66] While the ideal breaking stress value is expected to be one-ninth of the Young's modulus, for single-layer MoS$_2$, it reaches ~1/8.[64] This almost ideal behaviour is attributed to a low defect density in the fabricated devices, probably due to the lack of dangling bonds or other surface defects, and their high crystallinity. This is in clear contrast with conventional 3D semiconductors; while silicon typically breaks at strain levels of ~1.5%,[67,68] MoS$_2$ does not break until greater than 10% strain levels and it can be folded and wrinkled almost at will.[69,70]

Two main approaches have been employed to study the mechanical properties of atomically thin MoS$_2$: static deformation experiments and dynamic experiments. The static deformation approaches are based on the analysis of force versus deformation tests where a force load is applied to a freely suspended MoS$_2$ crystal while its deformation is recorded. In the dynamical experiments, the oscillation amplitude of suspended flakes subjected to a time-varying actuation (resonant or near resonant) is studied.

For both static deformation experiments and dynamic experiments, freely suspended MoS$_2$ samples have to be fabricated. Freely suspended MoS$_2$ 'drum heads' can be fabricated by transferring atomically thin flakes of MoS$_2$ onto a SiO$_2$/Si substrate pre-patterned with holes. Direct exfoliation of MoS$_2$ onto the pre-patterned substrates yields a low density of flakes, and rarely suspended single-layer MoS$_2$ devices could be fabricated using this method. An alternative method consists of exploiting an all-dry transfer technique to deposit the MoS$_2$ flakes onto the pre-patterned substrates.[71] **Figure 10.6** shows some of the steps employed to transfer single-layer MoS$_2$ onto holes to fabricate the mechanical resonators. Single-layer devices can be identified at a glance by means of optical microscopy[9,73] and confirmation of the exact layer thickness is performed by a combination of atomic force microscopy, Raman spectroscopy and photoluminescence.[45] **Figure 10.7** shows optical images of single-layer MoS$_2$ mechanical resonators.

Stamp surface   Looking through the stamp   After peeling off

Stamp almost in contact   Stamp in contact

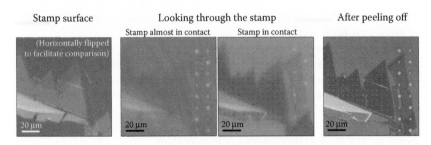

FIGURE 10.6 Optical microscopy images of the process to fabricate MoS$_2$ resonators. First, MoS$_2$ flakes are transferred onto a viscoelastic stamp (similar to those used in nanoimprinting). The holes can be seen through the stamp, the flake is aligned to the holes. The stamp is brought to gentle contact to the sample. By peeling off slowly the stamp, the flakes are transferred onto the pre-patterned substrate with holes. (Adapted from A. Castellanos-Gomez et al., *Adv. Mater.* **25**, 6719, 2013.)

FIGURE 10.7 Optical images of fabricated MoS$_2$ resonators with different thicknesses. Optical microscopy images of the MoS$_2$ flakes transferred onto the pre-patterned substrate with holes. The apparent colour of the flakes can be used to estimate the number of layers at a glance. (Adapted from A. Castellanos-Gomez et al., *Adv. Mater.* **25**, 6719, 2013.)

10.3.1 Static Deformation Experiments: Elastic Properties of MoS$_2$

Within the different approaches to study the elastic properties of 2D materials, the analysis of the force versus deformation traces measured at the centre of a freely suspended flake is one of the most used one. In these experiments, the tip of an atomic force microscope (AFM) applies a central point load. The produced deformation is simultaneously measured using the AFM. **Figure 10.8a** shows an AFM topography image of a freely suspended MoS$_2$ flake deposited on a SiO$_2$/Si substrate pre-patterned with holes (see a sketch in **Figure 10.8b**). Force versus deformation traces acquired on flakes with different thicknesses are compared in **Figure 10.6c**. When the tip and sample are in contact, the elastic deformation of the nanosheet (δ), the deflection of the AFM cantilever (Δz_c) and the displacement of the scanning piezotube of the AFM (Δz_{piezo}) are related by[12,65,74]

$$\delta = \Delta z_{piezo} - \Delta z_c \qquad (10.3)$$

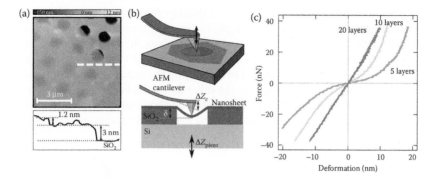

FIGURE 10.8 (a) Atomic force microscopy image of a MoS$_2$ flake transferred onto a SiO$_2$/Si substrate with pre-patterned holes, the thinner part of the flake is five layers thick. (b) Scheme of the central indentation measurement technique. The tip of the AFM is used to apply a point load at the centre of the suspended MoS$_2$ flake, the deflection of the cantilever is measured, along with the piezo displacement, making it possible to extract the deformation of the drum and the applied force. (c) Force versus deformation traces acquired on three different MoS$_2$ flakes with 5, 10 and 20 layers in thickness. (Adapted from A. Castellanos-Gomez et al., *Adv. Mater.* **24**, 772, 2012.)

The force applied is related to the cantilever deflection as $F = k_c \cdot \Delta z_c$, where k_c is the spring constant of the cantilever. In the case of an MoS$_2$ flake suspended over a hole (forming a circular drum-like structure), the force versus deformation relationship is expected to follow[64,65,75,76]:

$$F = \left[\frac{4\pi t^3}{3(1 - v^2)R^2} E + \pi T \right] \delta + \frac{tE}{(1.05 - 0.15v - 0.16v^2)^3 R^2} \delta^3 \quad \textbf{(10.4)}$$

where R and t are the radius and thickness of the 'drum', E is the Young's modulus, T is the initial pre-tension and v is the Poisson's ratio. The linear part in **Equation 10.4** includes a term that accounts for the bending rigidity of the layer (the first term, proportional to the Young's modulus) and a second term that accounts for the initial pre-tension (proportional to T). Therefore, the analysis of the linear term of the F versus δ traces does not allow one to unambiguously determine the Young's modulus and pre-tension of the suspended flakes. The cubic term in **Equation 10.4** that accounts for the stiffening in the layer due to the tension induced by the deflection, only depends on the Young's modulus and geometrical factors. Therefore, by fitting experimental non-linear F versus δ traces to **Equation 10.4**, one can determine the Young's modulus and the pre-tension independently from each other. **Figure 10.9a** shows an example of a F versus δ trace fitted to **Equation 10.4** to determine the Young's modulus $E = 0.35 \pm 0.02$ TPa and the pre-tension $T = 0.05 \pm 0.02$ N m^{-1} of an 8 nm thick MoS$_2$ layer. The pre-tension and Young's modulus values obtained following the same procedure for 13 other MoS$_2$ layers have been summarised in the histograms shown in **Figure 10.9b** and **c**.

According to Expression **10.4**, for very thin layers, the bending rigidity term (proportional to t^3) may become negligible in comparison to the pre-tension and the deformation-induced tension terms. This is the case for single- and bilayer MoS_2 in which the F versus δ traces are found to be highly non-linear, allowing them to be modelled by neglecting the bending rigidity term in **Equation 10.4**. For thick flakes, on the other hand, the bending rigidity term can become much larger than the other terms and consequently dominates the F versus δ traces. These then remain linear even for very large deformations.

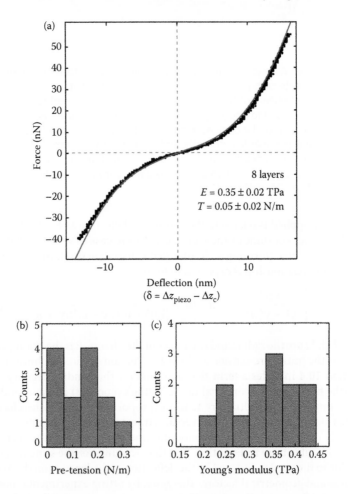

FIGURE 10.9 (a) Force versus deformation trace obtained on an eight layers thick MoS_2 flake suspended over a 1.1 µm hole. The solid line is the fit to Expression **10.4**, employed to obtain the Young's modulus $E = 0.35 \pm 0.02$ TPa and the pre-tension $T = 0.05 \pm 0.02$ N m^{-1} of this nanolayer. (b) and (c) Histogram of the initial pre-tension and Young's modulus obtained from the fit to Expression **10.4** for 13 sheets with different thicknesses ranging from 5 to 10 layers. (Adapted from A. Castellanos-Gomez et al., *Adv. Mater.* **24**, 772, 2012.)

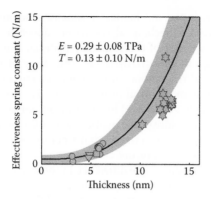

FIGURE 10.10 Thickness dependence of the effective spring constant of MoS$_2$ nanosheets with thicknesses ranging from 25 down to 5 layers. The relationship k_{eff}, circular drum versus t calculated with Expression **10.5** for $E = 0.29 \pm 0.08$ TPa and $T = 0.13 \pm 0.10$ N m^{-1} has been plotted as the drawn black line for comparison. (Adapted from A. Castellanos-Gomez et al., *Adv. Mater.* **24**, 772, 2012.)

Therefore, it is expected that a non-linear to linear transition occurs in the F versus δ traces as the thickness of the suspended flakes increases. While thinner flakes behave as membranes (tension dominated, negligible bending rigidity), thick flakes should show a plate-like behaviour (bending rigidity dominated, negligible tension). For intermediate thicknesses, the mechanical behaviour of the flakes should be described by a combination of membrane and plate mechanical behaviour. This crossover can be clearly seen in the force versus deformation traces shown in **Figure 10.8c**.

An alternative way to determine the Young's modulus and pre-tension of freely suspended MoS$_2$ flakes consists of measuring the linear term of the F versus δ traces (i.e. the effective spring constant, k_{eff}) acquired at the centre of suspended flakes with different thicknesses. In fact, according to Expression **10.4**, the effective spring constant depends on the sample geometry as[65,76]

$$k_{eff,\text{circular-drum}} = \frac{4\pi t^3}{3(1 - \nu^2)R^2}E + \pi T \qquad (10.5)$$

The line in **Figure 10.10** shows the fit of the experimental data to Expression **10.5** to determine the Young's modulus ($E = 290$ GPa) and the pre-tension ($T = 0.13$ N m^{-1}) of several few-layer MoS$_2$ flakes.

10.3.2 Dynamic Experiments: MoS$_2$ Mechanical Resonators

In contrast to the experiments detailed in the previous section, in dynamical experiments the oscillation amplitude of suspended flakes subjected to a time-varying actuation (resonant or near resonant) is studied.[72,77] The mechanical resonances of freely suspended flakes can be identified as peaks of large oscillation amplitude, occurring when the frequency of the actuation signal is swept across the mechanical resonator resonance frequency. The resonance frequency of mechanical resonators depends on their geometry and on physical properties of the resonator material (e.g. pre-tension, mass density and Young's modulus). Therefore, the study of the resonance frequency of mechanical resonators with different geometries can be exploited to determine intrinsic mechanical properties of a 2D material, complementing the static approaches described in the previous section. The width of the resonance peaks is a measure of the quality factor Q, which gives information about the damping processes in the mechanical resonator.

In order to drive the resonator, one can use a modulated laser signal, focussed on the mechanical resonators. The difference in thermal expansion coefficients between the flake and the substrate generates a time-varying strain at the laser modulation frequency that drives the resonator.[78,79] The read-out of the mechanical resonator motion can be done by optical methods as well.[79] The optical read-out exploits the change in reflectivity produced by the displacement of the suspended 2D material, which forms an optical cavity between the flake and the substrate; the 2D layer then acts as a semi-transparent mirror. The optical path in the cavity, and thus the phase difference between the incoming and reflected light beams, depends on the deflection of the resonator and strongly modifies the overall reflectivity. This facilitates the detection of the oscillation amplitude.

Figure 10.11 shows the resonance spectra obtained for single-layer and multilayer MoS$_2$ resonators with different diameters. The resonance frequency of the single-layer devices is in the order of 10–30 MHz which is comparable to that of graphene resonators with similar geometries (16–30 MHz).[80] The quality factor of single-layer MoS$_2$ resonators has been obtained by fitting the frequency response to a Lorentzian curve (dashed lines in **Figure 10.11**).

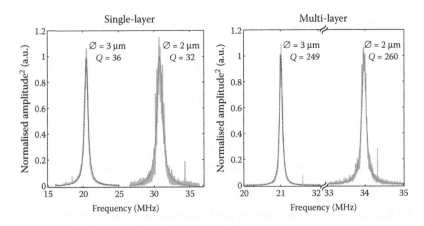

FIGURE 10.11 Mechanical resonance spectra measured for two different single-layer MoS$_2$ mechanical resonators (left) and two different multilayer resonators (right). Drumheads with both 2 and 3 μm diameters have been studied. The experimental data have been fitted to a Lorentzian curve (dashed lines) to extract their resonance frequency and quality factor.

The Q factors of the studied single-layer devices are between 17 and 105, with an average value of $Q = 54 \pm 30$, which is about a factor 3–4 lower than graphene drums with similar geometries[80] ($Q = 195 \pm 15$), and in the same order as those recently reported by Lee et al. for thicker MoS$_2$ resonators.[81]

For very thin MoS$_2$ resonators, the dynamics is dominated by their initial tension (membrane limit). For a membrane-like circular resonator, the fundamental frequency is given by[82]

$$f = \frac{2.4048}{2\pi R} \sqrt{\frac{T}{\rho \cdot t}} \tag{10.6}$$

where R is the resonator radius, T its initial pre-tension (in N m^{-1}), ρ its mass density and t is the thickness. For thick MoS$_2$ resonators, on the other hand, the dynamics is expected to be dominated by their bending rigidity (plate limit) since their pre-tension is negligible in comparison with the bending rigidity. For a plate-like circular resonator, the frequency is given by[82]

$$f = \frac{10.21}{4\pi} \sqrt{\frac{E}{3\rho(1 - \nu^2)}} \cdot \frac{t}{R^2} \tag{10.7}$$

where E is the Young's modulus and ν is the Poisson's ratio ($\nu = 0.125$[34]). A crossover from the membrane-to-plate behaviour is also expected, as seen from the nanoindentation experiments discussed in the previous sub-section.[65] The resonance frequency of the mechanical resonators in that crossover regime can be approximately calculated as

$$f = \sqrt{f_{\text{membrane}}^2 + f_{\text{plate}}^2} \tag{10.8}$$

where f_{membrane} and f_{plate} are calculated according to **Equations 10.6 and 10.7**, respectively.

Figure 10.12 shows the measured resonance frequency as a function of the thickness for 94 MoS$_2$ resonators (with a diameter of 2 and 3 µm). A noticeable dispersion of the frequency and the quality factor is found for devices with analogous geometry which is attributed to device-to-device variation of the pre-strain and clamping conditions (such as non-isotropic pre-tension).

The calculated resonance frequencies in the membrane and plate limit, as well as in the crossover regime are plotted as dotted, dashed and solid lines, respectively. A Young's modulus of $E = 300$ GPa and a pre-tension $T = 0.015$ N m^{-1} has been used for the calculation of the resonance frequencies, using Expressions **10.6** through **10.8**, in agreement with the values found by nanoindentation experiments on drums fabricated using the same technique.[65,74] The pre-tension value depends on the fabrication technique and it may be tuned by modifying the MoS$_2$ transfer process. Note that the calculated frequency versus thickness relationship is not a fit to the experimental values but a guide to the eye to point out the different thickness dependences expected for membranes and plates. From the analysis of the frequency versus thickness relationship (**Figure 10.12**), one can see how the membrane term in

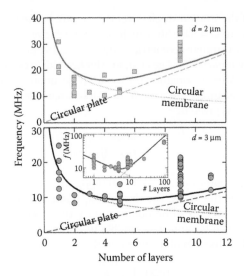

FIGURE 10.12 Measured resonance frequency as a function of the number of layers for MoS_2 mechanical resonators 2 μm in diameter (squares in the top panel) and 3 μm in diameter (circles in the lower panel). The lines indicate the calculated frequency versus thickness relationship for different cases: circular membrane (dotted), circular plate (dashed) and the combination of these two cases (solid). The inset in the lower panel in (a) includes data measured for seven thick MoS_2 resonators (more than 30 layers), which are in the plate limit. (Adapted from A. Castellanos-Gomez et al., *Adv. Mater.* **25**, 6719, 2013.)

Equation 10.8 can be neglected for flakes thicker than 10–12 layers, whereas the plate term can only be neglected for devices thinner than 3–4 layers. MoS_2 flakes 4–10 layers thick are in the crossover regime where membrane and plate terms in **Equation 10.8** are comparable. The inset in **Figure 10.12** (lower panel) includes measured resonance frequencies for much thicker MoS_2 devices, which follow the trend expected for plate-like resonators. This membrane-to-plate crossover is in good agreement with previous works based on nanoindentation in freely suspended MoS_2 layers.[65,74] While for flakes thicker than 10 layers, the force versus deformation relationship could be modelled as elastic plates, flakes 5–10 layers in thickness required a mechanical model including both membrane and plate terms.

10.4 Strain Engineering in MoS_2

Strain engineering is an interesting strategy to tune the material's electronic properties by subjecting its lattice to a mechanical deformation.[83–85] Conventional straining approaches, used for 3D materials, include epitaxial growth on a substrate with a lattice parameter mismatch, the use of a dielectric capping layer or heavy-ion implantation. However, these approaches are typically limited to strains lower than 2% due to the low maximum strains of the brittle bulk semiconducting materials used. Bulk silicon, for example, can be strained only up to 1.5% before breaking.

The isolation of atomically thin 2D semiconductors promises new avenues in the field of strain engineering.[86] In 3D semiconductors, the ultimate strain is limited by both bulk defects and surface imperfections. 2D semiconductors can circumvent these limitations because they lack dangling bonds or other imperfections at their surfaces and it is possible to apply highly anisotropic strain to 2D sheets (limiting the strain to a small area inside a single crystalline domain). Indeed atomically thin materials are particularly suited to withstand unprecedented non-homogeneous deformations before rupture: you can literally bend them and fold them up like a piece of paper. For instance, single-layer graphene and MoS$_2$ have shown very large breaking stress, only expected for ideal materials whose rupture is dominated by the interatomic bond strength and not by the presence of defects. Deformations of up to 10%–20% before rupture have been reported.

The band gap of single-layer MoS$_2$ can potentially be tuned from 1.8 eV (0% strain) all the way down to 0 eV (at ~10% biaxial strain) in contrast to the tunability of only 0.25 eV achieved for strained silicon (at the maximum strain experimentally reached, 1.5% biaxial strain).[67] An additional difference with respect to 3D semiconductors is that 2D semiconductors can be locally deformed, creating inhomogeneous strain profiles that could be used to generate a funnel-shape potential landscape enabling for instance the trapping of excitons for quantum optics experiments, or facilitating the collection of photogenerated charge carriers in solar cells.[87] In this section, the use of local deformations to spatially modify the optoelectronic properties of atomically thin MoS$_2$ will be reviewed.[88]

Large localised uniaxial strain (up to 2.5% tensile) in few-layers MoS$_2$ samples (3–5 layers) has been achieved in the following way: first MoS$_2$ flakes are deposited onto an elastomeric substrate which is pre-stretched by 100%. Subsequently, the tension in the elastomeric substrate is suddenly released generating well-aligned wrinkles in the MoS$_2$ layers perpendicular to the initial uniaxial strain axis in the substrate (see **Figure 10.13**). The mechanism behind the formation of these wrinkles is buckling-induced delamination.[89] This fabrication procedure reproducibly generates large-scale wrinkles (microns in height, separated by tens of microns) for bulk MoS$_2$ flakes (more than 10–15 layers), while thin MoS$_2$ layers (3–5 layers) exhibit wrinkles that are between 50 and 350 nm in height, separated by few microns. **Figure 10.13** shows optical images of wrinkled MoS$_2$ flakes in regions with different number of layers. Wrinkles in single- and bilayer MoS$_2$ samples are not stable as they tend to collapse forming folds, probably due to their reduced bending rigidity.

The maximum uniaxial tensile strain ε is accumulated at the top of the winkles and can be estimated as[89]

$$\varepsilon \sim \pi^2 t \delta / (1 - \sigma^2) \lambda^2 \qquad (10.9)$$

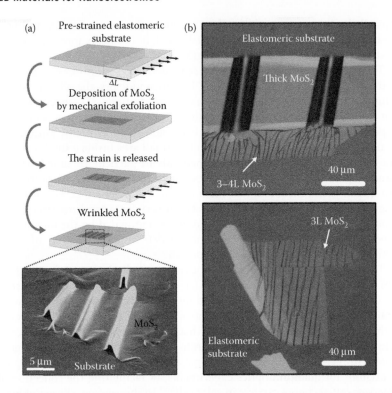

FIGURE 10.13 (a) Scheme of the method employed to fabricate wrinkled MoS$_2$ nanolayers. MoS$_2$ is deposited onto a pre-stretched elastomeric substrate by means of mechanical exfoliation. The strain is released afterwards producing buckling-induced delamination of the MoS$_2$ flakes. The last panel shows a scanning electron micrograph of a few-layer MoS$_2$ wrinkled in this method. (b) Optical microscopy images of two wrinkled MoS$_2$ flake fabricated by the buckling-induced delamination process. (Adapted from A. Castellanos-Gomez et al., *Nano Lett.* **13**, 5361, 2013.)

where v is the MoS$_2$ Poisson's ratio (0.125), t is the thickness of the flake and δ and λ are the height and width of the wrinkle, respectively. The values for δ and λ can be extracted from atomic force microscopy characterisation of the wrinkle geometry. To accurately determine the thickness, a combination of atomic force microscopy, quantitative optical microscopy,[9,73] Raman spectroscopy[45,46] and photoluminescence[24,25,45] can be employed. For the wrinkled thin MoS$_2$ flakes (three to five layers), the estimated uniaxial strain ranges from 0.2% to 2.5%. Interestingly, despite the large strain values, the wrinkles are stable in time and no slippage has been found. Note that slippage is usually a limiting factor in experiments applying uniform uniaxial strain to atomically thin crystals (graphene, MoS$_2$) in substrate bending geometries.[90,91]

Scanning photoluminescence measurements can be used to study the effect of localised tensile strain on the band structure of few-layer MoS$_2$. Although few-layer MoS$_2$ is an indirect band gap semiconductor, its photoluminescence spectrum is dominated by the direct transitions, at the K-point of the Brillouin zone, between the valence band (which is split by interlayer and spin–orbit coupling) and the conduction band.[24,25] These direct transitions appear in the photoluminescence spectra as two peaks, known as the A and B excitons (**Figure 10.14**).[24,25] The indirect band gap transition also contributes to the photoluminescence spectra, emerging as a weak peak at lower energy than the A and B excitons. Interestingly, when the photoluminescence spectrum is acquired on top of a wrinkle, the A and B excitons as well as the indirect band gap peak are red-shifted with respect to the spectrum measured on a flat region of the same flake. This indicates that the uniaxial strain localised on the top of the wrinkle modifies the band structure. Note that the band gap determined by photoluminescence spectroscopy (usually referred to as the optical band gap) differs from that determined by electronic transport due to the exciton binding energy.[87,92] Recent theoretical works have estimated that the exciton binding energy in MoS$_2$ (three to five layers thick) is on the order of 100 meV.[93] While this value is indeed large, it should not be significantly influenced by the applied strain, as was recently demonstrated theoretically in Reference [87] which showed that the exciton binding energy is expected to shift by less than 6 meV per % of strain. As these corrections are small compared to the shifts of the photoluminescence peak, one can neglect their contribution, and thus ascribe the observed changes in the photoluminescence spectra to the changes in the band structure of MoS$_2$. The band structure induced by this non-uniform strain is determined by performing scanning photoluminescence (see **Figure 10.14b**). **Figure 10.14b** shows how the A exciton energy decreases right at the wrinkle location. This observation underlines the potential of non-uniform strain to engineer a spatial variation of the optoelectronic properties of atomically thin MoS$_2$ crystals.

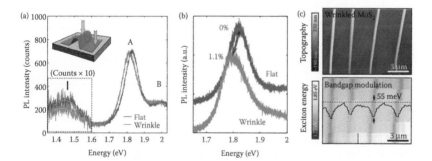

FIGURE 10.14 (a) Two photoluminescence spectra acquired on the flat region of a MoS$_2$ flake (blue) and on top of a wrinkle (red spectra). The region between 1.3 and 1.6 eV has been zoomed in to show the photoluminescence peak associated to the indirect transition (labelled as I). (b) Detail of the A exciton shift due to the uniaxial straining on top of the wrinkle. (c) Atomic force microscopy image (top) of the topography of a trilayer MoS$_2$ flake with four wrinkles. (Bottom) Colourmap image of the exciton energy acquired on the same region shown in the top panel. (Adapted from A. Castellanos-Gomez et al., *Nano Lett.* **13**, 5361, 2013.)

As reported by Rice et al., the E^1_{2g} Raman peak shifts with uniaxial strain by −1.7 cm^{-1} per % strain, and thus, it can be used to calibrate the amount of applied strain with higher accuracy.[91] **Figure 10.15** shows the shift of the direct band gap as a function of the strain (deduced from the E^1_{2g} Raman peak shift) obtained on more than 50 different wrinkles (three to five layers in thickness). The direct gap transition energy decreases for increasing uniaxial strain values; for an ~2.5% tensile strain, the change is about −90 meV, which corresponds to a reduction of the direct band gap transition energy of 5%. The change in the direct band gap of MoS$_2$ induced by the non-uniform strain generated by the wrinkles is comparable to that achieved in semiconducting nanowires by using a straining dielectric envelope,[94] and about five times larger than the change induced in quantum dots by biaxial straining with piezoelectric actuators.[95] Nonetheless, unlike these previous strain engineering approaches, this procedure using non-uniform strain allows one to locally modify the band structure of semiconductors at the nanometre scale.

Tight-binding calculations predict that the direct band gap should decrease linearly upon uniaxial strain.[88] However, the experimental datapoints shown in **Figure 10.15** show a marked sub-linear trend and a rather large dispersion. This difference between the measured and calculated strain tuning of the

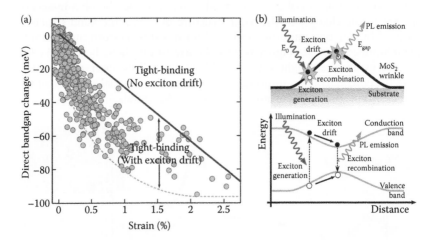

FIGURE 10.15 (a) Change in the energy of the direct band gap transition as a function of the strain measured by scanning the laser across more than 50 wrinkles. The change in the direct band gap is obtained from the shift of the *A* exciton in the photoluminescence spectra while the strain can be estimated from the E^1_{2g} Raman mode shift. The solid line in (a) is the band gap versus strain relationship calculated for wrinkled MoS$_2$ ribbons with different levels of maximum strain employing the tight-binding model discussed in the text. After accounting for the funnel effect in the tight-binding model, the band gap versus strain values are expected to be between the solid line and the dashed line. (b) Schematic diagram explaining the funnel effect due to the non-homogeneous strain in the wrinkled MoS$_2$. (Adapted from A. Castellanos-Gomez et al., *Nano Lett.* **13**, 5361, 2013.)

band gap is due to the movement of the photogenerated excitons which drift towards the position with lower band gap. The non-uniform band gap profile induced by the local strain of the wrinkles generates a trap for photogenerated excitons with a depth of up to 90 meV, referred to as the funnel effect.[87] When including this funnel effect to the tight-binding model, the theory predicts that the experimental datapoints should lay between the solid and the dashed lines in **Figure 10.15**; good agreement is found.

In this section, it has been shown how the local electronic band structure of atomically thin layers of MoS_2 can be tuned by strain engineering. The capability to engineer a local confinement potential for excitons using strain provides a unique tool to design and control the optoelectronic properties of atomically thin MoS_2-based devices.

10.5 Conclusions

This chapter summarises recent progress on the study of optoelectronic and mechanical properties of atomically thin MoS_2. In the first part of the chapter, the mechanism behind the photocurrent generation in unbiased MoS_2 phototransistors is discussed. Unlike for biased devices, in unbiased MoS_2, the photogenerated current is dominated by the photo-thermoelectric effect and not by photovoltaic or photoconductive effects. This interesting feature of MoS_2 devices could be exploited to fabricate broadband photo-detectors. In the second part of the chapter, different approaches to study the mechanical properties of MoS_2 are reviewed. Static measurements as well as dynamic methods are discussed. In the last section of the chapter, the possibility of modifying the optical and electronic properties of MoS_2 by applying deformations is described. In particular, the use of localised strains to generate a spatially varying potential landscape for excitons is reviewed.

Acknowledgements

This work was supported by the European Union (FP7) through the FP7-Marie Curie Project PIEF-GA-2011-300802 (STRENGTHNANO) and the Dutch organisation for Fundamental Research on Matter (FOM) and NWO/OCW.

References

1. K.S. Novoselov, A.K. Geim, S.V. Morozov, D. Jiang, Y. Zhang, S.V. Dubonos, I.V. Grigorieva and A.A. Firsov, *Science* **306**, 666, 2004.
2. K.S. Novoselov, D. Jiang, F. Schedin, T.J. Booth, V.V. Khotkevich, S.V. Morozov and A.K. Geim, *Proc. Natl. Acad. Sci. USA* **102**, 10451, 2005.
3. Q.H. Wang, K. Kalantar-Zadeh, A. Kis, J.N. Coleman and M.S. Strano, *Nat. Nanotechnol.* **7**, 699, 2012.

4. S.Z. Butler, S.M. Hollen, L. Cao, Y. Cui, J.A. Gupta, H.R. Gutiérrez, T.F. Heinz et al., *ACS Nano* 7, 2898, 2013.

5. M. Xu, T. Liang, M. Shi and H. Chen, *Chem. Rev.* 113, 3766, 2013.

6. J.N. Coleman, M. Lotya, A. O'Neill, S.D. Bergin, P.J. King, U. Khan, K. Young et al., *Science* 331, 568, 2011.

7. N. Staley, J. Wu, P. Eklund, Y. Liu, L. Li and Z. Xu, *Phys. Rev. B* 80, 184505, 2009.

8. H. Zhang, C.-X. Liu, X.-L. Qi, X. Dai, Z. Fang and S.-C. Zhang, *Nat. Phys.* 5, 438, 2009.

9. A. Castellanos-Gomez, N. Agraït and G. Rubio-Bollinger, *Appl. Phys. Lett.* 96, 213116, 2010.

10. C.R. Dean, A.F. Young, I. Meric, C. Lee, L. Wang, S. Sorgenfrei, K. Watanabe et al., *Nat. Nanotechnol.* 5, 722, 2010.

11. A. Castellanos-Gomez, M. Wojtaszek, N. Tombros, N. Agrait, B.J. van Wees and G. Rubio-Bollinger, *Small* 7, 2491, 2011.

12. A. Castellanos-Gomez, M. Poot, A. Amor-Amorós, G.A. Steele, H.S.J. van der Zant, N. Agraït and G. Rubio-Bollinger, *Nano Res.* 5, 550, 2012.

13. M.S. El-Bana, D. Wolverson, S. Russo, G. Balakrishnan, D.M. Paul and S.J. Bending, *Supercond. Sci. Technol.* 26, 125020, 2013.

14. B. Radisavljevic, A. Radenovic, J. Brivio, V. Giacometti and A. Kis, *Nat. Nanotechnol.* 6, 147, 2011.

15. M. Han, B. Özyilmaz, Y. Zhang and P. Kim, *Phys. Rev. Lett.* 98, 206805, 2007.

16. M. Moreno-Moreno, A. Castellanos-Gomez, G. Rubio-Bollinger, J. Gomez-Herrero and N. Agraït, *Small* 5, 924, 2009.

17. X. Li, X. Wang, L. Zhang, S. Lee and H. Dai, *Science* 319, 1229, 2008.

18. J.B. Oostinga, H.B. Heersche, X. Liu, A.F. Morpurgo and L.M.K. Vandersypen, *Nat. Mater.* 7, 151, 2008.

19. E. Castro, K. Novoselov, S. Morozov, N. Peres, J. dos Santos, J. Nilsson, F. Guinea, A. Geim and A. Neto, *Phys. Rev. Lett.* 99, 216802, 2007.

20. R. Balog, B. Jørgensen, L. Nilsson, M. Andersen, E. Rienks, M. Bianchi, M. Fanetti et al., *Nat. Mater.* 9, 315, 2010.

21. D.C. Elias, R.R. Nair, T.M.G. Mohiuddin, S. V Morozov, P. Blake, M.P. Halsall, A.C. Ferrari et al., *Science* 323, 610, 2009.

22. Y. Liu, F. Xu, Z. Zhang, E.S. Penev and B.I. Yakobson, *Nano Lett.* 14, 6782, 2014.

23. A. Castellanos-Gomez, M. Wojtaszek, Arramel, N. Tombros and B.J. van Wees, *Small* 8, 1607, 2012.

24. A. Splendiani, L. Sun, Y. Zhang, T. Li, J. Kim, C.-Y. Chim, G. Galli and F. Wang, *Nano Lett.* 10, 1271, 2010.

25. K. Mak, C. Lee, J. Hone, J. Shan and T. Heinz, *Phys. Rev. Lett.* 105, 2, 2010.

26. F.H.L. Koppens, T. Mueller, P. Avouris, A.C. Ferrari, M.S. Vitiello and M. Polini, *Nat. Nanotechnol.* 9, 780, 2014.

27. F. Xia, H. Wang, D. Xiao, M. Dubey and A. Ramasubramaniam, *Nat. Photonics*, 8, 899, 2014.

28. G. Fiori, F. Bonaccorso, G. Iannaccone, T. Palacios, D. Neumaier, A. Seabaugh, S.K. Banerjee and L. Colombo, *Nat. Nanotechnol.* 9, 768, 2014.

29. L. Liu, S.B. Kumar, Y. Ouyang and J. Guo, *IEEE Trans. Electron Devices* 58, 3042, 2011.

30. S. Ghatak, A.N. Pal and A. Ghosh, *ACS Nano* 5, 7707, 2011.

31. H. Li, Z. Yin, Q. He, H. Li, X. Huang, G. Lu, D.W.H. Fam, A.I.Y. Tok, Q. Zhang and H. Zhang, *Small* 8, 63, 2012.

32. Q. Hao, L. Pan, Z. Yao, J. Li, Y. Shi and X. Wang, *Appl. Phys. Lett.* **100,** 123104, 2012.
33. B. Radisavljevic, M.B. Whitwick and A. Kis, *ACS Nano* **5,** 9934, 2011.
34. S. Das, H.-Y. Chen, A.V. Penumatcha and J. Appenzeller, *Nano Lett.* **13,** 100, 2013.
35. P. Lu, X. Wu, W. Guo and X.C. Zeng, *Phys. Chem. Chem. Phys.* **14,** 13035, 2012.
36. Z. Yin, H. Li, H. Li, L. Jiang, Y. Shi, Y. Sun, G. Lu, Q. Zhang, X. Chen and H. Zhang, *ACS Nano* **6,** 74, 2012.
37. H. Wang, L. Yu, Y.-H. Lee, Y. Shi, A. Hsu, M.L. Chin, L.-J. Li, M. Dubey, J. Kong and T. Palacios, *Nano Lett.* **12,** 4674, 2012.
38. W. Zhang, J.-K. Huang, C.-H. Chen, Y.-H. Chang, Y.-J. Cheng and L.-J. Li, *Adv. Mater.* **25,** 3456, 2013.
39. O. Lopez-Sanchez, D. Lembke, M. Kayci, A. Radenovic and A. Kis, *Nat. Nanotechnol.* **8,** 497, 2013.
40. H.S. Lee, S.-W. Min, Y.-G. Chang, M.K. Park, T. Nam, H. Kim, J.H. Kim, S. Ryu and S. Im, *Nano Lett.* **12,** 3695, 2012.
41. W. Choi, M.Y. Cho, A. Konar, J.H. Lee, G.-B. Cha, S.C. Hong, S. Kim et al., *Adv. Mater.* **24,** 5832, 2012.
42. M. Shanmugam, C.A. Durcan and B. Yu, *Nanoscale* **4,** 7399, 2012.
43. M. Fontana, T. Deppe, A.K. Boyd, M. Rinzan, A.Y. Liu, M. Paranjape and P. Barbara, *Sci. Rep.* **3,** 1634, 2013.
44. M. Buscema, M. Barkelid, V. Zwiller, H.S.J. van der Zant, G.A. Steele and A. Castellanos-Gomez, *Nano Lett.* **13,** 358, 2013.
45. M. Buscema, G.A. Steele, H.S.J. van der Zant and A. Castellanos-Gomez, *Nano Res.* **7,** 561, 2015.
46. C. Lee, H. Yan, L.E. Brus, T.F. Heinz, Ќ.J. Hone and S. Ryu, *ACS Nano* **4,** 2695, 2010.
47. I. Popov, G. Seifert and D. Tománek, *Phys. Rev. Lett.* **108,** 156802, 2012.
48. G. Buchs, M. Barkelid, S. Bagiante, G.A. Steele and V. Zwiller, *J. Appl. Phys.* **110,** 074308, 2011.
49. F. Xia, T. Mueller, R. Golizadeh-Mojarad, M. Freitag, Y. Lin, J. Tsang, V. Perebeinos and P. Avouris, *Nano Lett.* **9,** 1039, 2009.
50. J. Park, Y.H. Ahn and C. Ruiz-Vargas, *Nano Lett.* **9,** 1742, 2009.
51. X. Xu, N.M. Gabor, J.S. Alden, A.M. van der Zande and P.L. McEuen, *Nano Lett.* **10,** 562, 2010.
52. N.M. Gabor, J.C.W. Song, Q. Ma, N.L. Nair, T. Taychatanapat, K. Watanabe, T. Taniguchi, L.S. Levitov and P. Jarillo-Herrero, *Science* **334,** 648, 2011.
53. R. Mansfield and S.A. Salam, *Proc. Phys. Soc. Sect. B* **66,** 377, 1953.
54. L. Dobusch, M.M. Furchi, A. Pospischil, T. Mueller, E. Bertagnolli and A. Lugstein, *Appl. Phys. Lett.* **105,** 253103, 2014.
55. J. Wu, H. Schmidt, K.K. Amara, X. Xu, G. Eda and B. Özyilmaz, *Nano Lett.* **14,** 2730, 2014.
56. S.N.F. Mott and H. Jones, *The Theory of the Properties of Metals and Alloys* (Dover Publications, Mineola, New York, USA, 1958), p. 326.
57. B. Bhushan, *Springer Handbook of Nanotechnology* (Springer Science and Business Media, Heidelberg, Germany, 2010), p. 354.
58. J.P. Heremans, V. Jovovic, E.S. Toberer, A. Saramat, K. Kurosaki, A. Charoenphakdee, S. Yamanaka and G.J. Snyder, *Science* **321,** 554, 2008.
59. J.H. Seol, I. Jo, A.L. Moore, L. Lindsay, Z.H. Aitken, M.T. Pettes, X. Li et al., *Science* **328,** 213, 2010.

60. J. Small, K. Perez and P. Kim, *Phys. Rev. Lett.* **91**, 256801, 2003.
61. K.P. Pernstich, B. Rössner and B. Batlogg, *Nat. Mater.* **7**, 321, 2008.
62. J.P. Fleurial, L. Gailliard, R. Triboulet, H. Scherrer and S. Scherrer, *J. Phys. Chem. Solids* **49**, 1237, 1988.
63. A. Li Bassi, A. Bailini, C.S. Casari, F. Donati, A. Mantegazza, M. Passoni, V. Russo and C.E. Bottani, *J. Appl. Phys.* **105**, 124307, 2009.
64. S. Bertolazzi, J. Brivio and A. Kis, *ACS Nano* **5**, 9703, 2011.
65. A. Castellanos-Gomez, M. Poot, G.A. Steele, H.S.J. van der Zant, N. Agraït and G. Rubio-Bollinger, *Adv. Mater.* **24**, 772, 2012.
66. A.A. Griffith, *Philos. Trans. R. Soc. Lond.* **221**, 63, 1921.
67. J. Munguiʹa, G. Bremond, J.M. Bluet, J.M. Hartmann and M. Mermoux, *Appl. Phys. Lett.* **93**, 102101, 2008.
68. O. Marty, T. Nychyporuk, J. de la Torre, V. Lysenko, G. Bremond and D. Barbier, *Appl. Phys. Lett.* **88**, 101909, 2006.
69. A. Castellanos-Gomez, H.S.J. van der Zant and G.A. Steele, *Nano Res.* **7**, 572, 2015.
70. D.-M. Tang, D.G. Kvashnin, S. Najmaei, Y. Bando, K. Kimoto, P. Koskinen, P.M. Ajayan et al., *Nat. Commun.* **5**, 3631, 2014.
71. A. Castellanos-Gomez, M. Buscema, R. Molenaar, V. Singh, L. Janssen, H.S.J. van der Zant and G.A. Steele, *2D Mater.* **1**, 011002, 2014.
72. A. Castellanos-Gomez, R. van Leeuwen, M. Buscema, H.S.J. van der Zant, G.A. Steele and W.J. Venstra, *Adv. Mater.* **25**, 6719, 2013.
73. H. Li, G. Lu, Z. Yin, Q. He, H. Li, Q. Zhang and H. Zhang, *Small* **8**, 682, 2012.
74. A. Castellanos-Gomez, M. Poot, G.A. Steele, H.S. van der Zant, N. Agraït and G. Rubio-Bollinger, *Nanoscale Res. Lett.* **7**, 233, 2012.
75. C. Lee, X. Wei, J.W. Kysar and J. Hone, *Science* **321**, 385, 2008.
76. A. Castellanos-Gomez, V. Singh, H.S.J. van der Zant and G.A. Steele, *Ann. Phys.* **527**, 27, 2015.
77. R. van Leeuwen, A. Castellanos-Gomez, G.A. Steele, H.S.J. van der Zant and W.J. Venstra, *Appl. Phys. Lett.* **105**, 041911, 2014.
78. J.S. Bunch, A.M. van der Zande, S.S. Verbridge, I.W. Frank, D.M. Tanenbaum, J.M. Parpia, H.G. Craighead and P.L. McEuen, *Science* **315**, 490, 2007.
79. J.S. Bunch, S.S. Verbridge, J.S. Alden, A.M. van der Zande, J.M. Parpia, H.G. Craighead and P.L. McEuen, *Nano Lett.* **8**, 2458, 2008.
80. R.A. Barton, B. Ilic, A.M. van der Zande, W.S. Whitney, P.L. McEuen, J.M. Parpia and H.G. Craighead, *Nano Lett.* **11**, 1232, 2011.
81. J. Lee, Z. Wang, K. He, J. Shan and P.X.-L. Feng, *ACS Nano* **7**, 6086, 2013.
82. T. Wah, *J. Acoust. Soc. Am.* **34**, 275, 1962.
83. R.S. Jacobsen, K.N. Andersen, P.I. Borel, J. Fage-Pedersen, L.H. Frandsen, O. Hansen, M. Kristensen et al., *Nature* **441**, 199, 2006.
84. J. Cao, E. Ertekin, V. Srinivasan, W. Fan, S. Huang, H. Zheng, J.W.L. Yim et al., *Nat. Nanotechnol.* **4**, 732, 2009.
85. J.H. Park, J.M. Coy, T.S. Kasirga, C. Huang, Z. Fei, S. Hunter and D.H. Cobden, *Nature* **500**, 431, 2013.
86. V. Pereira and A. Castro Neto, *Phys. Rev. Lett.* **103**, 046801, 2009.
87. J. Feng, X. Qian, C.-W. Huang and J. Li, *Nat. Photonics* **6**, 866, 2012.
88. A. Castellanos-Gomez, R. Roldán, E. Cappelluti, M. Buscema, F. Guinea, H.S.J. van der Zant and G.A. Steele, *Nano Lett.* **13**, 5361, 2013.
89. D. Vella, J. Bico, A. Boudaoud, B. Roman and P.M. Reis, *Proc. Natl. Acad. Sci. USA* **106**, 10901, 2009.

90. T. Mohiuddin, A. Lombardo, R. Nair, A. Bonetti, G. Savini, R. Jalil, N. Bonini et al., *Phys. Rev. B* **79**, 205433, 2009.
91. C. Rice, R. Young, R. Zan, U. Bangert, D. Wolverson, T. Georgiou, R. Jalil and K. Novoselov, *Phys. Rev. B* **87**, 081307, 2013.
92. H. Shi, H. Pan, Y.-W. Zhang and B. Yakobson, *Phys. Rev. B* **87**, 155304, 2013.
93. H.-P. Komsa and A.V. Krasheninnikov, *Phys. Rev. B* **86**, 241201, 2012.
94. M. Bouwes Bavinck, M. Zieliński, B.J. Witek, T. Zehender, E.P.A.M. Bakkers and V. Zwiller, *Nano Lett.* **12**, 6206, 2012.
95. R. Trotta, P. Atkinson, J.D. Plumhof, E. Zallo, R.O. Rezaev, S. Kumar, S. Baunack, J.R. Schröter, A. Rastelli and O.G. Schmidt, *Adv. Mater.* **24**, 2668, 2012.

11

Device Physics and Device Mechanics for Flexible TMD and Phosphorene Thin-Film Transistors

Hsiao-Yu Chang, Weinan Zhu and Deji Akinwande

Contents

2D Materials for Nanoelectronics edited by Michel Houssa, Athanasios Dimoulas and Alessandro Molle © 2016 CRC Press/Taylor & Francis Group, LLC. ISBN: 978-1-4987-0417-5.

11.1 Introduction to 2D Flexible Electronics

Future ubiquitous smart electronic systems are envisioned to afford arbitrary form factors, robust elasticity, high-speed charge transport and low power consumption, a combined set of attributes which demonstrate uniform electronic properties across a wide range of applied strains.[1–3] A major contemporary challenge concerns the choice of a semiconducting material suitable for high-performance field-effect transistors (FETs) on a flexible substrate.[1,2,4] Using pliable electronic materials, such as semiconducting polymers and organic molecules, to fabricate thin-film transistors (TFTs) on soft substrates has had limited applications due to low field-effect mobilities.[1,5,6] Enhanced device performance has been achieved by utilizing thin films of conventional semiconductor materials, including crystalline and polycrystalline Si and III–V semiconductors, that offer improved electronic properties albeit at the cost of overall device flexibility and thickness scalability.[7–10] More recently, graphene has attracted substantial interest for high-performance flexible electronics owing to its high carrier mobility (>10,000 cm^2/Vs) and outstanding radio-frequency properties[11–16]; however, its lack of a band gap is a major drawback since low-power switching or digital transistors cannot be realized.[17] This drawback has consequently motivated the search for other layered atomic sheets with substantial band gaps such as the semiconducting transition metal dichalcogenides (TMDs).[18,19] Molybdenum disulphide (MoS$_2$) is a prototypical TMD that has been attracting rapidly growing interest owing to its large semiconducting band gap (~1.8 eV for monolayer and ~1.3 eV for bulk films), which is ideal for low-power electronics on hard and soft substrates.[18,20–24] In addition to electronic properties, two-dimensional (2D) materials possess clear advantages, both in material thickness and elastic limit, in comparison with traditional semiconductors for flexible nanoelectronics, which demonstrate high electronic performance as well as high flexibility (see **Figure 11.1**).

Table 11.1 summarises the basic optical, electrical, mechanical and thermal properties of several 2D atomic sheets that span the range from semimetals to insulators.[10] This table is a useful reference for guiding the selection of material(s) and device design for flexible nanoelectronics. As a general material, graphene offers the fastest charge transport, stiffness and thermal conductivity. MoS$_2$, phosphorene and other semiconducting TMDs are well suited to serve as the semiconducting channel for digital electronics. In addition, direct

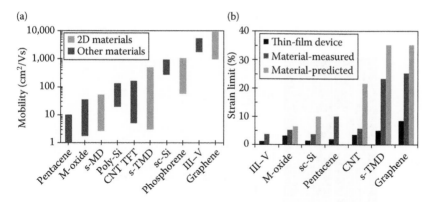

FIGURE 11.1 Mobility and strain comparison of candidate materials for flexible TFTs. (a) Comparison of experimentally reported FET mobilities, including mobility variations reported across a wide variety of experimental samples at room temperature. (b) Maximum elastic strain limits of candidate materials for flexible TFTs (From Akinwande, D., Petrone, N. and Hone, J. *Nat. Commun.* 2014, 5, 5678.)

band gap monolayer TMDs can be used for optoelectronics. For atomically thin insulating sheets, multilayer hBN has been proven to be the ideal dielectric for 2D TFTs, owing to its atomically smooth surface, large phonon energies, high dielectric breakdown field and high in-plane thermal conductivity compared to conventional dielectrics.[25–27]

11.2 Fabrication of MoS$_2$ Device on Both Rigid and Flexible Substrate

For the flexible devices, we used commercially available polyimide (PI) film (Kapton) as the flexible substrate, and spin coated an additional liquid PI film (PI-2574 from HD Micro Systems) on the surface to reduce the surface roughness (total substrate thickness ~100 μm). The liquid PI was cured at 300°C for 1 h. Ti/Pd (2/50 nm) deposited by electron beam evaporation was used as the bottom-gate electrode, and Al$_2$O$_3$ or HfO$_2$ (25 nm) deposited at 200°C by atomic layer deposition (ALD) method as the gate dielectric. MoS$_2$ devices were prepared by mechanical exfoliation from commercial crystal (SPI Supplies, Inc.). Flakes with thickness around 5–25 nm were selected by optical microscope and confirmed by atomic force microscope (AFM). Source/drain contacts were defined by electron beam lithography, and Ti/Au (2/50 nm) or Au (50 nm) were deposited by electron beam evaporation followed by the lift-off process. **Figure 11.2** shows the schematic depiction of the flexible MoS$_2$ device made on PI. In addition to the flexible devices, we also made MoS$_2$ devices on rigid substrates. Highly doped Si substrate was used as the bottom gate, and 270 nm SiO$_2$ deposited by thermal oxidation or 25 nm Al$_2$O$_3$ by ALD method was used as the gate dielectric. The remaining steps are the same as the flexible samples. Channel length is fixed to

Table 11.1   Room Temperature Solid-State Properties of Selected 2D Crystalline Materials

2D Material	Optical		Electrical		Mechanical		Thermal	
	Band Gap (eV)	Band Type	Device Mobility (cm^2/V-s)	v_{sat} (cm/s)	Young's Mod. (GPa)	Fracture Strain (%) Theor (Meas)	κ (W/m-K)	CTE (10^{-6}/K)
Graphene	0	D	10^3–5×10^4	1–5×10^7	1,000	27–38 (25)	600–5,000	−8
1L MoS_2	1.8	D	10–130	4×10^6	270	25–33 (23)	40	N/A
Bulk MoS_2	1.2	I	30–500	3×10^6	240	N/A	50 (II), 4($\perp$)	1.9 (II)
1L WSe_2	1.7	D	140–250	4×10^6	195	26–37 (N/A)	N/A	N/A
Bulk WSe_2	1.2	I	500	N/A	75–100	N/A	9.7 (II), 2($\perp$)	11 (II)
h-BN	5.9	D	N/A	N/A	220–880	24 (3–4)	250–360 (II) 2($\perp$)	−2.7
Phosphorene	0.3–2*	D	50–1,000	N/A	35–165	24–32	10–35 (II)	N/A

Source: From D. Akinwande, N. Petrone and J. Hone, *Nat Commun* 2014, 5, 5678.

Note: All listed values should be considered estimates. In some cases, experimental or theoretical values are not available (N/A).

* The precise value for the bandgap, which is a maximum for a monolayer is a matter of ongoing research.

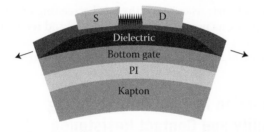

FIGURE 11.2   The schematic depiction of the flexible bottom gate device structure. (From Chang, H.-Y. et al. *ACS Nano* 2013, 7(6), 5446–5452.)

1 μm, except for the device structures of transfer length method (TLM), while channel width is determined by the width of each flake, typically in the range of 0.7 ~ 4 μm.

Electrical characteristics of the flexible MoS_2 FETs were evaluated under ambient conditions. Representative transfer $(I_D–V_G)$ characteristics are shown in **Figure 11.3a**. The extracted low-field mobility is 30 cm²/Vs using the Y-function method which is defined as $I_D/\sqrt{g_m}$ (I_D is the drain current and g_m is the transconductance) and is especially suitable for studying device physics because it excludes the contact resistance effect on the mobility.[28,29] The details of the low-field mobility extraction will be provided in the next chapter. The on/off switching ratio is more than 10^7, and the sub-threshold slope is ~82 mV/decade. Output $(I_D–V_D)$ characteristics show negligible Schottky barrier in the linear region, and current saturation at high fields as shown in **Figure 11.3b**. These results are comparable with unpassivated MoS_2 FETs realized on Si substrates,[20,30] indicating that its unique electrical properties can be accessed on hard and soft substrates alike which is a welcome benefit for flexible electronics. Further improvement of the device performance can be achieved by mobility enhancement through passivation

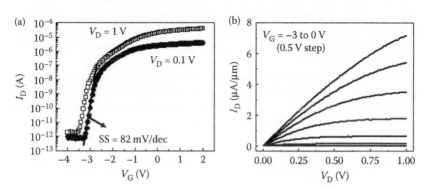

FIGURE 11.3   A representative MoS_2 FET made with Al_2O_3 as gate dielectric on flexible PI. (a) I_D-V_G characteristics in log scale. (b) I_D-V_D characteristics indicates negligible Schottky barrier in the linear region, and current saturation at high fields.

with a high-k dielectric to enhance the local screening effect and suppress Coulomb scattering as previously reported,[20,30–32] and reducing the contact resistance by using low work function metals such as scandium[23] or the metallic 1T phase of MoS_2[33] to minimise the Schottky barrier at the contact.

11.3 Y-Function Method for Extracting Mobility and Contact Resistance

11.3.1 Introduction to the Y-Function Method

The major challenge in understanding the intrinsic charge transport properties arises from the substantial R_c that is typical of metal interfaces to low-dimensional systems such as 2D atomic layers,[23,34–36] which is often a Schottky barrier interface when the 2D layer is a semiconductor as is the case for MoS_2, and similar TMDs. As a result, established methods based on the transconductance or two-point conductance of field-effect devices can result in mobility underestimation in the presence of typical R_c profile, which is either gate independent (Ohmic contact) or inversely proportional to the magnitude of the gate voltage (Schottky contact). In the unusual case of R_c that is directly proportional to gate voltage, we show later that mobility overestimation can even occur. While this challenge has been long recognised, and indeed discussed in the first reports of (one-dimensional) carbon nanotube transistor devices (in 1998),[37] there remains a need for a routine technique that can decouple the mobility evaluation from the contact resistance and yield accurate estimates for both, particularly for understanding the properties and prospects of semiconducting atomic crystals based on TMDs, which has only recently attracted broad attention.[38]

The Y-function method[28] was first proposed by Ghibaudo for low-field mobility (μ_o) and threshold voltage (V_T) extraction in Si transistor devices for its accuracy and simplicity considering first-order mobility attenuation coefficient and constant contact resistance. We show in this chapter that the Y-function method remains robust and accurate even in the presence of gate voltage dependent (Schottky barrier) contact resistance, which is typical for FETs based on TMDs.[39–42]

The Y-function method is based on the straightforward analysis of the drain current (I_D) in the linear region. Considering that Schottky barrier induced contact resistance at source and drain end will result in additional voltage drops, the I–V equation can be expressed as[43]

$$I_D = \left(\frac{\mu_o}{1 + \theta_o(V_G - V_T)}\right) C_{ox} \frac{W}{L}(V_G - V_T - 0.5V_D)(V_D - I_D R_c) \quad (11.1)$$

where first-order mobility attenuation coefficient, θ_o, introduced by remote phonon scattering and surface roughness is included to better depict the realistic device performance.[44] C_{ox} is the gate capacitance and W and L are the device width and length, respectively. V_G and V_D are the gate and drain voltages, respectively. $L = 1\ \mu m$, and $V_D = 0.1$ V was chosen to evaluate the mobility in the low-field bias condition. In the low-field limit, $V_G - V_T \gg 0.5\ V_D$ under strong inversion, hence, the $0.5\ V_D$ factor can be ignored to further simplify **Equation 11.1**. For convenience, we combine both θ_o and R_c effects as one effective mobility attenuation factor, $\theta = \theta_o + \mu_o C_{ox} R_c W/L$. Therefore, **Equation 11.1** can be rewritten as

$$I_D = \left(\frac{\mu_o}{1 + \theta(V_G - V_T)}\right) C_{ox} \frac{W}{L}(V_G - V_T)V_D \qquad (11.2)$$

The Y-function is defined as $I_D/\sqrt{g_m}$, where g_m is the transconductance defined as dI_D/dV_G. Owing to the existence of Schottky barriers at the source and drain contact, we consider the case, where R_c is also V_G dependent when calculating dI_D/dV_G. It follows that in the general case, the Y-function is given by

$$Y = \frac{I_D}{\sqrt{g_m}} = \frac{\sqrt{\mu_o C_{ox} V_D (W/L)}}{\sqrt{1 - \mu_o C_{ox} R_c'(W/L)(V_G - V_T)^2}}(V_G - V_T) \qquad (11.3)$$

where R_c' is dR_c/dV_G. We first assume that R_c is not V_G dependent, resulting in the simplified expression $Y = (\mu_o C_{ox} V_D W/L)^{0.5}(V_G - V_T)$ for the extraction of the low-field mobility and the threshold voltage.[28] Subsequently, we will examine the validation of this assumption afterwards in order to understand the impact of voltage dependent R_c on mobility extraction.

Figure 11.4a through **c** is a representative example for extracting μ_o, V_T and R_c of an experimental MoS$_2$ FET from the simplified Y-function. First, we plot the Y-function with respect to V_G as shown in **Figure 11.4a**. From the linear fit in the strong inversion region, $V_T = -32.1$ V and $\mu_o = 55.7\ cm^2/Vs$ can be extracted from the x-intercept and the slope respectively, given that the capacitance of the gate dielectrics used in this work were measured separately by standard C-V techniques on fabricated test structures. Strong agreement to a linear profile is observed. **Figure 11.4b** shows θ versus V_G. In the absence of estimates for θ_o, which is the case for MoS$_2$, an upper bound can be placed on R_c in the limit of negligible θ_o. With $\theta \approx \mu_o C_{ox} R_c W/L$, $R_c \approx 20.4\ \Omega$-mm is extracted from the strong inversion region. **Figure 11.4c** shows the transfer characteristic (I_D–V_G) for MoS$_2$ FET in both log and linear scales. The on/off switching ratio is ~10^7 at $V_D = 0.1$ V from the log scale, and the model (**Equation 11.2**) offers strong agreement to the experimental data shown in the linear scale.

11.3.2 Mobility Underestimation from $G_{m,max}$

Arguably, the most common method typically chosen to extract the field-effect mobility is from the maximum or peak transconductance ($g_m = \mu_{FE} C_{ox} V_D W/L$, where μ_{FE} is the field-effect mobility extracted from g_m) in the transistor linear region. However, this extracted mobility (μ_{FE}) can significantly underestimate the true low-field mobility (μ_o) in the presence of substantial contact

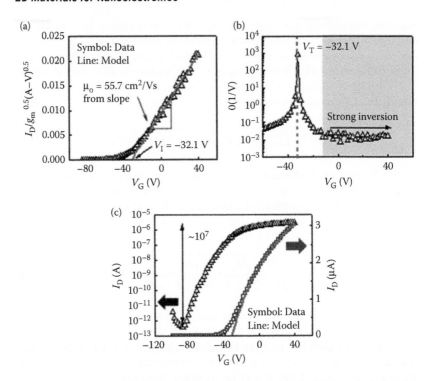

FIGURE 11.4 A representative example for extracting μ_o, V_T and R_c of an experimental MoS$_2$ FET from the simplified Y-function. (MoS$_2$ FET fabricated with 270 nm SiO$_2$ as gate dielectric) (a) From the linear fitting in the strong inversion region, which shows good agreement, both V_T and μ_o can be extracted from the x-intercept and the slope respectively. (b) With $\theta \sim \mu_o C_{ox} R_c W/L$, $R_c \sim 20.4$ Ω-mm is extracted from the strong inversion region. (c) The transfer characteristic (I_D–V_G) for MoS$_2$ FET in both log and linear scales at $V_D = 0.1$ V. The model shows strong agreement to the experimental data. (From Chang, H.-Y., Zhu, W. and Akinwande, D. *Appl. Phys. Lett.* 2014, 104(11), 113504.)

resistance. In the strong inversion region, we can derive the dependence of the mobility underestimation error as a function of the normalised contact resistance (R_c/R_{tot}), which can be expressed as

$$\frac{\mu_{FE} - \mu_o}{\mu_o} = \left(\frac{R_c}{R_{tot}} \right) \left(\frac{R_c}{R_{tot}} - 2 \right) \tag{11.4}$$

We note that the total (channel plus contact) resistance, $R_{tot} = V_D/I_D$ is a function of $V_G - V_T$, and so is μ_{FE}. With **Equation 11.4** (which is $\approx -2R_c/R_{tot}$ when $R_c/R_{tot} \ll 2$), we can quantitatively estimate the mobility underestimation error arising from finite contact resistance. **Figure 11.5** shows device statistics of the mobility underestimation from μ_{FE} extracted from g_m and $g_{m,max}$ compared to μ_o extracted from the Y-function as a function of the significance

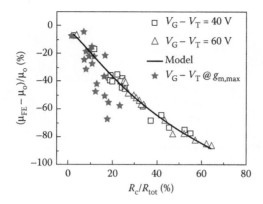

FIGURE 11.5 Electrical data statistics from 19 MoS_2 devices highlighting the mobility underestimation error from the g_m method compared to the Y-function method. Relation between the mobility underestimation extracted from g_m (μ_{FE}) compared to μ_o as a function of R_c/R_{tot}. (From Chang, H.-Y., Zhu, W. and Akinwande, D. *Appl. Phys. Lett.* 2014, 104 (11), 113504.)

of the contact resistance (R_c/R_{tot}). Data extracted from 19 devices made on 270 nm SiO_2 as the gate dielectric and Ti/Au or Au as the contact metal show strong agreement with the model in the strong inversion region. However, for devices with larger R_c, $g_{m,max}$ happened at small gate overdrive which is outside (not fully applicable for) the fitting range of the Y-function. As a result, the extraction from $g_{m,max}$ gives additional underestimation of μ_o due to this deviation from the model. For 16 out of the 19 devices in our statistics, R_c/R_{tot} is in the range of 7%–23% corresponding to the mobility underestimation of 15%–60%.

11.3.3 Y-Function Robust Analysis for Contact Resistance Extraction

In order to independently corroborate the Y-function contact resistance extraction, we fabricated two test structures for TLM made with 25 nm Al_2O_3 as the gate dielectric and Au as the metal contacts. Each TLM device structure consists of 1, 0.5 and 0.25 μm channel lengths. FETs with $L = 0.5$ μm shown in **Figure 11.6a** and **b** are selected to represent each TLM device structure. The device shown in **Figure 11.6a** has a negligible Schottky barrier at room temperature as indicated by the linear response from the low-field I_D–V_D profile, while the other one shown in **Figure 11.6b** has an obvious Schottky barrier. Robustness of the Y-function method is discussed and validated in the following paragraphs.

As shown in **Figure 11.6c** and **d**, the experimental R_c data extracted from the TLM structure demonstrate a clear dependence on gate overdrive, which is in similar trend with previous published results.[39–41] In other words, the assumption of constant R_c in μ_o extraction using the Y-function method should be re-examined. According to the original equation (**Equation 11.3**),

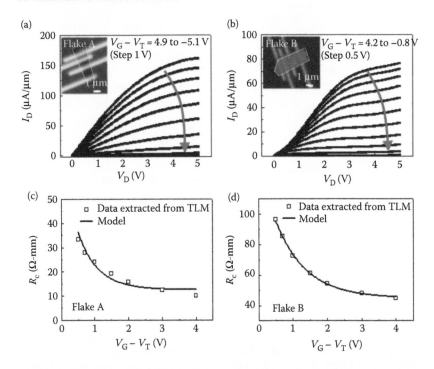

FIGURE 11.6 (a) and (b) TLM structure made with 25 nm Al_2O_3 as the gate dielectric and Au as the metal contact. MoS_2 crystals are outlined in the optical microscope images, and the thicknesses are 7.7 and 21.6 nm for flake A and B respectively. Each device structure includes 1, 0.5 and 0.25 mm channel lengths. Devices with $L = 0.5$ mm are selected to represent each TLM structure. Devices made with flake A show negligible Schottky-barrier as evident by the linear low-field I_D–V_D profile, in contrast, noticeable Schottky-barrier can be seen for flake B. (c) and (d) Gate voltage dependence of R_c extracted from TLM data for flake A and flake B are presented and fitted with the exponential FN R_c model (solid line). (From Chang, H.-Y., Zhu, W. and Akinwande, D. *Appl. Phys. Lett* 2014, 104 (11), 113504.)

the additional contact resistance component in the denominator does not affect the V_T extraction, that is, the threshold voltage can still be accurately determined from the linear intercept. However, the extracted mobility (μ_{ext}) from the slope of Y in the presence of gate-dependent contact resistance can in general depend on the contact resistance value (note: when R_c is constant, $\mu_{ext} = \mu_o$). One key insight from **Equation 11.3** is that the Y-function extracted mobility can underestimate or overestimate the true low-field mobility (μ_o) depending on whether R'_c is negative or positive, respectively.

The gate-dependent source and drain contact resistance is attributed primarily to the Schottky barrier formed at the metal/MoS_2 interface.[23,34] Electron transport across the Schottky barrier can be well described by Fowler–Nordheim tunnelling current, $I = [A^*AT^2\exp(-\Phi_B/KT)][\exp(q(V_G - V_T)/nKT) - 1]$,[45] where A^* is the 2D Richardson coefficient, A is the contact area, T is the temperature, Φ_B is the Schottky barrier height (SBH) in unit of eV, K is the Boltzmann's constant and q is the electron charge. n is the gate voltage coupling coefficient and the metal/MoS$_2$ contact resistance is evaluated in the linear region rather than the sub-threshold region. For convenience, a composite fitting parameter is defined here, $a = q/nKT$. The Schottky barrier induced metal/MoS$_2$ contact resistance, therefore, can be interpreted as

$$R_c = R_m + R_o \exp[-a(V_G - V_T)], \tag{11.5}$$

where the former term (R_m), a voltage independent contribution arising from the finite metal Ohmic resistance, is the limiting series resistance under high gate overdrive; and the latter term is the gate voltage dependent term derived from the Schottky interface, where R_o is the maximum value of the Schottky contribution when $V_G - V_T = 0$. As shown in **Figure 11.6c** and **d**, the gate dependence of the experimental metal/MoS$_2$ R_c measured via TLM method can be well fitted by the exponential model. The SBH, $\Phi_B = kT/q$ ($\ln(a/R_o) - \ln(AA^*T^2)$) was extracted from the exponential fit to be ~54 meV under zero gate overdrive for the Au/MoS$_2$ contact, which is within the published range given by recent articles (20–120 meV).[40,41]

With the Schottky-barrier-based R_c model, the robustness of the Y-function method in the presence of gate-dependent contact resistance effect in μ_o extraction can now be analysed. Experimentally, μ_{ext} is achieved from the linear fitting of Y-function in the strong inversion regime corresponding to relatively high gate overdrive. In order to analyse the impact of the V_G dependence of R_c on the linearity of the Y-function under high overdrive bias, we applied the first-order approximation of the exponential R_c equation, which yield an inverse dependence between the gate overdrive and R_c, $R_c \approx R_m + R_o/(1 - a(V_G - V_T))$. This simplification makes the analysis more straightforward without impacting the validity of ensuing conclusions. Substituting the simplified R_c expression into the Y-function equation (**Equation 11.3**), under high overdrive bias, where $a(V_G - V_T) \gg 1$, yields

$$Y = \frac{\sqrt{\mu_o C_{ox} V_D (W/L)}}{\sqrt{1 - \mu_o C_{ox}(W/L)(R_o/a)}}(V_G - V_T) \tag{11.6}$$

The influence of V_G bias in the denominator of **Equation 11.3** is eliminated in **Equation 11.6**, therefore, this first-order approximation of R_c helps to retain the linearity of the Y-function despite the underestimation of the extracted mobility. The deviation between μ_{ext} and μ_o can be expressed as $(\mu_{ext} - \mu_o)/\mu_o = \mu_{ext} C_{ox}(W/L)(R_o/a)$, which is dependent on the value of R_o/a from the R_c fitting for a measured μ_{ext}. With the aid of numerical simulations, we observed that increasing the value of R_o/a from 0 to 10^3 merely increased the deviation to 0.1%. Based on the fitting results in this work of the experimental data, a reasonable range for R_o/a is 10–100, which validates that the Y-function method is robust in μ_o and V_T extraction for realistic MoS$_2$ FETs.

11.3.4 Comparison between Contact Resistance Extraction Methods

From the theoretical analysis shown above, we have demonstrated that the simplified Y-function method is robust for μ_o, and V_T extraction for both Ohmic contact and modest Schottky barrier devices typical of reported TMD transistors[23,39,41,46,47]. In this paragraph, we are going to verify that extracting the contact resistance from the simplified Y-function method can be adopted as a reliable way to provide a close estimation for R_c in strong inversion even in the presence of Schottky barriers. First, we examined the contact resistance by comparing the extraction from the Y-function and TLM methods (**Figure 11.7**), and noticed a clear discrepancy. From the assumption of $\theta \sim \mu_o C_{ox} R_c W/L$, R_c is extracted by neglecting θ_o, which results in an upper bound for R_c. However, it is shown that the R_c extracted from the Y-function is even higher than the value extracted from TLM. We examined our data more closely, and then found out that μ_o extracted from the Y-function method showed degradation with channel length from 1 μm to 250 nm, and similar observation was also mentioned in a previous published result for MoS$_2$ bottom-gate transistors.[41,46] This L-dependent mobility effect needs to be better understood and certainly considered for R_c extraction from the TLM structure. For the original extracted data, we assumed μ_o is constant over different channel lengths. Ignoring the mobility degradation introduced by channel length scaling will lead to an overestimation of R_c. Since μ_o is extracted in the strong inversion region, we only apply the length dependent correction to the high gate overdrive region (shaded region in **Figure 11.7**). With this correction, there is good agreement between R_c extracted from the Y-function

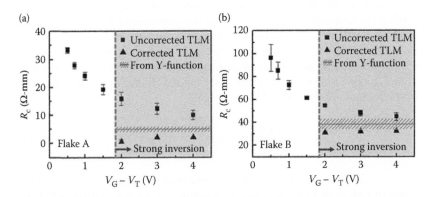

FIGURE 11.7 (a) and (b) The same TLM structures as shown in **Figure 11.6**. R_c extracted from the Y-function method shows the uncertainty range for 3 different channel lengths for each TLM structure. With the corrected μ_o for each channel length, a good agreement between R_c extracted from the Y-function method and the corrected TLM method can be observed. The overestimation of R_c extracted from the Y-function method was due to neglecting the effect of θ_o, which relates to additional mobility degradation factors such as surface roughness and remote phonon scattering. (From Chang, H.-Y., Zhu, W. and Akinwande, D. *Appl. Phys. Lett* 2014, 104 (11), 113504.)

method and the corrected TLM method. The overestimation of R_c extracted from the Y-function method was due to neglecting the effect of θ_o, which relates to additional mobility degradation factors such as surface roughness and remote phonon scattering.[44]

11.3.5 Conclusion

In summary, the Y-function method for FET mobility extraction has been shown to be accurate and robust in the presence of gate-dependent or Schottky barrier contact resistance frequently encountered in semiconducting TMD transistor devices. Furthermore, independent TLM contact resistance studies corroborate the Y-function contact resistance extracted in the strong inversion region even for Schottky barrier TMD FETs. By comparing the Y-function method and the corrected TLM method, we verify that the Y-function method can be adopted as a convenient way to provide a close estimation of R_c. The main conclusion of this combined experimental and analytical study is that the Y-function method is accurate for the evaluation of the low-field mobility, contact resistance and threshold voltage for Ohmic and Schottky barrier TMD transistor devices and is a straightforward unambiguous technique for experimental FET studies of 2D atomic sheets.

11.4 Device Mechanics: Failure Mechanisms under Strain

11.4.1 Failure Mechanisms from Bending Test

To study the mechanical properties of flexible MoS_2 device, tensile strain was applied to the devices by convexly bending the flexible substrate using a home-built mechanical bending fixture. The sample was held for 10 s at each bending radius, and then released for measurement. During the bending process, tensile strain was applied directly to the device. Owing to the sliding of the MoS_2 while bending to the smaller bending radius, the compressive strain would able to be induced during the releasing process.

Figure 11.8a and **b** shows the dependence of the normalised OFF current and normalised ON current on the bending radius, respectively. Device characteristics are robust down to a bending radius of 2 mm, which we attribute to the high deformability of MoS_2,[48] and the relatively low strain placed on the dielectric thin films. At or below 2 mm bending radius, MoS_2 devices show significant OFF current increase owing to structural damage to the gate dielectric. Similarly, the ON current degrades ≤ 2 mm radius. However, buckling delamination of MoS_2 is mainly responsible.

11.4.2 Cracks in Inorganic Dielectrics: Stretching Test

To unambiguously identify the mechanism responsible for device failure after severe bending, the gate dielectric structural integrity was investigated under varying tensile strains. For this purpose, HfO_2 and Al_2O_3 films are deposited

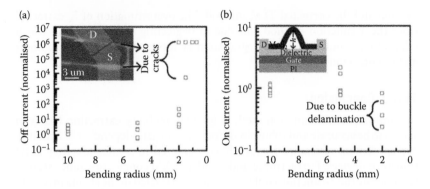

FIGURE 11.8 Mechanical studies of flexible M_oS_2 FET with Al_2O_3 as the gate dielectric. (a) Below 2 mm bending radius, the exponential increase in OFF current is due to crack formation in the gate dielectric shown in the inset. (b) The ON current degrades ≤ 2 mm radius. Buckling delamination of MoS_2 (shown in inset) is mainly responsible. (From Akinwande, D., Petrone, N. and Hone, *J. Nat. Commun.* 2014, 5, 5678.)

on 26-μm-thick rectangular PI strips, with a sample cross-section similar to the device sample as illustrated in **Figure 11.9a**. Without the Kapton substrate, the thinner 26 μm PI affords a greater range of tensile strain to be studied, and maintained the same surface property as the device structure. We did not perform our own measurements of the Young's modulus of each dielectric material but materials fabricated in very similar conditions are measured to have E_{HfO_2} = 73.4 GPa, and $E_{Al_2O_3}$ = 163.3 GPa, respectively.[49] Stretch tests were subsequently done using a home-built mechanical test fixture in situ under optical microscope (**Figure 11.9b**). The stretch tests revealed formation of channel cracks aligned perpendicular to the stretch direction in the dielectric materials as shown in **Figure 11.9c**. The growing density of dielectric cracks lead to increased gate leakage and subsequent device failure. A quantitative count of the crack density as a function of the applied tensile strain can be seen in **Figure 11.9d**. The critical strain and saturation crack density are extracted using an empirical model that is applicable to this work.[50] The result suggests a slightly higher critical strain for HfO_2 (1.72%) compared to Al_2O_3 (1.69%). We found the crack density of HfO_2 saturates at slightly higher values (~10%) compared to that of Al_2O_3, which is consistent with the expectations that films with lower strength $\sigma_{max} = E\varepsilon_{cr}$, where $\varepsilon_{cr}^{Al_2O_3} \approx \varepsilon_{cr}^{HfO_2}$ but $E_{Al_2O_3} > E_{HfO_2}$ exhibit lower saturation crack spacing.[51]

11.4.3 Approaches to Improve Device Flexibility

Structure design is one of the approaches which can effectively improve flexibility. When we consider the case of bending, strain on the surface of a material subject to flexural bending decreases linearly with substrate thickness;

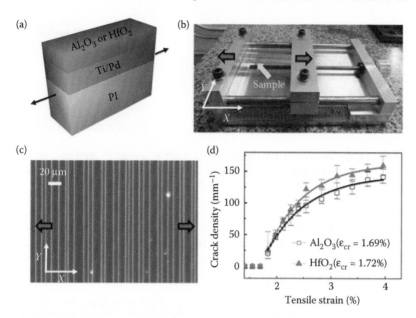

FIGURE 11.9 (a) Test structure for the stretching experiments to elucidate the mechanical reliability of selected gate dielectrics on flexible PI. (PI 26 μm/ Ti 2 nm/Pd 50 nm/Al_2O_3 or HfO_2 25 nm) (b) Photograph of the stretcher test fixture. The stretching direction was along the x-direction. (c) Optical microscope image of the sample of HfO_2 at strain ~2.5%. The parallel cracks aligned to the y-direction are due to tensile stress. (d) The dependence of the crack density on tensile strain for Al_2O_3 and HfO_2. The stretch test shows that the critical crack onset strain is around 1.69% and 1.72%, and the crack density saturates at 145 and 164 mm^{-1} for Al_2O_3 and HfO_2 respectively. (From H.-Y. Chang et al., *ACS Nano* 2013, 7(6), 5446–5452.)

hence, even materials that are brittle in bulk form can be flexed to a degree when produced as a thin film. Under pure bending, the strain, ε, at any given point in the substrate is a function of both the bending radius, ρ, and the perpendicular distance, z, from the neutral axis, given by the relationship $\varepsilon = z/\rho$. Strategies to minimise this distance include thinning the substrate or moving the device from the substrate surface (where strains reach a maximum) nearer to the neutral axis (where strains vanish), for instance by two-sided encapsulation. Although practical design requirements may complicate the ability to move the TFT device plane to coincide with the neutral axis of the substrate, in principle it is possible to design a highly flexible device out of a relatively brittle material. Indeed, TFTs fabricated from crystalline Si on 25 μm thick substrates and encapsulated to maintain the device plane within 2 μm of the substrate's neutral axis have achieved bending radii below 400 μm while maintaining tensile strains of less than 0.2% in the active devices.[52] For the

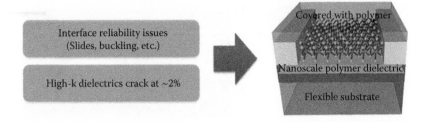

FIGURE 11.10 Replacing the conventional high-k dielectric with the polymer dielectric to prevent the cracks in dielectric, and covering the surface with polymer to avoid buckling delamination of MoS_2.

stretch induced strain, ε is expressed as the ratio of total deformation to the initial dimension of the material body in which the forces are being applied. Similarly, by structure designs, the non-uniform stress can be enhanced to avoid the strain applied to the device region which can effectively improve the stretchability.

Besides the structure design, the other approach remaining is to increase the strain limit of all the components in the device. The failure mechanisms under strain have been discussed in Section 11.4.1. The critical strain of high-k dielectric is around 2%. As shown in **Figure 11.10**, by replacing the conventional high-k dielectric with the nanoscale polymer dielectric, and covering the surface to avoid sliding between the interfaces which may result in buckling delamination, will enable us to study the strain effect beyond 2%. The nanoscale PI with thickness around 60 nm can serve as the gate dielectric was demonstrated by MoS_2 FET, with mobility round 20–30 cm²/Vs which is comparable to other dielectric, including SiO_2, and high-k dielectrics with the same device structure and fabrication process (**Figure 11.11**).

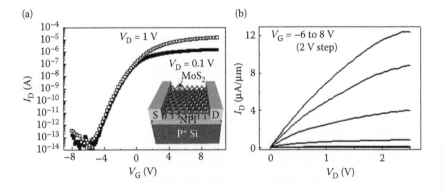

FIGURE 11.11 The MoS_2 device was successfully made with nanoscale polyimide (NPI). (a) I_D–V_G characteristics in log scale. (b) I_D–V_D characteristics.

11.5 Flexible Phosphorene Transistors and Circuits

11.5.1 Introduction to Phosphorene

Black phosphorus is a thermal dynamically stable and the least reactive form of phosphorus allotropes, which was first synthesised from red phosphorus under high temperature and high pressure a century ago.[53] In **Figure 11.12**, three common phosphorus allotropes are shown: white, red and black phosphorus.

Single crystal bulk black phosphorus exhibits layer stacked crystal structure similar to graphite but with a natural direct band gap of 0.31 eV,[54,55] which was favourable to optical electronics such as the infrared detector.[56] Owing to this van der Waals crystalline nature (**Figure 11.13**), the monolayer of black phosphorus atomic sheets, known as phosphorene, can be mechanically cleaved from the bulk crystal, a practical method for obtaining 2D atomic sheets, for instance, graphene and monolayer TMDs.[30, 47,57] By stacking more layers of phosphorene, the direct band gap nature will not be destroyed but reduced from 2 eV for the monolayer to 0.31 eV for thickness exceeding 10 nm.[58] The puckered crystal structure of phosphorene results in strong in-plane anisotropy

FIGURE 11.12 Phosphorus allotropes: white (left), red and black phosphorus (right).

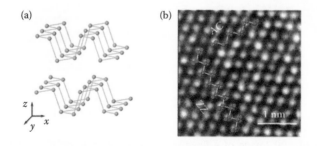

FIGURE 11.13 The crystal structure and band structure of phosphorene. (a) The illustration of puckered crystal structure of bilayer phosphorene. (b) HRTEM of few-layer phosphorene. AC and ZZ edges are identified accordingly.

in electron effective mass, therefore, electron transport along zig-zag and armchair directions and high sensitivity in strain engineering. Direct to indirect band gap change, semi conductive to metallic phase change in monolayer phosphorene has been predicted under certain compressive/tensile strain and can be experimentally observed in PL, Fourier transform infrared spectroscopy (FTIR) and even in Raman spectroscopy. All these interesting phenomena that lie upon its unique crystal structure make this new 2D material a fascinating candidate for advanced flexible electronics.[59]

11.5.2 Flexible Phosphorene Transistors

Few-layer phosphorene was mechanically exfoliated onto PI substrate as the functional channel material. Flakes with uniform thickness around 5–15 nm were targeted for the study and with thickness verified by AFM. Flexible bottom-gate phosphorene (BP) devices are shown in **Figure 11.14**. For BP FETs, source and drain contact were defined by electron beam lithography and followed by electron beam evaporation of Ti/Au metal stack. This is the standard nanofabrication process adopted from flexible MoS$_2$ FETs.[47] In comparison with MoS$_2$ devices, however, phosphorene is much more chemically reactive with moisture and oxygen from the ambient environment, therefore, special encapsulation processes need to be developed to help against the degradations, including immediate poly(methyl methacrylate) (PMMA) coating after exfoliation and high-k dielectric capping after source/drain metal lift-off.[60,61]

As a newly joint candidate in the 2D layered semiconductors family, phosphorene quickly gained worldwide attention for its outstanding optical and electrical properties, including high room temperature hole mobility (~1000 cm^2/Vs),[62] efficient drain current modulation (10^5),[63] strong current saturation[62] and symmetric ambipolar transport characteristics.[64] The superior

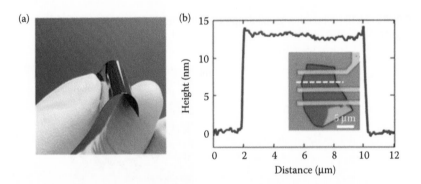

FIGURE 11.14 (a) Image of a flexible BP device sample on highly bendable PI substrate. (b) AFM data of a typical BP flake with thickness of 13 nm used as channel material. Inset is the optical image of bottom gate devices fabricated. (From Zhu, W. et al. *Nano lett.* 2015, 15, 1883–1890.)

(a)

(b)

(c)

(d)

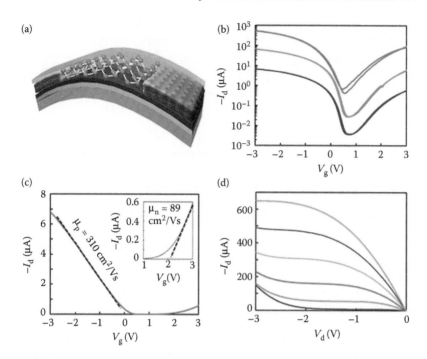

FIGURE 11.15 The device performance of flexible BP FETs. (a) The illustration of encapsulated bottom gated BP FETs on PI substrate. (b) The transfer characteristics of the device showing highest mobility. $V_d = -10$ mV, -100 mV and -1 V. (c) Transfer characteristics in linear plot with low field mobility extracted as μ_p ~310 cm²/Vs. Insert is the n-branch transport with electron low field mobility μ_n ~90 cm²/Vs. (d) Output characteristics with strong current saturation. (From Zhu, W. et al. *Nano lett.* 2015, 15, 1883–1890.)

electrical performances for few-layer phosphorene were well preserved on a flexible substrate of PI, see **Figure 11.15**.

Owing to the effective encapsulation strategy, negligible hysteresis were obtained for the flexible BP device which validates the mobility extraction from transfer curves using the Y-function method as discussed in Chapter 3.[65] The highest hole mobility achieved from flexible BP devices is 310 cm²/Vs. Ambipolar transport behaviour was observed for a flexible BP device with clear on-state of n-branch, which yielded an electron mobility of 89 cm²/Vs. Strong current saturation was obtained from this typical device together with high transconductance, which are two key parameters for successful accomplishment of high gain voltage amplifiers.

11.5.3 Mechanical Robustness

Mechanical robustness is another core criteria of evaluating flexible nanoelectronics. One standard and efficient method to test mechanical robustness is to apply strain to the device through either bending or stretching the substrate. When tensile strain was applied to the device through bending, the local strain

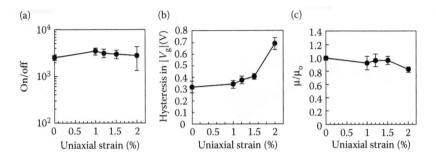

FIGURE 11.16 The statistics of strain effect on transport properties of flexible BP FETs, including (a) normalized hole mobility, (b) drain current modulation and (c) hysteresis. (From Zhu, W. et al. *Nano lett.* 2015, 15, 1883–1890.)

can be accordingly calculated from the bending radius. When no slippage was observed between the flake and the substrate, the mechanical robustness of the flexible device is in fact limited by the robustness of the high-k dielectric layer instead of the 2D semiconductor. As shown in **Figure 11.16**, before reaching 2% of tensile strain, the device performance of flexible BP-FETs remain stable with only slight variation observed.

Another interesting yet critical criteria is the number of repeated cycles of strain (smaller than but close to the critical strain of the dielectric layer) the device is able to endure. An RSA-G2 dynamic mechanical analyser was applied for multi-cycle three-point bending measurements. About 1.5% of tensile strain was repeated for 5,000 cycles onto the flexible sample to verify its mechanical robustness. Impressively, the electrical properties were well preserved after this test, as shown in **Figure 11.17**, which indicated strong mechanical robustness of flexible BP-FETs.

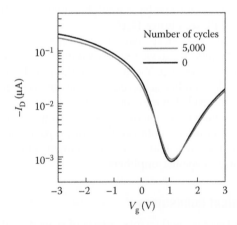

FIGURE 11.17 Transfer characteristics of one typical device before and after 5,000 cycles of bending at 1.5% tensile strain. (From Zhu, W. et al. *Nano lett.* 2015, 15, 1883–1890.)

11.5.4 Flexible Phosphorene Circuits

With all the superior electrical and mechanical performances obtained from flexible BP devices, it is natural to expect outstanding performance from the flexible BP-FET based circuit units, including both digital and analogue applications. The current results validate this expectation. Digital inverters, inverting and non-inverting voltage amplifiers and frequency doublers have been successfully realized with state-of-the-art functionality obtained. Circuits functionalities are shown in **Figure 11.18**.

The inverter device structure enabled by the ambipolar transport characteristics and the high drain current modulation was successfully achieved by splitting the minimum conduction points of two series transistors, similar in concept to graphene ambipolar inverters.[66] As shown in the inserted schematic, two identical FETs were combined into a complementary inverter circuit, with the global bottom gate as the input, and the centre terminal as the output. Clear inverting functionality was observed with peak inverter gain of

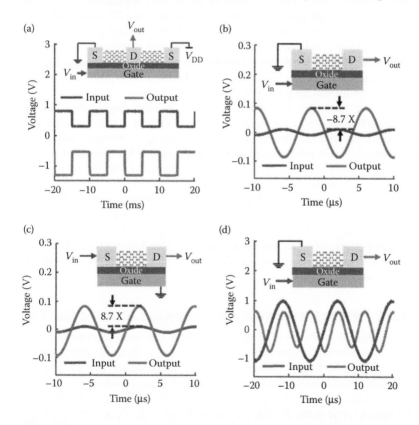

FIGURE 11.18 Basic digital and analog circuits based on flexible phosphorene transistors. (a) Digital inverter. (b) and (c) are the inverting and non-inverting amplifiers. (d) Frequency doubler. (From Zhu, W. et al. *Nano lett.* 2015, 15, 1883–1890.)

~|4.6| for $V_g = -0.46$ V and $V_{dd} = -2$ V. Functioning as a digital inverter, an input pulse signal with peak-to-peak amplitude (V_{pp}) of 0.5 V was connected to the gate input, and the amplified inverted signal was monitored at the output terminal with a voltage gain of ~1.68. This inverter operates at a maximum frequency of 100 Hz, which is limited by the very large parasitic capacitance from the bottom-gate structure. Higher frequency operation will be accessible by optimising the device structure, for example, utilizing top-gate or patterned bottom-gate electrodes.[15,67–69]

Inverting and non-inverting single-transistor amplifiers (**Figure 11.18b and c**) are among the most important circuit units for analogue signal processing. In order to achieve high voltage gain, both output current saturation and high transconductance (g_m) are required.[47,70] Source (gate) terminal was grounded for the inverting (non-inverting) amplifier, while the output was collected at the drain terminal. Absolute voltage gain ($|A_v| = |V_{out}/V_{in}|$) of 8.7 with a small-signal input of $V_{pp} = 20$ mV applied was observed for both configurations. The identical amplifier gains with opposite polarity reflects the symmetry of the gate–source terminal afforded in a 2D semiconductor, in contrast to a bulk semiconductor where the bulk terminal can introduce significant asymmetry between the common-gate and common-source configurations due to the body effect.[71] This present phosphorene amplifiers have the highest circuit amplification factor reported compared to previous 2D flexible amplifiers with voltage gains less than 4.[69,72,73]

A single-transistor frequency doubler is another highly desirable analogue circuit owing to the benefits of low transistor count, reduced power consumption and the absence of device matching, a stringent requirement for frequency multipliers based on transistor pairs.[74] Even based on a similar concept of symmetry near the minimum conduction point, the main distinction between ambipolar phosphorene and ambipolar graphene (under ideal electron–hole symmetry) is that the ambipolar symmetry point reflects a diffusive (exponential) transport for phosphorene and a drift (linear) transport for graphene.[26,36,75] A higher conversion gain requires more optimum square-law properties, whereas phosphorene can offer lower power and higher power efficiency owing to its lower off current and DC power dissipation, respectively.

11.5.5 Flexible Phosphorene AM Demodulator and Radio Receiver

The amplitude modulation (AM) demodulator and radio receiver are of great importance for communication and connectivity in future flexible smart systems.[10] The first flexible phosphorene AM demodulator was accomplished based on the successful demonstration of the frequency doubler.

A 55 kHz AC carrier signal was AM modulated by music baseband signal before input to the flexible AM demodulator. The flexible BP single FET-based AM demodulator successfully demodulated the baseband music, which was played via a loudspeaker directly connected to the output drain terminal of the BP-FET. To systematically evaluate its AM demodulation efficiency, a 5 kHz sinusoid signal was chosen as the baseband signal for AC voltage gain analysis and modulation index dependence analysis. The illustration of the AM demodulator system is shown in **Figure 11.19a**, where the BP-FET functions as the active component, which is biased near its minimum conduction point, as analysed in the frequency doubler. **Equation 11.7** presents the case for an AM input signal,

$$V_{in} = (1 + m \cos(\omega_m t))V_i \cos(\omega_c t) \tag{11.7}$$

where ω_m represents the frequency of the modulated signal, ω_c is the carrier frequency, V_i is the amplitude of the carrier and m is the modulation index. The general equation for frequency multiplier output is as follows:

$$V_d \approx V_{DD} - 2I_o R_L \left[1 + \frac{\beta^2 V_g^2}{2} \right] \tag{11.8}$$

where V_{DD} is the DC power supply, I_o is the coefficient current in subthreshold region which depends on fixed parameters including transistor device geometry, mobility, drain bias and threshold voltage.[45] R_L is the load resistance, β is a composite parameter comprising the field-effect capacitive coupling and the thermal energy (~0.026 eV at room temperature) and V_g is gate bias.

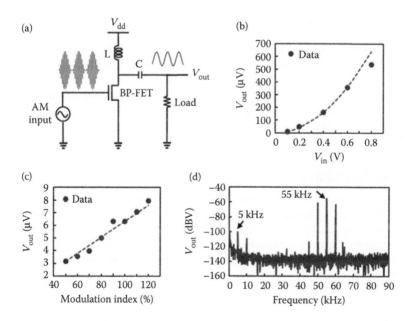

FIGURE 11.19 Flexible AM demodulator based on BP FETs. (a) System schematic. (b) Quadratic dependence of demodulated signal on AC carrier amplitude. (c) Linear dependence of demodulated signal on modulation index. (d) A typical output spectrum. (From Zhu, W. et al. *Nano lett.* 2015, 15, 1883–1890.)

Substituting **Equation 11.7** into **Equation 11.8**, we get

$$V_D \approx V_{DD} - 2I_o R_L \left[1 + \frac{\beta^2[(1 + m\cos(\omega_m t))V_i \cos(\omega_c t)]^2}{2} \right] \quad (11.9)$$

Expanding the output equation yields

$$V_D \approx V_{DD} - 2I_o R_L - I_o R_L \beta^2 \left(\frac{V_i^2}{2} + \frac{mV_i^2}{4} \right) - I_o R_L \beta^2 \left[mV_i^2 \cos\omega_m t \right.$$

$$+ \frac{mV_i^2}{4}\cos2\omega_m t + \left(\frac{V_i^2}{2} + \frac{mV_i^2}{4} \right)\cos2\omega_c + \frac{mV_i^2}{2}\cos((2\omega_c + \omega_m)t)$$

$$+ \frac{mV_i^2}{2}\cos((2\omega_c - \omega_m)t) + \frac{m^2 V_i^2}{8}\cos((2\omega_c + 2\omega_m)t)$$

$$\left. + \frac{m^2 V_i^2}{8}\cos((2\omega_c - 2\omega_m)t) \right] \quad (11.10)$$

where the equation contains the demodulated output signal V_{out} at frequency ω_m, higher order harmonics and intermodulation terms and a DC component. The demodulated baseband output V_{out} can be isolated as

$$V_{out} \approx I_o R_L m\beta^2 V_i^2 \cos(\omega_m t) \quad (11.11)$$

From this analysis, it is evident that the output voltage V_{out} is linearly dependent on the modulation index m, and is quadratically dependent on the amplitude V_i. Both trends have been observed in the experimental results shown in **Figure 11.19b** and **c**. In the output spectrum, odd harmonics at the output including the carrier feed-through can be observed due to asymmetry in the ambipolar curve.

11.6 Summary

Few-layer phosphorene, as a new 2D semiconductor, has demonstrated interesting physics aspects and outstanding electrical and mechanical properties. Flexible devices and circuits haven been successfully realized using this new material with impressive functionality obtained. Based on its superior performance, phosphorene can be the new promising candidate for advanced flexible nanoelectronics.

References

1. Nathan, A.; Ahnood, A.; Cole, M. T.; Sungsik, L.; Suzuki, Y.; Hiralal, P.; Bonaccorso, F. et al. Flexible electronics: The next ubiquitous platform. *Proc. IEEE* 2012, 100, 1486–1517.

2. Reuss, R. H.; Chalamala, B. R.; Moussessian, A.; Kane, M. G.; Kumar, A.; Zhang, D. C.; Rogers, J. A. et al. Macroelectronics: Perspectives on technology and applications. *Proc. IEEE* 2005, 93, 1239–1256.

3. Lee, J.; Tao, L.; Parrish, K. N.; Hao, Y.; Ruoff, R. S.; Akinwande, D., Multi-finger flexible graphene field effect transistors with high bendability. *Appl. Phys. Lett.* 2012, 101, 252109.

4. Chason, M.; Brazis, P. W.; Zhang, J.; Kalyanasundaram, K.; Gamota, D. R., Printed organic semiconducting devices. *Proc. IEEE* 2005, 93, 1348–1356.

5. Dodabalapur, A., Organic and polymer transistors for electronics. *Mater. Today* 2006, 9, 24–30.

6. Rogers, J. A.; Someya, T.; Huang, Y., Materials and mechanics for stretchable electronics. *Science* 2010, 327, 1603–1607.

7. Wang, C.; Chien, J. C.; Fang, H.; Takei, K.; Nah, J.; Plis, E.; Krishna, S.; Niknejad, A. M.; Javey, A., Self-aligned, extremely high frequency III–V metal–oxide–semiconductor field-effect transistors on rigid and flexible substrates. *Nano Lett.* 2012, 12, 4140–4145.

8. Zhou, H.; Seo, J. H.; Paskiewicz, D. M.; Zhu, Y.; Celler, G. K.; Voyles, P. M.; Zhou, W.; Lagally, M. G.; Ma, Z., Fast flexible electronics with strained silicon nano-membranes. *Sci. Rep.* 2013, 3, 1291.

9. Yoon, C.; Cho, G.; Kim, S., Electrical characteristics of GaAs nanowire-based MESFETs on flexible plastics. *IEEE Trans. Electron Devices* 2011, 58, 1096–1101.

10. Akinwande, D.; Petrone, N.; Hone, J., Two-dimensional flexible nanoelectronics. *Nat. Commun.* 2014, 5, 5678.

11. Geim, A. K.; Novoselov, K. S., The rise of graphene. *Nat. Mater.* 2007, 6, 183–191.

12. Petrone, N.; Dean, C. R.; Meric, I.; van der Zande, A. M.; Huang, P. Y.; Wang, L.; Muller, D.; Shepard, K. L.; Hone, J., Chemical vapor deposition-derived graphene with electrical performance of exfoliated graphene. *Nano Lett.* 2012, 12, 2751–2756.

13. Petrone, N.; Meric, I.; Hone, J.; Shepard, K. L., Graphene field-effect transistors with gigahertz-frequency power gain on flexible substrates. *Nano Lett.* 2012, 13, 121–125.

14. Lin, Y. M.; Dimitrakopoulos, C.; Jenkins, K. A.; Farmer, D. B.; Chiu, H. Y.; Grill, A.; Avouris, P., 100-GHz transistors from wafer-scale epitaxial graphene. *Science* 2010, 327, 662.

15. Ramón, M. E.; Parrish, K. N.; Chowdhury, S. F.; Magnuson, C. W.; Movva, H. C. P.; Ruoff, R. S.; Banerjee, S. K.; Akinwande, D., Three-gigahertz graphene frequency doubler on quartz operating beyond the transit frequency. *IEEE Trans. Nanotechnol.* 2012, 11, 877–883.

16. Lee, J.; Parrish, K. N.; Chowdhury, S. F.; Ha, T.-J.; Hao, Y.; Tao, L.; Dodabalapur, A.; Ruoff, R. S.; Akinwande, D., State-of-the-art graphene transistors on hexagonal boron nitride, high-k, and polymeric films for GHz flexible analog nanoelectronics, *Tech. Dig. Int. Electron Devices Meeting*, San Francisco, CA, 2012, 343–346.

17. Schwierz, F., Graphene transistors. *Nat. Nanotechnol.* 2010, 5, 487–496.

18. Neto, A. H. C.; Novoselov, K., New Directions in science and technology: Two-dimensional crystals. *Rep. Prog. Phys.* 2011, 74, 082501.

19. Xu, M.; Liang, T.; Shi, M.; Chen, H., Graphene-like two-dimensional materials. *Chem. Rev.* 2013, 113, 3766–3798.

20. Radisavljevic, B.; Radenovic, A.; Brivio, J.; Giacometti, V.; Kis, A., Single-layer MoS_2 transistors. *Nat. Nanotechnol.* 2011, 6, 147–150.

21. Kim, S.; Konar, A.; Hwang, W.-S.; Lee, J. H.; Lee, J.; Yang, J.; Jung, C. et al. High-mobility and low-power thin-film transistors based on multilayer MoS_2 crystals. *Nat. Commun.* 2012, 3, 1011.

22. Pu, J.; Yomogida, Y.; Liu, K. K.; Li, L. J.; Iwasa, Y.; Takenobu, T., Highly flexible MoS$_2$ thin-film transistors with ion gel dielectrics. *Nano Lett.* 2012, 12, 4013–4017.

23. Das, S.; Chen, H.-Y.; Penumatcha, A. V.; Appenzeller, J., High performance multi-layer MoS$_2$ transistors with scandium contacts. *Nano Lett.* 2013, 13, 100.

24. Yoon, J.; Park, W.; Bae, G. Y.; Kim, Y.; Jang, H. S.; Hyun, Y.; Lim, S. K. et al. Highly flexible and transparent multilayer MoS$_2$ transistors with graphene electrodes. *Small* 2013, 9, 3295–3300.

25. Lee, G.-H.; Yu, Y.-J.; Cui, X.; Petrone, N.; Lee, C.-H.; Choi, M. S.; Lee, D.-Y.; Lee, C.; Yoo, W. J.; Watanabe, K., Flexible and transparent MoS$_2$ field-effect transistors on hexagonal boron nitride–graphene heterostructures. *ACS Nano* 2013, 7, 7931–7936.

26. Lee, J.; Ha, T.-J.; Parrish, K. N.; Fahad Chowdhury, S.; Tao, L.; Dodabalapur, A.; Akinwande, D., High-performance current saturating graphene field-effect transistor with hexagonal boron nitride dielectric on flexible polymeric substrates. *IEEE Electron Device Lett.* 2013, 34, 172–174.

27. Meric, I.; Dean, C. R.; Petrone, N.; Wang, L.; Hone, J.; Kim, P.; Shepard, K. L., Graphene field-effect transistors based on boron–nitride dielectrics. *Proc. IEEE* 2013, 101, 1609–1619.

28. Ghibaudo, G., New method for the extraction of MOSFET parameters. *Electron. Lett.* 1988, 24, 543–545.

29. Fleury, D.; Cros, A.; Brut, H.; Ghibaudo, G., New Y-function-based methodology for accurate extraction of electrical parameters on nano-scaled MOSFETs, *Proc. Conf. Microelectron. Test Struct.*, Edinburgh, UK, 2008, pp. 160–165.

30. Wang, H.; Yu, L.; Lee, Y.-H.; Shi, Y.; Hsu, A.; Chin, M. L.; Li, L.-J.; Dubey, M.; Kong, J.; Palacios, T., Integrated circuits based on bilayer MoS$_2$ transistors. *Nano Lett.* 2012, 12, 4674–4680.

31. Jena, D.; Konar, A., Enhancement of carrier mobility in semiconductor nanostructures by dielectric engineering. *Phys. Rev. Lett.* 2007, 98, 136805.

32. Liu, H.; Ye, P. D., MoS$_2$ dual-gate MOSFET With atomic-layer-deposited as top-gate dielectric. *IEEE Electron Device Lett.* 2012, 33, 546–548.

33. Kappera, R.; Voiry, D.; Yalcin, S. E.; Branch, B.; Gupta, G.; Mohite, A. D.; Chhowalla, M., Phase-engineered low-resistance contacts for ultrathin MoS$_2$ transistors. *Nat. Mater.* 2014, 13, 1128–1134.

34. Liu, W.; Kang, J.; Sarkar, D.; Khatami, Y.; Jena, D.; Banerjee, K., Role of metal contacts in designing high-performance monolayer n-type WSe$_2$ field effect transistors. *Nano Lett.* 2013, 13, 1983–1990.

35. Xia, F.; Perebeinos, V.; Lin, Y.-M.; Wu, Y.; Avouris, P., The origins and limits of metal–graphene junction resistance. *Nat. Nano* 2011, 6, 179–184.

36. Parrish, K. N.; Akinwande, D., Impact of contact resistance on the transconductance and linearity of graphene transistors. *Appl. Phys. Lett.* 2011, 98, 183505.

37. Martel, R.; Schmidt, T.; Shea, H. R.; Hertel, T.; Avouris, P., Single- and multi-wall carbon nanotube field-effect transistors. *Appl. Phys. Lett.* 1998, 73, 2447–2449.

38. Wang, Q. H.; Kalantar-Zadeh, K.; Kis, A.; Coleman, J. N.; Strano, M. S., Electronics and optoelectronics of two-dimensional transition metal dichalcogenides. *Nat. Nanotechnol.* 2012, 7, 699–712.

39. Das, S.; Appenzeller, J., Where does the current flow in two-dimensional layered systems? *Nano Lett.* 2013, 13, 3396–3402.

40. Chen, J.-R.; Odenthal, P. M.; Swartz, A. G.; Floyd, G. C.; Wen, H.; Luo, K. Y.; Kawakami, R. K., Control of Schottky barriers in single layer MoS$_2$ transistors with ferromagnetic contacts. *Nano Lett.* 2013, 13, 3106–3110.

41. Liu, H.; Neal, A. T.; Ye, P. D., Channel length scaling of MoS₂ MOSFETs. *ACS Nano* 2012, 6, 8563–8569.

42. Larentis, S.; Fallahazad, B.; Tutuc, E., Field-effect transistors and intrinsic mobility in ultra-thin MoSe₂ layers. *Appl. Phys. Lett.* 2012, 101, 223104–223104-4.

43. Schroder, D. K., *Semiconductor Material and Device Characterization.* John Wiley & Sons, New York, 2006.

44. Reichert, G.; Ouisse, T., Relationship between empirical and theoretical mobility models in silicon inversion layers. *IEEE Trans. Electron Devices* 1996, 43, 1394–1398.

45. Sze, S. M.; Ng, K. K., *Physics of Semiconductor Devices.* John Wiley & Sons, New York, 2006.

46. Kim, S.; Konar, A.; Hwang, W.-S.; Lee, J. H.; Lee, J.; Yang, J.; Jung, C. et al. High-mobility and low-power thin-film transistors based on multilayer MoS₂ crystals. *Nat. Commun.* 2012, 3, 1011.

47. Chang, H. Y.; Yang, S.; Lee, J.; Tao, L.; Hwang, W. S.; Jena, D.; Lu, N.; Akinwande, D., High-performance, highly bendable MoS₂ transistors with high-k dielectrics for flexible low-power systems. *ACS Nano* 2013, 7, 5446–5452.

48. Bertolazzi, S.; Brivio, J.; Kis, A., Stretching and breaking of ultrathin MoS₂. *ACS Nano* 2011, 5, 9703–9709.

49. Ilic, B.; Krylov, S.; Craighead, H., Young's modulus and density measurements of thin atomic layer deposited films using resonant nanomechanics. *J. Appl. Phys.* 2010, 108, 044317–044317-11.

50. Jen, S. H.; Bertrand, J. A.; George, S. M., Critical tensile and compressive strains for cracking of Al₂O₃ films grown by atomic layer deposition. *J. Appl. Phys.* 2011, 109, 084305–084305-11.

51. Leterrier, Y.; Wyser, Y.; Månson, J.; Hilborn, J., A method to measure the adhesion of thin glass coatings on polymer films. *J. Adhes.* 1994, 44, 213–227.

52. Viventi, J.; Kim, D.-H.; Vigeland, L.; Frechette, E. S.; Blanco, J. A.; Kim, Y.-S.; Avrin, A. E.; Tiruvadi, V. R.; Hwang, S.-W.; Vanleer, A. C., Flexible, foldable, actively multiplexed, high-density electrode array for mapping brain activity *in vivo. Nat. Neurosci.* 2011, 14, 1599–1605.

53. Bridgman, P., Two new modifications of phosphorus. *J. Am. Chem. Soc.* 1914, 36, 1344–1363.

54. Maruyama, Y.; Suzuki, S.; Kobayashi, K.; Tanuma, S., Synthesis and some properties of black phosphorus single-crystals. *Physica B and C* 1981, 105, 99–102.

55. Akahama, Y.; Endo, S.; Narita, S.-I., Electrical properties of black phosphorus single crystals. *J. Phys. Soc. Jpn.* 1983, 52, 2148–2155.

56. Baba, M.; Takeda, Y.; Shibata, K.; Ikeda, T.; Morita, A., Optical properties of black phosphorus and its application to the infrared detector. *Jpn. J. Appl. Phys.* 1989, 28, L2104.

57. Novoselov, K.; Geim, A.; Morozov, S.; Jiang, D.; Zhang, Y.; Dubonos, S.; Grigorieva, I.; Firsov, A., Electric field effect in atomically thin carbon films. *Science* 2004, 306, 666–669.

58. Castellanos-Gomez, A.; Vicarelli, L.; Prada, E.; Island, J. O.; Narasimha-Acharya, K.; Blanter, S. I.; Groenendijk, D. J.; Buscema, M.; Steele, G. A.; Alvarez, J., Isolation and characterization of few-layer black phosphorus. *2D Mater.* 2014, 1, 025001.

59. Zhu, W.; Yogeesh, M. N.; Yang, S.; Aldave, S. H.; Kim, J.-S.; Sonde, S.; Tao, L.; Lu, N.; Akinwande, D., Flexible black phosphorus ambipolar transistors, circuits and AM demodulator. *Nano Lett.* 2015, 15, 1883–1890.

60. Wood, J. D.; Wells, S. A.; Jariwala, D.; Chen, K.-S.; Cho, E.; Sangwan, V. K.; Liu, X.; Lauhon, L. J.; Marks, T. J.; Hersam, M. C., Effective passivation of exfoliated black phosphorus transistors against ambient degradation. *Nano Lett.* 2014, 14, 6964–6970.

61. Kim, J.-S.; Liu, Y.; Zhu, W.; Kim, S.; Wu, D.; Tao, L.; Dodabalapur, A.; Lai, K.; Akinwande, D., Toward air-stable multilayer phosphorene thin-films and transistors. *Sci. Rep.* 2015, 5, 8989.

62. Li, L.; Yu, Y.; Ye, G. J.; Ge, Q.; Ou, X.; Wu, H.; Feng, D.; Chen, X. H.; Zhang, Y., Black phosphorus field-effect transistors. *Nat. Nanotechnol.* 2014, 9, 372–377.

63. Liu, H.; Neal, A. T.; Zhu, Z.; Luo, Z.; Xu, X.; Tománek, D.; Ye, P. D., Phosphorene: An unexplored 2D semiconductor with a high hole mobility. *ACS Nano* 2014, 8, 4033–4041.

64. Das, S.; Demarteau, M.; Roelofs, A., Ambipolar phosphorene field-effect transistor. *ACS Nano* 2014, 8, 11730–11738.

65. Chang, H. Y.; Zhu, W. N.; Akinwande, D., On the mobility and contact resistance evaluation for transistors based on MoS_2 or two-dimensional semiconducting atomic crystals. *Appl. Phys. Lett.* 2014, 104, 113504.

66. Li, S.-L.; Miyazaki, H.; Kumatani, A.; Kanda, A.; Tsukagoshi, K., Low operating bias and matched input–output characteristics in graphene logic inverters. *Nano Lett.* 2010, 10, 2357–2362.

67. Wang, H.; Wang, X.; Xia, F.; Wang, L.; Jiang, H.; Xia, Q.; Chin, M. L.; Dubey, M.; Han, S.-J., Black phosphorus radio-frequency transistors. *Nano Lett.* 2014, 14, 6424–6429.

68. Lee, J.; Ha, T.-J.; Li, H.; Parrish, K. N.; Holt, M.; Dodabalapur, A.; Ruoff, R. S.; Akinwande, D., 25 GHz embedded-gate graphene transistors with high-K dielectrics on extremely flexible plastic sheets. *ACS Nano* 2013, 7, 7744–7750.

69. Cheng, R.; Jiang, S.; Chen, Y.; Liu, Y.; Weiss, N.; Cheng, H. C.; Wu, H.; Huang, Y.; Duan, X., Few-layer molybdenum disulfide transistors and circuits for high-speed flexible electronics. *Nat. Commun.* 2014, 5, 5143.

70. Schwierz, F., Graphene transistors: Status, prospects, and problems. *Proc. IEEE* 2013, 101, 1567–1584.

71. Gray, P. R.; Meyer, R. G., *Analysis and Design of Analog Integrated Circuits.* John Wiley & Sons, Inc., Hoboken, NJ, 2008.

72. Nayfeh, O. M., Graphene transistors on mechanically flexible polyimide incorporating atomic-layer-deposited gate dielectric. *IEEE Electron Device Lett.* 2011, 32, 1349–1351.

73. Yeh, C.-H.; Lain, Y.-W.; Chiu, Y.-C.; Liao, C.-H.; Moyano, D. R.; Hsu, S. S.; Chiu, P.-W., Gigahertz flexible graphene transistors for microwave integrated circuits. *ACS Nano* 2014, 8, 7663–7670.

74. Razavi, B., CMOS technology characterization for analog and RF design. *IEEE J. Solid-State Circuits* 1999, 34, 268–276.

75. Wang, H.; Nezich, D.; Kong, J.; Palacios, T., Graphene frequency multipliers. *IEEE Electron Device Lett.* 2009, 30, 547–549.

SECTION III

Novel 2D Materials

12

Structural, Electronic and Transport Properties of Silicene and Germanene

Michel Houssa, Valery Afanas'ev and André Stesmans

Contents

12.1 Introduction

The recent progress in the growth, characterisation and understanding of the physical properties of graphene has paved the way to the study of other two-dimensional (2D) materials, for example, hexagonal boron nitride, transition metal dichalcogenides (TMD), phosphorene and the so-called van der Waals heterostructures of these different materials [1–6].

2D Materials for Nanoelectronics edited by Michel Houssa, Athanasios Dimoulas and Alessandro Molle © 2016 CRC Press/Taylor & Francis Group, LLC. ISBN: 978-1-4987-0417-5.

The possible growth of single atomic layers of the other group-IV elements (Si, Ge and Sn) has also recently emerged [7–11]. As an example, silicene and germanene, the Si and Ge counterpart of graphene, respectively, are gaining much interest in the scientific community. A natural advantage of silicene and germanene (compared to graphene) for nanoelectronic applications, is their better compatibility and expected integration with existing Si nanotechnology. These new 2D materials have been extensively studied theoretically [12–16], and very recently, the likely formation of silicene on several metallic substrates, like Ag(111) [17–19], ZrB$_2$(0001) [20] and Ir(111) [21] surfaces has been reported. The possible growth of germanene on Au(111) surfaces has also been highlighted very recently [22].

In this chapter, we give an overview of recent theoretical works on these novel 2D materials. We first discuss the structural, electronic and vibrational properties of free-standing silicene and germanene, as predicted from first-principles calculations. We next briefly review theoretical studies on the interaction of silicene and germanene with different substrates. The computed ballistic transport properties of these novel 2D materials are also discussed.

12.2 Free-Standing Silicene and Germanene

Free-standing silicene (germanene) consists of a single layer of Si (Ge) atoms with a hexagonal arrangement. But contrary to graphene, which is flat, with its carbon atoms in a pure sp^2 hybridised state, silicene and germanene are predicted to be slightly buckled in their lowest energy configurations [13], consisting of top and bottom Si or Ge atoms, separated by a vertical distance ΔZ of about 0.44 and 0.69 Å, respectively, as shown in **Figure 12.1**. The origin of the buckling of these 2D materials arises from the larger Si–Si and Ge–Ge bond lengths, as compared to the C–C bond length, which prevents the Si and Ge atoms to be 'purely' sp^2 hybridised and to form π bonds [14]. The buckling of the Si and Ge atoms enables a larger overlapping of their orbitals and results in a mixed sp^2–sp^3 hybridisation. The computed structural parameters

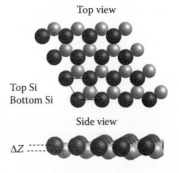

Top view

Top Si
Bottom Si

Side view

ΔZ

FIGURE 12.1 Top and side views of the atomic configuration of relaxed free-standing silicene. Dark and light grey spheres correspond to top and bottom Si atoms, respectively.

Table 12.1 Computed Structural Parameters of Silicene and Germanene, Obtained within the Generalised Gradient Approximation (PBE) for the DFT Exchange–Correlation Functional		
	In-Plane Lattice Parameter (Å)	Buckling Distance ΔZ (Å)	Bond Length (Å)
Silicene	3.87	0.44	2.28
Germanene	4.06	0.69	2.44

of free-standing silicene and germanene, obtained from density functional theory (DFT) simulations, are summarised in **Table 12.1**.

Slightly buckled silicene and germanene are predicted to be gapless semiconductors, like graphene, with Dirac cones at the K-points of their hexagonal Brillouin zone [13], as shown in **Figure 12.2**. Very interestingly, the 'natural' buckling of silicene and germanene potentially allows the tailoring of their electronic properties. For instance, applying an electric field perpendicular to their planes is predicted to open an energy gap at the K-points [23–26], as illustrated in the inset of **Figure 12.2**; the opening of the energy gap results from the charge transfer between the top and bottom Si or Ge atoms, breaking the inversion symmetry in the system. The electric-field induced gap openings in silicene and germanene are potentially very interesting for logic applications. When spin–orbit coupling is included in the DFT calculations, silicene and germanene are predicted to present a topological phase transition from a quantum spin Hall state to a trivial insulator, when the out-of-plane electric field is increased, silicene and germanene thus potentially being 2D topological insulators [15]. Note, though, that the spin–orbit coupling is much larger in stanenene (the Sn counterpart of graphene), making this material a better candidate of 2D topological insulators [27], as discussed in the last chapter of this book.

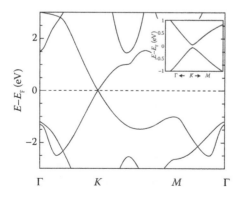

FIGURE 12.2 Calculated electronic energy band structure of free-standing silicene. The Fermi level corresponds to the reference (zero) energy. The inset shows the energy band structure of silicene in presence of an out-of-plane electric field of $E_z = 1$ V/Å.

The chemical functionalisation of silicene and germanene with different ad-atoms and molecules has also been reported recently [28–33]. The adsorption of various atoms on silicene and germanene usually results in the opening of an energy gap, due to the sp³ hybridisation of the Si or Ge atoms. Mechanical strain can also be used to tailor the electronic properties of silicene and germanene [34,35]. For example, biaxial tensile strain larger than about 5% induces hole doping in silicene, due to the weakened Si–Si bonds [34]. On the other hand, uniaxial strain results in the breaking of the lattice symmetry and incudes an energy band gap at the *K*-points [35].

The vibrational properties of silicene and germanene were recently computed [36,37] within the density-functional perturbation theory. The calculated Raman spectra of H-passivated armchair silicene and germanene nanoribbons with N dimer lines across their width (N-ASiNR and N-AGeNR) is shown in **Figure 12.3**. Two major Raman active modes are observed, corresponding to the so-called G- and D-like peaks, in analogy with graphene.

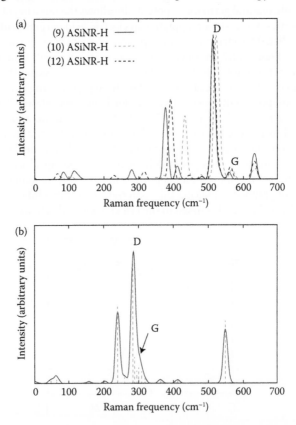

FIGURE 12.3 Computed Raman spectra of (a) hydrogenated armchair silicene nanoribbons with different widths *W* and (b) hydrogenated armchair germanene nanoribbon (*W* = 10), obtained by the calculated vibrational spectrum convoluted with a uniform Gaussian broadening having 10 cm⁻¹ width.

Table 12.2 Computed Raman Modes Corresponding to the G- and D-Like Peak in Silicene and Germanene

	G-Like Peak (cm⁻¹)	D-Like Peak (cm⁻¹)
Silicene	575	515
Germanene	290	270

The frequencies of these modes are summarised in **Table 12.2**. The G- and D-like modes of silicene and germanene nanoribbons can be identified from the analysis of the eigendisplacements of the phonon modes at Γ-point [36]. The eigendisplacements corresponding to the frequency of the G-like peak can be identified as the E_{2g} vibrational mode, which is due to the bond stretching of all pairs of mixed sp^2–sp^3 hybridised Si or Ge atoms. The D-like peak corresponds to the A_{1g} breathing mode of the hexagonal rings of Si or Ge atoms. This Raman active mode is forbidden in 'disordered-free' (perfect) layers, but appears in the presence of disorder, and is located near the K-point zone boundary, like in disordered graphene. Note that the frequency of the D mode corresponds to that of the higher frequency phonon branch around the K-point, where a discontinuity of its derivative can also be observed; this discontinuity is related to the so-called Kohn anomaly at the K-point of silicene and germanene [37], in analogy with graphene. Note that the computed frequencies of the G- and D-like Raman modes in silicene and germanene could potentially contribute to their experimental identification and characterisation, see, for example, Reference 38.

12.3 Interaction with Metallic Surfaces

The growth of silicene on different metallic substrates has been recently successfully achieved, as on Ag(111) [17–19], ZrB_2(0001) [20] and Ir(111) [21] surfaces. Different reconstructions of Si monolayers on Ag(111) surfaces have been recently identified using scanning tunnelling microscopy [17–19], as also discussed in more detail in the next chapter. One of these reconstructions is the so-called (4×4) silicene/Ag structure, which corresponds to a (3×3) silicene supercell on a (4×4) Ag supercell. This structure consists of 18 Si atoms arranged in two sublattices, as shown in **Figure 12.4** [17,39–41]: six atoms at top position, at about 2.93 Å from the Ag surface and 12 atoms at bottom positions, at about 2.18 Å from the Ag surface. The electronic density of states (DOS) and energy band structure of the (3×3) silicene supercell, stripping off the underlying Ag(111) substrate is shown in **Figure 12.5a** [42]. The silicene layer is predicted to be semiconducting, with a calculated energy band gap of about 0.3 eV. This energy gap opening originates from the breaking of the inversion symmetry in the silicene layer, due to the different number of top (6) and bottom (12) Si atoms in the supercell, as well as from the distribution of Si–Si buckling distances and Si–Si bond lengths and angles [42]. The inversion symmetry breaking and disorder in the silicene layer induce a charge transfer

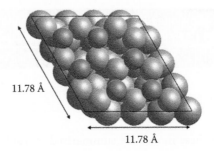

11.78 Å

11.78 Å

FIGURE 12.4 Slab model of the (4 × 4) silicene/(111)Ag structure. Dark purple, light purple and grey spheres are top Si, bottom Si and Ag atoms, respectively.

between the top and bottom atoms, leading to the gap opening; the bottom Si atoms (from the planar hexagons) are mainly contributing to the valence (π) band and the top Si atoms are essentially contributing to the (π*) conduction band of the system, as shown in **Figure 12.5a**. The computed electronic DOS of the (4 × 4) silicene/(111)Ag system is presented in **Figure 12.5b**. The mixing between the 3p orbitals of Si atoms and the 5s orbitals of Ag atoms results in an overall metallic system [42], in agreement with other recent theoretical calculations [40,41]. The distribution of the bond lengths, bond angles and buckling

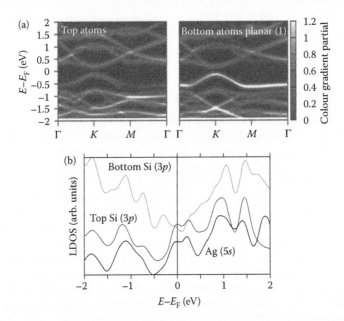

FIGURE 12.5 (a) Energy band population of the top and bottom Si atoms of the (3 × 3) silicene supercell. The highest electron density is shown in white and the lowest one in grey. (b) Projected electronic DOS of the (4 × 4) silicene/(111)Ag supercell.

distances in the Si sheet also results in the distribution of the projected DOS of the different Si atoms [42], which could explain the distribution in the local electronic DOS recently measured by scanning tunnelling spectroscopy on different silicene/Ag(111) superstructures [19]. The electronic structure of the silicene/Ag(111) surface is discussed in more detail in the next chapter.

So far, the possible growth of germanene has been only reported on Au(111) surfaces [22]. Germanene was grown in an ultra-high vacuum system, from the evaporation of Ge atoms on a Au(111) surface at a typical temperature of 200°C. Combining scanning tunneling microscopy (STM) images and synchrotron radiation core-level spectroscopy experiments with DFT simulations, the formation of a monolayer of Ge atoms arranged on an hexagonal lattice was inferred, most likely corresponding to a $\sqrt{3} \times \sqrt{3}$ germanene layer on top of a $\sqrt{7} \times \sqrt{7}$ Au(111) supercell.

12.4 Interaction with Non-Metallic Surfaces

Layered non-metallic substrates, with strong intra-layer covalent bonding and weak interlayer van der Waals bonding, are expected to interact weakly with silicene and germanene, potentially preserving their peculiar electronic properties. As an example, the interaction of silicene and germanene with semiconducting layered TMD, namely MoX_2 and GaX with $X = S$, Se, Te has been recently reported [43]. When a flat layer of silicene or germanene is placed on top of a TMD substrate, both Si and Ge keep their hexagonal arrangement, but tend to buckle; the buckling distance and in-plane lattice parameter mismatch is given in **Tables 12.3** and **12.4** for silicene on TMD and germanene on TMD, respectively. From these results, the buckling distance is clearly correlated to the in-plane lattice mismatch. Note that the typical interlayer distance between the TMD and silicene or germanene lies between 3 and 3.5 Å, indicating a weak van der Waals like interaction between the 2D material and the substrate; the typical adhesion energy between silicene (germanene) and the top layer of MoX_2 or GaX is about 0.2–0.3 eV.

Table 12.3 Computed Silicene Buckling Distance and In-Plane Lattice Mismatch between Silicene and the (Di)Chalcogenide Substrate

	In-Plane Lattice Mismatch (%)	Buckling Distance (Å)	Electronic Properties
MoS_2	18.3	1.9	Metallic
$MoSe_2$	14.7	1.0	Metallic
$MoTe_2$	9	0.7	Gapless semiconductor
GaS	7.5	0.7	Gapless semiconductor
GaSe	3.4	0.5	Gapless semiconductor

Note: The predicted electronic properties of the silicene layer is also indicated, that is, metallic or gapless semiconductor (with preserved Dirac cones at the *K*-points).

Table 12.4 Computed Germanene Buckling Distance and In-Plane Lattice Mismatch between Germanene and the (Di)Chalcogenide Substrate

	In-Plane Lattice Mismatch (%)	Buckling Distance (Å)	Electronic Properties
MoTe$_2$	12	1.2	Metallic
GaSe	6.5	1	Metallic
GaTe	3.5	0.7	Gapless semiconductor

Note: The predicted electronic properties of the germanene layer is also indicated, that is, metallic or gapless semiconductor (with preserved Dirac cones at the *K*-points).

Both the Si buckling distance and the Si–MoS$_2$ interlayer distance are in very good agreement with recently reported experimental STM results on silicene/MoS$_2$ systems [44] as shown in **Figure 12.6**. The step profile between a Si domain and the MoS$_2$ substrate amounts to 3 Å and exhibits a feature at about 2 Å, consistently with the highly buckled silicene arrangement. The relatively good agreement between the experimental and computed structural properties of silicene on MoS$_2$ indicates that the weak van der Waals

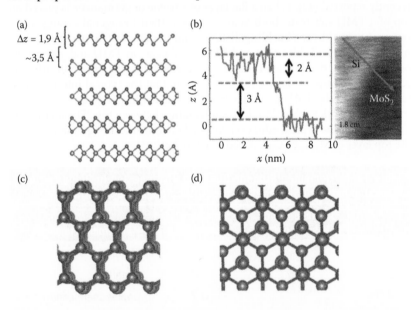

FIGURE 12.6 (a) Relaxed atomic configuration of silicene on bulk MoS$_2$ (side view). In (b), the STM image (VB = 0.2 V, IS = 2 nA) of a partially covered MoS$_2$ surface by Si atoms is shown (right side). A line profile taken across the two terraces allows measuring the amplitude of the step which amounts to about 5 Å (left side of b). In (c) and (d), the top view of the most stable silicene/MoS$_2$ slab models (obtained after atomic relaxations) is shown.

interaction between silicene and MoS$_2$ likely favours the formation of the metastable highly buckled silicene layer [13].

The predicted electronic structure of the silicene or germanene layer on the MoX$_2$ template largely depends on the buckling parameter in the 2D material. Highly buckled silicene on MoS$_2$ is predicted to be metallic [43,44]. On the other hand, low-buckled silicene on MoTe$_2$ or germanene on GaTe are predicted to preserve their Dirac cones at the K-points, as illustrated in **Figure 12.7**. Consequently, by increasing the in-plane lattice parameter of the (di) chalcogenide substrates, it is found out that the buckling distance in the silicene or germanene layer can be reduced and the 2D material can eventually preserves its gapless semiconducting behaviour. Note that very recent theoretical studies on the weak van der Waals interactions between silicene and (111) CaF$_2$ surfaces [45] or layered Ga chalcogenides [46] also predict the persistence of Dirac cones in the electronic structure of silicene.

The formation of covalent bonds between silicene or germanene and an underlying substrate can result in the partial or complete sp^3 hybridisation of the Si or Ge atoms, and consequently, in the opening of an energy gap in its electronic structure, for example, in silicene or germanene functionalised by the adsorption of ad-atoms [28–33]. We discuss here the covalent bonding between germanene and ZnSe surfaces [47], as a typical example of a stronger interaction (compared to the weak van der Waals bonding) between germanene and a non-metallic substrate. Note that very similar results are obtained on silicene/ZnS surfaces [48].

ZnSe crystallises in the Wurtzite phase [49,50] and is a semiconductor, with a direct energy band gap of about 2.9 eV. Interestingly, its in-plane lattice constant (3.98 Å) is very close to the computed one of free-standing germanene (about 4.06 Å). A (0001) polar ZnSe surface was thus considered as a possible substrate for germanene. A slab model with eight atomic layers (64 atoms) and with a 15 Å vacuum layer was used for DFT simulations. Displacements of the top and bottom ZnSe layers were observed during atomic relaxation, resulting in a surface reconstruction very similar to the one of the non-polar (1010) ZnS surfaces [51,52], as discussed in more detail in Reference 47. Similar results were reported on the (0001) ZnS surface, the reconstructed surface being predicted to be more stable than the non-reconstructed polar surface for layers

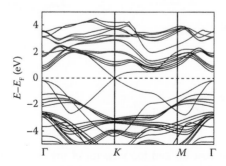

FIGURE 12.7 Energy band structure of the germanene/GaTe slab model.

up to about 6.6 nm [48,53]. The reconstructed (0001) ZnSe surface is semi-conducting, with a computed energy gap of about 2.1 eV. The interaction of germanene with the reconstructed (0001) ZnSe surface was then investigated. The most energetically stable structure is presented in **Figure 12.8**, and corresponds to a hexagonal arrangement of the Ge atoms placed at intermediate positions between top and hollow sites of the ZnSe hexagons. Two Ge–Se bonds and two Ge–Zn bonds are formed, with a charge transfer essentially involving the $4p_z$ orbitals of the Ge atoms and the $4s$ states of Zn and $4p$ states of Se, the bonded Ge atoms thus adopting a sp³-like character. Four other Ge atoms are not bonded to the ZnSe surface, two of these atoms lying at about 2.9 Å from the surface (marked 'intermediate' on **Figure 12.8**) and two other atoms lying at about 3.71 Å from the surface (marked 'top' on **Figure 12.8**). The charge transfer at the germanene/(0001) ZnSe interface leads to an excess of negative (Mulliken) charge of about 0.15 |e| on the top Ge atoms, with respect to the intermediate Ge atoms, resulting in the formation of a dipole at this interface. The average Ge–Ge distance (2.49 Å) is very similar to the one of free-standing germanene.

The germanene/(0001) ZnSe interface is predicted to be semiconducting, with a computed indirect energy band gap of about 0.4 eV, as shown in **Figure 12.9**. The energy gap opening in germanene is due to the charge transfer and partial sp³ hybridisation of the Ge atoms bonded to the ZnSe surface [47,48]. The effect of an out-of-plane electric field on the energy band structure of the system is also illustrated in **Figure 12.9** (dashed lines). The electric field has a substantial effect on the conduction band near the Γ-point, leading to a transition from an indirect (Γ to L-point) to direct (at Γ-point) energy band gap in germanene, for an electric field of about 0.4 V/Å, as indicated in **Figure 12.10**. The electric-field dependence of the energy band gap of the germanene layer is related to the modulation of the electric dipole at the germanene/ZnSe interface [47,48]. The predicted tuning of the energy band gap of germanene on ZnSe by an out-of-plane electric field is potentially very interesting for field-effect nanoelectronic devices.

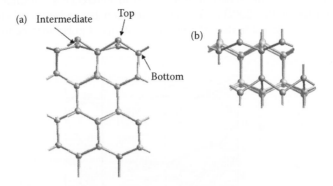

FIGURE 12.8 Side view (a) and top view (b) of the relaxed germanene/(0001) ZnSe slab model. Yellow, green and magenta spheres are Se, Zn and Ge atoms, respectively.

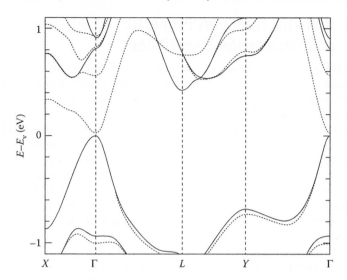

FIGURE 12.9 Computed energy band structure of the germanene/(0001) ZnSe slab model, without (solid lines) and with (dashed lines) an external electric field of 0.6 V/Å in the direction perpendicular to the interface. The reference (zero) energy level corresponds to the top of the valence band E_v of germanene.

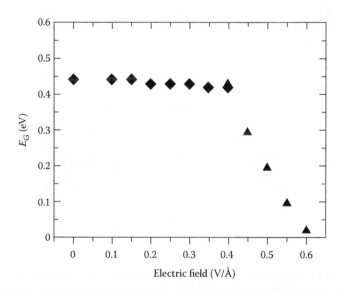

FIGURE 12.10 Computed direct (filled triangles) and indirect (filled diamonds) energy band gaps of the germanene/(0001) ZnSe slab model, as a function of the external electric field applied to the system.

12.5 Ballistic Transport Properties of Silicene and Germanene Nanoribbons

The ballistic transport properties of silicene and germanene nanoribbons, computed using DFT within the non-equilibrium Green's function (NEGF) method, were reported recently [23,54]. For these simulations, the system is partitioned into a left lead (L), a central scattering region (CS) and a right lead (R) along the z-(transport) direction, as shown in **Figure 12.11** [55]. Hydrogen passivated silicene and germanene armchair nanoribbons (SiANRs and GeANRs) were considered. The width W of the nanoribbon refers to the number of parallel Si- or Ge-chains (dimers) along the x-direction (see **Figure 12.11**). The ribbon width has a large impact on the resulting electronic structure, since a decreasing width increases quantum confinement and results in a larger band gap. In addition, in analogy with graphene [56,57], the threefold coordination for the Si or Ge atoms for armchair edges suggests partitioning the system into three different widths, that is, $W = \{3m - 1, 3m, 3m + 1\}$, where m is an integer. The band structures and transmission functions (transmission probability times the number of energy modes) for zero bias armchair SiANRs and GeNRs are shown in **Figure 12.12**. All the NRs exhibit a direct band gap originating from the quantum confinement. The values are summarised in **Table 12.5**. Those with widths $3m$ and $3m + 1$ have larger band gaps than those with width $3m - 1$, as in graphene [56,57].

The value of the transmission spectra at a certain energy corresponds to the number of energy modes which are available for electronic transport at that energy. Accordingly, the transmission spectra exhibits a 'transmission gap' in the energy range of the band gap. When applying a bias voltage V_b, the bands of the left and right contacts are shifted with respect to each other. The shift of the bands can also be seen in the transmission spectra, which is shown in **Figure 12.13** for the case of SiANR ($W = 6$), for V_b ranging between 0 V (bottom curve) and 1.6 V (top curve). When increasing the bias voltage, the transmission gap

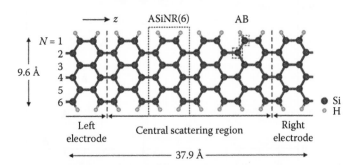

FIGURE 12.11 Schematic of an armchair silicene nanoribbon (ASiNR) with $W = 6$ used for ballistic transport simulations. The system is partitioned into a semi-infinite left electrode, a central scattering region and a semi-infinite right electrode along the transport (z) direction. The ribbon unit cell is indicated by the dashed box.

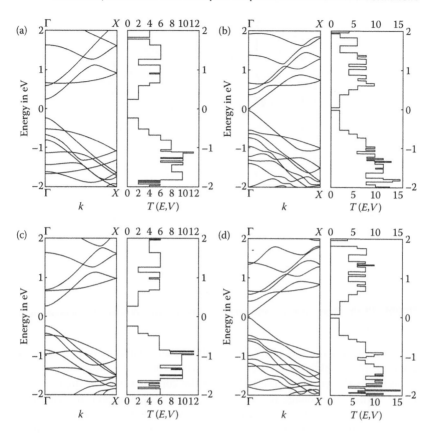

FIGURE 12.12 Energy band structure and transmission function for the silicene nanoribbons with $W = 6$ (a) and $W = 8$ (b) and for the germanene nanoribbons with $W = 6$ (c) and $W = 8$ (d).

splits in two and is shifted to lower and higher energies. A current starts to flow in the system when available states from the left electrode can 'tunnel' though an energy mode in the channel and reach an empty state at the right electrode, that is, when eV is equal to the band gap energy. At this voltage, a transport channel appears in the transmission spectra. The computed ballistic current–voltage

Table 12.5 Computed Energy Band Gap of Group-IV ANRs, with Two Different Widths ($W = 6$ and 8)

	$W = 6$ E_g (eV)	$W = 8$ E_g (eV)
Graphene	1.2	0.2
Silicene	0.49	0.06
Germanene	0.46	0.05
Stanene	0.37	0.01

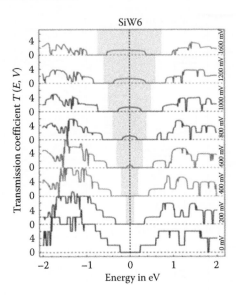

FIGURE 12.13 Transmission function of the SiANR (*W* = 6), for bias voltages ranging from 0 to 1.6 V.

characteristics of the W6 and W8 nanoribbons are compared in **Figure 12.14**. For bias voltages higher than the band gap, a current starts to flow and increases almost linearly with the bias, indicating the contribution of a single energy mode within this voltage range. The corresponding graphene ANRs, which are also shown for comparison, exhibit the same behaviour. Very interestingly, the current flowing through the silicene, germanene and stanene nanoribbon is higher than the one flowing in the graphene one (at a fixed bias) [54]. This is

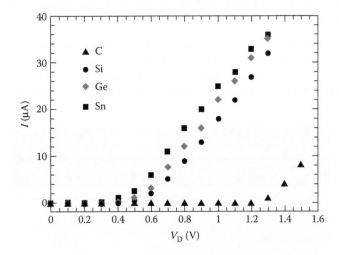

FIGURE 12.14 Computed ballistic current–voltage characteristics for W6 ANRs.

due to the reduction of the energy band gap of the group-IV nanoribbon, going down from C to Sn in the periodic table (see **Table 12.5**), which leads to ballistic current flow at a lower onset voltage.

Very interestingly, all the nanoribbons have a very similar conductance (about 50 µS), as extracted from the slope of their I–V characteristics. The conductance in a ballistic conductor is related to the injection velocity [58,59]

$$v_{inj} = \sqrt{\frac{2k_B T}{\pi m^*}}, \tag{12.1}$$

where k_B is the Boltzmann constant, T is the temperature and m^* is the carrier effective mass. The calculated electron and hole effective masses for all the group-IV nanoribbons are comparable (within 1%) and equal $m_e^* \sim m_h^* \sim 3 \times 10^{-3} m_0$, with m_0 the free electron mass, leading to a similar injection velocity and ballistic conductance.

12.6 Concluding Remarks

2D materials like graphene could enable the fabrication of novel low-power nanoelectronic devices. However, the integration of graphene with the Si-based nanotechnology platform is challenging. In that respect, alternative 2D materials like silicene and germanene could be potentially better integrated with the existing technology. The recent progress in the growth and characterisation of silicene and germanene on different substrates, combined with theoretical predictions on their structural, vibrational and electronic properties are very encouraging. As a matter of fact, the likely growth of silicene on various metallic surfaces has been successfully achieved, and its electronic structure is consistent with theoretical predictions. The recently reported possible growth of silicene on (semiconducting) MoS_2 surfaces is also very promising. In addition, the very recent report of silicene-based field-effect transistors operating at room temperature and presenting ambipolar current–voltage characteristics [60] paves the way to the possible realisation of silicene-based nanoelectronic devices. Fundamental and technological challenges still lie ahead, like the growth and characterisation of silicene and germanene over large areas, their possible growth and/or transfer to non-metallic substrates and further integration into functional devices.

Acknowledgements

We are thankful to E. Scalise (Max Planck Institute), B. van den Broek and K. Iordanidou (KU Leuven) as well as G. Pourtois (IMEC) for their valuable contributions to this work. Fruitful discussions and collaborations with A. Dimoulas (NCSR Demokritos) and A. Molle (CNR-IMM) are also gratefully acknowledged. This work has been financially supported by the European Project 2D-NANOLATTICES, within the Future and Emerging Technologies

(FET) programme of the European Commission, under the FET-grant number 270749 as well as by the KU Leuven Research Funds, project GOA/13/011.

References

1. J.C. Meyer, A. Chuvilin, G. Algara-Siller, J. Biskupek and U. Kaiser, *Nano Lett.* **9**, 2683, 2009.
2. Q.H. Wang, K. Kalantar-Zadeh, A. Kis, J.N. Coleman and M.S. Strano, *Nat. Nanotechnol.* **7**, 699, 2012.
3. A.K. Geim and I.V. Grigorieva, *Nature* **499**, 419, 2013.
4. S.Z. Butler et al. *ACS Nano* **7**, 2898, 2013.
5. G. Fiori, F. Bonaccorso, G. Iannaccone, T. Palacios, D. Neumaier, A. Seabaugh, S. Bannerjee and L. Colombo, *Nat. Nanotechnol.* **9**, 768, 2014.
6. H. Liu, A.T. Neal, Z. Zhu, Z. Luo, X. Xu, D. Tománek and P.D. Ye, *ACS Nano* **8**, 4033, 2014.
7. A. Kara, H. Enriquez, A.P. Seitsonen, L.C.L.Y. Voon, S. Vizzini, B. Aufray and H. Oughaddou, *Surf. Sci. Rep.* **67**, 1, 2012.
8. D. Jose and A. Datta, *Acc. Chem. Res.* **47**, 593, 2014.
9. L.C. Lew Yan Voon and G.G. Guzman-Verri, *MRS Bull.* **39**, 366, 2014.
10. S. Balendhran, S. Walia, H. Nili, S. Sriram and M. Bhaskaran, *Small* **11**, 640, 2015.
11. M. Houssa, A. Dimoulas and A. Molle, *J. Phys.: Condens. Matter* **27**, 253002, 2015.
12. G.G. Guzman-Verri and L.C. Lew Yan Voon, *Phys. Rev. B* **76**, 075131, 2007.
13. S.S. Cahangirov, M. Topsakal, E. Aktürk, H. Sahin and S. Ciraci, *Phys. Rev. Lett.* **102**, 236804, 2009.
14. M. Houssa, G. Pourtois, V.V. Afanas'ev and A. Stesmans, *Appl. Phys. Lett.* **97**, 112106, 2010.
15. M. Ezawa, *Phys. Rev. Lett.* **109**, 055502, 2012.
16. L. Matthes, O. Pulci and F. Bechstedt, *New J. Phys.* **16**, 105007, 2014.
17. P. Vogt, P. De Padova, C. Quaresima, J. Avila, E. Frantzeskakis, M.C. Asensio, A. Resta, B. Ealet and G. Le Lay, *Phys. Rev. Lett.* **108**, 155501, 2012.
18. B. Feng, Z. Ding, S. Meng, Y. Yao, X. He, P. Cheng, L. Chen and K. Wu, *Nano Lett.* **12**, 3507, 2012.
19. D. Chiappe, C. Grazianetti, G. Tallarida, M. Fanciulli and A. Molle, *Adv. Mater.* **24**, 5088, 2012.
20. A. Fleurence, R. Friedlein, T. Ozaki, H. Kawai, Y. Wang and Y. Takamura, *Phys. Rev. Lett.* **108**, 245501, 2012.
21. L. Meng et al., *Nano Lett.* **13**, 685, 2013.
22. M.E. Davila, L. Xian, S. Cahangirov, A. Rubio and G. Le Lay, *New J. Phys.* **16**, 095002, 2014.
23. Z. Ni et al., *Nano Lett.* **12**, 113, 2012.
24. H.H. Gurel, V.O. Ozcelik and S. Ciraci, *J. Phys.: Condens. Matter* **25**, 305007, 2013.
25. W.F. Tsai, C.Y. Huang, T.R. Chang, H. Lin, H.T. Jeng and A. Bansil, *Nat. Commun.* **4**, 1500, 2013.
26. V. Vargiamidis, P. Vasilopoulos and G.Q. Hai, *J. Phys.: Condens. Matter* **26**, 345303, 2014.
27. Y. Xu, B. Yan, H.-J. Zhang, J. Wang, G. Xu, P. Tang, W. Duan and S.-C. Zhang, *Phys. Rev. Lett.* **111**, 136804, 2013.

28. L.C. Lew Yan Voon, E. Sandberg, R.S. Aga and A.A. Farajian, *Appl. Phys. Lett.* **97**, 163114, 2010.
29. M. Houssa, E. Scalise, K. Sankaran, G. Pourtois, V.V. Afanas'ev and A. Stesmans, *Appl. Phys. Lett.* **98**, 223107, 2011.
30. R. Quhe et al., *Sci. Rep.* **2**, 853, 2012.
31. Y. Ding and Y. Wang, *Appl. Phys. Lett.* **100**, 083102, 2012.
32. B. van den Broek, M. Houssa, E. Scalise, G. Pourtois, V.V. Afanas'ev and A. Stesmans, *Appl. Surf. Sci.* **291**, 104, 2014.
33. T.P. Kaloni, N. Singh and U. Schwingenschlögl, *Phys. Rev. B* **89**, 035409, 2014.
34. T.P. Kaloni, Y.C. Cheng and U. Schwingenschlögl, *J. Appl. Phys.* **113**, 104305, 2013.
35. H. Zhao, *Phys. Lett. A* **376**, 3546, 2012.
36. E. Scalise, M. Houssa, G. Pourtois, B. van den Broek, V.V. Afanas'ev and A. Stesmans, *Nano Res.* **6**, 19, 2013.
37. E. Scalise, *Vibrational Properties of Defective Oxides and 2D Nanolattices* (Springer, Berlin, 2014).
38. E. Cinquanta, E. Scalise, D. Chiappe, E. Grazianetti, B. van den Broek, M. Houssa, M. Fanciulli and A. Molle, *J. Phys. Chem. C* **117**, 16719, 2013.
39. D. Kaltsas, L. Tsetseris and A. Dimoulas, *J. Phys: Condens. Matter* **24**, 442001, 2012.
40. C.L. Lin, R. Arafune, K. Kawahora, M. Kanno, N. Tsukahara, E. Minamitani, K. Yousoo, M. Kawai and N. Takagi, *Phys. Rev. Lett.* **110**, 076801, 2013.
41. S. Cahangirov, M. Audiffred, P.Z. Tang, A. Iacomino, W.H. Duan, G. Merino and A. Rubio, *Phys. Rev. B* **88**, 035432, 2013.
42. M. Houssa et al., *Appl. Surf. Sci.* **291**, 98, 2014.
43. E. Scalise, M. Houssa, E. Cinquanta, C. Grazianetti, B. van den Broek, G. Pourtois, A. Stesmans, M. Fanciulli and A. Molle, *2D Mater.* **1**, 011010, 2014.
44. D. Chiappe, E. Scalise, E. Cinquanta, C. Grazianetti, B. van den Broek, M. Fanciulli, M. Houssa and A. Molle, *Adv. Mater.* **26**, 2096, 2014.
45. S. Kokott, P. Pflugradt, L. Matthes and F. Bechstedt, *J. Phys.: Condens. Matter.* **26**, 185002, 2014.
46. Y. Ding and Y. Wang, *Appl. Phys. Lett.* **103**, 043114, 2013.
47. M. Houssa, E. Scalise, B. van den Broek, A. Lu, G. Pourtois, V.V. Afanas'ev and A. Stesmans, *ECS Trans.* **64** (8), 111, 2014.
48. M. Houssa, B. van den Broek, E. Scalise, G. Pourtois, V.V. Afanas'ev and A. Stesmans, *Phys. Chem. Chem. Phys.* **15**, 3702, 2013.
49. M.J. Weber, Ed., *Handbook of Laser Science and Technology* (CRC Press, Cleveland, OH, 1986).
50. Y.-N. Xu and W.Y. Ching, *Phys. Rev. B* **48**, 4335, 1993.
51. J. E. Northrup and J. Neugebauer, *Phys. Rev. B* **53**, R10477, 1996.
52. A. Filippetti, V. Fiorentini, G. Cappellini and A. Bosin, *Phys. Rev. B* **59**, 8026, 1999.
53. X. Zhang, H. Zhang, T. He and M. Zhao, *J. Appl. Phys.* **108**, 064317, 2010.
54. B. van den Broek, M. Houssa, G. Pourtois, V.V. Afanas'ev and A. Stesmans, *Phys. Status Solidi RRL* **8**, 931, 2014.
55. M. Brandbyge, J.L. Mozos, P. Ordejon, J. Taylor and K. Stokbro, *Phys. Rev. B* **65**, 165401, 2002.
56. V. Barone, O. Hod and G.E. Scuseria, *Nano Lett.* **6**, 2748, 2006.
57. L.E.F. Foa Torres, S. Roche and J.C. Charlier, *Introduction to Graphene-Based Nanomaterials: From Electronic Structure to Quantum Transport* (Cambridge University Press, Cambridge, 2014).

58. S. Datta, *Lessons from Nanoelectronics: A New Perspective on Transport* (World Scientific, Singapore, 2012).

59. M. Lundstrom and C. Jeong, *Near Equilibrium Transport: Fundamentals and Applications* (World Scientific, Singapore, 2013).

60. L. Tao, E. Cinquanta, D. Chiappe, C. Grazianetti, M. Fanciulli, M. Dubey, A. Molle and D. Akinwande, *Nat. Nanotechnol.* **10**, 227, 2015.

Group IV Semiconductor 2D Materials

The Case of Silicene and Germanene

Alessandro Molle, Dimitra Tsoutsou and Athanasios Dimoulas

Contents

2D Materials for Nanoelectronics edited by Michel Houssa, Athanasios Dimoulas and Alessandro Molle © 2016 CRC Press/Taylor & Francis Group, LLC. ISBN: 978-1-4987-0417-5.

13.1 Introduction: Motivation and Technology Drivers

Silicene and germanene are conceptualised as monatomic two-dimensional (2D) crystals with a honeycomb lattice resembling that of graphene. The interest in silicene and germanene is motivated by the need to answer fundamental scientific questions. The most intriguing question is why silicene and germanene do not exist in nature in free-standing forms just like graphene? A related question is why silicon and germanium behave so differently compared to carbon despite the fact they belong to the same group IV of the periodic table? Is it possible to engineer silicene and germanene on suitable substrates and if so, what would be their structure? More specifically, are they going to be flat on the microscopic scale indicating a purely sp²-type of hybridisation or buckled instead meaning that a mixed sp²–sp³ hybridisation is favoured? Even more importantly, there are questions regarding their electronic properties. Are they semimetals like graphene featuring a Dirac cone, metallic or semiconducting/insulating instead and what would be the role of the substrate in determining their electronic identity?

The effort in realising silicene and germanene is not merely a matter of scientific interest or curiosity but it is also driven by potential applications. The continuous lateral scaling of Si nanoelectronic devices and circuits suffers from insufficient electrostatic control which is mitigated mainly by reducing the gate equivalent oxide thickness and the channel thickness in fully depleted devices. Reducing channel thickness reduces also the physical (or geometrical) screening length, allowing lateral scaling without adverse short-channel effects. In fact, the industry has already moved to thin film channel devices, a good example being the FinFET, a thin film silicon channel placed vertically (the 'fin') in a field-effect transistor (FET) structure as illustrated in **Figure 13.1**. While the FinFET is already in production, it is not clear what the device solutions for future generations would be beyond the 7 nm node [1]. A natural evolution of the FinFET may lead to the nanowire as the main option for next generation devices. On the other hand, the ultrathin body silicon on insulator that is essentially a thin film Si channel placed horizontally (**Figure 13.1**) is an attractive alternative because of its benefits for low power operation and its architectural simplicity. At present, the Si body is around 10 nm and it is foreseen that it will scale down to 8 or 7 nm, however, getting lower will be a major challenge. One of the reasons is that scattering from imperfect Si/BOX interface and surface roughness will limit mobility.

Moreover, very thin films may be highly non-uniform or even discontinuous, while quantum confinement effects will unfavourably increase the effective band gap. With silicene, an inherently 2D material consisting of a single layer of Si atoms, one can reach the ultimate limit of channel thickness. This is a new Si allotrope with radically different electronic properties compared to Si

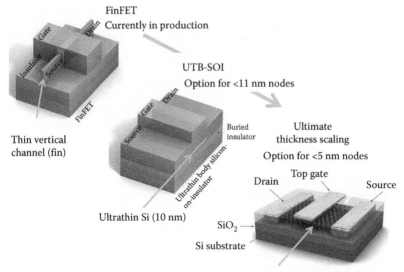

FinFET
Currently in production

UTB-SOI
Option for <11 nm nodes

Thin vertical
channel (fin)

Buried
insulator

Ultimate
thickness scaling

Option for <5 nm nodes

Ultrathin Si (10 nm)

SiO₂

Si substrate

Top gate

Drain

Source

Extremely thin channels/one or few atomic layers

FIGURE 13.1 Scaling of nanoelectronic devices in terms of channel thickness reduction. Ultimate thickness scaling beyond the 5 nm technology could be achieved with atomically thin layers offering excellent electrostatic control. (Reproduced from A. Dimoulas, *Microelectron. Eng.* 131, 68, 2015.)

ultrathin films scaled down from the bulk. The main idea behind the effort for silicene and germanene is that they will maintain some of the good properties of graphene, namely its stability in 2D form, the reduced reactivity with the environment and the relatively high mobility, while at the same time, they will present some new physical properties such as the energy band gap (lacking in graphene) and spin functionality due to larger spin–orbit coupling which could give silicene nontrivial topological properties. Silicene and germanene offer compatibility with semiconductor processing which gives them the advantage to rival not only graphene but also other prospective candidates such as MoS_2 and other transition metal dichalcogenide 2D semiconductors in the race for ultimate thickness scaling of nanoelectronic devices.

13.2 Free-Standing Silicene and Germanene

The concept of a graphite-like Si nanosheet originally dates back to pioneers Takeda and Shiraishi [3] who based on first-principles total energy calculations, argued that the aromatic stage of 2D Si and Ge layers is expected to be corrugated rather than planar (PL) as in the carbon counterpart. The reduction to a single 2D layers of Si and Ge, otherwise termed as silicene and germanene in analogy with the benzene ring stage and ultimately with the graphene analogue, was more recently addressed by Cahangirov et al. [4], and Liu et al. [5]

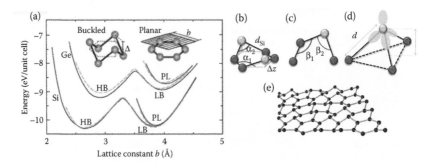

FIGURE 13.2 (a) Plot of the total energy versus hexagonal lattice constant for a total energy diagram of 2D Si and Ge layers calculated with PL, LB and HB structures. (Reproduced from S. Cahangirov et al., *Phys. Rev. Lett.* 102, 236804, 2009.) Schematic pictures of a buckled hexagonal ring (b), a buckled dimer (c) usually observed in (2 × 1) reconstructed Si(001) and Ge(001) surfaces, a Si or Ge dangling bond with a tetrahedral geometry (d) and a buckled silicene lattice reproducing the experimentally observed topography (e).

who revived the originally proposed corrugation of the graphite-like phase in a pseudo-PL, 2D buckled structure alternating upper and downer atoms throughout the hexagonal rings of the sheet lattice.

Total energy minimisation resulted in the buckled configurations of **Figure 13.2a**, known as low-buckled (LB) and high-buckled (HB), that basically differs from the PL one in terms of the vertical distortion ΔZ (see the calculated structural parameters for the three configurations **Table 13.1** as extracted from Reference 4). The nature of the buckling can be figured out in the sketch of the partially distorted hexagonal ring in **Figure 13.2b**. As a consequence of the buckling, two bond angles can be identified, one (120°), corresponding to the PL sp² disposition and another one approaching the tetrahedral-like value (109.5°) of a conventional sp³ bonding. The PL and HB structures are unstable as

Table 13.1   Structural Parameters for the Bulk and 2D Buckled Si and Ge Crystal

	a_{bulk}	Δ_{HB}	Δ_{LB}	b_{LB}	d_{LB}
Si	5.41	2.13	0.44	3.83	2.25
Ge	5.64	2.23	0.64	3.97	2.38

Source: S. Cahangirov et al., *Phys. Rev. Lett.* 102, 236804, 2009.

Note: a_{bulk}, Δ, b and d stand for bulk cubic lattice constant, buckling distance, hexagonal lattice constant and corresponding nearest neighbour distance; HB and LB denote the high-buckled and low-buckled silicene.

they have phonon modes with imaginary frequencies in the Brillouin zone. As such, the HB structure does not correspond to a real local minimum of the total energy, whereas the LB counterpart is stable, and as graphene it exhibits a semimetallic character with linearly dispersed π and π^* bands that mutually cross at the Dirac points to form the well-known Dirac cone features and preserve the chiral symmetry. It deserves to notice that according to the structural parameters in **Table 13.1**, the LB configuration is predicted to have a lattice constant of $b_{LB} = 3.83$ Å and a bond length of $d_{LB} = 2.25$ Å, whereas the vertical distortion translates as a buckling distance $\Delta_{LB} = 0.47$ Å and a bond angle $\theta_{LB} = 101°$ that implicates the interplay of sp^2- and sp^3-hybridised states in the lattice.

In analogy with graphene, the electronic structure of the LB silicene lattice can be rationalised with a graphene-like Hamiltonian where two (vertically displaced) sublattices can be clearly identified in the preserved hexagonal symmetry of the lattice. Following Reference 5, in its simplest form, the low-energy Hamiltonian around the Dirac points K looks like as

$$H_{eff}^{K} \approx \begin{pmatrix} -\xi\sigma_z & v_f(k_x - k_y) \\ v_f(k_x - k_y) & \xi\sigma_z \end{pmatrix} \tag{13.1}$$

where v_f is the Fermi velocity, k_x and k_y are the in-plane projections of the momentum and σ_z is the out-of-plane projection of the Pauli matrix. While the non-diagonal terms originate from the nearest neighbour hopping exchange that take place in a graphene-like hexagonal lattice, relation 13.1 differs from the graphene Hamiltonian in that the diagonal terms account for the actual spin orbit coupling (SOC) contribution (ξ being the exchange SOC energy). Strictly speaking, the buckled nature of silicene and germanene should result in an additional staggered potential once an external electric field is applied and in intrinsic Rashba SOC terms [6], both neglected in **Equation 13.1** for simplicity. However, even in its approximate form, the Hamiltonian in **Equation 13.1** implicates a nearly linear energy dispersion in proximity of the K points,

$$E(\mathbf{k}) = \pm\sqrt{(v_f k)^2 + \xi^2} \tag{13.2}$$

leading to the SOC induced band gap opening of 2ξ. The resulting energy gap is expected to be proportional to the ion radius. As such, compared to graphene the stronger SOC contribution in 2D materials of group IV semiconductors with larger atomic volume, such as silicene, germanene and stanene, that is, the honeycomb analogue of tin [7], is expected to open an energy band gap large enough to be measured at room temperature thus paving the way to the detection of a quantum spin Hall (QSH) state according to the model proposed by Kane and Mele [8]. Along this direction, a topological phase diagram for the electronic state of silicene can be derived which includes transitions from a trivial insulator to QSH states and to quantum anomalous Hall states as a function of the (intrinsic or extrinsic) magnetisation and of the applied electric field [6]. In this scenario, the buckling introduces a staggered potential in the Hamiltonian in **Equation 13.1** that can be leveraged to drive topological phase transitions in silicene and germanene as a function of an externally applied electric field [6]. However, in practice, the observation of the characteristic electronic states in silicene is jeopardised by the natural tendency of the Si atom to sp^3 hybridise due to the energetic balance resulting from the core–core Coulomb repulsion between nuclei. Only few configurations in nature permit the sp^2 hybridisation between Si atoms, namely the constitutive prerequisite for a silicene sheet to exist. In this respect, sp^2 hybridisation naturally occurs in the Si–Si bond of the disilene, that is, the inorganic Si_2H_4 compound molecule incorporating one Si–Si and four Si–H bonds, as well as in Si dangling bonds in reconstructed Si(001) and Si(111) surfaces [9]. The latter case provides an interesting analogy with the buckled nature of the silicene/germanene lattice since a likewise configuration takes place as due to the Jahn–Teller distortion of the Si/Ge dimers (see the sketch in **Figure 13.2c and d**). Such a distortion implies that the bonding character in the dimer is forced to vary from the sp^3-hybridised bonding in the bulk to an sp^2 hybrid in the downer atom and a p-like character in the upper atom [10]. Following Takagi et al. [11], the mixed sp^2/sp^3 hybridisation character in the LB silicene structure can be elucidated by taking the geometrical feature of Si and Ge dangling bonds as a paradigmatic example (see the sketch in **Figure 13.2c**). In there, the bonding state energy E_{db} of the dangling bond is expressed as Reference 12

$$E_{db} = \frac{2\Delta^2 E_s + (d^2 - 3\Delta^2)E_p}{d^2 - \Delta^2} \tag{13.3}$$

that is, in terms of the energy of the valence s and p states (E_s and E_p), and of the geometrical parameters Δ and d, the vertical displacement of the partially unbound atom and the atom–atom distance in analogy with the buckled silicene. Different geometrical configurations of relevance for the silicene lattice bonding can be singled out depending on the ratio $q = \Delta/d$. One borderline case is the PL configuration, obtained when $\theta = 90°$ and $\Delta = 0$, that corresponds to a flat honeycomb silicene lattice with pure sp^2 bonding, and the fully sp^3 configuration, obtained for a tetrahedral symmetry with $q = 1/3$ and $\theta = 109.5°$ resulting in the interplay of one s orbital and three p orbitals. The LB silicene corresponds to the intermediate situation, where $q \approx 0.2$ and $\theta \approx 101°$ therein implicating a mixed contribution of sp^2 and sp^3 hybrid states. The collateral built-in of the sp^2 hybridisation in the Si and Ge dangling bonds is driven by the surface free energy minimisation and is accommodated by the buckled dimer. As will be discussed in the following sections, an *ad hoc* hosting template is similarly requested for the epitaxial silicene and germanene to set although the influence of a substrate can have nontrivial implications in the atomistic arrangement and in the electronic character of the 2D layer.

13.3 Epitaxial Silicene on Metal Substrates

13.3.1 Case of Ag(111)

As previously anticipated, silicene cannot be synthesised as a free-standing layer, but in recent years, an increasing number of scientific reports proved that epitaxial silicene can be grown on a limited number of hosting substrates somehow acting as 'catalysers' for the silicene accommodation and having an intrinsic metallic character in common.

The case of Ag(111) as template for the silicene epitaxy is by far the more intensively studied and it was curiously anticipated by the evidence of silicon nanoribbons with hexagonal symmetry epitaxially grown on Ag(001) surfaces [13,14]. The growth methodology consists of an ultra-high vacuum molecular beam of Si atoms supplied by sublimation of ultra-pure Si fragments via Joule heating or by e-beam evaporator of a Si matrix. Si adatoms self-arrange in a nominally flat Ag(111) surface in a carefully tailored range of temperatures so as to form a regular honeycomb-like pattern (see the sequential illustration of the process in **Figure 13.3**) [15].

Unlike the theoretically predicted LB silicene descending from alternated adjacent Si atoms, buckling in epitaxial silicene results in a number of super-structures (also denoted as reconstructions or phases) that mutually distinguish in terms of periodicity and structural parameters (see the picture of a truly buckled silicene lattice in **Figure 13.2e**) extracted by stripping off the Ag(111) support from the reconstructed silicene overlayer). The atomistic details of the silicene superstructures were convincingly probed by high-resolution scanning probe microscopy (scanning tunneling microscopy, STM) supported by *ab initio* calculation of the stable atomic topography. Two representative cases are here taken into account in **Figure 13.4** to illustrate this 'polymorphic nature' of silicene, that is, (I) the (3×3) silicene on (4×4) Ag(111) (well-known as 4×4 silicene) and (II) the $\sqrt{7} \times \sqrt{7}$ silicene on $(2\sqrt{3} \times 2\sqrt{3})$R30° Ag(111). According to the reported notation, the superstructure of a silicene monolayer (ML) phase is equivalently denoted in terms of the periodicity of its coincidence cell with respect to the free-standing silicene or with respect to the underlying Ag(111) surface, for instance, the (3×3) silicene on (4×4) Ag(111) has a coincidence cell with (3×3) periodicity with respect to the free-standing silicene cell and (4×4) periodicity with respect to the Ag(111) surface cell. The rotational index is often added to mark the cell orientation from the high symmetry Ag(111) surface directions. The *ab initio* calculated atomic structure (both as a top view and as a side view) and the simulated STM images as reported in

Step 1: Place silicon wafer above crystalline silver (Ag) in a vacuum.

Step 2: Heat the silicon until it begins to sublimate.

Step 3: Silicon atoms spontaneously organize on the silver surface as silicene.

FIGURE 13.3 Process flow for the silicene epitaxy on Ag(111) substrates. (Reproduced from G. Brumfiel, *Nature* 495, 152–153, 2013.)

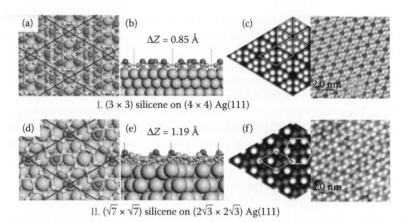

I. (3 × 3) silicene on (4 × 4) Ag(111)

II. ($\sqrt{7}$ × $\sqrt{7}$) silicene on (2$\sqrt{3}$ × 2$\sqrt{3}$) Ag(111)

FIGURE 13.4 Atomic structures and simulated STM images of two representative allotropes of Ag(111) supported silicene, the (3 × 3) silicene on (4 × 4) Ag(111) (top) and the (2$\sqrt{3}$ × 2$\sqrt{3}$)R30° Ag(111) (bottom). (a), (d): Top views; (b), (e): side views (with the buckling parameters ΔZ included); (c), (f): simulated STM images (left) and experientially recorded STM topographies (right). (*Ab initio* simulations in panels (a) through (f) are adapted from J. Gao and J. Zhao, *Sci. Rep.* 2, 861, 2012.)

Reference 16 are compared in **Figure 13.4** with the experimentally measured STM topography of the three above-mentioned superstructures. As follows from the substantial matching between simulated and experimental patterns in **Figure 13.5**, the buckling distribution in the Ag(111)-supported epitaxial silicene significantly differs from that of the free-standing LB silicene basically because of two constraints, one structural in character (the lattice match between silicene and the substrate) and one chemical in character (the strong hybridisation taking place between the silicene band structure and the Ag electronic surface states). Such a discrepancy can be clarified in terms of a commensurability relation between the epitaxial silicene superstructures and its substrate. In detail, type I silicene results in a unit cell aligned with the high symmetry $[110]_{Ag}$ direction, where only 6 of the 18 available atoms are in the upper position (corresponding to 1/3 of the total Si atoms of the silicene lattice) and the cell parameter is fourfold larger than that of the Ag(111) surface cell. These features confer the characteristic flower-like fashion to the STM topography (see the STM simulation and the observed STM topography in **Figure 13.4**). Conversely, type II silicene is tilted by 30° with respect to the $[110]_{Ag}$ direction and more closely resembles an ideally flat silicene layer as only two buckled atoms are present in the 14 atoms made supercell (i.e. 1/7 of the total number of Si atoms are on top positions). Not only the buckling distribution dictates the different periodicity of the two superstructures, but also the buckling parameter changes from one another, being $\Delta \approx 0.74$ and 1.05 Å, respectively. Detailed structural parameters of the superstructures in **Figure 13.4** are reported in **Table 13.2** for clarity. Each silicene superstructure

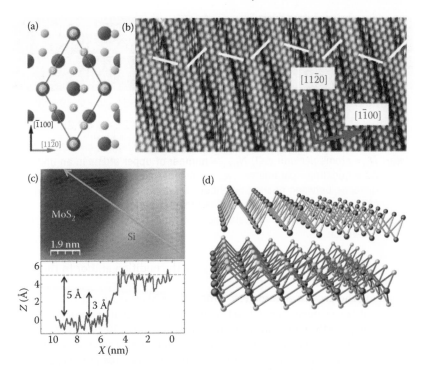

FIGURE 13.5 (a) Model of the Si honeycomb structure on the topmost Zr layer of $ZrB_2(0001)$. Chemically different types of Si atoms 'A', 'B' and 'C' are indicated according to their position with respect to the underlying lattice. (b) Large scale (20 nm × 9.5 nm) STM topography of 2 × 2 reconstructed epitaxial silicene on ZrB_2 with characteristic stripe domains (the orientation of the domains are marked with white lines). (Panels (a) and (b) are adapted from A. Fleurence et al. *Phys. Rev. Lett.* 108, 245501, 2012.) (c) STM topography (top image) of a Si ML domain (left) that partially covers the underlying MoS_2 surface (right). The line profile (bottom diagram) across two adjacent terraces allows for the measurement of the step height and for the extraction of the buckling parameter. (d) Pictorial view of the Si/MoS_2 heterosheet junction where the highly buckled silicene is shown to replicate the MoS_2 surface structure. (Panels (c) and (d) are adapted from D. Chiappe et al., *Adv. Mater.* 26, 2096, 2014.)

is self-consistently inferred from the combined use of atomically resolved evidences and theoretical models and is qualified by a characteristic buckling distribution that determine a specific electronic character if we assume that interaction with the substrate takes place. Indeed, the larger and more diffuse buckling of the type I silicene is predicted to result in parabolic bands with a gap of 0.3–0.4 eV [17], whereas the more PL morphology of the type II silicene is consistent with the Dirac cones features in analogy with the free-standing LB silicene. For an exhaustive view, other silicene superstructures have been

Table 13.2 Structural Parameters of the Two More Representative Silicene Phases

Silicene Phase	N_{Si} ($N_{Si,top}$)	ΔZ (Å)	a_{cell} (Å)	d_{Si} (Å)	C_{buckl} (%)
Type I: $3 \times 3/4 \times 4$	18 (6)	0.74–0.85	11.78	$2.34 < d < 2.39$	33.34
Type II: $\sqrt{7} \times \sqrt{7}/2\sqrt{3} \times 2\sqrt{3}$	14 (2)	1.1–1.2	10.20	$2.28 < d < 2.36$	14.3

Note: N_{Si} = atoms per unit cell, $N_{Si,top}$ = number of upper atoms in an unit cell, ΔZ = buckling parameter, a_{cell} = silicene lattice parameter, d_{Si-Si} = distance between nearest neighbour Si atoms and C_{buckl} = concentration of buckled atoms in a phase. The range of parameters is deduced from experimental values in Reference 55 and theoretical outcomes from Reference 16.

extensively reported depending on the growth condition and/or in the same silicene sheet [18,19].

Among the growth parameters, temperature plays a critical role. Indeed, the stability of a silicene superstructure is deeply impacted by the slight variation of the growth temperature in the range between 220°C and 290°C. As a consequence, consecutive transitions between majority silicene superstructures were observed with increasing growth temperatures [20], types I and II silicene positioning in the lowest and highest temperature edge, respectively. It is interesting to notice that the type II silicene has been recently argued to undergo intermixing with the Ag substrate atoms and dewetting in clusters because its formation temperature is close to the activation of the Si desorption from Ag or Si precipitation into Ag [21]. Unlike other coexisting phases, the higher temperature threshold causes the type II silicene to be largely dominant over the other minority [22] thus resulting in more extended domains. The overall scenario depicts the silicene growth as a kinetics-limited self-organisation of silicon adatoms, where the silicene superstructure are metastable points in the surface phase diagram as a function of the temperature and of chemical potential during growth [17]. From this point of view, each silicene superstructure can be referred to as a metastable 2D phase that results from the local excess or deficiency of silicon adatoms at a given growth temperature. Local variations of the silicon adatom population in the narrow temperature range where phase transition occurs are thus responsible for the observed phase coexistence.

As a matter of fact, coexistence of different silicene phases as individual domains of a single silicene sheet has been initially reported by Chiappe et al. [19]. The intrinsic multiphase character of a silicene sheet has been subsequently consolidated by Moras et al. [22] in a thorough electron diffraction study as a function of the growth temperature where different silicene phases were proved to coexist in proportions dictated by the thermal constraint of the growth. As such, the silicene phases in **Figure 13.4** can be regarded as majority (but not unique) superstructures for a given growth temperature. It is not clear at the moment if a majority phase can be forced to homogeneously cover the whole silicene sheet. Nonetheless, given the metastable character of the silicene phases, silicene domains with a specific phase can be forced to become the majority structure and their lateral extension can be increased with respect to the as-grown configuration (up to width of the order of 100 nm or anyway limited by the terrace width of the underlying Ag surface) upon carefully tailoring the post-growth isothermal treatments as reported in Reference 23.

13.3.2 Case of Other Metal Substrates

While the earliest attempts and the current reports on the synthesis of the epitaxial silicene are more intensively focussed on the Ag(111) substrate, other metallic substrates were also successfully investigated as growth templates. To date, an epitaxial silicene lattice was observed on $ZrB_2(0001)$ [24], Ir(111) [25] and more recently on ZrC(111) [26]. Generally speaking, these surfaces provide a compliant lattice match as they are commensurate with a stable structure of a buckled silicene. Similarly to the case of the Ag(111), a pseudo-PL-like phase must adapt to the epitaxial constraints of the substrate thus deviating from the free-standing stable form. In this respect, it is particularly noteworthy to mention the epitaxial silicene on $ZrB_2(0001)$/Si substrates for the thorough and comprehensive understanding of the atomistic and chemical details. The growth process basically descends from the segregation of Si atoms from the Si bottom substrate on top of the commensurate ZrB_2 buffer layer, and results in a $\sqrt{3} \times \sqrt{3}$ silicene structure superimposed onto a (2×2) $ZrB_2(00001)$ surface where three different fashions of atomic positions can be distinctly identified in relation to their proximity to the Zr atoms of the ZrB_2 surface (hollow, top and bridge sites denoted as A, B and C atoms in the scheme of **Figure 13.5a**) [24]. The overall morphology of this silicene structure interestingly exhibits a striped pattern (see the STM topography in **Figure 13.5b**) as a consequence of periodic deformations at the boundaries which are in turn induced by epitaxial strain on the in-plane structure [27]. Here also, the interaction with the substrate is critical in dictating the atomic arrangement of the silicene lattice as far as the silicene derived electronic bands are hybridised with the Zr d electronic states [27], therein resulting in a bucking distribution and in a sp^3/sp^2 bonding interplay qualitatively similar to those observed in the case of the Ag(111) surface.

Assuming the crystallographic matching to be a driving force for the silicene epitaxy, template engineering for the silicene accommodation was so far tackled in substrates with a hexagonal symmetry and with a commensurate surface lattice with silicene such as hex-BN and hydrogenated Si(111) [28] or layered chalcogenides (MoS_2, $MoSe_2$, $MoTe_2$, GaS and GaSe) [29]. Among the latter ones, despite the substantial lattice mismatch, MoS_2 proved to host epitaxially grown 2D silicon domains with highly puckered Si bonding [30]. In detail, combining atomically resolved topography and density functional theory (DFT) calculations, Si atoms were observed to self-arrange in an unreconstructed (1×1) highly buckled lattice that is perfectly matched with the underlying MoS_2 surface lattice. The topographic contrast between a 2D Si domain and the underlying MoS_2 surface is illustrated in the STM topography of **Figure 13.5c**. A buckling parameter of 2 Å is deduced from the linear cut profile throughout the STM topography in **Figure 13.5c**. This value is consistent with the atomic model of relative minimum energy represented in **Figure 13.5d**. As follows from the general consideration of Section 13.1, the high-buckling results in a metallic character of the silicene domain. It is not clear whether this growth mode can be rationalised as a van der Waal epitaxy, namely in the epitaxial growth of a 2D crystal lattice with an extremely weak interaction with the hosting substrate, or if the substrate constraints have a leading role in locally dictating the psedomorphic character of the Si ML growth. Nonetheless, the Si/MoS_2 heterosheet structure is of technological relevance for optoelectronic and photonic MoS_2-based applications as interfacing with Si was proved to result in an enhanced photoresponse [31].

13.3.3 Case of the Silicene Multilayer

Special consideration should be taken for the growth of multiple silicene layers on Ag(111) substrates because of the potentially outstanding advantages in terms of chemical stability and electronic properties. According to the former point, it is expected that the silicene multilayer is more robust against oxidation or degradation in air while preserving Dirac fermions as charge carriers. The epitaxy of the silicene multilayer has been observed to follow a Stranski–Krastanov growth [32], where 2D islands grow over a homogeneous 3×3 silicene structural template (see type I silicene in **Figure 13.4**) and pile up before totally completing a whole layer as follows from the STM morphology in **Figure 13.6**. It is relevant to notice that each overlying island from the bilayer stage further on exhibits a unique $\sqrt{3} \times \sqrt{3}$ R30° unit cell (with respect to the 1×1 silicene) as follows from the combined low energy electron diffraction and STM information extracted from Reference 32. Such a structure is consistent with a honeycomb-like arrangement of buckled silicon atoms with an adjacent hexagon distance of 0.64 nm and is preserved well through the multilayer regime of growth up to the 60th layer [33]. As reported by Vogt et al. [32], the step height of the silicene bilayer islands from the 3×3 silicene plane amounts to 0.2 nm that differs from the height of 0.24 nm of the Ag(111) surface steps.

A similar structure was also reported as the high temperature stage of the temperature driven phase transitions in the silicene ML by Feng et al. [20]. Nonetheless, there is wide consensus considering that the $\sqrt{3} \times \sqrt{3}$ R30° stage as due to the activation of a dewetting mechanism leads to the formation of a second layer. On the other hand, a debate is currently ongoing on what is the true stacking between the piled-up silicene layers. In this respect, Shirai et al. [34] argue that the silicene multilayer consists of a thin film of bulk diamond-like silicon terminated as Si(111) with an additional $\sqrt{3} \times \sqrt{3}$ superstructure induced by Ag segregation from the substrate. Accordingly, Si dissolution in the near-surface region of the Ag and consequent intermixing of Si and Ag are inferred by Mannix et al. [35] for the $\sqrt{3} \times \sqrt{3}$ phase in the multilayer regime and growth temperatures (340–360°C) comparatively higher than those reported in Reference 32 or used for the silicene ML growth. In the

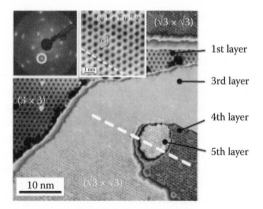

FIGURE 13.6 Structural properties of multilayer silicene grown on Ag(111). Insets: RHEED pattern of the multilayer silicene (left), high-resolution STM topography on a multilayer silicene terrace showing a √3 × √3 registry. (After P. Vogt et al., *Appl. Phys. Lett.* 104, 021602, 2014.)

same study, these authors concluded that the observed $\sqrt{3} \times \sqrt{3}$ phase is a sp³ allotrope in character that according to local scanning tunnelling spectroscopy (STS) features a semiconductor-like energy gap down to the extreme 2D level. This result opens a controversy with the angle resolved photoemission spectroscopy (ARPES) study by De Padova et al. [36] which reports a linear energy dispersion of massless quasiparticles in the silicene multilayer that is suggestive of a carrier population of Dirac fermions. De Padova et al. [36] identified the hallmark of the Dirac cones in the silicene multilayer directly from the signature of the valence and conduction bands extracted from the ARPES. A linear dispersion of the energy–momentum relation is measured in the Γ points of the extended Brillouin zone scheme of the $\sqrt{3} \times \sqrt{3}$ silicene, which is related to band folding from K points of the 1 × 1 silicene. The physical picture is rationalised as zero-gap Dirac cones originating from the π and π* silicene bands, with a partially filled π* band as due to charge transfer from the substrate, and an inferred Fermi velocity at room temperature amounting to 0.3×10^6 ms⁻¹. This behaviour is corroborated by similar evidences of linear bands in the parental system made of silicene multilayer nanoribbons grown on Ag(110) [14] and by the oscillatory behaviour of the quasiparticle interference pattern throughout the $\sqrt{3} \times \sqrt{3}$ silicene terraces reported by Feng et al. [37] to prove the existence of Dirac fermions chirality. The controversy in the exact electronic band structure identification of the $\sqrt{3} \times \sqrt{3}$ phase is still open and demands a careful analysis of the structural and electronic properties of the Si multilayer growth as a function of the growth temperature so as to distinguish a graphite-like phase from a diamond-like one in different thermal regime of growth or to rule out the existence of one of them in the limit of ultrathin Si film growth. It should be finally added that the direct observation of Dirac cones at the K points was reported from multilayer silicene intercalated in $CaSi_2$ crystals incorporating hexagonal Si bilayer (silicene) lattices in between tetragonal Ca MLs [38]. The two lattices are alternately stacked along the [0001] axis of the hexagonal crystal, and the Ca lattices play a stabilising role for the honeycomb-like silicene bilayer thus effectively resembling a graphite-like interacted compound.

From the theoretical point of view, the $\sqrt{3} \times \sqrt{3}$ R30° pattern of the silicene bilayer takes its origin from the exothermic formation of atomic dumbbell structures resulting from the adsorption of excess adatoms over the 3 × 3 silicene phase [39]. The stacking of successive layers is modelled as a consecutive sequence of dumbbell structures alternating honeycomb and trigonal arrangements. The resulting layered structure, termed as silicite in Reference 39, is expected to be thermodynamically stable as no negative phonon frequency is inferred, and it is not influenced by the underlying Ag(111) surface in its lattice features (despite a 5% mismatch between the two) and electronic band structure (a substantial indirect energy band gap opens that is predicted to range in between 0.98 and 1.26 eV). In fair accordance, a much smaller energy gap opening of 100 meV is independently deduced in between linearly dispersed Dirac bands by modelling the band structure of the $\sqrt{3} \times \sqrt{3}$ R30° silicene bilayer with a honeycomb arrangement of the buckled atoms [40].

It is relevant to notice that the electronic structure of the silicene multilayer dramatically deviates from that reported for the silicene ML. In particular, local electronic probing measurements under magnetic field ruled out the emergence of Landau levels in 3 × 3 silicene ML thus claiming for symmetry breaking and hybridisation induced suppression of the Dirac cones in the silicene ML [41]. ARPES measurements of the silicene ML were apparently consistent with a linear energy band dispersion [42] that was subsequently attributed to Si-enhanced Ag surface states and to silicene–Ag hybrid states [43]. A more detailed discussion on this matter will be addressed in the following section. Overall, it seems as though while the ML suffers from a strong interaction with the substrates that warps its native electronic signature, the electronic band structure of the silicene multilayer is rid of any Ag-related hybridisation and it may possess a pure Dirac-like character of its charge carriers as well as exhibits a semiconducting behaviour depending on the growth conditions. For these qualities and for its intrinsic resistance against oxidation, the silicene multilayer is supposed to be a good alternative to the silicene ML for functional applications and device integration. Nonetheless, a careful consideration of the $\sqrt{3} \times \sqrt{3}$ R30° silicene phase is still necessary both in the bilayer and in the multilayer configurations in order to definitely elucidate the intrinsic electronic character and hence its exploitation in the electronic field.

13.4 Electronic Band Structure of Silicene and Hybrid States

The electronic properties of the silicene phase are still a controversial issue, mainly due to the fact that epitaxial silicene grown on various substrates exhibits sp^2–sp^3 orbital hybridisation and a buckling, which could affect its electronic band structure. The possible existence of a Dirac cone, which was recently hotly debated, is the way to clarify whether silicene on Ag(111) is semimetallic like graphene, metallic or semiconducting/insulating. In a pioneering work, Vogt et al. [42] observed a linear energy–momentum dispersion near E_F by ARPES measurements providing a hint for the existence of a Dirac cone in 4 × 4 silicene. Signatures of a Dirac cone with Fermi velocity of ~1.3 × 10^6 m/s have been observed along ΓK [42] and ΓM [44] directions of Ag(111) surface Brillouin zone (SBZ), indicating a gaped (semiconducting) silicene. Further evidence for the existence of Dirac fermions in a honeycomb lattice based on silicon was provided by STS measurements, where a linear dispersion above the Fermi level was revealed [45].

Subsequently, additional evidence combined with *ab initio* calculations provides a different picture. The absence of discrete Landau levels in STS measurements under high magnetic field, do not support the idea of Dirac fermions in (4 × 4) silicene [41]. Towards this direction, there have also been a number of DFT calculations that further supported the absence of a Dirac cone. In particular, the experimentally observed linear dispersions have been related either to the Ag(111) substrate [28,46–48], or to a new state that emerges due to the strong hybridisation between Si and Ag orbitals [49], or more recently to a substrate state that is modified by the interaction with silicene [50]. The hybridisation scenario was further supported by ARPES experiments where a new surface band (SB) located near the bulk Ag sp-band was observed along both ΓK and ΓM (**Figure 13.7a** and **b**) directions of the BZ [43]. It was proved that the new band has no dependence on photon excitation energy and consequently no dispersion with $k_\perp$, only with $k_{//}$, which means that it is a true 2D surface band.

The new surface band was not associated with the π band of (1 × 1) silicene, since it disperses at higher rather than lower k// values with respect to the Ag sp band. Instead, the appearance of the new band was attributed to the formation of hybrid metallic states between Si and Ag orbitals. The fact that such a band is not visible in the ARPES spectra of amorphous Si, implies that indeed the buckling and hybridisation play a crucial role in the electronic properties of the (4 × 4) silicene/Ag(111) system (namely the type I silicene in **Figure 13.4**).

Moreover, it should be noted that the hybrid band exhibits a Λ-shaped along the ΓM direction, which resembles a gaped Dirac cone originating from (4 × 4) silicene (**Figure 13.7b**). Nevertheless, the absence of a Dirac cone was verified by the constant energy k_x–k_y contour plots where a non-vortical ')(' instead of a circularly shaped behaviour was observed (**Figure 13.7c**). Moreover, the band imaging along the perpendicular k//y (M_{Ag}–K_{Ag}) direction reveals an upward parabolic-like branch (**Figure 13.7d**). This branch is part of the surface band SB forming a saddle point at M_{Ag} which is located 0.3–0.4 eV below the Fermi surface as schematically illustrated in **Figure 13.7e**.

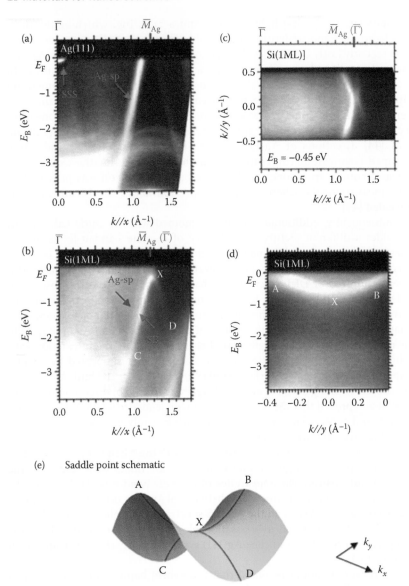

FIGURE 13.7 Valence band imaging by ARPES of (a) bare Ag(111) and (b) Ag(111)/Si(1ML) structure (corresponding to the formation of 4×4 silicene), recorded along the ΓM direction of Ag(111). The new Λ-shaped hybrid surface band is indicated as SB by the red arrow, (c) constant energy k_x–k_y contour plots of the Ag(111)/Si(1ML) structure, showing a non-vortical ')(' behaviour, (d) Energy dispersion along the perpendicular $k//y$ (M_{Ag}–K_{Ag}) direction exhibiting an upward parabolic-like branch and (e) schematic illustration of the saddle point at formed at M_{Ag} in the Ag(111)/ Si(1ML) structure. (After D. Tsoutsou et al., *Appl. Phys. Lett.* 103, 231604, 2013.)

Based on the ARPES data, the hybrid surface band is reconstructed schematically and compared with graphene in **Figure 13.8**. A steep linearly varying dispersion near K_{Ag} and a saddle point near M_{Ag} reveal a striking similarity with the π-band behaviour in graphene which shows similar characteristics. It could be said that silicene converts the surface of Ag(111) into a graphene-like system with regard to the electronic structure.

Further ARPES and x-ray absorption spectroscopy experiments as well as DFT calculations have also confirmed the hybridisation between Si and Ag, which results in the metallic nature of epitaxial silicene on Ag(111) [51–53]. Interestingly, the σ bands of silicene are clearly resolved in one of these works (**Figure 13.9**), confirming the successful silicene growth, but without the Dirac π bands. Another ARPES work [52] reports the presence of a hybrid band in addition to the bulk Ag band confirming the earlier work [43] and shows that the Si–Ag hybridisation gradually decays due to oxygen adsorption, resulting in the revival of the Ag(111) Shockley surface state.

Overall, it can be concluded that a Dirac cone in the (4×4) silicene/Ag(111) system has not been confirmed experimentally. Nevertheless, silicene hybridising with Ag produces new 2D surface metallic states resembling the graphene π bands. This opens a new way to explore other possible metal/semiconductor hybrid systems with graphene-like bands. It is worth noticing also that the saddle point of the hybrid band lying only 0.5 eV below the Fermi level produces a van Hove singularity with a high density of states near the E_F. In principle, such a singularity near E_F could enhance any possible weak electron–electron correlation effects and give rise to magnetism or

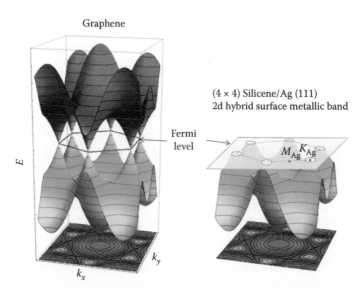

FIGURE 13.8 Schematic illustration of the 2D hybrid surface band of 4×4 silicene on Ag(111) (right), which resembles the π band of graphene (left). (Reproduced from A. Dimoulas, *Microelectron. Eng.* 131, 68, 2015.)

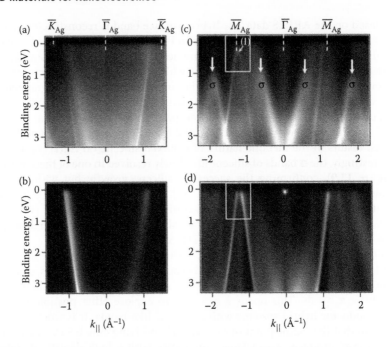

FIGURE 13.9 Valence band imaging by ARPES recorded along the Ag(111) ΓK direction of (a) 4 × 4 silicene on Ag(111) and (b) bare Ag(111). Valence band imaging by ARPES recorded along the ΓM direction of (c) 4 × 4 silicene on Ag(111) and (d) bare Ag(111), showing the clearly resolved silicene σ-bands. (After S.K. Mahatha et al., *Phys. Rev. B* 89, 201416(R), 2014.)

superconductivity increasing the critical point. The presence of such a singularity near E_F may be a possible explanation for the reported superconducting transition in the silicene/Ag(111) system at a relatively high $T_C \sim 35$ K [54].

13.5 Stability of Silicene and Process Issues

Although the identification of a honeycomb silicon structure on commensurate metal surfaces opened exceptional opportunities in the fundamental physics and technology playground, a key issue remains as to how to utilise the silicene in practical devices. Indeed, two obstacles stand up against any feasible exploitation of the silicene at first glance. One is the stability under environmental conditions because silicene is prone to oxidise in air [55]. The other one is the unavoidable symbiosis with the metallic support template (whatever template it is) that makes the silicene epitaxy possible, but causes a strong hybridisation (see Section 13.4) and prevents silicene from being transferred in any practical device platform. The former issue was addressed by engineering a non-destructive encapsulation, where an Al ML is first grown onto the silicene in ultra-high condition at room temperature and then Al is sequentially co-deposited with ultra-pure O_2 [55]. This process is illustrated in its physical details in **Figure 13.10a** and it consists of an ultrathin Al_2O_3 capping that preserves the intrinsic chemical and structural environment of the underlying silicene. To support the non-destructive character of the Al_2O_3 capping, an *in situ* photoemission study was carried out on a sufficiently thin Al_2O_3 overlayer proving the survival of a Si–Si bonding (see **Figure 13.10b**). In analogy with the graphene characterisation, a key-probe allowing for the *ex situ* identification of the silicene ML was found in the Raman spectroscopy (see **Figure 13.10c**). However, unlike graphene, the Raman spectrum of the silicene is quite elusive owing to the apparent resemblance with the

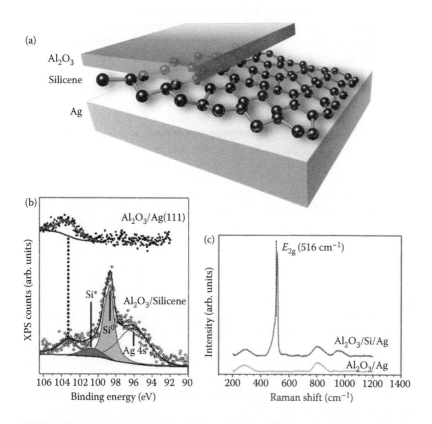

FIGURE 13.10 (a) Schematic picture of the Al_2O_3 encapsulation of the Ag-supported silicene; (b) physical characterisation of the silicene encapsulation: (b) Si 2p XPS core-level line taken after deposition of Al_2O_3 capping layer onto silicon/Ag(111) (bottom) and directly onto the Ag(111) surface for comparison (top) making evidence of the chemical integrity of the silicene when sandwiched in between the Al_2O_3 encapsulation and the Ag substrate; (c) comparative Raman spectra taken for the same configurations as in panel (b) therein making evidence of the characteristic E_{2g} peak coming from the epitaxial silicene. (Panels (a) and (b) are adapted from A. Molle et al., *Adv. Funct. Mater.* 23, 4340, 2013.)

spectrum of the bulk diamond-like silicon [56]. Nonetheless, several features allow for a nontrivial but definite discrimination between the two spectra. First of all, a narrow and intense peak was measured from the (3 × 3) silicene at a frequency of 516 cm^{-1}, that is, red shifted with respect to the position of the bulk Si.

Based on parallel *ab initio* simulations, such a red shift along with the presence of an asymmetric shoulder in the low frequency tail were self-consistently interpreted as due to the phonon modes of buckled silicene lattice in fine agreement with the true epitaxial silicene phases. As such, the narrow peak was attributed to an in-plane E_{2g} mode, whereas any other asymmetry of the peak was rationalised as a result of buckling-induced breathing modes. Other clues also hint at the Raman signature of a honeycomb lattice therein including the frequency dispersive behaviour of the Raman overtone peak [56]. It is relevant to notice that the interpretation of the Raman spectrum of the encapsulated silicene has been recently confirmed by *in situ* Raman spectroscopy measurements that definitely ruled out any influence on the silicene structure coming from the encapsulation layer (e.g. stress and intermixing) [57].

The second, perhaps tougher, hurdle is the awkward presence of the substrate. Restricting the consideration to the case of the Ag(111) substrates, an Ag monocrystal as supporting template is unpractical for device integration and it turns out to detrimentally impact the electronic character of the silicene ML as discussed in Section 13.4. One route to bypass the Ag template is the synthesis of silicene in device friendly platforms such as insulating or semi-conducting substrates. Despite the intensive interest in templates like hex-BN or H-passivated Si(111), where silicene is supposed to weakly interact with the substrate and conserve its inherent Dirac cones [28], to date only MoS_2 bulk substrates proved to accommodate 2D Si domains [30]. This outcome can be potentially extended to MoS_2 nanosheets opening a route to the fabrication of heterosheet transistors. However, atomically resolved topography of the 2D silicon domains are consistent with a dramatically HB arrangement leading to a metallic character. Although the metallicity of the 2D silicon domains can be hardly applicable to device functionality, it is interesting to elucidate the electronic details of the Si/MoS_2 interface as well as the Si growth on MoS_2 because of the potential impact in several fields of technological and scientific interest such as photodioded with enhanced response, extrinsic doping and quantum confined states at the MoS_2 surface.

On the other hand, a revolutionary approach to the handling of Ag-supported silicene has been proposed by Tao et al. [58] who performed an extraction process denoted as 'silicene encapsulated delamination with native electrode' (SEDNE) and illustrated in **Figure 13.11**. The SEDNE process starts from the epitaxy of silicene on commercial 300 nm-thick Ag(111)-on-mica substrates. The silicene sandwiched in between the encapsulation Al_2O_3 layer and the residual Ag is mechanically cleaved from the mica (see the second step of the process flow in **Figure 13.11**). The free-standing silicene sandwich (as depicted in the third step of **Figure 13.11**) is then placed upside down on an inert SiO_2/Si platform, where the heavily doped Si can be used as bottom gate electrode (see the fourth step of **Figure 13.11**). An iodine-based solvent is then

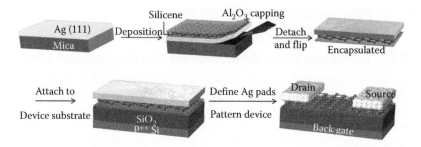

FIGURE 13.11 SEDNE process enabling silicene integration into a FET operating at room temperature and manifesting an ambipolar behaviour in its transfer characteristics. (After L. Tao et al., *Nat. Nanotechnol.* 10, 227, 2015.)

dropped on the top Ag side of the silicene sandwich in order to selectively etch the native Ag away from patterned electrodes which serve as source and drain contacts in the resulting bottom gate field effect transistor (FET) (fifth step in **Figure 13.11**). A transfer curve is measured from the silicene transistor which exhibits an ambipolar behaviour thereby suggesting of a graphene-like transport. Indeed, a diffusive transport model describing the transport through a FETs based on graphene nicely fits to the transfer curve thus attesting to the semimetallic nature of the silicene. Control group devices incorporating oxidised silicene or amorphous silicon in place of the silicene channel as well as Ag residue rule out any process artefact and confirm that the observed transport feature is silicene specific [58]. The groundbreaking character of the room temperature operation of the silicene transistor is anyway jeopardised by the limited stability of the silicene once the native Ag is etched off as the silicene is observed to degrade in the timescale of a few minutes. This apparent drawback can be anyway tackled by *ad hoc* passivation strategies (e.g. by tuning the Ag removal down to the atomic scale or by performing the Ag in a controlled environment) that will be mandatory to set the silicene in a reliable technological mainstream. What actually matters is that silicene has been manipulated away from its native template and its electrical signature finely matches with an ambipolar behaviour in close accordance with a Dirac fermions population [58]. From the ambipolar behaviour of the FET transfer characteristic, a carrier mobility of 129 cm^2 V^{-1}s^{-1} was extracted that is significantly lower than that of graphene but competing with other 2D atomically thin materials such as MoS$_2$ and phosphorene. The quantitative discrepancy from the graphene mobility can be likely due to structural inhomogeneity as described in Section 13.3 as well as from the comparatively stronger electron–phonon coupling. Despite the poor stability of the silicene when decoupled from the native Ag, this finding opens the route to future exploitation of the material in electronic applications which may take benefit from the natural compatibility of silicene with the ubiquitous Si technology and from the inherent flexibility of silicene to embody externally tunable electronic states (such as QSH insulator states, band gap opening, etc.).

13.6 Progress in the Search for Germanene: Metallic Substrates

13.6.1 Germanium on Ag: An Ordered Surface Substitutional Alloy

The Ge counterpart of silicene, namely germanene has proven to be more difficult to realise. Similar to silicene, Ag(111) was regarded as a candidate substrate for germanene growth. Nevertheless, it turns out that contrary to the silicene case, Ge on Ag(111) hybridises strongly with the substrate forming an ordered Ag_2Ge surface alloy with a $(\sqrt{3} \times \sqrt{3})R30°$ superstructure at 1/3 Ge coverage [59,60]. *In situ* ARPES experiments revealed a rich surface hybrid band structure consisting of cone like, linearly dispersing features at the superstructure zone centre Γ, which coincides with the K-point of Ag (**Figure 13.12**).

A characteristic split-band behaviour has been observed for one of the bands that exhibits a saddle point near the M point of the SBZ. This behaviour distinguishes the Ge surface alloy from other similar systems such as Ag_2Sn and has been attributed to the deviation from the ideal $(\sqrt{3} \times \sqrt{3})R30°$ periodicity [61]. Although germanene Dirac fermions have not been confirmed, the band structure of the Ag(111)/Ge system with steep linearly varying bands near K_{Ag} and saddle point near M_{Ag} overall resembles the band structure of the (4×4) silicene on Ag(111) which is similar to the π bands of graphene as already discussed in Section 13.4. This demonstrates that the formation of new graphene-like bands is a more general occurrence in hybrid 2D materials/Ag systems.

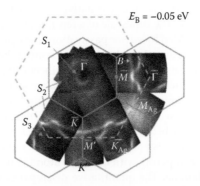

FIGURE 13.12 Valence band imaging by ARPES showing the constant energy contour k_x–k_y plots of the Ge/Ag(111) structure near Fermi level ($E_B = -0.05$ eV). Dashed blue and solid orange hexagons indicate the (1×1) Ag(111) and $(\sqrt{3} \times \sqrt{3})R30°$ superstructure, respectively. B marks the bulk Ag sp band, while S1–S4 are surfaces states of the superstructure. (After E. Golias et al., *Phys. Rev. B* 88, 075403, 2013.)

13.6.2 Evidence for Germanene on Various Metal Substrates Other than Ag(111)

2D lattices of germanium interpreted as germanene have been demonstrated on metallic substrates other than Ag(111). In particular, highly distorted and corrugated Ge honeycomb lattices are grown on Pt(111) substrate, taking advantage of the template's hexagonal symmetry as well as its reduced reactivity with other adsorbed 2D honeycomb lattices such as graphene. It is claimed that a 2D continuous germanene layer is successfully fabricated on Pt(111), exhibiting a $(\sqrt{19} \times \sqrt{19})$ superstructure (**Figure 13.13a**) in coincidence with (3×3) buckled germanene [62]. The reported buckling is 0.6 Å, comparable to that of silicene (0.75 Å), whereas the Ge–Pt interaction did not seem to affect the formation of the germanene sheet.

In another work, it is shown that a large number of ordered structures can be grown on Au(111). In one of these structures, Ge is arranged in a nearly flat honeycomb lattice with a superstructure $(\sqrt{3} \times \sqrt{3})$ with respect to (1×1) germanene or $(\sqrt{7} \times \sqrt{7})R19.1°$ with respect to the Au(111) surface, attributed to germanene (**Figure 13.13b**) [63]. The conclusions are made on the basis of STM and synchrotron-excited core-level spectroscopy measurements as well as DFT calculations.

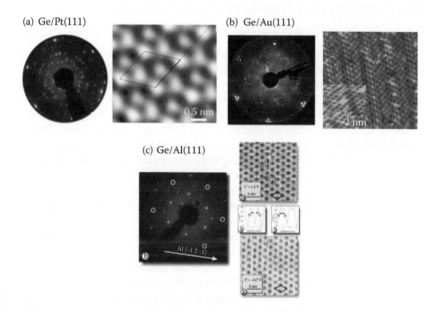

(a) Ge/Pt(111)

(b) Ge/Au(111)

(c) Ge/Al(111)

FIGURE 13.13 LEED and STM partners indicate different Ge superstructures for the germanene layer grown on (a) Pt(111). (After L. Li et al., *Adv. Mater.* 26, 4820, 2014.) (b) Au(111). (After M.E. Davila et al., *New J. Phys.* 16, 095002, 2014.) (c) Al(111) substrates. (After M. Derivaz et al., *Nano Lett.* 15, 2510, 2015.)

More recently, a continuous 2D honeycomb germanene layer has been also fabricated onto an Al(111) substrate [64]. A (3 × 3) surface periodicity with respect to the Al(111) surface was confirmed by low energy electron diffraction (LEED) and STM measurements (**Figure 13.13c**). In addition, first-principles calculations indicate the formation of a buckled structure, in which two of the Ge atoms are displaced upwards with respect to the remaining six Ge atoms of the unit cell. The absence of covalent bonds at the Ge–Al interface has been confirmed both experimentally and theoretically by x-ray photoemission spectroscopy/diffraction measurements and the calculation of the electron localisation function, respectively.

Although there is substantial evidence for germanium ordered lattices on metal substrates, which could be considered as buckled germanene, there is need for better identification of germanene by ARPES or other methods based on the existence of Dirac fermions and verification that possible interaction (or hybridisation) with the substrate does not disturb their electronic band structure. Most importantly, the same as silicene, germanene needs to be directly grown on crystalline insulators or transferred on insulating substrates to electrically characterise them and fully exploit their expected unique properties in electronic devices.

13.7 Alternative Non-Metallic Substrates

13.7.1 2D Crystalline Dielectric Substrates

Most of the work regarding silicene and germanene has been demonstrated on metallic substrates. As already discussed in Sections 13.4 and 13.6 above, silicene and germanene, in most cases, hybridise with the metallic substrates essentially losing their electronic identity, and essentially transforming to 2D crystals without π bands and Dirac fermions. Moreover, the presence of a metallic substrate screens the electric field inhibiting the fabrication of functional field effect electronic devices. According to Section 13.5, the recent success in transferring silicene from Ag onto insulating SiO_2-based substrates and the demonstration of the first silicene FET [58] creates new hopes for applications of silicene devices in nanoelectronics. However, one of the biggest challenges at this moment is to directly grow silicene epitaxially on crystalline non-metallic substrates in order to obtain good control of the growth, upscale the process to larger area wafers and avoid the hurdles of layer transferring (limited stability, complexity and yield) for device fabrication. First attempts to grow on non-metallic substrates concern the silicene growth on exfoliated flakes of MoS_2 2D semiconductor [29]. A highly buckled (2 Å) Si honeycomb lattice is observed by STM measurements, in agreement with DFT calculations. Given the high-buckling, it is rather difficult to claim silicene in this case, however, the structural and chemical stability and the absence of extensive reaction and mixing with the substrate creates hopes that single-layer crystals made of silicon can be obtained on suitable non-metallic substrates.

For the growth of a 2D material such as silicene, 2D graphite-like crystalline insulating substrates (e.g. BN) may be a good choice for lab-based research. However, BN is only available as small size crystals and it is rather difficult to obtain it epitaxially on a given substrate. For this reason, BN is less attractive for electronics applications requiring fab-based processing on large area wafers. Other materials like AlN which can be grown epitaxially on Si(111) or other large area substrates may be a better option. However, AlN on Si adopts the wurtzite crystal structure which may not be ideal for the overgrowth of silicene. It is envisaged that a flat 2D sp^2-hybridised dielectric in the hexagonal structure (like BN) would be more appropriate as a growth template for silicene since it will interact weekly via van der Waals forces maintaining the integrity of the interface. Indeed, it has been predicted by Reference 65 that silicene remains stable when encapsulated between two ML-thick 2D hex AlN dielectrics. In such a case, the 2D AlN/silicene (or germanene)/2D AlN heterostructure depicted in **Figure 13.14**, could be an ideal device layer structure ensuring good electrostatic control and allowing for charge transport through a single layer of Si atoms with minimum dissipation and leakage current.

To implement the device layout in **Figure 13.14**, it is important to realise first a hexagonal sp^2-hybridised AlN with a flat geometry. As shown by first principles [66] on a large number of wurtzite materials, this hexagonal metastable phase is the precursor of the more stable wurtzite one. As wurtzite is polar material with a large built-in electrostatic energy, the 2D hexagonal configuration is favoured at small film thickness to minimise the electrostatic energy.

Remarkably, the AlN dielectric is predicted to maintain the 2D hexagonal crystal structure for much higher thickness of up to 22 MLs and for this reason is a good candidate template for silicene and germanene overgrowth.

ZnO was the first wurtzite material where a hexagonal structure was experimentally demonstrated for ultrathin layers of two to three MLs epitaxially grown on Ag(111) [67]. Recently, experimental evidence has been reported [68] by plasma assisted molecular beam epitaxy (MBE) that AlN (~4 ML) also grows epitaxially on Ag(111) adopting a 2D hexagonal structure with a

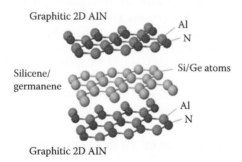

Graphitic 2D AlN

Al
N

Silicene/
germanene

Si/Ge atoms

Al
N

Graphitic 2D AlN

FIGURE 13.14 Graphical illustration of silicene or germanene sandwiched between two graphite-like sp^2-hybridised AlN layers. (Reproduced from A. Dimoulas, *Microelectron. Eng.* 131, 68, 2015.)

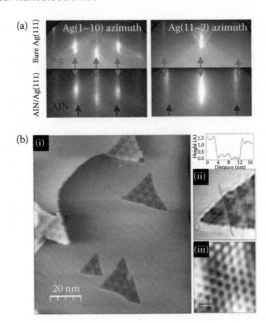

FIGURE 13.15 Experimental evidence for the formation of graphite-like 2D AlN layers serving as insulating substrates for silicene and germanene overgrowth: (a) RHEED patterns indicating the high epitaxial quality of hexagonal AlN grown on Ag(111) and (b) STM images indicating the formation of a hexagonal honeycomb AlN lattice: (i) large scale topography of triangle-like AlN islands", (ii) magnification (down) and height profile (top) of a single island, (iii) atomically-resolved structure of a AlN island. (After P. Tsipas et al., *Appl. Phys. Lett.* 103, 251605, 2013.)

modified electronic valence band structure indicative of a smaller band gap. The latter is in agreement with theory [69] which predicts that hex AlN has a smaller band gap compared to wurtzite. By combining reflection high energy electron diffraction (RHEED) (**Figure 13.15a**) and STM data (**Figure 13.15b**), the lattice constant of AlN on Ag is estimated to be 3.14 Å, which is larger than the bulk wurtzite one, supporting the conclusion that a 2D hexagonal unit cell of AlN is formed on Ag(111) [68]. STM images (**Figure 13.15b**) show triangular shaped AlN layers with a long period contrast modulation which is attributed to the Moiré pattern associated with a coincidence lattice similar to that previously observed for ZnO [67].

13.7.2 Epitaxial Germanene on 2D Hexagonal AlN Substrates

Very recently [70], evidence has been obtained that germanene is grown epitaxially on 2D graphite-like hexagonal AlN insulating layers on Ag(111). RHEED data indicate a faint 4×4 superstructure with respect to (1×1) Ag(111), which reflects a coincidence lattice arrangement according to which three lattice constants of germanene are nearly equal to four lattice spacing on Ag(111) surface. The extended x-ray absorption fine structure (EXAFS) spectroscopy provides compelling evidence of 2D Ge layer formation with an interatomic distance of $d_{Ge-Ge} = 2.38$ Å, characteristic of free-standing germanene [4]. First-principles calculations predict that almost symmetrically buckled germanene with buckling ~0.705 Å and a bond length of 2.38 Å situated 3.83 Å from the AlN surface is the most stable configuration supporting the experimental observations. Ultraviolet photoelectron spectroscopy (UPS) data reveal a faint feature at binding energy of ~1 eV which is associated to the σ bond of germanene predicted by DFT to be ~0.5 eV below the Fermi level. While a full proof of germanene would require the observation of π bands and Dirac fermions, the recent results create hope that germanene can be synthesised on crystalline dielectrics such as hexagonal AlN without hybridising with the substrate.

13.8 Summary

We introduced this chapter in Section 13.1 with some basic questions on the nature of silicene. Now, we are in a position to give some answers and propose research perspectives that were not even considered few years ago when silicene was merely a concept without any practical implication. Silicene is no longer a concept as appears from its elusive free-standing form described in Section 13.2. In Section 13.3, we extensively report on compelling evidences on the formation of an epitaxial silicene on substrates. Are these substrates suitable? It depends for what. For sure, substrates that are commensurate with the theoretically stable free-standing silicene are the only candidates to accommodate a silicene lattice as far as we presently know. Perhaps, a van der Waals epitaxy of silicene can be envisioned but at present there is no solid and unambiguous evidence of it. On the other hand, it becomes clear that substrates hosting silicene have a dramatic role in dictating the structural and electronic properties of silicene. Indeed, silicene grows buckled with a mixed sp^2/sp^3 bonding and the details of the buckling determine a number of silicene superstructures that can be rationalised in a surface phase diagram. Moreover, silicene suffers from a strong hybridisation with its substrate that is responsible for the suppression of the Dirac cones and for the built-in characteristic hybrid bands. These two facts respond to the preliminary questions about the actual hybridisation in the epitaxial silicene and the character of its electronic band structure. Although these aspects bear a nontrivial complexity on the handling of silicene, outstanding opportunities concomitantly rise for silicene exploitation in electronic devices. Indeed, silicene constitutes the ultimate frontier of silicon at the 2D limit in nanoelectronics. In this respect, the present review reports on the recent integration of silicene in a FET operating at room temperature. Opposed to the semiconductor roadmap targeting a post-silicon technology, silicene may bring a renaissance of silicon in nanoelectronics, though in a different allotropic form than the one conventionally known. For sure, the silicene debut opens the routes to a class of epitaxial materials that may fundamentally expand the actual knowledge and functionalities of currently know 2D materials. 2D lattices of group IV semiconductors, not only silicene but also germanene and stanene (i.e. the 2D analogues of Ge and Sn), may actually access new states of matter at room temperature such as the QSH insulator state thus pointing out to the realisation of 'ballistic electron highways' or quantum topological transistors. Although many issues concerned with the material stability and handling are still open, a new era of X-enes (with X being Si, Ge, Sn, etc.) is on the way in the realm of the emerging class of 2D materials beyond graphene.

References

1. The International Technology Roadmap for Semiconductors, http://www.itrs. net/.
2. A. Dimoulas, *Microelectron. Eng.* 131, 68, 2015.
3. K. Takeda and K. Shiraishi, *Phys. Rev. B* 50, 14916, 1994.
4. S. Cahangirov, M. Topsakal, E. Akturk, H. Sahin and S. Ciraci, *Phys. Rev. Lett.* 102, 236804, 2009.
5. C.-C. Liu, W. Feng and Y. Yao, *Phys. Rev. Lett.* 107, 076802, 2011.
6. M. Ezawa, *Phys. Rev. Lett.* 109, 055502, 2012.
7. Y. Xu, B. Yan, H.-J. Zhang, J. Wang, G. Xu, P. Tang, W. Duan and S.-C. Zhang, *Phys. Rev. Lett.* 111, 136804, 2013.
8. C.L. Kane, and E. J. Mele, *Phys. Rev. Lett.* 95, 226801, 2005.
9. K. C. Pandey, *Phys. Rev. Lett.* 47, 1913, 1981.
10. H. Ibach, *Physics of Surfaces and Interfaces*, Springer-Verlag, Berlin, Heidelberg, 2006.
11. N. Takagi, C.-L. Lin, K. Kawahara, E. Minamitani, N. Tsukahara, M. Iwai and R. Arafune, *Prog. Surf. Sci.* 90, 1–20, 2015.
12. M.-C. Desjonquères, and D. Spanjaard, *Concepts in Surface Physics*, 2nd edition, Springer-Verlag, Berlin, Heidelberg, 1996.
13. P. De Padova, C. Quaresima, C. Ottaviani, P.M. Sheverdyaeva, P. Moras, C. Carbone, D. Topwal et al., *Appl. Phys. Lett.* 96, 261905, 2010.
14. P. De Padova, O. Kubo, B. Olivieri, C. Quaresima, T. Nakayama, M. Aono and G. Le Lay, *Nano Lett.* 12, 5500, 2012.
15. G. Brumfiel, *Nature* 495, 152–153, 2013.
16. J. Gao, and J. Zhao, *Sci. Rep.* 2, 861, 2012.
17. P. Pflugradt, L. Matthes and F. Bechstedt, *Phys. Rev. B* 89, 035403, 2014.
18. H. Jamgotchian, Y. Colignon, N. Hamzaoui, B. Ealet, J.Y. Hoarau, B. Aufray and J.P. Biberian, *J. Phys.: Condens. Matter* 24, 172001, 2012.
19. D. Chiappe, C. Grazianetti, G. Tallarida, M. Fanciulli and A. Molle, *Adv. Mater.* 24, 5088, 2012.
20. B. Feng, Z. Ding, S. Meng, Y. Yao, X. He, P. Cheng, L. Chen and K. Wu, *Nano Lett.* 12, 3507, 2012.
21. Z.-L. Liu, M.-X. Wang, C. Liu, J.-F. Jia, P. Vogt, C. Quaresima, C. Ottaviani, B. Olivieri, P. De Padova and G. Le Lay, *APL Mater.* 2, 092513, 2014.
22. P. Moras, T.O. Mentes, P.M. Sheverdyaeva, A. Locatelli and C. Carbone, *J. Phys.: Condens. Matter* 26, 185001, 2014.
23. C. Grazianetti, D. Chiappe, E. Cinquanta, M. Fanciulli and A. Molle, *J. Phys.: Condens. Matter* 27, 255005, 2015.

24. A. Fleurence, R. Friedlein, T. Ozaki, H. Kawai, Y. Wang and Y. Takamura, *Phys. Rev. Lett.* 108, 245501, 2012.
25. L. Meng, Y. Wang, L. Zhang, S. Du, R. Wu, L. Li, Y. Zhang et al., *Nano Lett.* 13, 685, 2013.
26. T. Aizawa, S. Suehara and S. Otani, *J. Phys. Chem. C* 118, 23049, 2014.
27. C.-C. Lee, A. Fleurence, R. Friedlein, Y. Yamada-Takamura and T. Ozaki, *Phys. Rev. B* 90, 241402, 2014.
28. Z.-X. Guo, S. Furuya, J.-I. Iwata and A. Oshiyama, *Phys. Rev. B* 87, 235435, 2013.
29. E. Scalise, M. Houssa, E. Cinquanta, C. Grazianetti, B. van den Broek, G. Pourtois, A. Stesmans, M. Fanciulli and A. Molle, *2D Mater.* 1, 011010, 2014.
30. D. Chiappe, E. Scalise, E. Cinquanta, C. Grazianetti, B. van den Broek, M. Fanciulli, M. Houssa and A. Molle, *Adv. Mater.* 26, 2096, 2014.
31. M.R. Esmaeili-Rad, and S. Salauddin, *Sci. Rep.* 3, 2345, 2013.
32. P. Vogt, P. Capiod, M. Berthe, A. Resta, P. De Padova, T. Bruhn, G. Le Lay and B. Grandidier, *Appl. Phys. Lett.* 104, 021602, 2014.
33. P. De Padova, C. Ottaviani, C. Quaresima, B. Olivieri, P. Imperatori, E. Salomon, T. Angot et al., *2D Mater.* 1, 021003, 2014.
34. T. Shirai, T. Shirasawa, T. Hirahara, N. Fukui, T. Takahashi and S. Hasegawa, *Phys. Rev. B* 24, 891403R, 2014.
35. A.J. Mannix, B. Kiraly, B.L. Fisher, M.C. Hersam and N.P. Guisinger, *ACS Nano* 7, 7538, 2014.
36. P. De Padova, J. Avila, A. Resta, I. Razado-Colambo, C. Quaresima, C. Ottaviani, B. Olivieri et al., *J. Phys.: Condens. Matter* 25, 382202, 2013.
37. B. Feng, H. Li, C.-C. Liu, T.-N. Shao, P. Cheng, Y. Yao, S. Meng, L. Chen and K. Whu, *ACS Nano* 7, 9049, 2013.
38. E. Noguchi, K. Sugawara, R. Yaokawa, T. Hitosugi, H. Nakano and T. Takahashi, *Adv. Mater.* 27, 856, 2015.
39. S. Cahangirov, V. Ongun Ozcelik, A. Rubio and S. Ciraci, *Phys. Rev. B* 90, 085426, 2014.
40. L. Chen, H. Li, B. Feng, Z. Ding, J. Qiu, P. Cheng, K. Wu and S. Meng, *Phys. Rev. Lett.* 110, 085504, 2013.
41. C.-L. Lin, R. Arafune, K. Kawahara, M. Kanno, N. Tsukahara, E. Minamitani, Y. Kim, M. Kawai and N. Takagi, *Phys. Rev. Lett.* 110, 076801, 2013.
42. P. Vogt, P. De Padova, C. Quaresima, J. Avila, E. Frantzeskakis, M.C. Asensio, A. Resta, B. Ealet and G. Le Lay, *Phys. Rev. Lett.* 108, 155501, 2012.
43. D. Tsoutsou, E. Xenogiannopoulou, E. Golias, P. Tsipas and A. Dimoulas, *Appl. Phys. Lett.* 103, 231604, 2013.
44. J. Avila, P. De Padova, S. Cho, I. Colambo, S. Lorcy, C. Quaresima, P. Vogt, A. Resta, G. Le Lay and M.-C. Asensio, *J. Phys.: Condens. Matter* 25, 262001, 2013.
45. L. Chen, C. Liu, B. Feng, X. He, P. Cheng, Z. Ding, S. Meng, Y. Yao and K. Wu, *Phys. Rev. Lett.* 109, 056804, 2012.
46. Z.-X. Guo, S. Furuya, J. Iwata and A. Oshiyama. *J. Phys. Soc. Jpn.* 82, 063714, 2013.
47. Y.-P. Wang, and H.-P. Cheng, *Phys. Rev. B* 87, 245430, 2013.
48. P. Gori, O. Pulci, F. Ronci, S. Colonna and F. Bechstedt, *J. Appl. Phys.* 114, 113710, 2013.
49. S. Cahangirov, M. Audiffred, P. Tang, A, Iacomino, W. Duan, G. Merino and A. Rubio, *Phys. Rev. B* 88, 035432, 2013.
50. M. X. Chen, and M. Weinert, *Nano Lett.* 14, 5189, 2014.
51. S.K. Mahatha, P. Moras, V. Bellini, P.M. Sheverdyaeva, C. Struzzi, L. Petaccia and C. Carbone, *Phys. Rev. B* 89, 201416(R), 2014.

52. X. Xu, J. Zhuang, Y. Du, H. Feng, N. Zhang, C. Liu, T. Lei et al., *Sci. Rep.* 4, 7543, 2014.
53. N. W. Johnson, P. Vogt, A. Resta, P. De Padova, I. Perez, D. Muir, E.Z. Kurmaev, G. Le Lay and A. Moewes, *Adv. Funct. Mater.* 24, 5253, 2014.
54. L. Chen, B. Feng and K. Wu, *Appl. Phys. Lett.* 102, 081602, 2012.
55. A. Molle, C. Grazianetti, D. Chiappe, E. Cinquanta, E. Cianci, G. Tallarida and M. Fanciulli, *Adv. Funct. Mater.* 23, 4340, 2013.
56. E. Cinquanta, E. Scalise, D. Chiappe, C. Grazianetti, B. van den Broek, M. Houssa, M. Fanciulli and A. Molle, *J. Phys. Chem. C* 117, 16719, 2013.
57. J. Zhuang, X. Xu, Y. Du, K. Wu, L. Chen, W. Hao, J. Wang, W. K. Yeoh, X. Wang and S. X. Dou, *Phys. Rev. B* 91, 161409(R), 2015.
58. L. Tao, E. Cinquanta, D. Chiappe, C. Grazianetti, M. Fanciulli, M. Dubey, A. Molle and D. Akinwande, *Nat. Nanotechnol.* 10, 227, 2015.
59. H. Oughaddou, S. Sawaya, J. Goniakowski, B. Aufray, G. Le Lay, J. M. Gay, G. Treglia et al., *Phys. Rev. B* 62, 16653, 2000.
60. E. Golias, E. Xenogianopoulou, D. Tsoutsou, S. Giamini, P. Tsipas and A. Dimoulas, *Phys. Rev. B* 88, 075403, 2013.
61. W. Wang, H.M. Sohail, J.R. Osiecki and R.I.G. Uhrberg, *Phys. Rev. B* 89, 125410, 2014.
62. L. Li, S.-Z. Lu, J. Pan, Z. Qin, Y.-Q. Wang, Y. Wang, C.-Y. Cao, S. Du and H.-J. Gao, *Adv. Mater.* 26, 4820, 2014.
63. M.E. Davila, L. Xian, S. Cahangirov, A. Rubio and G. Le Lay, *New J. Phys.* 16, 095002, 2014.
64. M. Derivaz, D. Dentel, R. Stephan, M.-C. Hanf, A. Mehdaoui, P. Sonnet and C. Pirri, *Nano Lett.* 15, 2510, 2015.
65. M. Houssa, G. Pourtois, V.V. Afanas'ev and A. Stesmans, *Appl. Phys. Lett.* 97, 112106, 2010.
66. C.L. Freeman, F. Claeyssens, N.L. Allan and J.H. Harding, *Phys. Rev. Lett.* 96, 066102, 2006.
67. C. Tusche, H.L. Meyerheim and J. Kirschner, *Phys. Rev. Lett.* 99, 026102, 2007.
68. P. Tsipas, S. Kassavetis, D. Tsoutsou, E. Xenogiannopoulou, E. Golias, S.A. Giamini, C. Grazianetti et al., *Appl. Phys. Lett.* 103, 251605, 2013.
69. C. Zhang, and F. Zheng, *J. Comput. Chem.* 32, 3122, 2011.
70. F. d'Acapito, S. Torrengo, E. Xenogiannopoulou, P. Tsipas, J. Marquez Velasco, D. Tsoutsou and A. Dimoulas, *ImagineNano 2015 Conference*, March 10–13, 2015, BEC (Bilbao Exhibition Centre), Bilbao, Spain.

14

Stannene
A Likely 2D Topological Insulator

*William Vandenberghe, Ana Suarez Negreira
and Massimo Fischetti*

Contents

2D Materials for Nanoelectronics edited by Michel Houssa, Athanasios Dimoulas and Alessandro
Molle © 2016 CRC Press/Taylor & Francis Group, LLC. ISBN: 978-1-4987-0417-5.

Following investigations into monolayer silicon (silicene) and germanium (germanene), interest has recently extended towards monolayer tin ('stannene') [1–3]. In this chapter, we will discuss the interesting physical and electronic features of monolayer tin compared to the other hexagonal lattices of group IV elements. Currently, there are no experimental results on 'stannene' so we limit our discussion to theoretical insights.

In Section 14.1, we analyse the atomic structure of monolayer tin and define 'stannene', 'stannanane' and 'halostannanane'; in Section 14.2, we describe the electronic structure of these monolayers; in Section 14.3, we discuss the topological insulating nature of stannene and related materials; in Section 14.4, we discuss conductivity and mobility in iodostannanane and in Section 14.5, we analyse the electronic properties of stannanane grown on a substrate. In Section 14.6, we present our conclusion and in the appendix, we specify the computational methodology we use. Some sections of text and figures have been taken from References 4 and 5, especially in Sections 14.5 and 14.6.

14.1 Stannene

Bulk tin (lat: *stannum*) has two common allotropes: diamond-like α-tin, with a two-atom basis in a face-centred cubic lattice and β-tin with a two-atom basis in a face-centred tetragonal lattice. β-tin is the most commonly found allotrope but it undergoes a phase transition to α-tin below 13.1°C [6]. In thin layers, however, it was shown experimentally that the α-tin phase is also stable at higher temperatures [7]. To study the possibility of a graphene-like tin allotrope 'stannene', we employ density functional theory (DFT) to determine the lowest-energy hexagonal monolayer tin state [8]. In **Figure 14.1**, we show the

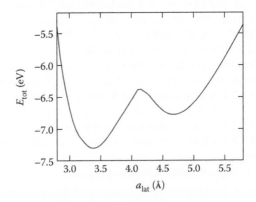

FIGURE 14.1 DFT ground-state energy for relaxed ion positions for hexagonal monolayer tin. Two minima are observed, the first minimum with $a_0 = 3.38$Å corresponds to a high-buckled phase while the second minimum $a_0 = 4.65$Å corresponds to a low-buckled phase.

(a) (b)

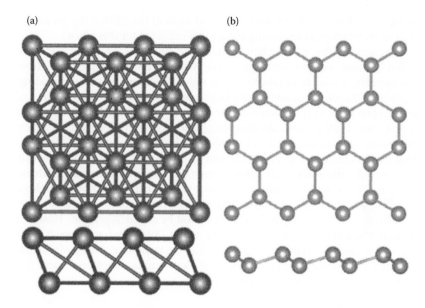

FIGURE 14.2 Top and side view of high-buckled stannene structure (a) and low-buckled stannene structure (b).

energy of the relaxed buckled hexagonal lattice as a function of the lattice constant. The two minima in **Figure 14.1** can be attributed to a low-buckled ($a = 4.65$ Å) and a high-buckled lattice ($a = 3.38$ Å) structure illustrated in **Figure 14.2**. To further verify the stability of these structures, we compute the phonon spectra for the low- and high-buckled phase and present them in **Figure 14.3**.

None of the phonon spectra exhibit negative frequencies indicating that both structures are stable or meta-stable [9]. This is contrary to silicon and germanium, structures for which the high-buckled phase is reported to be unstable [6]. Nevertheless, we mainly focus our discussion on the low-buckled

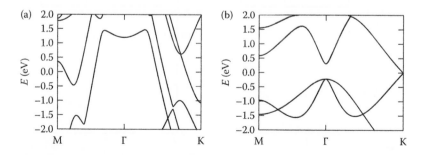

FIGURE 14.3 Band structure of high-buckled (a) and low-buckled stannene (b). The Fermi level lies at $E = 0$.

phase, since most of the current work and most of the interesting properties such as its topological insulating nature, are expected for this phase.

Considering the nomenclature used, the low-buckled phase has been called 'stanene' because each tin is bound to three neighbouring tin atoms and appears similar to graphene at first sight. However, because of buckling, the interpretation of stannene as a lattice of sp^2-bonded tin atoms fails and an interpretation as sp^3-bonded tin atoms with an empty orbital seems more appropriate. Furthermore, the absence of a double bond makes the classification as an '-ene' in the strict sense unjustified but to maintain compatibility with the literature and given the relaxed attribution of the 'ene' suffix within the scientific community (cfr. phosphorene [10]), we will maintain 'stannene' as a name for the low-buckled carbon phase in this chapter. Following the logic of calling low-buckled monolayer tin stannene, functionalised monolayer tin with hydrogen on the top and bottom should be called 'stannane'. However, stannane is a common gas (SnH_4), like silane (SiH_4) and germane (GeH_4). For this reason, we refer to monolayer tin functionalised with hydrogen as 'stannanane', stannene functionalised with halogens as 'halostannanane' and stannene functionalised with iodine as 'iodostannanane'.

14.2 Electronic Structure

Figure 14.4 shows the high-buckled hexagonal monolayer tin and stannene electronic band structure calculated using DFT without incorporating the effects of spin–orbit coupling (SOC). This crucial interaction will be considered shortly.

The high-buckled phase is metallic. Inspecting the crystal structure in **Figure 14.2**, it can be seen that in the high-buckled phase, each tin atom has six nearest neighbours in the same layer and three nearest neighbours in the opposite layer, making each tin atom have a nine-fold coordination.

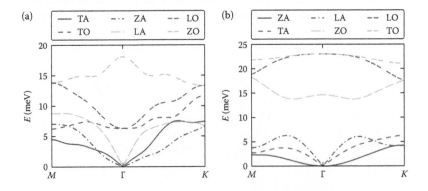

FIGURE 14.4 Phonon dispersion of the high-buckled phase (a) and the low-buckled phase (b).

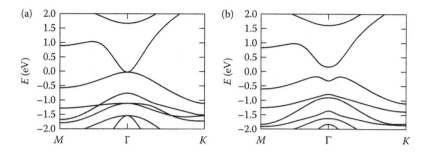

FIGURE 14.5 Band structure of iodine-functionalized monolayer tin (iodostannanane) calculated without spin-orbit coupling (a) and with spin-orbit coupling (b).

The nine-fold rather than the four-fold sp^3 coordination explains the metallic nature of high-buckled tin.

The low-buckled phase exhibits a dispersion resembling the graphene band structure with its electronic properties near the Fermi level determined by a Dirac cone at the K-point. The stannene Fermi velocity is calculated to be 5×10^6 m/s, about half of the graphene Fermi velocity.

The effects of SOC are known to lift the degeneracy at the K-point in graphene but this effect is small [11] because carbon is a light atom and the SOC is weak. For stannene, which is instead composed of heavy tin atoms, the effect is large and a gap of 0.07–0.1 eV is predicted by DFT accounting for SOC [1,2]. Moreover, graphene was the first material to be shown to be a quantum spin Hall insulator, a topological insulator. Like graphene, stannene is a topological insulator, as will be discussed in the following section.

The calculated band structure of stannanane and iodostannanane is shown in **Figures 14.5** and **14.6**. Stannanane is a trivial insulator with a 0.25 eV band gap making it an interesting low band-gap two-dimensional (2D) material. Iodostannanane is gapless with a two-fold degenerate state at Γ in the absence of SOC and a topological gap of 0.35 eV when accounting for SOC. An overview of the calculated band gaps for different functionalisations is given in **Figure 14.7**.

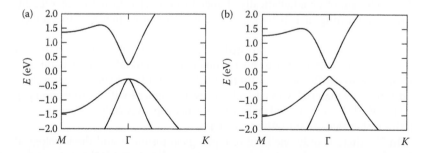

FIGURE 14.6 Band structure of hydrogen-functionalized monolayer tin (stannanane) calculated without spin-orbit coupling (a) and with spin-orbit coupling (b).

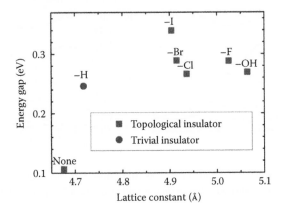

FIGURE 14.7 Band gap for different functionalized stannanane ribbons. (Reprinted with permission from Y. Xu et al., *Phys. Rev. Lett.* 111, 136804, 2013. Copyright 2013, by the American Physical Society.)

14.3 Topological Classification of Stannene and Stannananes

In this section, we introduce the concept of a topological insulator and show how stannene and halostannanane are topological insulators.

To understand the concept of a topological insulator, consider the gradual (adiabatic) change of one material to another or the gradual change of the crystalline structure of a material, and the effect this has on its band structure. Theoretically, a silicon lattice can be changed into a germanium lattice by gradually replacing each silicon atom by a germanium atom and the band gap of the lattice will evolve gradually from that of silicon to that of germanium. Alternatively, a silicon lattice can be theoretically strained gradually until the bonding between the silicon atoms vanishes and the band structure resembles that of vacuum. This is only possible because silicon and germanium are topologically classified as 'trivial insulators'. However, there are also materials which cannot be adiabatically changed into trivial insulators or vacuum while maintaining their band gap. These materials are known as 'topological insulators'.

For our purpose, we are interested in the classification of 2D materials with time-reversal symmetry (e.g. no ferromagnetic materials) and a band gap. Also, to avoid confusion, we want to clarify that for topological-classification purposes, any material with a band gap, even graphene with its tiny band gap (≈ 24 μeV) [11], is considered an insulator, since topology only deals with qualitative rather than quantitative properties.

In 2D, there is only one kind of topological insulator and the topological classification is done based on a quantity ν equaling 0 or 1 [12]. ν is more formally referred to as a Z_2 invariant and materials with $\nu = 0$ are trivial insulators and materials with $\nu = 1$ are topological insulators.

If a description of the bulk band structure is available and the Bloch functions $|u_n(\mathbf{k})\rangle$ are known throughout the 2D Brillouin zone, ν can be calculated explicitly using

$$(-1)^\nu = \prod_{a=1}^{4} \delta_a$$

where $\delta_a = \text{Pf}[w(\Delta_a)]/\sqrt{\text{Det}[w(\Delta_a)]}$ with $\text{Pf}[w(\Delta_a)]$ the Pfaffian and $\text{Det}[w(\Delta_a)]$ the determinant of the unitary matrix $w_{mn} = \langle u_m(\mathbf{k})|\Theta|u_n(\mathbf{k})\rangle$, where the index m and n runs over all occupied bands [13]. Θ is the time-reversal operator and Δ_a are the four points in the general 2D Brillouin zone, where $\mathbf{k} = -\mathbf{k}$. These are the three M-points and the Γ-point in a hexagonal lattice. The evaluation of ν is greatly simplified when inversion symmetry is present in the system, in this case,

$$\delta_a = \prod_m \xi_m(\Delta_a)$$

with $\xi_m(\Delta_a) = \pm 1$, the parity with respect to inversion of each Bloch function.

The most interesting and characteristic property of topological insulators is their edge states. These are topologically protected because any transition to a trivial insulator such as vacuum, has to be accompanied by the closing of the band gap. Furthermore, time-reversal symmetry mandates that edge states with opposite momenta also have opposite spin polarisation. This has led to the measurement of quantised conductance in CdTe/HgTe/CdTe topological insulators [14]. However, while the presence of edge states is topologically protected, the quantised conductance is not and scattering with magnetic impurities, the hyperfine interaction, inter-edge scattering, inelastic scattering and several other processes can destroy the quantised conductance.

To prove that graphene, stannene and iodostannanane are topological insulators, whereas the stannanane band structure is topologically trivial, we analyse the symmetry of the different bands obtained from DFT at Γ, M and K and show the result in **Table 14.1**. DFT only treats the four valence electrons of carbon and tin and the seven valence electrons of iodine and this results in eight electrons per unit cell for graphene and stannene, 10 for stannanane and 22 for iodostannanane. Knowing that the representations subscripted by u and g are odd and even under inversion $(\delta_\Gamma, \delta_M)_{\text{graphene}} = (\chi(A_{1g}) \times \chi(A_{2u}) \times \chi(E_{2g}) \times \chi(E_{2g}), \chi(B_{2u}) \times \chi(A_g) \times \chi(B_{3u}) \times \chi(B_{3g})) = (-1,1)$ and similarly $(\delta_\Gamma, \delta_M)_{\text{stannene}} = (-1,1)$ making $\delta_\Gamma \delta_M^3 = -1$ and $\nu_{\text{graphene}} = \nu_{\text{stannene}} = 1$. For stannanane, $(\delta_\Gamma, \delta_M)_{\text{stannanane}} = (-1,-1)$ and $\nu_{\text{stannanane}} = 0$ and for iodostannanane, $(\delta_\Gamma, \delta_M)_{\text{halostannanane}} = (-1,1)$ and $\nu_{\text{iodostannanane}} = 1$.

An alternative way to show that these materials are topological insulators is to compute the band structure of a ribbon. An iodostannanane ribbon band structure is shown in **Figure 14.8** and the ribbon band structure resembles the bulk band structure shown in **Figure 14.5** but has no band gap because of the two edge states traversing the band gap.

A last question we investigate is whether the classification of the topological nature is maintained when improved band structure calculation methods are

Table 14.1 Symmetry of the Bloch Functions of Graphene, Stannene, Stannanane and Halostannanane

Graphene (8 Electrons)

Γ (D_{6h})	A_{1g}	A_{2u}	E_{2g}	E_{2g}	A_{1g}
M (D_{2h})	B_{2u}	A_g	B_{3u}	B_{3g}	B_{1u}
K (D_{3h})	E'	E'	A_1'	E''	E''

Stannene (8 Electrons)

Γ (D_{3d})	A_{1g}	A_{2u}	E_g	E_g	A_{2u}
M (C_{2h})	B_u	A_g	A_u	A_g	B_u
K (D_3)	E	E	A_1	E	E

Stannanane (10 Electrons)

Γ (D_{3d})	A_{1g}	A_{2u}	A_{1g}	E_g	E_g	A_{2u}
M (C_{2h})	B_u	A_g	A_g	B_u	A_u	A_g
K (D_3)	E	E	E	E	A_1	E

Iodostannanane (22 Electrons)

Γ (D_{3d})	A_{1g}	A_{2u}	A_{1g}	A_{2u}	A_{1g}	E_g	E_g	E_u	E_u	A_{2u}	E_g	E_g
M (C_{2h})	B_u	A_g	B_u	A_g	A_g	B_u	A_u	B_u	A_g	B_g	A_u	A_g
K (D_3)	E	E	E	E	A_1	E	E	A_2	E	E	A_1	E

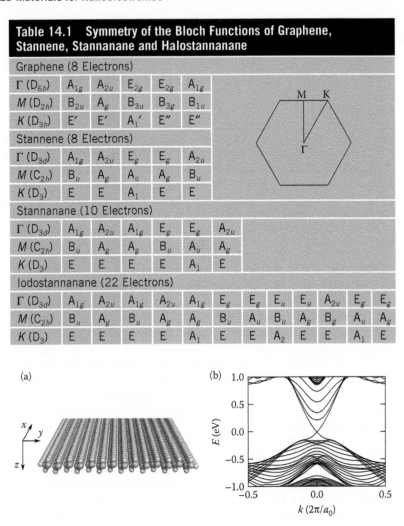

FIGURE 14.8 Illustration of the 14-unitcell zigzag iodostannanane ribbon (a) and its band structure (b). The ribbon band structure resembles the bulk band structure shown in **Figure 14.5** folded along the Γ–M direction apart from a 2-fold degenerate band associated with both edges closing the bandgap. (Reprinted with permission from W. Vandenberghe and M. Fischetti, Calculation of room temperature conductivity and mobility in tin-based topological insulator nanoribbons. *J. Appl. Phys.*, 116, 173707, 2014. Copyright 2014, American Institute of Physics.)

used, since it is well-known that band structure results obtained using DFT do not reliably predict the band gap. We perform a G_0W_0 analysis for unsupported stannanane and halostannanane and observe that the band order is maintained while the gap at Γ increases from 0.48 to 0.9 eV for iodostannanane and from 0.27 to 1.0 eV for stannanane. These results indicate that the use of generalised-gradient approximation (GGA) DFT is a good indicator of the topological nature of a structure.

14.4 Conductivity and Mobility in Iodostannanane Ribbons

In this section, we present our results obtained in Reference 15 where the conductivity and mobility are calculated in iodostannanane ribbons.

As outlined in Section 14.3, iodostannanane's topologically non-trivial band structure leads to topologically protected edge states. Thanks to time-reversal symmetry, intra-edge elastic backscattering is prohibited, whereas inelastic backscattering is very weak because of the spin polarisation of the edge states [16,17]. The absence of backscattering leads to a quantised conductance (G) [14,18] corresponding to a very high conductivity ($\sigma_{1D} = GL$, L being the length of the insulator) which makes topological insulators (TIs) interesting for electronic applications. Of course, in practice scattering with magnetic impurities, the hyperfine interaction [19], scattering with bulk or other edge states, inelastic scattering or scattering to the opposite edge will lead to a finite conductivity.

In addition to quantum computing applications [20,21], TIs have been envisioned as active elements in classical electronics in two possible ways: (1) as highly conductive interconnects and (2) as a spin-polarising materials for spintronics [22].

Here, we perform a theoretical study of the upper limit of the conductivity of stannanane as limited by phonon-mediated inter-edge scattering at room temperature. We show that ribbons of TIs have a high conductivity and mobility when the Fermi level lies in the bulk band gap, but the conductivity decreases as the Fermi level moves out of the bulk band gap. This high conductivity and the ability to modulate it suggest that 2D TIs can also be of high interest for classical computing, even without the need to of a non-zero band gap [23].

To visualise how backscattering occurs for the edge states, we show their wave functions in **Figure 14.9**. In **Figure 14.10**, we show the overlap of the wavefunctions at opposite edges for different ribbon widths. The reduced overlap for wider ribbons gives rise to a high conductivity and mobility. To obtain a quantitative estimate of the conductivity, we wish to use the Kubo–Greenwood formalism [24–26]. However, to enable the calculation of the lifetime required for the Kubo–Greenwood formalism, we first require an expression for the electron–phonon interaction.

14.4.1 Electron–Phonon Interaction

An estimate of the strength of the electron–phonon interaction can be obtained by calculating the deformation potentials for the conduction band of bulk iodostannanane: the Hartree potential for all independent displacements of each atom in the unit cell is calculated from the Vienna *ab initio*

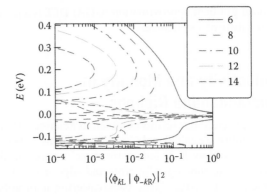

FIGURE 14.9 The wavefunction overlap as a function of energy for different ribbon width. For most E, the wavefunction overlap decreases exponentially as the ribbon width increases but in a narrow region near $E = 0$ ($k = 0$) where $k \approx -k$, there is no distinction between left- and right-edge waves and the overlap is close to unity. On the other hand, the states near $k = 0$ will not affect the conductivity since the electron velocity dE/dk vanishes. (Reprinted with permission from W. Vandenberghe and M. Fischetti, Calculation of room temperature conductivity and mobility in tin-based topological insulator nanoribbons. *J. Appl. Phys.*, 116, 173707, 2014. Copyright 2014, American Institute of Physics.)

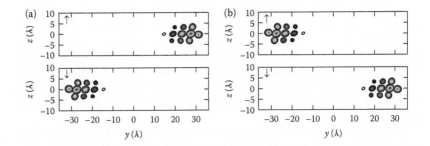

FIGURE 14.10 Illustration of the magnitude of the spin components for both left and right edge states at $k^+ = 0.0025 \times 2\pi/a > 0$ in the conduction band state ($E > 0$) associated with the left edge (a) and both the left and right edge state for $k^- = -k^+ < 0$ (b) for the 14-cell wide ribbon. The quantity plotted is the squared magnitude averaged along the x-direction (as indicated in **Figure 14.8**) and only areas where the averaged squared magnitude exceeds 2% of the maximum averaged squared magnitude are coloured. The centre of the ribbon coincides with the centre of the figure. Back-scattering between the left-edge states is prohibited because of time-reversal symmetry but back-scattering between opposite edge states is allowed although strongly suppressed for wide ribbons. (Reprinted with permission from W. Vandenberghe and M. Fischetti, Calculation of room temperature conductivity and mobility in tin-based topological insulator nanoribbons. *J. Appl. Phys.*, 116, 173707, 2014. Copyright 2014, American Institute of Physics.)

simulation package (VASP), as done in References 27, 28 and 29. Using the electronic wavefunctions and the ionic displacement (polarisation) vectors $(e_{q,i}^v)$ (v is the phonon branch index, q is the 2D phonon wave vector and i is the ion index), from the phonon dispersion calculation, the deformation potential is calculated from

$$DK_{k,q}^v = \sum_{\substack{r,i \\ s=\uparrow,\downarrow}} \phi_{k+q,s}^*(r)e^{iq\cdot(r-R_i)}\phi_{k,s}(r)\sqrt{\frac{M}{M_i}}e_{q,i}^v \cdot \frac{\partial V_H(r)}{\partial R_i}$$

where M_i and M are the masses of the ions and of the unit cell, respectively, $V_H(r)$ is the Hartree part of the local potential, and the sum is performed over all positions in a suitable real-space mesh and over both spin components. The largest deformation potential at small momenta is due to the phonon branch corresponding to the out-of-phase vibrations of the tin–iodine bond with an energy of 24.5 meV.

However, since both the magnitude of the displacement vector and the phonon occupation number are inversely proportional to the phonon frequency, it is the ratio $|DK^v|^2/\omega^2$ that determines which phonons cause the largest scattering rate. In the limit $q \to 0$, this ratio is two orders of magnitude larger for the transverse and longitudinal acoustic phonons than for the optical phonons and for the out-of-plane acoustic phonons. Thus, one has only to consider scattering with transverse and longitudinal acoustic phonons and the computed value for the deformation potentials $D_{LA,TA} = dDK/dq = 27$ eV.

The bulk 2D electron–phonon Hamiltonian in the deformation potential approximation reads

$$H_{ep} = \sum_q D|q|\sqrt{\frac{\hbar}{WL2\rho\omega_q}}e^{iq\cdot r}(\hat{a}(q) + \hat{a}^\dagger(-q))$$

where D is the deformation potential, W and L are the ribbon width and length, respectively, ρ is the 2D mass density, ω_q is the phonon frequency and $\hat{a}(q)$ and $\hat{a}^\dagger(-q)$ are the phonon annihilation and creation operators. In the elastic approximation, the transition probability due to phonon scattering is

$$W_{k_xk_x'} = \frac{2\pi}{\hbar}\sum_q \frac{\hbar D^2|q|^2}{WL2\rho\omega_q}I_{nkn'k'q}(1 + 2v(\omega_q))\delta(E_{nk} - E_{n'k'})$$

with $I_{nkq} = |<nk|e^{iq\cdot r}|n'k'>|^2$. For the acoustic phonons $\omega_q \approx v_s|q|$ and the Bose–Einstein distribution $v(\hbar\omega_q) \approx (k_BT/\hbar\omega_q) \gg 1$ which yields

$$W_{nk_xn'k_x'} \approx \frac{2\pi}{\hbar}\sum_q \frac{D^2 2k_BT}{WL2\rho v_s^2}I_{nkn'k'q}\delta(E_{nk} - E_{n'k'}).$$

The momentum relaxation rate is computed by

$$\tau_n^{-1}(k) = \sum_{n'k_x'} W_{nk_x n'k_x'} \frac{v_n(k) - v_{n'}(k')}{v_n(k)}$$

$$= \sum_{\alpha} \frac{2\pi D^2 k_B T}{\hbar \rho v_s^2} \left(\frac{dE_n(k_x)}{dk_x} \right)^{-1} \Bigg|_{nk_x = \alpha} \frac{v_n(k) - v_{n'}(k')}{v_n(k)} \sum_{q_y} \left| \int dr^3 u_\alpha^*(r) e^{iq_y y} u_{nk_x}^*(r) \right|^2$$

where α is an index and is summed over all $n'k_x' \neq nk_x$ which satisfy $E_\alpha = E_{nk}$ and $\rho_{1D} = \rho W = MN_y/a$ with M is the mass of the unit cell, N_y is the number of unit cells in the ribbon and a is the unit cell length along the x-direction. For our ribbons, we sum over $q_y = n2\pi/W$ with n going from $-\lfloor(N_y - 1)/2\rfloor \to \lfloor N_y/2 \rfloor$; this corresponds to imposing Born–von Kármann boundary conditions on the phonon displacement.

In our calculations, we use $a = 4.9$ Å as the ribbon lattice constant. The momentum relaxation rate is calculated as the sum of a LA and TA scattering rate: $\tau^{-1} = \tau_{LA}^{-1} + \tau_{TA}^{-1}$. Each scattering rate $\tau_{TA,LA}^{-1}$ is calculated with its own deformation potential, $D_{TA,LA} = 27$ eV, and its own sound velocity, $v_{LA} = 1.4$ km/s and $v_{TA} = 2.3$ km/s.

A more rigorous approach to deal with the phonon dispersion and electron-phonon coupling would either involve the calculation of the ribbon phonons from first principles or to impose free-floating boundary conditions, as done in Reference 30. However, the main physics will not be drastically affected and the use of bulk phonons and imposing Born–von Kármann boundary conditions suffices here.

14.4.2 Calculation of Conductivity and Mobility

In order to evaluate the electron conductivity, we use the Kubo–Greenwood formula:

$$\sigma = \frac{2e}{k_B T} \sum_j dk \tau_j(k) v_j(k) f(E_j(k))(1 - f(E_j(k)))$$

where $f(E)$ is the Fermi–Dirac distribution whose Fermi level E_F is taken as a parameter, the temperature T is taken as 300 K, j is the band index, $v_j(k)$ is the electron-velocity computed from the band structure as $v_j(k) = dE_j/d(\hbar k)$ and the factor 2 results from considering left- and right-edge degeneracy.

In **Figure 14.11**, we show the conductivity as a function of the Fermi level, E_F. The conductivity increases as the ribbon width increases due to the reduced overlap between the edge states reaching a maximum when the Fermi level is in the bulk band gap for all ribbons wider than eight unit cells. However, since the group velocity is higher near the conduction-band minimum than near the valence-band maximum, the largest conductivity is seen when E_F is closer to the bulk conduction-band minimum than to the bulk valence-band maximum.

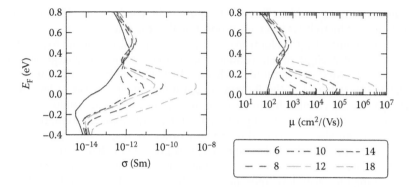

FIGURE 14.11 One-dimensional conductivity and mobility as a function of Fermi level for different ribbon widths, width is expressed in number of unit cells. The conductivity and mobility reach a maximum when the Fermi level lies in the bulk bandgap. For wide ribbons, the mobility is very high and exceeds 10^6 cm²/ Vs. (Reprinted with permission from W. Vandenberghe and M. Fischetti, Calculation of room temperature conductivity and mobility in tin-based topological insulator nanoribbons. *J. Appl. Phys.*, 116, 173707, 2014. Copyright 2014, American Institute of Physics.)

A more familiar figure-of-merit for electronic performance is the mobility. This can be obtained from the conductivity and the conduction-band charge density,

$$\mu = \frac{\sigma}{E_F \rho(E_F)} \quad \text{and} \quad \rho(E_F) = \int \frac{dk}{2\pi} f(E_c(k)).$$

Figure 14.11 shows that the phonon-limited mobility of iodostannanane for the widest ribbon exceeds the graphene mobility $\approx 10^6$ cm²/Vs [31].

With this high conductivity and mobility, it is important to remember that competing scattering processes such as impurity scattering or edge-roughness scattering, will also be strongly suppressed for wide ribbons. But the current in iodostannanane will also never exceed the ballistic limit and for short ribbons, conductance rather conductivity will limit the current. Nevertheless, as the ribbon lengths increase, diffusive transport will always dominate.

The physical processes that control the electron–phonon-limited conductivity in 2D TI ribbons differ sharply from those that control the conductivity of conventional semiconductors. In ordinary semiconductors, the conductivity is approximately linearly proportional to the density of free carriers, n, via $\sigma = ne\mu$, and the mobility is mainly determined by the electron velocity (in addition to the scattering rates, obviously). In 2D TIs, instead, the conductivity is determined by inter-edge scattering and so by the overlap factor, resulting in a behaviour that is extremely (exponentially) sensitive to the width of the ribbon. Furthermore, as the Fermi level increases towards the conduction-band minimum, the free-carrier density in the conduction band increases but, simultaneously, the carrier velocity decreases and scattering towards bulk states becomes possible, effects that suppress the conductivity. Therefore, in TIs, the conductivity is inversely proportional to the charge density.

When considering TIs for transistor applications, it is important to remember the great strength of semiconductors: the ability to modify charge density over many orders of magnitude with a small change in bias. Ideally, for every $\log(10)kT/e$ (where k is the Boltzmann constant) change in Fermi level, the charge density changes by an order of magnitude. Since the conductivity of semiconductors is linearly proportional to the charge concentration, the conductivity can also be changed over many orders of magnitude. Thus, viewing this from a topological perspective, it is possible to change an ordinary semiconductor from an insulator to a metal by applying an external bias and so moving the Fermi level into the conduction band. This is exactly the basic working principle of the metal–oxide–semiconductor field-effect transistor (MOSFET).

Here, we have demonstrated a similar concept: the mobility of TI ribbons is also strongly (exponentially) dependent on the position of the Fermi level, although in the opposite way, and it is possible to modulate the conductivity over three orders of magnitude. This finding opens the idea of a TI-based MOSFET in which a transition from a low- to high-conductivity state can be obtained by shifting the Fermi level towards the middle of the TI bulk band gap via the gate bias. A possible advantage of this alternative gating scheme consists in the fact that only a small amount of charge needs to be displaced to obtain the highly conductive state, thus suggesting high-speed switching thanks to the high velocity of the charge carriers.

14.5 Supported Stannene and Halostannanane

In this section, we present our analysis presented in Reference 5 of the stability of stannene and halostannanane and the effect of the support on the band structure.

14.5.1 Geometry and Band Structure

The optimisation of unsupported stannene (Sn-ML) and halostannanane (Sn-ML-X) systems is performed first to obtain a reference state for the supported monolayers. The crystal structure of the stannene and halostannananes with their respective band structures is shown in **Figure 14.12**. The optimisation of the 2D hexagonal lattice of stannanane results in a lattice constant of 4.70 Å, a Sn–Sn bond distance of 2.84 Å and a buckling distance of 0.84 Å, values that are in agreement with previous DFT studies [2,32,33]. As shown previously, the SOC induced minimum band gap is at the K symmetry point for unfunctionalised stannanane, whereas upon functionalisation with halogens it is at the Γ symmetry point and has an increased magnitude (from 0.07 to 0.3 eV), due to the saturation of the π-orbital [1,2].

The stability of the halostannananes is studied through the formation energies calculated as

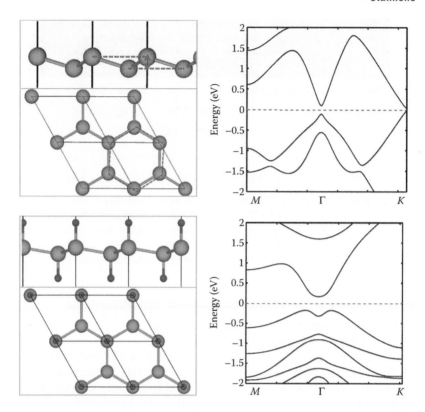

FIGURE 14.12 Top: Side- and top-view of the crystal structure (left) and band structure (right) calculated accounting for SOC for pristine unsupported stannanane. Bottom: The same, but for stannanane functionalized with I. The horizontal dashed lines (red online) in the right frames indicate the Fermi level. (Reprinted with permission from A. Suarez Negreira, W. Vandenberghe and M. Fischetti, *Phys. Rev. B* 91, 245103, 2015. Copyright 2015, by the American Physical Society.)

$$E_{\text{form}} = \frac{E_{\text{Sn-ML-X}} - (E_{\text{Sn-ML}} + N_X E_X)}{N_X}$$

where $E_{\text{Sn-ML-X}}$, $E_{\text{Sn-ML}}$ and E_X are the total energy of a double-side halogenated Sn-ML (as shown in **Figure 14.8**), the energy of the pristine Sn-ML and the binding energy per atom of a halogen molecule, respectively. The formation energies, changes of the lattice constant of the Sn-MLs and of the SOC-induced band gaps upon halogen chemisorption are summarised in **Table 14.2**.

As shown in **Table 14.2**, the binding strength of halogens on unsupported Sn-MLs and the geometrical strain generated on the film structure (as indicated by the variation of its lattice constant) decrease from F to I. For the electronic structure, I-chemisorption leads to the largest band gap, in agreement with previous theoretical results [1,2].

Table 14.2   Calculated Structural Parameters and Binding Energies of Pure and Halogenated Sn-ML Systems

System	Bandgap (eV)	E_{form} (eV)	Lattice Constant (Å)
Sn-ML	0.07(0.07)	NA	4.70(4.67)
Sn-ML-F	0.29(0.37)	−3.38(−3.34)	5.03(4.94)
Sn-ML-Cl	0.29(0.35)	−1.80(−1.68)	4.94(4.93)
Sn-ML-Br	0.29(0.39)	−1.58(−1.46)	4.89(4.91)
Sn-ML-I	0.34(0.48)	−1.25(−1.13)	4.89(4.89)

Source: Y. Ma et al., *The Journal of Physical Chemistry* C, vol. 116, no. 23, pp. 12977–12981, 2012; A. Suarez Negreira, W.G. Vandenberghe and M.V. Fischetti, *Physical Review B*, vol. 91, p. 245103, 2015.

Note: The data in parentheses correspond to previous calculations.

14.5.2 Impact of the Support on Stannene and Stannanane

14.5.2.1 Optimisation and Stability of Substrates: CdTe(111), InSb(111) and Si(111)

The three materials chosen as possible substrates for the Sn-ML films are Si(111), CdTe(111) and InSb(111). We choose the (111) surface because it is commensurate with the hexagonal lattice of the Sn-ML, as illustrated in **Figure 14.12**. The lattice constants resulting from the optimisation of the bulk substrate (6.63 Å for CdTe, 6.65 Å for InSb and 5.47 Å for Si) are all in agreement with previous DFT studies (6.63 Å [34], 6.64 Å [35] and 5.47 Å [36] for CdTe, InSb and Si, respectively).

The determination of the minimum number of atomic layers needed to obtain an accurate representation of a condensed phase must be established by judging the change of interlayer distances of the (111) slab with respect to the bulk value. The interlayer spacing of the three-, four-, five- and six-layer slabs and their difference with respect bulk distances (in %) is summarised in **Table 14.3**. We conclude that the use of slabs consisting of five layers (schematically represented in **Figure 14.13**) yields sufficient accuracy, since they exhibit an interlayer difference smaller than 5%.

The CdTe and InSb(111) surfaces can be terminated in two different ways: Cd- and Te-terminated surfaces for CdTe(111) and In- and Sb-terminated surfaces for InSb(111). The surface free energy of each termination dictates their relative stability and it is used as a selection factor. The surface free energy is proportional to the Gibbs free energy, and the latter can be approximated by the total DFT energy, as described in Reference 37. The surface free energy, γ, of the CdTe and InSb semi-infinite slabs with one surface is given by

$$\gamma = \frac{1}{A}\left[E_{syst} - N_{Cd}E_{CdTe}^{bulk} - (N_{Te} - N_{Cd})\mu_{Te} - N_H\mu_H\right] \quad (14.1)$$

Table 14.3 Change of the Interlayer Distance (in %) of Substrate Slabs with Increasing Number of Layers with Respect the Bulk Distances for InSb(111), CdTe(111) and Si(111) Slabs

Interlayer	InSb(111)				CdTe(111)				Si(111)			
Change (%)	6 lyrs.	5 lyrs.	4 lyrs.	3 lyrs.	6 lyrs.	5 lyrs.	4 lyrs.	3 lyrs.	6 lyrs.	5 lyrs.	4 lyrs.	3 lyrs.
1st–2nd lyr.	−0.3	−0.27	−0.3	−0.33	0.01	0.03	0.02	0.03	0.75	0.4	2.85	0.21
2nd–3rd lyr.	0.05	−0.05	−0.16	−0.45	0.1	−0.16	−0.17	−0.02	0.31	0.39	5.19	−3.84
3rd–4th lyr.	0.03	−0.06	−0.48	–	0.14	0.09	0.23	–	0.29	0.53	−2.26	–
4th–5th lyr.	−0.19	−0.4	–	–	0.26	0.27	–	–	0.61	−3.45	–	–
5th–6th lyr.	−0.53	–	–	–	0.46	–	–	–	−3.59	–	–	–

Source: Adapted from A. Suarez Negreira, W.G. Vandenberghe and M.V. Fischetti, *Physical Review B*, vol. 91, p. 245103, 2015.

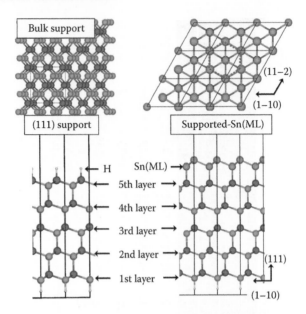

FIGURE 14.13 Left: Atomic structure of bulk and the (111) surface of a 5-layer-slab of the substrate materials: CdTe, InSb and Si. The pink and green balls represent the substrate atoms: Cd and Te for CdTe and In and Sb for InSb. For the case of Si, there is only one colour. Hydrogen atoms (small white balls) saturate the dangling bonds of top and bottom layer of the substrates. Right: Top- and side-view of the supported Sn-ML systems. The dashed hexagonal lattices correspond to the Sn monolayer (grey balls). (Reprinted with permission from A. Suarez Negreira, W. Vandenberghe and M. Fischetti, *Phys. Rev. B* 91, 245103, 2015. Copyright 2015, by the American Physical Society.)

and

$$\gamma = \frac{1}{A}\Big[E_{\text{surf}} - N_{\text{Cd}}E_{\text{InSb}}^{\text{bulk}} - (N_{\text{Sb}} - N_{\text{In}})\mu_{\text{Sb}} - N_{\text{H}}\mu_{\text{H}}\Big]\gamma \qquad (14.2)$$

where E_{surf} is the total energy of the five-layer slab, $E_{\text{CdTe}}^{\text{bulk}}$ and $E_{\text{InSb}}^{\text{bulk}}$ are the energy of bulk CdTe and bulk InSb per formula unit and μ_{Te}, μ_{Sb} and μ_{H} are the chemical potentials of Te, Sb and H, respectively. Since stoichiometric slabs are used as a substrate ($N_{\text{Te}} = N_{\text{Cd}}$ and $N_{\text{In}} = N_{\text{Sb}}$), these terms cancel in **Equations 14.1** and **14.2**. For the case of zero temperature, the μ_{H} simplifies to 1/2 E_{H2}.

The surface free energies obtained are −0.07 eV/Å² for an In-terminated InSb substrate (InSb(In)), 0.05 eV/Å² for a Sb-terminated InSb substrate (InSb(Sb)), −0.17 eV/Å² for a Cd-terminated CdTe substrate (CdTe(Cd)) and 0.01 eV/Å² for a Te-terminated CdTe substrate (CdTe(Te)). Owing to their lower surface energy and the resulting higher stability, we choose the InSb(In) and CdTe(Cd) terminations as substrates, along with Si(111).

14.5.2.2 Characterisation of Supported Stannene
14.5.2.2.1 Geometry and Thermodynamic Stability

For thick layers, the different interactions between ad-species and substrate atoms, and the buildup of strain in the overlayer as the film thickness increases, produce dislocations in the supported film. However, since our aim is to study monolayer Sn films, an epitaxial growth is assumed in which the structures of the CdTe(111), the InSb(111) and the Si(111) substrates are extended to a Sn(111) termination (i.e. the atoms in the Sn-ML follow the structure of the underneath substrate). When the lattice constant mismatch between the film and the substrate is small, even multiple layers can be grown homoepitaxially and dislocation free [38]. The relative stability of each supported Sn-ML structure is analysed based on the binding energies, using the (1 × 1) five-layer H-passivated bottom-layer substrate and an unsupported Sn-ML as references. In the supported Sn-ML structure, only the bottom layer is passivated with fictitious H atoms, since the dangling bonds of the atoms of the top layer of the substrate are saturated with the Sn atoms. The binding energy of these supported Sn structures, $E_{binding}$, is defined as

$$E_{binding} = E_{sys} - (E_{subs} + E_{Sn\text{-}ML})$$

where E_{sys}, E_{subs} and $E_{Sn\text{-}ML}$ are the total energies of the supported system, the H-passivated bottom-layer substrate and the unsupported monolayer Sn, respectively. The binding energies and the structural parameters of the unsupported and supported Sn-ML systems are summarised in **Table 14.4**.

This table shows the predicted lattice strain on the Sn-ML film during epitaxial growth on each substrate. The binding energy is used to predict the relative stability of the different supported systems, suggesting InSb(In)-Sn ($E_{binding} = -0.91$ eV) as the most promising candidate. The much larger lattice mismatch between the Si(111) substrate and the Sn-ML is responsible for the lower stability of the epitaxial growth of Sn on Si(111). The predicted lattice mismatch ($\approx 18\%$) is in good agreement with the lattice mismatch measured experimentally (19.5%) [39]. The in-plane compression induced by the Si(111)

Table 14.4 Calculated Lattice Constant a_0, Lattice Strain ε, Buckling Distance and Binding Energy of Supported Sn-ML Systems

System	a_0(Å)	ε (%)	Bucking (Å)	$E_{binding}$ (eV)
Unsupported Sn	4.70	–	0.84	–
Si-Sn	3.86	−17.92	1.96	−0.10
CdTe(Cd)-Sn	4.69	−0.36	0.87	−0.83
InSb(In)-Sn	4.70	−0.06	0.89	−0.91

Source: Adapted from A. Suarez Negreira, W.G. Vandenberghe and M.V. Fischetti, *Physical Review B*, vol. 91, p. 245103, 2015.

substrate on the Sn-ML leads to a large buckling distance displacement (out-of plane) of the Sn atoms compared to the unsupported Sn-ML (increasing from 0.84 to 1.95 Å). Despite the large lattice mismatch between Si(111) and Sn and the low stability of the supported system, there is some experimental evidence suggesting the existence of stable supported systems with up to four mono-layers in thickness [39]. For this reason, in the following, we further analyse the Si–Sn-ML system, along with the CdTe(Cd)-Sn-ML and InSb(In) Sn-ML systems.

14.5.2.2.2 Electronic Structure

The band structures of CdTe(Cd)-Sn, InSb(In)-Sn and Si–Sn with and without accounting for SOC are shown in **Figure 14.14**.

The Si–Sn band structure is greatly distorted compared with the band structures of the two other systems. This can be attributed to the large lattice mismatch and the geometrical constraint imposed by the Si(111) substrate and the band structure resembles that of stannene in its high-buckled phase [9]. When thicker substrates (with up to 12 atomic layers) are used for the Si–Sn system, negligible changes are observed in the band structure, indicating that the distortion of the band structure is not due to an excessively thin Si(111) substrate.

A first observation in **Figure 14.14** for the InSb and CdTe-supported mono-layers is that the Fermi level does not fall between the conduction and valence band. This is because of the fractional charge transfer of the substrate. In this

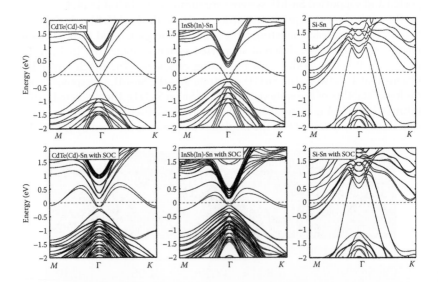

FIGURE 14.14 Top: Band structures for CdTe(Cd)-Sn, InSb(In)-Sn and Si-Sn without spin-orbit-coupling effect. Bottom: Band structures with spin-orbit-coupling effect. The dashed line indicates the energy of the Fermi level. (Reprinted with permission from A. Suarez Negreira, W. Vandenberghe and M. Fischetti, *Phys. Rev. B* 91, 245103, 2015. Copyright 2015, by the American Physical Society.)

sense, monolayer Sn supported on a polar substrate can never be a topological insulator since it is not an insulator. However, in applications where gating is possible, charge can be depleted/accumulated and the Fermi level can be moved towards the centre of the gap.

But further inspection of **Figure 14.14** reveals that even when moving the Fermi level, supported stannene does not have a gap throughout the entire Brillouin zone and it is therefore metallic.

An additional observation we made is that when the symmetry of our structure is slightly broken, the fractional occupation of the band results in convergence towards a ferromagnetic state for the calculations accounting for SOC. However, experimental observation of this ferromagneticity is unlikely, given that the obtained energies of the ferromagnetic state do not differ significantly from that of the non-ferromagnetic state.

14.5.2.3 Characterisation of Supported Halostannanane

Since the largest reported gap is observed for halogen-functionalised Sn-MLs, a similar effect is expected in the supported system. A thorough analysis is performed in this section in order to quantify the effect of surface functionalisation on the geometry, stability and band structure of the InSb-Sn-ML systems (and, to a lesser extent, to halogenated CdTe-Sn-ML systems).

14.5.2.3.1 Geometry and Thermodynamic Stability

The first step in the analysis of the thermodynamic stability of the halogenated InSb-Sn-ML systems is the estimation of the corresponding halogen binding energies. As done above, the binding energies are calculated as $E_{binding} = E_{(InSbSn-X)} - [E_{(InSbSn)} + E_X]$, an expression whose terms represent the total energy of the halogenated InSb-Sn system, of the pristine InSb-Sn system, and the binding energy per atom of a halogen molecule, respectively. The resulting binding energies are shown in **Table 14.5**, which also reports the Sn-X, Sn–Sn and Sn-In distances on the InSb(In)-Sn-ML-X system in order to quantify the geometrical changes upon functionalisation.

Table 14.5 Calculated Structural Parameters and Binding Energy of Clean and Halogenated InSb(In)-Sn-ML Systems

System	$E_{binding}$ (eV)	Sn-X (Å)	Sn-Sn (Å)	Sn-In (Å)
InSb-Sn-ML	NA	NA	2.85	2.92
InSb-Sn-ML-F	−3.57 (−3.37)	1.95	2.87	2.87
InSb-Sn-ML-Cl	−2.00 (−1.80)	2.37	2.88	2.85
InSb-Sn-ML-Br	−1.79 (−1.60)	2.51	2.89	2.85
InSb-Sn-ML-I	−1.47 (−1.29)	2.73	2.89	2.87

Source: A. Suarez Negreira, W.G. Vandenberghe and M.V. Fischetti, *Physical Review B*, vol. 91, p. 245103, 2015.

Note: Values in parentheses correspond to the CdTe-Sn-ML systems.

The large binding energies of halogens on InSb-Sn-ML highlight the stability of these systems, which are slightly higher than those determined for CdTe-Sn-ML systems. The interaction of F with the topmost Sn atoms of the InSb-Sn-ML system leads to the strongest binding energy (−3.57 eV), whereas I has the lowest (−1.47 eV), following the same trend observed in the halogenated unsupported Sn-MLs. However, contrary to what was found for unsupported Sn-MLs, the adsorption of F on the InSb-Sn-ML leads to the smallest perturbation of the Sn-ML film geometry, which is measured through the changes in the Sn–Sn and Sn-In distances. A slight increase of the Sn–Sn bond distance and a contraction of the Sn-ML-substrate interlayer distance (Sn-In distance) are observed upon halogen chemisorption.

The adsorption energies presented in **Table 14.5** are calculated using DFT and do not give any indication of the effect of temperature or pressure on the stability of these systems. The implementation of *ab initio* thermodynamics is needed in order to include the entropic effects, important at high temperature, on the adsorption energies of halogens on the InSb-Sn-ML and CdTe-Sn-ML systems. The *ab initio* thermodynamic methodology has been explained in detail in previous studies, so only the basic equations are presented here for the case of the InSb-Sn-ML system (similar equations are derived for the halogenated CdTe-Sn-ML surfaces) [37,40–42].

The stability of phases resulting from adsorption of different species on a given surface can be judged in terms of the energetic cost required to create a modified surface starting from the pristine surface [40]. The stability of different surface terminations can be evaluated from the Gibbs free energy of adsorption which, for a semi-infinite surface slab in equilibrium with a gas-phase reservoir for a given temperature and pressure, is defined as

$$\Delta G^{abs}(T, p) = \frac{1}{A}\left(G_{\text{InSb-Sn-X}} - G_{\text{InSb-Sn}} - \sum N_X \mu_X\right) \qquad (14.3)$$

where $G_{\text{InSb-Sn-X}}$ is the Gibbs free energy of the halogenated system, $G_{\text{InSb-Sn}}$ is the Gibbs free energy of the pristine InSb-Sn-ML surface, N_X is the number of gas-phase halogen species, and μ_X the chemical potential of the gas-phase halogen (μ_F, μ_{Cl}, μ_{Br} and μ_I). The Gibbs free energy terms in **Equation 14.3** are defined as $G = E_{\text{total}} + F^{vib} + F^{conf} + pV$, where the first three terms, which represents the Helmholtz free energy, are the energy at constant volume calculated with DFT, the vibrational free energy and the configurational free energy, respectively. The last term (pV) is the free energy contribution from the pressure–volume expansion. Previous DFT studies [40,43–45] consider that for $p < 100$ atm and $T < 1000$ K, the contributions from pressure–volume expansion (pV) and configurational free energy (F^{conf}) may be considered negligible (<10⁻³ meV/Å²). Rogal and Reuter [40] also suggested that vibrational free energies of the bulk solid for both the functionalised and the pristine surfaces cancel, so that only the vibrational contribution from the adsorbed species requires consideration. In this study, the vibrations associated with the Sn monolayer and adsorbed halogens are included in the surface free energy calculations. The vibrational contributions are calculated using the harmonic oscillator approximation [40] as

$$F_{H_{ad}}^{vib} = \sum_{k}^{3N} \left[\frac{\hbar\omega_k}{2} + k_B T \ln(1 - e^{(\hbar\omega_k/k_B T)}) \right]$$

in which the sum is over the vibrational modes, ω_k, of each of the N adsorbed halogen atoms and Sn atoms. The vibrational modes are obtained by running the vibrational frequently calculation, where only the halogen and Sn atoms are allowed to vibrate in the direction perpendicular to the surface. The temperature and pressure dependence of the chemical potential of the gas-phase halogen species [40] are described as

$$\mu_i = E_i^{tot}(DFT) + E_i^{ZPE} + \mu_i(T, p^0) + k_B T \ln\left(\frac{p_i}{p^0}\right)$$

where $\mu_i(T,p^0)$ can be obtained from the NIST-JANAF thermochemical tables at standard pressure p_0, 1 atm [46]; E_i^{ZPE} arises from the zero-point vibrations and E_i^{tot} is the total energy obtained through the DFT calculations of a halogen molecule in a $20 \times 20 \times 20(\text{Å}^3)$ periodic box.

The thermodynamic stability of the halogenated InSb-Sn-ML and CdTe-Sn-ML surfaces is studied under two extreme environments: Ultra-high vacuum (UHV) (for partial pressures of any gas-phase species, $p_X = 10^{-10}$ atm) and rich-halogen conditions ($p_X = 10\%$ vol. for the halogen species). The first case is important because a great number of the spectroscopy techniques (e.g. x-ray photo-electron spectroscopy, Auger electron spectroscopy or time of flight secondary ion mass spectroscopy) used for surface characterisation of these materials in the laboratory require UHV conditions. Under these conditions, weak adsorbed species (physisorbed) are no longer stable on the surface so that they are not detected during the surface analysis. Therefore, it is essential to predict the stability of these halogenated systems when exposed to these common environmental conditions. On the other hand, a higher stability of these halogenated systems could be obtained by exposing these materials to a halogen-rich atmosphere, therefore preserving the surface properties caused by halogen chemisorption. The thermodynamic stability of halogenated InSb-Sn-ML and CdTe-Sn-ML surfaces as a function of temperature (100–1000 K) under these two extreme environments are presented in **Figure 14.15**, where the lowest energy corresponds to the most thermodynamically stable system.

Note how F-adsorption on both types of surfaces leads to the most stable functionalised surfaces at any temperature under UHV and halogen-rich conditions, thanks to the strong adsorption-energy of F, as shown in **Table 14.5**. As the adsorption energies decrease from F to I, the same effect is observed on the thermodynamic stability as a function of temperature and pressure. The weakest adsorption energy corresponds to the case of I, which becomes unstable under UHV conditions for temperatures above 600 and 500 K for

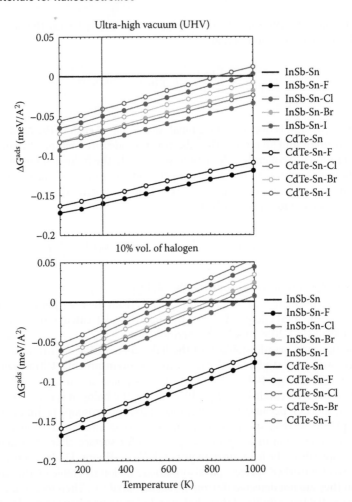

FIGURE 14.15 Temperature dependence of the adsorption energies of halogens on the surfaces of the InSb-Sn-ML (filled symbols) and CdTe-Sn-ML (empty symbols) systems under ultra-high vacuum (top) and halogen-rich (bottom) conditions. The vertical line shows room temperature (298 K), while black horizontal line represents pristine InSb-Sn-ML and CdTe-Sn-ML surfaces. The more negative adsorption energies correspond to more thermodynamically stable systems. (Reprinted with permission from A. Suarez Negreira, W. Vandenberghe and M. Fischetti, *Phys. Rev. B* 91, 245103, 2015. Copyright 2015, by the American Physical Society.)

InSb-Sn-ML and CdTe-Sn-ML, respectively, while under halogen-rich conditions, the same I-terminated surfaces are stable at temperatures below 900 and 800 K. Overall, at room temperature (marked by a vertical line), all halogenated InSb-Sn-ML and CdTe-Sn-ML systems are predicted to be stable for both types of environmental conditions.

14.5.2.3.2 Electronic Structure

The band structure of InSb-Sn-ML-I is compared with the band structure of CdTe-Sn-ML-I, with and without inclusion of SOC in **Figure 14.16**.

From **Figure 14.16**, it is immediately clear that InSb-Sn-ML-X is a topological insulator and that the two-fold degenerate E-band lies above the non-degenerate A_1-band similar to the topological insulating iodostannanane in Section 14.3. For CdTe-Sn-ML-X, the two-fold degenerate E-band is degenerate and lies below the non-degenerate A_1-band in the absence of SOC. However, when analysing the symmetry of the wavefunctions after SOC, it is clear that the first conduction band is an E-band and the last valence band is an A_1-band making CdTe-Sn-ML-X a topological insulator as well as InSb-Sn-ML-X.

The band gap of InSb-Sn-ML-X is 0.15, 0.14, 0.15 and 0.17 eV for F-, Cl-, Br- and I-adsorption on InSb-Sn-ML, respectively. The same values were obtained for the band gaps of halogenated CdTe-Sn-ML (0.15, 0.14, 0.15 and 0.17 eV for F-, Cl-, Br- and I-adsorption on CdTe-Sn-ML, respectively). This trend with

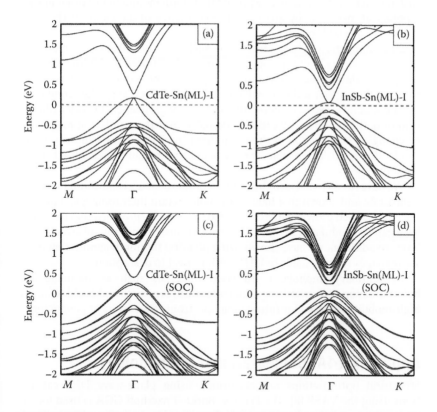

FIGURE 14.16 Band structures for: (a) CdTe(Cd)-Sn-I, (b) InSb(In)-Sn-I, (c) CdTe(Cd)-Sn-I with SOC, (d) InSb(In)-Sn-I with SOC. The dashed line indicates the energy of the Fermi level. (Reprinted with permission from A. Suarez Negreira, W. Vandenberghe and M. Fischetti, *Phys. Rev. B* 91, 245103, 2015. Copyright 2015, by the American Physical Society.)

respect to the functionalising agent is similar to the one observed in unsupported halostannanane, where I-chemisorption also resulted in the largest gap. Although, these topological band gap values are still smaller than those found in the pristine unsupported Sn-MLs (0.34 eV for Sn-ML), these band gaps exceed the thermal energy at room temperature, rendering their use in nanoelectronics possible.

14.6 Conclusion

Using DFT, we have shown that monolayer tin is stable or meta-stable in a high-buckled and a low-buckled hexagonal phase. The high-buckled phase is metallic, whereas the low-buckled phase has a band gap and is a topological insulator. Naming hexagonal tin in its low-buckled phase *stannene*, stannene functionalised with hydrogen *stannanane* and stannene functionalised with halogens *halostannane*, we proved that stannene and halostannanane are topological insulators by analysing the symmetry of the bands at the high-symmetry M and Γ points.

Since stannene and halostannananes such as iodostannane are topological insulators, they exhibit topologically protected edge states with high conductivity. Calculating the conductance using the Kubo–Greenwood formalism, we have shown that iodostannanane ribbons exhibit a very high conductivity and mobility, even exceeding the mobility of graphene. For wider ribbons (~10 nm), the conductivity can be modulated over three orders of magnitude by changing the position of the Fermi level from inside the band gap to the conduction band. This enables the use of iodostannanane for classical transistor applications.

We have further analysed the stability of stannene and halostannananes on a substrate and shown that halostannananes retain their topological insulating nature when grown epitaxially on CdTe and InSb, provided that the Fermi level can be modulated to offset the charge transfer from the substrate to the layer. In addition, we have shown using *ab initio* thermodynamics that halostannananes remain stable in UHV with respect to desorption.

In summary, monolayer tin-based materials such as stannene and halostannanane are promising avenues for future research and could serve as a high-mobility material for future nano-electronic transistors or conductors.

Appendix 14A: Computational Methodology

Structural optimisations are performed using plane-wave DFT calculations using the VASP [8]. The Perdew–Burke–Enzerhoff GGA is used for the exchange–correlation functional [47]. For the CdTe, InSb, Si and Sn bulk calculations, a projector augmented wave [48] pseudopotential is used with an energy cut-off of 500 eV. The number of k-points for the Brillouin-zone integration is chosen according to a Monkhorst–Pack [49] grid of $8 \times 8 \times 8$ with a convergence criterion of 10^{-4} eV.

The surfaces of the CdTe(111), InSb(111) and Si(111) substrates and the monolayer Sn(111) are studied using an asymmetric 1×1 unit cell (i.e. the slab does not have a mirror symmetry plane parallel to the surface due to the difference between the top and bottom layers). The slabs are separated by a 15 Å thick vacuum padding to prevent interaction between adjacent super cells and to minimise the impact of any dipole moment present. To study the slabs, an $8 \times 8 \times 1$ k-mesh is used. The positions of all atoms in the substrate are allowed to relax with the exception of the atoms in the bottom layer, which are kept fixed in their bulk positions throughout the calculations.

The bonds in CdTe are formed with 0.5 electrons from the Cd atom and 1.5 electrons from the Te atom, whereas the bonds in InSb are formed with 0.75 electrons from the In atom and 1.25 electrons from the Sb atom. In order to saturate the dangling bonds of the atoms in the top and bottom layers of the CdTe and InSb substrates, fictitious hydrogen atoms (hydrogen atoms with a fractional nuclear and electronic charge) are used in the following way: H atoms with a +1.5 electron charge saturate the Cd-terminated surfaces, whereas H atoms with a +0.5 electron charge are used for the Te-terminated surfaces. For the InSb substrate, H atoms with +0.75 and +1.25 electron charge are used to saturate the Sb- and In-terminated surfaces, respectively [50,51]. As for the Si(111) surface, the dangling bonds are passivated with 'conventional' +1 electron charge H atoms.

Whenever information was needed about gas-phase species (F_2, Cl_2, Br_2 and I_2), these are modelled as isolated molecules in a $20 \times 20 \times 20$ (Å^3) periodic box.

To assist with the symmetry classification, we employed Quantum ESPRESSO [52] and we use the notation from Reference 53 for the irreducible representations of the point groups describing the symmetry of the different bands under study.

Phonon band structures are calculated using the small displacement method using PHONOPY [54].

References

1. Y. Xu, B. Yan, H.-J. Zhang, J. Wang, G. Xu, P. Tang, W. Duan and S.-C. Zhang, Large-gap quantum spin Hall insulators in tin films, *Physical Review Letters*, vol. 111, no. 13, p. 136804, 2013.

2. Y. Ma, Y. Dai, M. Guo, C. Niu and B. Huang, Intriguing behavior of halogenated two-dimensional tin, *The Journal of Physical Chemistry C*, vol. 116, no. 23, pp. 12977–12981, 2012.

3. B. van den Broek, M. Houssa, E. Scalise, G. Pourtois, V. Afanas'ev and A. Stesmans, Two-dimensional hexagonal tin: *Ab initio* geometry, stability, electronic structure and functionalization, *2D Materials*, vol. 1, no. 2, p. 021004, 2014.

4. W. Vandenberghe and M. Fischetti, Calculation of room temperature conductivity and mobility in tin-based topological insulator nanoribbons, *Journal of Applied Physics*, vol. 116, p. 173707, 2014.

5. A. Suarez Negreira, W. G. Vandenberghe and M. V. Fischetti, DFT study of the electronic properties and thermodynamic stability of supported 2D-Sn (stannanane) films, *Physical Review B*, vol. 91, p. 245103, 2015.

6. S. Cahangirov, M. Topsakal, E. Aktürk, H. C. Sahin and S. Ciraci, Two-and one-dimensional honeycomb structures of silicon and germanium, *Physical Review Letters*, vol. 102, no. 23, p. 236804, 2009.

7. X. Yi, J. Hao, X. Zhang and G. Wong, Alpha-Sn thin film grown on GaAs substrate by MBE and investigation of its multiquantum well structure, *Science in China Series A: Mathematics*, vol. 41, no. 4, pp. 399–404, 1998.

8. G. Kresse and J. Furthmüller, Efficiency of ab-initio total energy calculations for metals and semiconductors using a plane-wave basis set, *Computational Materials Science*, vol. 6, no. 1, pp. 15–50, 1996.

9. P. Rivero, J.-A. Yan, V. M. Garcia-Suarez, J. Ferrer and S. Barraza-Lopez, Stability and properties of high-buckled two-dimensional tin and lead, *Physical Review B*, vol. 90, no. 24, p. 241408, 2014.

10. H. Liu, A. T. Neal, Z. Zhu, Z. Luo, X. Xu, D. Tomanek and P. D. Ye, Phosphorene: An unexplored 2D semiconductor with a high hole mobility, *ACS Nano*, vol. 8, no. 4, pp. 4033–4041, 2014.

11. M. Gmitra, S. Konschuh, C. Ertler, C. Ambrosch-Draxl and J. Fabian, Band-structure topologies of graphene: Spin–orbit coupling effects from first principles, *Physical Review B*, vol. 80, no. 23, p. 235431, 2009.

12. M. Z. Hasan and C. L. Kane, Colloquium: Topological insulators, *Reviews of Modern Physics*, vol. 82, no. 4, p. 3045, 2010.

13. L. Fu and C. L. Kane, Topological insulators with inversion symmetry, *Physical Review B*, vol. 76, no. 4, p. 045302, 2007.

14. M. König, S. Wiedmann, C. Brüne, A. Roth, H. Buhmann, L. W. Molenkamp, X.-L. Qi and S.-C. Zhang, Quantum spin Hall insulator state in HgTe quantum wells, *Science*, vol. 318, no. 5851, pp. 766–770, 2007.

15. W. G. Vandenberghe and M. V. Fischetti, Calculation of room temperature conductivity and mobility in tin-based topological insulator nanoribbons, *Journal of Applied Physics*, vol. 116, no. 17, p. 173707, 2014.

16. T. L. Schmidt, S. Rachel, F. von Oppen and L. I. Glazman, Inelastic electron backscattering in a generic helical edge channel, *Physical Review Letters*, vol. 108, no. 15, p. 156402, 2012.

17. J. C. Budich, F. Dolcini, P. Recher and B. Trauzettel, Phonon-induced backscattering in helical edge states, *Physical Review Letters*, vol. 108, no. 8, p. 086602, 2012.

18. B. A. Bernevig and S.-C. Zhang, Quantum spin Hall effect, *Physical Review Letters*, vol. 96, no. 10, p. 106802, 2006.

19. A. M. Lunde and G. Platero, Hyperfine interactions in two-dimensional HgTe topological insulators, *Physical Review B*, vol. 88, no. 11, p. 115411, 2013.

20. L. Fu and C. L. Kane, Probing neutral Majorana fermion edge modes with charge transport, *Physical Review Letters*, vol. 102, no. 21, p. 216403, 2009.

21. J. E. Moore, The birth of topological insulators, *Nature*, vol. 464, no. 7286, pp. 194–198, 2010.

22. D. Pesin and A. H. MacDonald, Spintronics and pseudospintronics in graphene and topological insulators, *Nature Materials*, vol. 11, no. 5, pp. 409–416, 2012.

23. J. Chang, L. F. Register and S. K. Banerjee, Topological insulator Bi_2Se_3 thin films as an alternative channel material in metal–oxide–semiconductor field-effect transistors, *Journal of Applied Physics*, vol. 112, no. 12, p. 124511, 2012.

24. D. Greenwood, The Boltzmann equation in the theory of electrical conduction in metals, *Proceedings of the Physical Society*, vol. 71, no. 4, p. 585, 1958.

25. B. Soree, W. Magnus and W. Vandenberghe, Low-field mobility in ultrathin silicon nanowire junctionless transistors, *Applied Physics Letters*, vol. 99, no. 23, p. 233509, 2011.

26. J. Kim, M. V. Fischetti and S. Aboud, Structural, electronic and transport properties of silicane nanoribbons, *Physical Review B*, vol. 86, no. 20, p. 205323, 2012.

27. K. M. Borysenko, J. T. Mullen, E. A. Barry, S. Paul, Y. G. Semenov, J. M. Zavada, M. B. Nardelli and K. W. Kim, First-principles analysis of electron–phonon interactions in graphene, *Physical Review B*, vol. 81, p. 121412, 2010.

28. K. Kaasbjerg, K. S. Thygesen and K. W. Jacobsen, Phonon-limited mobility in n-type single-layer MoS_2 from first principles, *Physical Review B*, vol. 85, no. 11, p. 115317, 2012.

29. W. G. Vandenberghe and M. V. Fischetti, Deformation potentials for band-to-band tunneling in silicon and germanium from first principles, *Applied Physics Letters*, vol. 106, no. 1, p. 013505, 2015.

30. N. Bannov, V. Mitin and M. Stroscio, Confined acoustic phonons in a free-standing quantum well and their interaction with electrons, *Physica Status Solidi (b)*, vol. 183, no. 1, pp. 131–142, 1994.

31. K. I. Bolotin, K. Sikes, Z. Jiang, M. Klima, G. Fudenberg, J. Hone, P. Kim and H. Stormer, Ultrahigh electron mobility in suspended graphene, *Solid State Communications*, vol. 146, no. 9, pp. 351–355, 2008.

32. C.-C. Liu, H. Jiang and Y. Yao, Low-energy effective Hamiltonian involving spin–orbit coupling in silicene and two-dimensional germanium and tin, *Physical Review B*, vol. 84, no. 19, p. 195430, 2011.

33. J. C. Garcia, D. B. de Lima, L. V. Assali and J. F. Justo, Group IV graphene- and graphane-like nanosheets, *The Journal of Physical Chemistry C*, vol. 115, no. 27, pp. 13242–13246, 2011.

34. J. Wang, X. Wu and D. Bai, Structural and electronic properties of Zn on a CdTe (001) surface, *Solid State Communications*, vol. 149, no. 25, pp. 982–985, 2009.

35. Y.-S. Kim, K. Hummer and G. Kresse, Accurate band structures and effective masses for InP, InAs and InSb using hybrid functionals, *Physical Review B*, vol. 80, no. 3, p. 035203, 2009.

36. G. I. Csonka, J. P. Perdew, A. Ruzsinszky, P. H. T. Philipsen, S. Lebègue, J. Paier, O. A. Vydrov and J. G. Angyan, Assessing the performance of recent density functionals for bulk solids, *Physical Review B*, vol. 79, p. 155107, 2009.

37. K. Reuter and M. Scheffler, Composition, structure and stability of $RuO_2(110)$ as a function of oxygen pressure, *Physical Review B*, vol. 65, no. 3, p. 035406, 2001.

38. J. Evans, P. Thiel and M. Bartelt, Morphological evolution during epitaxial thin film growth: Formation of 2D islands and 3D mounds, *Surface Science Reports*, vol. 61, no. 1, pp. 1–128, 2006.

39. L. Wang, X. Ma, S. Ji, Y. Fu, Q. Shen, J. Jia, K. Kelly and Q. Xue, Epitaxial growth and quantum well states study of Sn thin films on Sn induced Si (111)-(2 3$\times$ 2 3) R 30Â° surface, *Physical Review B*, vol. 77, no. 20, p. 205410, 2008.

40. J. Rogal and K. Reuter. Ab initio atomistic thermodynamics for surfaces: A primer. Max-Planck-Gesellschaft zur Foerderung Der Wissenschaften ev Berlin (Germany Fr) Fritz-Haber-Inst, 2006.

41. A. S. Negreira, S. Aboud and J. Wilcox, Surface reactivity of $V_2O_5(001)$: Effects of vacancies, protonation, hydroxylation and chlorination, *Physical Review B*, vol. 83, no. 4, p. 045423, 2011.

42. A. Suarez Negreira and J. Wilcox, DFT study of Hg oxidation across Vanadia–Titania SCR catalyst under flue gas conditions, *Journal of Physical Chemistry C*, vol. 117, no. 4, pp. 1761–1772, 2013.

43. C. Lo, K. Tanwar, A. Chaka and T. Trainor, Density functional theory study of the clean and hydrated hematite (1102) surfaces, *Physical Review B*, vol. 75, no. 7, p. 075425, 2007.

44. R. Blum, H. Niehus, C. Hucho, R. Fortrie, M. Ganduglia-Pirovano, J. Sauer, S. Shaikhutdinov and H. Freund, Surface metal–insulator transition on a vanadium pentoxide (001) single crystal, *Physical Review Letters*, vol. 99, no. 22, p. 226103, 2007.

45. X. Wang, A. Chaka and M. Scheffler, Effect of the environment on $Al_2O_3(0001)$ surface structures, *Physical Review Letters*, vol. 84, no. 16, pp. 3650–3653, 2000.

46. NIST-JANAF, Fluorine, chlorine, bromine, iodine. In J. Chase (ed.), *NIST-JANAF Thermochemical Tables*, 4th edn, American Chemical Society, Washington, DC, 1998.

47. J. Perdew, K. Burke and M. Ernzerhof, Generalized gradient approximation made simple, *Physical Review Letters*, vol. 77, no. 18, pp. 3865–3868, 1996.

48. P. Blochl, Projector augmented-wave method, *Physical Review B*, vol. 50, no. 24, pp. 17953–17979, 1994.

49. H. Monkhorst and J. Pack, Special points for Brillouin-zone integrations, *Physical Review B*, vol. 13, no. 12, pp. 5188–5192, 1976.

50. J. Wang, G. Tang, X. Wu and M. Gu, The adsorption of O on (001) and (111) CdTe surfaces: A first-principles study, *Thin Solid Films*, vol. 520, no. 11, pp. 3960–3964, 2012.

51. T. Ohno and K. Shiraishi, First-principles study of sulfur passivation of GaAs (001) surfaces, *Physical Review B*, vol. 42, no. 17, p. 11194, 1990.

52. P. Giannozzi, S. Baroni, N. Bonini, M. Calandra, R. Car, C. Cavazzoni, D. Ceresoli et al., QUANTUM ESPRESSO: A modular and open-source software project for quantum simulations of materials, *Journal of Physics: Condensed Matter*, vol. 21, no. 39, p. 395502 (19pp), 2009.

53. M. S. Dresselhaus, G. Dresselhaus and A. Jorio, *Group Theory: Application to the Physics of Condensed Matter*, Springer-Verlag, Berlin, Heidelberg, 2007.

54. A. Togo, F. Oba and I. Tanaka, First-principles calculations of the ferroelastic transition between rutile-type and $CaCl_2$-type SiO_2 at high pressures, *Physical Review B*, vol. 78, no. 13, p. 134106, 2008.

15

Phosphorene
A Novel 2D Material for Future Nanoelectronics and Optoelectronics

Yexin Deng, Zhe Luo, Han Liu, Yuchen Du,
Xianfan Xu and Peide D. Ye

Contents

15.1 Introduction

The discovery of graphene has attracted extensive research interest in two-dimensional (2D) materials in the past 10 years [1,2]. The unique physical and chemical properties of graphene make it a promising material for many different applications [3]. The outstanding transport properties of graphene have

2D Materials for Nanoelectronics edited by Michel Houssa, Athanasios Dimoulas and Alessandro Molle © 2016 CRC Press/Taylor & Francis Group, LLC. ISBN: 978-1-4987-0417-5.

drawn the attention of the semiconductor industry, which is now a hundreds-of-billions dollar business – the foundation of all the electronics devices in our daily life. The basic elements of electronic circuits are electronic devices, and the most famous one is the metal–oxide–semiconductor field-effect transistor (MOSFET). The state-of-art MOSFET is scaled down to a 10 nm regime. The limited mobility of silicon, which is the dominant material in the semiconductor industry, drives the research interest in looking for materials with higher carrier mobility. With an extremely high mobility of up to 200,000 cm^2/V s at low temperature, as well as 10,000 cm^2/V s in graphene MOSFETs, it can in principle lead to a higher on-current in MOSFET devices [4]. Moreover, the electrostatic control of the ultimately scaled graphene MOSFET can be better than the Si MOSFET due to its ultra-thin 2D nature. Besides, its high thermal conductivity, high optical damage threshold and third-order optical non-linearity also make it promising for use in photonic devices [5]. Although many efforts have been made in exploring graphene for nanoelectronics, its zero band gap limits its applications. Without a band gap, the MOSFET cannot be turned off, and the current cannot saturate when turned on, which is crucial for the MOSFET. On the other hand, the zero band gap also makes the lifetime of the photo-generated carriers very short, which limits its efficiency for photonics applications such as when used as photodetectors.

The researcher then tried to find 2D materials with a finite band gap. Transition metal dichalcogenides (TMD) such as semiconducting MoS_2, have caught the eyes of researchers [6,7]. With a 1.8 eV band gap in monolayer form, MoS_2 has been proved to be a good semiconductor which can achieve a large on/off ratio of up to 10^8 in a MOSFET. Its direct band gap in monolayer form also makes it good for photodetection and other photonics applications [8]. Other TMD materials have also been studied for their electronics and optoelectronics applications [9]. As an insulator, hBN has been proved to be an ultra-flat substrate for graphene quantum transport studies. It also can be used as a gate dielectric in the MOSFET, which is suitable for flexible electronics [10]. However, the band gap of most of these materials is more than 1 eV, which limits the possibility for higher mobility and large drive current in the MOSFET, as well as the applications in infrared optoelectronic and photonic devices. Recently, researchers from China and United States have found a new 2D material which shows a 0.33 eV narrow direct band gap: black phosphorus [11–13].

Black phosphorus is the most stable allotrope of phosphorus. It was first discovered in 1914 by applying high pressure on white phosphorus. However, just from the beginning of 2014, this material has drawn much more interest from the 2D research community. This layered material shows a 0.33 eV direct band gap in bulk, and its band gap increases to more than 1 eV in monolayer form, which is called 'phosphorene'. It shows a mobility up to order of 100,000 cm^2/V s cm in bulk (at low temperature) and 1000 cm^2/V s in a MOSFET made from few-layer phosphorene at room temperature, while maintaining an on/off ratio of 10^6 thanks to its ultra-thin 2D nature. Its high mobility and reasonable on/off ratio makes it promising for use in thin film transistors and radio-frequency (RF) transistors. Based on its thickness-dependent direct band gap, black phosphorus is also suitable for infrared optoelectronics and photonics. In this chapter, we will first briefly introduce the early research results of black phosphorus from earlier days, including the synthesis method, atomic structure, electrical properties and other results. This can give us a basic background about bulk black phosphorus. Then, we will focus more on the 2D research results on phosphorene. We will cover the materials properties, including electrical, optical, thermal and mechanical properties. Based on the understanding of these 2D materials, we will then give an introduction about the research of applications of phosphorene on electronics and optoelectronics.

15.2 Brief History of Black Phosphorus

In this section, we will discuss about the synthesis, atomic structure, electrical properties at normal condition, under high pressure and superconductivity. These include most of the knowledge people had about bulk black phosphorus, which may be helpful for us to get a deeper understanding of the current 2D research on this material.

Black phosphorus was first synthesised in 1914 by applying high pressure on white phosphorus [14]. White phosphorus was placed in a high-pressure cylinder under kerosene. High pressure of up to 0.6 GPa was applied at room temperature. The temperature of the cylinder was increased to 200°C, and the pressure was increased to 1.2 GPa. It took 5–30 min for the transition from white phosphorus to black phosphorus. Bridgman found that black phosphorus showed a higher density of 2.69 g/cm³ compared to 1.83 g/cm³ of white phosphorus and 2.05–2.34 g/cm³ of red phosphorus. Unlike white or red phosphorus, it did not catch fire in air or ignite by fire, and could sustain up to 400°C in air without spontaneous ignition. A similar result was reproduced by Keyes under the pressure of 1.3 GPa at 200°C. The ingot was polycrystalline, and the grain size was roughly around 0.1 mm [15]. He found that black phosphorus was a good conductor of both electrons and heat, but was less diamagnetic than white or red phosphorus. The structure of black phosphorus prepared by the Bridgman method has been known to change from amorphous to polycrystalline form, depending on the applied pressure and temperature [16]. Black phosphorus was also synthesised using mercury or the bismuth-flux method [17,18]. Brown and Rundqvist found that black phosphorus crystallises in orthorhombic form, and has an infinite puckered layer structure. Silicon or germanium doped black phosphorus was achieved by the bismuth-flux method later on Reference 19. Large single crystals of black phosphorus were first obtained in 1981 from red phosphorus melted at high temperature and high pressure using a wedge-type cubic high-pressure apparatus. Samples grown under 3.8 GPa higher than 270°C were crystallised in a single crystal [20].

It was not until very recently that black phosphorus could be prepared under lower pressures. Lange and his colleagues [21] reported a low-pressure method to produce high-quality black phosphorus by using a mineraliser as reaction promoter under non-toxic conditions. Black phosphorus was prepared by the reaction of Au, Sn, red phosphorus and SnI4 in evacuated silica ampules. The starting materials were heated to 600°C, and gradually cooled down to room temperature. This reaction usually took dozens of hours and even up to 5–10 days. This process is further simplified by Köpf using Sn/SnI4 as the only mineralisation additive [22].

Based on synthesised black phosphorus, its basic physical properties were studied. At normal conditions, bulk black phosphorus shows a layered structure, which is much like graphite. The atomic structure is shown in **Figure 15.1**. The unit cell contains eight atoms, which gives a calculated density of 2.69 g/cm³. Using black phosphorus by the Bridgman method, the lattice constant was first determined by Hultgren et al. with an x-ray pattern. Each phosphorus atom is bonded with three neighbouring atoms at 2.18 Å. Two of them are in the plane of the layer at 99° from one another and the third is between the layers at 103° [23]. Similar results were obtained by Brown and Rundqvist using black phosphorus crystals from the bismuth-flux method [17]. Lattice parameter extractions at normal conditions were done by Brown and Rundqvist, as listed in **Table 15.1**.

The crystal structure of black phosphorus can be discriminated under high pressure [24]. This mostly originates from the anisotropic compressibility of black phosphorus due to the asymmetrical crystal structures. It can be expected that in the z-direction, the van der Waals bond can be greatly compressed. However, it also shows a strong variation in compressibility across the orthogonal x–y plane. The x-axis shows similar compressibility at relatively lower pressure (<2.66 GPa) with the z-axis, whereas it almost remains constant along the y-axis. This results in the structural changing of black phosphorus under high pressure [25,26]. Black phosphorus undergoes two reversible structural transitions under high pressure. The first transition from the orthorhombic to rhombohedral phase occurs around 5.5 GPa at room temperature, accompanied by displacement of the puckered layers and a volume change.

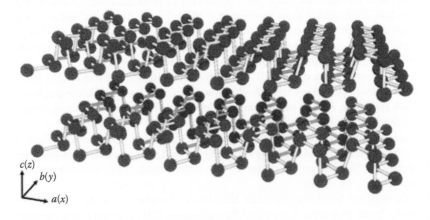

$c(z)$
$b(y)$
$a(x)$

FIGURE 15.1 Atomic structure of black phosphorus.

Table 15.1 Crystal Structural Parameters of Black Phosphorus under the Normal Condition

a	4.374 Å	Lattice constants in orthorhombic system
b	3.3133 Å	
c	10.473 Å	
u	0.0806 Å	Crystal structural parameters
v	0.1034 Å	
d1	2.222 Å	Bond length
d2	2.277 Å	
α1	96.5°	Bond angle between d1s
α2	102.09°	Bond angle between d1 and d2

Under higher pressure, it goes from the rhombohedral to simple cubic phase at around 10 GPa. Transition temperature is almost independent with temperature [27,28]. This simple cubic phase keeps stable even when the pressure is increased up to 60 GPa [29].

The electronic structure of black phosphorus was determined by various approaches, including the tight-bonding method, self-consistent pseudopotential method and local-orbital method [30–32]. From band structure calculation, it shows that the effective mass in the x–y plane is relatively large for both electrons and holes. The effective mass in the x-direction is lightest, and that in the z-direction is lighter than that in the in-plane y-direction due to its van der Waals structure. An average effective mass of $0.22m_0$ for electrons and $0.24m_0$ for holes was determined by the self-consistent pseudopotential method. Both were slightly smaller than those in silicon [31].

Electrical transport property of black phosphorus was first measured by Bridgman. The results showed the resistivity ranged from 0.48 to 0.77 Ω cm at 30°C [33]. It was later revisited by Keyes and Warchauer using polycrystalline crystals by the Bridgman method [15,34]. Both of them determined that the un-doped samples showed p-type conductivity and a positive Hall coefficient. This indicated that conductivity was more dominated by holes than by electrons. Temperature-dependent conductance study revealed its energy gap to be around 0.33–0.35 eV. In 1980s, both n- and p-type black phosphorus were successfully synthesised and studied. The n-type black phosphorus was realised by Te-doping. N-type conductivity was attributed to the substitution of Te atoms to the P atoms in black phosphorus, however, p-type conductivity is still not clear so far. The effective donor and acceptor concentration in these samples were determined to be around 2×10^{16} to 3×10^{16} and 2×10^{16} to 5×10^{15} cm^{-3}, respectively. The activation energy for n- and p-type black phosphorus was determined to be ~39 and ~18 meV [35]. Baba et al. observed two types of acceptors, of which activation energies were 26.1 and 11.8 meV in the bismuth-flux prepared samples. Effective concentrations of acceptors were typically 1.36×10^{15} cm^{-3} for the deeper level and 0.44×10^{15} cm^{-3} for the shallower level [36].

At room temperature, hole and electron mobilities are of 350 and 220 cm^2/V s, respectively. They can be characterised with the $T^{-2/3}$ relationship, indicating that they are mostly limited by lattice vibration in the measurement range [15]. The resistivity was observed to increase near liquid helium temperature, and the Hall coefficient peak occurred between 24 and 30 K [34]. Maruyama et al. measured the resistance between 1.4 and 400 K, and it turned out to be an impurity dominated semiconductor. A band gap of 0.31 eV was also estimated. Carrier mobilities extracted from the magneto-resistance coefficients showed the values of 2.7×10^4 cm^2/V s at 77 K and 1.5×10^4 cm^2/V s at 294 K, respectively, which are roughly one order of magnitude larger than those found from the Hall effect on polycrystalline samples. At 20 K, Hall mobility of holes reached the maximum value of 6.5×10^4 cm^2/V s, whereas the peak shifted to 50 K for electrons around 1.5×10^4 cm^2/V s [37]. Magneto-resistance was also studied in these samples down to 0.5 K in temperature and up to 6 T in magnetic fields, and 2D Anderson localisation was observed [38]. The electric conductivity showed a log T-like behaviour below about 5 K, and the Hall coefficient had its sign reversal at around 7 K. Magneto-transport properties observed at low temperatures in black phosphorus crystals originated from the 2D Anderson localisation in an inversion layer on the surface [39].

Bridgman first found that the electrical properties of black phosphorus were sensitive to pressure. Under 1.2 GPa, the resistivity was found to be only ~3% of its value under atmosphere pressure [40]. The effect pressure on electrical conductivity was further investigated by Okajima et al. [41] with single crystal black phosphorus. Akahama et al. measured the resistivity at room temperature in terms of the applied pressure. The resistivity decreased logarithmically at higher pressure. Two anomalies were observed occurring at 1.7 and 4.2 GPa. The first one was attributed to the change of the band gap and the second one was attributed to the phase transition from the orthorhombic to rhombohedral. It was found that the energy gap decreased at higher pressure, and finally became zero around 1.7 GPa, and resistivity turned out to be like the characteristic of a metal, which showed a pressure-induced semiconductor–metal transition [35]. Calculation of energy gap under pressure based on the self-consistent pseudopotential method indicated a change of −212 to −235 meV/GPa [31,42]. The linear change of band gap under pressure was attributed to the overlap between conduction and valence bands without any structural change.

Superconductivity was first observed by Wittig et al. [43] near 4.7 K at the simple cubic phase of black phosphorus. This constituted one of the four last missing links in the proof that superconductivity is normal behaviour for every truly metallic sp element besides arsenic, sulphur and iodine. The following three paths were studied on the dependence of the treatment methods of cooling and pressurising [42]. (a) Applied 15 GPa at room temperature to convert the samples to the simple cubic phase, and then cooled down to liquid helium temperature. T_c was measured to be around 6 K, and increased with increasing pressure. Even after removing the applied pressure, it remains superconducting at 6 K for several hours. (b) Pressure was applied at 8.7 GPa at room temperature where samples were transformed into the rhombohedral phase. The pressure cell was then cooled down to liquid helium temperature, and then pressure dependent T_c was measured. The transition observed at 8.7 GPa suggested the possibility of superconductivity in the rhombohedral phase. With increasing pressure, the transition curve at about 25 GPa resembles the previous path, suggesting the transformation from the rhombohedral to simple cubic may occur even at liquid helium temperature. (c) Samples were cooled down to liquid helium temperature at normal pressure. Pressure dependent T_c was measured by applying high pressure. T_c rose from 4 to 10.7 K when pressure increased from 11 to 30 GPa. Also, the transition temperature decreases rapidly by decreasing the applied pressure and becomes null before about 3 GPa in contrast with the first path [43,44].

15.3 Material Properties of 2D Phosphorene

Although the basic properties of black phosphorus have been studied since it was discovered about 100 years ago, the research on this material did not attract too much attention until 2014. The 2D research community found this layered material just fit their needs for nanoelectronics and optoelectronics. Moreover, its unique atomic structure also makes the researcher interested in its thermal and mechanical properties. However, the current research focusses more on the 2D counterpart of black phosphorus–phosphorene or few-layer phosphorene. Unlike the bulk material, its properties change as the thickness of black phosphorus decreases down to the nanometre regime. This makes the material properties even more interesting to study. In this section, we will give a brief review of the recent study of the material properties of phosphorene. We first talk about the basic characterisation method of 2D phosphorene, and we then discuss the electrical transport properties, optical properties, mechanical and thermal properties.

The earliest papers about phosphorene were published at the beginning of 2014. In these papers, the basic properties of phosphorene have been covered [11–13]. Single-layer phosphorene was obtained by using the standard scotch tape mechanical exfoliation method, much like the method used to obtain graphene from graphite [1]. The single-layer phosphorene was characterised by atomic force microscopy (AFM), and showed a thickness from 0.7 to 0.85 nm (**Figure 15.2**) [11,46]. This range is slightly larger than the 0.53 nm theoretical value but much less than the 1.06 nm bilayer black phosphorus, indicating its monolayer feature. Raman spectrum was used to characterise the phosphorene flake [11,13]. In graphene and other 2D materials study, Raman spectrum is a useful tool to identify the basic material properties without damaging the sample [1–3]. The Raman spectrum was obtained from monolayer to few-layer phosphorene by several groups [11,45,46]. It was found that the Raman spectrum was angle dependent due to the anisotropy of phosphorene in the x–y plane. By using the polarised Raman spectrum, the crystalline direction of phosphorene flakes can be effectively determined [47]. To observe the

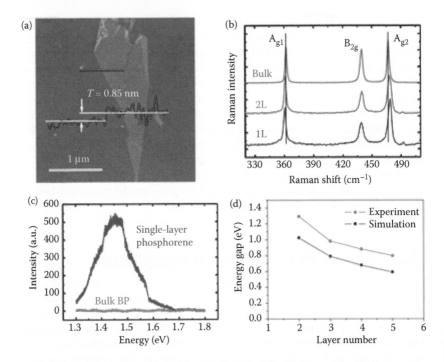

FIGURE 15.2 (a) AFM image of single-layer phosphorene. (b) Raman spectrum of single, bi-layer phosphorene and bulk black phosphorus. (c) PL spectrum of single-layer phosphorene and bulk black phosphorus. (d) Thickness-dependent energy gap of phosphorene obtained by PL from bi-layer to five layer. ((a)–(c) Reprinted with permission from H. Liu et al., *ACS Nano* 2014, 8, 4033–4041. (d) Reprinted with permission from S. Zhang et al., *ACS Nano* 2014, 8, 9590–9596. Copyright 2014 American Chemical Society.)

polarised Raman scattering, a linear polariser was placed at the spectrometer entrance. With the detection polarisation perpendicular (referred to as 'VH configuration', where 'V' stands for vertical laser polarisation and 'H' for horizontal detection polarisation) or parallel (VV configuration) to the incident laser polarisation, the optical phonon modes of different symmetries can be selected or eliminated when lattice principal axes are aligned with the laser polarisation. In the case of phosphorene, the A_g modes and the B_{2g} mode can be filtered out in VH and VV configurations, respectively, when either the armchair or zigzag axis is aligned with the laser polarisation, as shown in **Figure 15.3**. In this way, we were able to identify the armchair or zigzag axis by, for example, observing the B_{2g} mode Raman intensity in the VV configuration while rotating the black phosphorus (BP) flake. To further distinguish these two axes, we looked into the A_{g2}/A_{g1} Raman intensity ratio in the VV configuration. The armchair-oriented atomic vibrations of A_{g2} phonons lead to maximised A_{g2} Raman intensity when laser polarisation is along the armchair direction, while the A_{g1} Raman intensity remains unchanged because the

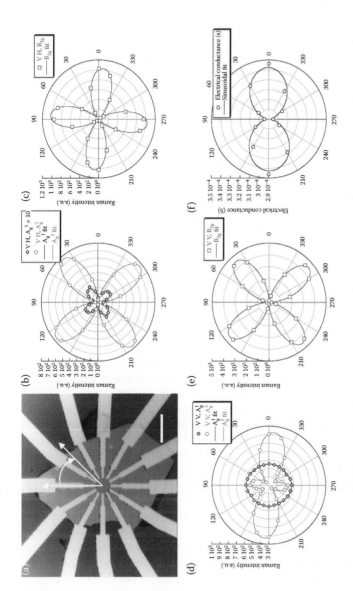

FIGURE 15.3 Polarised Raman and electrical conductance measurements on a 32-nm-thick phosphorene flake. (a) Optical image showing the flake and the electrodes. Scale bar is 10 µm. Angle-resolved polarised Raman intensity of (b) A_g modes (c) B_{2g} mode in VH configuration and (d) A_g modes (e) B_{2g} mode in VV configuration. In (b)–(e), solid lines are all curve fits using equation. (f) Angle-resolved electrical conductance measured by six pairs of electrodes using 50 mV voltage. Solid line is the sinusoidal curve fit. (Adapted from Z. Luo et al., arXiv:1503.06167, 2015.)

A_{g1} phonon vibrations are out-of-plane. Therefore, the A_{g2}/A_{g1} intensity ratio becomes larger (~2) with armchair-polarised laser excitation, and smaller (~1) with zigzag-polarised laser excitation, which serves as Raman signatures of armchair and zigzag lattice axes. However, we have to notice that the laser may damage the phosphorene flakes if the flakes are too thin (single layer or a few layers), or the laser power is high enough [11]. This is because the laser heating may accelerate the degradation of phosphorene in air, which will be discussed later in this chapter. Thus, a capping layer on top of phosphorene or a careful choice of laser power is necessary for any laser processing of phosphorene. Photoluminescence (PL) was studied on phosphorene with different thicknesses, as well as its angle dependence characteristics [11,45,46]. The PL wavelength increases as the thickness of phosphorene increases from monolayer to five layers, indicating that the energy gap decreases as the thickness increases. However, we note that the reported PL peak values on monolayer phosphorene are different in different papers, while there is one paper that reported no peak from encapsulation phosphorene [11,45,46,48]. It is still necessary to make more efforts to understand the reason for this phenomenon. Transmission electron microscopy and scanning tunnelling microscope images were also obtained from multi-layer phosphorene samples [49,50].

The earliest study on 2D phosphorene was focussed on its electrical properties and its promising use in thin film field-effect transistors (FETs). The electrical conductivity is anisotropy in phosphorene flake as determined through electrical measurements (**Figure 15.3**). By combining with the Raman spectrum, it is easy to determine the more conductive direction, and make FETs on it to achieve larger drive current. The fabricated transistor showed a calculated field-effect mobility from several hundred to more than one thousand (cm^2/V s). It was found that the field-effect mobility of few-layer phosphorene FET strongly depends on the thickness of the phosphorene. A 5–10 nm thickness may be beneficial for a larger field-effect mobility for FET when we consider the optimisation of the on/off ratio of FET [11]. The Hall mobility at low temperature was measured to be 1000 cm^2/V s in the light effective mass direction for multi-layer phosphorene (>10 nm). It was also found that the Hall mobility showed a strong dependence on phosphorene thickness, a thicker flake often showed a larger Hall mobility [48]. Inspired by the high mobility obtained from the hBN/graphene vertical van der Waals heterostructure, efforts have been made in combining the 2D hBN with phosphorene to realise a heterostructure to reach a higher mobility [48,51–53]. It was believed that a higher Hall mobility of up to more than 4000 cm^2/V s on multi-layer (>10 layer) phosphorene was due to the flatter surface of 2D hBN compared with SiO_2. Moreover, an hBN top layer was proved to be effective as an encapsulation layer for the phosphorene devices to survive for even several months [48]. This makes it possible to investigate the transport properties even in single-layer phosphorene. Note that the Hall mobilities decrease dramatically from tri-layer, bi-layer to monolayer, from 1200, 80 to 1 cm^2/V s, which is way lower than that in bulk black phosphorus [48]. Thanks to the high mobility of phosphorene, quantum oscillation was observed on multi-layer phosphorene, even down to tri-layers (**Figure 15.4**).

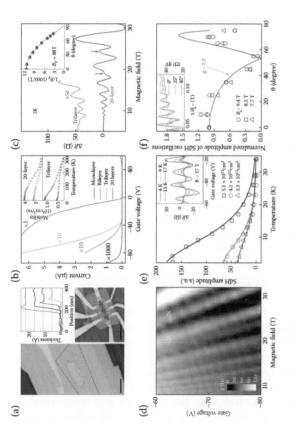

FIGURE 15.4 (a) Left: Tri-layer phosphorene (outlined in red and partially folded) is encapsulated with monolayer hBN outlined with the black lines. Right top: AFM measurements of thickness for several encapsulated BP samples. The dashed lines correspond to the interlayer spacing of 5.5 Å. Bottom: optical micrograph of a typical Hall bar devices. Scale bars: 5 μm. (b) Source–drain current at 10 K as a function of V_g for phosphorene devices of different thickness, bias voltage 30 mV. Inset: T dependences of μ found using Hall and field-effect measurements (red and blue curves, respectively) for 3- and 20-layer BP. (c) Changes in resistance (ΔR) for the devices in (b). The red arrows mark spin-splitting of Landau levels. Inset: angle dependence of the oscillation frequency. (d) Colour map $\rho(V_g, B)$ for the 20-layer device. Navy to white: ρ changes by 115 Ω. (e) T dependence of SdH oscillations. Inset: examples at different temperatures. Their amplitude can be fitted by the Lifshitz–Kosevich formula (solid curves in the main plot). (f) Amplitude of SdH oscillations in the 20-layer BP for different orientations of magnetic field. Examples are shown in the inset. (Adapted from Y. Cao et al., arXiv:1502.03755, 2015.)

Detailed studies on the optical properties of phosphorene have also been made. A simple model has been proposed to model the optical properties of multi-layer phosphorene (10–60 layers), in which the disorder effect was included [54]. It was found that the optical conductivity and the absorption spectra were sensitive to the thickness of phosphorene, doping and light polarisation, especially in the frequency range of 2500–5000 cm^{-1}, indicating the flexibility of multi-layer phosphorene for use in infrared optoelectronics. The screening properties of multi-layer phosphorene were also studied by the same group [55]. The optical properties of monolayer phosphorene were studied by researchers from the Yale University [46]. They found the highly anisotropic and tightly bound excitons in monolayer phosphorene using polarisation-resolved PL measurements at room temperature. It was found that the emitted light from the monolayer phosphorene was linearly polarised along the light effective mass direction regardless of the excitation laser polarisation, which is a clear signature of emission from highly anisotropic bright excitons. Moreover, PL excitation spectroscopy revealed a quasiparticle band gap of 2.2 eV, from which an exciton binding energy of around 0.9 eV was estimated, consistent with theoretical results based on first principles. The observation of highly anisotropic, bright excitons with large binding energy showed the potential for the future explorations of many-electron effects in 2D phosphorene, and also suggested its promising future for use in optoelectronic devices such as on-chip infrared light sources.

The anisotropic lattice structure attracted the interests of researchers to work on its thermal properties. The thermoelectric power of bulk black phosphorus has been reported, indicating that BP could be used as an efficient thermoelectric material at around 380 K [56]. First principles studies also showed the potential of phosphorene in thermoelectric applications due to its anisotropic lattice structure [57,58]. It was found that the armchair direction showed high electrical conductivity and low lattice thermal conductivity, which is desirable for thermoelectrics. On the contrary, such orthogonal electron and phonon transport of BP may not be favourable in a typical FET as low thermal conductivity in the channel direction can lead to thermal management issues. An experimental study was then performed to elucidate the anisotropic thermal transport in multi-layer phosphorene [47]. The anisotropic in-plane thermal conductivity of suspended few-layer phosphorene was measured by micro-Raman spectroscopy. The armchair and zigzag thermal conductivities were ~20 and ~40 W/mK for multi-layer phosphorene films thicker than 15 nm, respectively and decreased to ~10 and ~20 W/mK as the film thickness was reduced, showing significant anisotropy of in-plane thermal transport and strong surface scattering of acoustic phonons. The thermal conductivity anisotropic ratio was found to be ~2 for the thicker phosphorene film, and drops to ~1.5 for the thinner film (9.5 nm). First principles calculations showed that the observed anisotropy was related to the asymmetric phonon dispersion, whereas the intrinsic phonon scattering rates were found to be similar along the armchair and zigzag directions. Surface scattering in the phosphorene thin films was shown to strongly suppress the contribution of long-mean-free-path acoustic phonons (**Figure 15.5**).

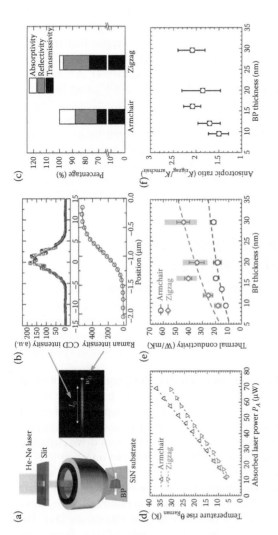

FIGURE 15.5 Thermal conductivity measurements of BP using micro-Raman technique. (a) Illustration of the experimental setup and an optical image of the produced laser focal line. (b) The lengthwise profile and the knife-edge-measured widthwise integrated profile of the laser focal line. The solid lines are Gaussian function and error function curve fits, respectively. (c) The optical absorptivity A, reflectivity R and transmissivity T of the 9.5-nm-thick suspended BP film upon armchair- and zigzag-polarised laser incidence. The uncertainty of absorptivity is ~0.2%. (d) Laser power-dependent temperature rise (θRaman) of the 16.1-nm-thick BP film determined by the micro-Raman spectroscopy along armchair and zigzag transport directions. The dashed lines are linear fits. (e) Extracted armchair and zigzag in-plane thermal conductivities of multiple BP films. Dashed lines are first principles-based calculation results. The grey error bars account for the uncertainty of SiN substrate thermal conductivity kSiN, whereas the blue/red error bars do not. (f) The anisotropic ratio kzigzag/karmchair at different BP thicknesses. The ratio at 12 nm is calculated using linearly interpolated armchair thermal conductivity from adjacent thicknesses. (Adapted from Z. Luo et al., arXiv:1503.06167, 2015.)

The mechanical properties of phosphorene are also interesting to study due to its anisotropy lattice structure. Most of the studies were based on modelling and simulations. The impact of strain on the band structure of single-layer phosphorene was studied. It was found that the energy ordering of the conduction band valleys changed with strain, so that it was possible to switch from a nearly direct band gap semiconductor to an indirect semiconductor, semimetal and metal with the compression along only one direction [59]. The Young's modulus was also studied based on first principles calculations [60]. The in-plane Young's modulus in the direction perpendicular to the pucker was found to be only half of that in the parallel direction, while the ultimate strain was much larger in the direction perpendicular to the pucker. The Poisson's ratio in phosphorene was studied by the same group [61]. They pointed out the existence of a negative Poisson's ratio in phosphorene. In contrast to engineered bulk auxetics, this behaviour was shown to be intrinsic for single-layer phosphorene, and originated from its puckered structure, where the pucker can be regarded as a re-entrant structure that was comprised of two coupled orthogonal hinges. A negative Poisson's ratio was observed in the out-of-plane direction under uniaxial deformation in the direction parallel to the pucker.

15.4 Nanoelectronic and Optoelectronic Applications of Phosphorene

As 2D phosphorene shows excellent electrical transport properties and optical properties, researchers are trying to explore its promising applications in nanoelectronics and optoelectronics. In this section, we will first discuss the recent progress in electronic devices based on phosphorene, as well as its device passivation study. Then, we will discuss optoelectronic devices based on phosphorene.

As the reason people started to work on 2D phosphorene was its high mobility and reasonable band gap for FET applications, it was easy to find researchers from different groups working on different aspects of phosphorene FET. The earliest papers demonstrated the p-FET based on multi-layer phosphorene on a SiO_2/Si back gate structure. The maximum drive current of the on state could reach 200 mA/mm, while maintaining an on/off ratio of 10^5 [11–13]. The effects of metal contacts on phosphorene FETs were then studied [62,63]. Similar to the FETs device made on other nanomaterials such as carbon nanotube, graphene and TMD, the FETs made on phosphorene by metal contacts also made them the typical Schottky barrier transistors [64]. This means the operation mechanisms of the FETs are quite different from traditional Si FETs, in which heavily doped Si S/D region and Ohmic contacts are made. In the cases of FETs made on nanomaterials, there is often a Schottky barrier between the metal and the channel material, as the nanomaterials are often hard to dope through the conventional doping method. This barrier can be modulated by the back gate if the device is made on a back gate structure, and can result in a quite different device operation phenomenon compared with normal Si FETs, as can be found in the cases of carbon nanotube FETs [65]. It was found that, unlike in the case of MoS_2, the metal with different work functions strongly influenced the performance of the FET. The metal with higher work function was found to align closer to the valence band, and could be beneficial for the holes conduction for the p-FET. This could lead to a higher on-state drive current [62,63]. If a metal with lower work function was used, the ambipolar conduction or even n-FET could be obtained. Moreover, the contact resistance was calculated for different metal contacts, it was proved that the FETs with higher work function metal contacts really showed a lower contact resistance for the p-FET [63]. The effect of channel length scaling was also studied [63]. The ambipolar characteristics were found to be more obvious as the channel lengths were scaled down, due to the stronger effect, the drain voltage could make on the Schottky barrier. The maximum current was found to be saturated if the channel length was scaled down to 500 nm due to relatively large contact resistance. It was also noted that the intrinsic mobility decreased as the channel length scaled down due to velocity saturation. The relatively large contact resistance was found to be an important limiting factor for the further scaling of phosphorene FETs. The scaling property of the dielectric was also studied based on a back gate structure. A 20 nm HfO_2 and Pd was used as a gate dielectric and gate, instead of SiO_2 and Si [66]. The scaled gate dielectric resulted in a maximum on-state current of 200 mA/mm at a channel length of 1 μm. The device also displayed extrinsic transconductance, g_m of 101 μS/μm at a source–drain bias $V_{ds} = -3$ V. Temperature-dependent analysis showed that the subthreshold slope (SS) was nearly ideal, with a minimum value of SS = 66 mV/decade at room temperature and $V_{ds} = -0.1$ V. Moreover, the device with 7-nm HfO_2 dielectrics and 0.3 μm channel length displayed g_m of up to 204 μS/μm at $V_{ds} = -1.5$ V. In addition to the back gate structure, the top gate structure, which is a more practical way of device and circuit fabrication, was used to fabricated FETs and invertors based on atomic layer deposition (ALD) [11]. The ratio frequency FET was also demonstrated based on the top gate structure [67]. The devices showed excellent current saturation with an on/off ratio exceeding 2×10^3. The drive current of 270 mA/mm and DC transconductance above 180 mS/mm were achieved for holes conduction. Using standard high frequency characterisation techniques, the short-circuit current-gain cut-off frequency f_T of 12 GHz and a maximum oscillation frequency f_{max} of 20 GHz in 300 nm channel length devices were obtained. This showed that phosphorene RF devices may offer advantages over graphene transistors for high frequency electronics in terms of voltage and power gain due to the good current saturation properties arising from their finite band gap (**Figure 15.6**).

As researchers were working on the phosphorene device, the degradation of phosphorene was found to be an important limiting factor of real applications phosphorene devices. Although black phosphorus is the most stable allotrope of phosphorus, and it is stable enough to survive through the standard lithography process, it still degrades as observed during the electrical measurements [11,49]. Different characterisation techniques were used to study the mechanism and the passivation method [68–70]. Through optical

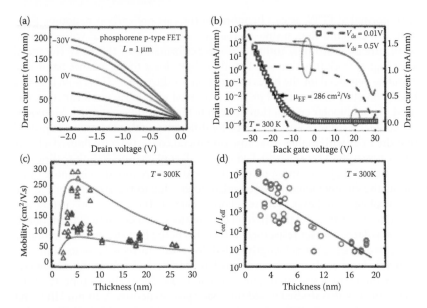

FIGURE 15.6 Device performance of p-type transistors based on few-layer phosphorene. Output (a) and transfer (b) curves of a typical few-layer phosphorene transistor with a film thickness of ~5 nm. The arrow directions are also back gate bias sweeping directions. (c) Mobility summary of few-layer phosphorene and black phosphorus thin film transistors with varying thicknesses. Red and green lines are models after Reference 14 with light and heavy hole masses for phosphorene, respectively. (d) Current on/off ratio summary of few-layer phosphorene and black phosphorus thin film transistors with varying thicknesses. (Reprinted with permission from H. Liu et al., *ACS Nano* 2014, 8, 4033–4041. Copyright 2014 American Chemical Society.)

microscopy and AFM, people could observe the degradation phenomenon. It was found that 'bubbles' were observed on the surface of the phosphorene flakes after the phosphorene sample was put in air for a short time. Then these bubbles become bigger, and finally, a huge bubble was formed on top of the whole flake, which makes the volume of the flake increase many times [68]. It was also found that the very thin flake could be etched away and finally disappear after a long time in air. Simulation and experiment suggested that the water and oxygen in air played an important role in the degradation of phosphorene. As phosphorene showed strong affinity for water, so the oxygen saturated water irreversibly reacts with phosphorene to form an oxidised phosphorus species or related acid, which formed the 'bubbles', and could be the main reason for the degradation [68]. Thus, the fabricated FET transistors also suffered from the degradation. Upon exposure to air, it was observed that after a short timescale (minutes), there was a shift in the threshold voltage that occurs due to physically absorbed oxygen and nitrogen. Then after a long timescale (hours), strong p-type doping occurred from water/oxygen absorption. Continuous measurements of phosphorene FETs in air revealed eventual degradation and breakdown of the channel material after several days due to the etching process. This shows that a passivation technique is really necessary for real applications (**Figure 15.7**).

Many groups have worked on a passivation method for phosphorene devices. Generally, ALD capping and hBN capping were studied as two passivation methods. ALD is widely used in the industry for device fabrication, which quickly came to our mind as a possible passivation method due to its easy process. Moreover, the ALD dielectric can be used as a top gate dielectric, which makes phosphorene devices with an ultra-thin gate dielectric possible.

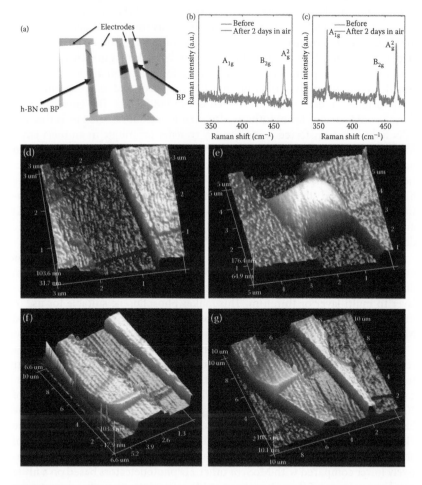

FIGURE 15.7 Effect of hBN passivation on phosphorene. (a) Optical image of device with and without h-BP on top. Raman spectrum of the device (b) without and (c) with hBN passivation of the as-fabricated device and put in air after 2 days. (d) and (e) show the AFM image of the phosphorene device without passivation, before and after 2 days in air. (f) and (g) show the AFM image of the phosphorene device with hBN passivation, before and after 2 days in air.

This makes phosphorene more compatible for the standard device fabrication process. Several groups have reported ALD passivation using Al_2O_3 or HfO_2 [67,69,70]. It was found that about 30 nm ALD Al_2O_3 grown at low temperature (150°C) could make the device stable for weeks [69]. However, it was also noted that a relatively thin dielectric (several nanometres) could not protect the phosphorene devices effectively [70]. By applying a fluoropolymer on top of the thick Al_2O_3 (25 nm), the fabricated device could be stable for months. This makes the phosphorene devices really promising for real applications. As the ALD process involves in different precursors, we need to point out that these precursors may react with phosphorene and degrade the quality of phosphorene. For example, water is used as one of the precursors of ALD Al_2O_3 process, which reacts with phosphorene as mentioned above. Thus, a low temperature growth or even room temperature growth at the beginning step was used [69,70]. However, the quality of the dielectric grown at low temperature may be a problem if we are looking for a high-quality dielectric for real device applications. In addition, hBN was also used as a capping layer for phosphorene devices as mentioned above [48,51–53]. The encapsulation device could survive for months as also mentioned above. The key is that the hBN flake was transferred using the dry transfer technique in an inert environment to avoid phosphorene degradation. Then, a 2D to 2D interface could be formed between hBN and phosphorene through van der Waals interactions. Although the hBN/phosphorene/hBN device structure could achieve high mobility and stable operation due to the ideal 2D interface, the fabrication process is complicated and hard to use for real applications. Thus, ALD and other possible easy-process method may be favourable for the electronic device applications of phosphorene.

Thanks to its thickness-dependent direct band gap, phosphorene was explored for use in optoelectronic devices. The band gap of phosphorene can decrease from around 1.4 to 0.3 eV as the thickness of it decreases from monolayer to multi-layer or thin film. This makes phosphorene suitable for use in infrared and near-infrared optoelectronics, compared with graphene and MoS_2 [71].

Several groups have worked on the photodetector application of phosphorene. The first demonstration was based on the structure which is very similar to back gate phosphorene FET [72]. Multi-layer phosphorene FETs were fabricated on a SiO_2/Si substrate and the device showed mobilities in the order of 100 cm²/V s and current on/off ratio larger than 10^3. Upon illumination, the black phosphorus FET showed response to excitation wavelengths from the visible wave length to 940 nm. The minimum rise time was measured to be of about 1 ms. However, the responsivity only reached 4.8 mA/W, which is relatively lower compared with MoS_2 or other 2D photodetectors based on a finite band gap material. Efforts have been made on optimisation of the responsivity through contact engineering [62]. It was found that by using a metal with higher work function, the responsivity could be substantially increased up to 223 mV/W under a back gate voltage of −30 V or 76 mA/W without a back gate voltage. The photodetector was also demonstrated to be used for image detecting both in the visible ($\lambda = 532$ nm) as well as in the infrared ($\lambda = 1550$ nm) spectral regime [73]. The multi-layer phosphorene device was demonstrated as a point-like detector in a confocal microscope setup. They acquired diffraction-limited optical images with sub-micron resolution. By integrating with silicon wave guide structure operating in the near-infrared telecom band, phosphorene photodetectors could operate under a bias with very low dark current and attain intrinsic responsivity up to 135 and 657 mA/W in 11.5 and 100 nm thick devices, respectively, at room temperature [74]. The photocurrent was found to be dominated by the photovoltaic effect with a high response bandwidth exceeding 3 GHz. However, the mechanism of photodetection in phosphorene is still under debate. Although several papers reported that the photovoltaic effect was the dominant factor, a researcher from IBM still found that the thermally driven thermoelectric and bolometric processes could be dominant [75]. Thus, more efforts should be made on elucidating the photodetection dominant mechanism in phosphorene.

Several approaches have been made to fabricate a p–n diode: one of the basic elements of optoelectronics. The earliest reported work was based on the van der Waals heterostructure of multi-layer phosphorene and monolayer MoS_2 [76]. The rectifying IV characteristics were demonstrated with a rectifying ratio of 10^5, which is tunable by the back gate. Upon illumination, these ultra-thin p–n diodes show a maximum photodetection responsivity of 418 mA/W at the wavelength of 633 nm, and photovoltaic energy conversion with an external quantum efficiency of 0.3%. The p–n diode was also demonstrated based on a local back gate structure [77]. They demonstrated electrostatic control of the local charge carrier type and density above each gate in the phosphorene device, tuning its electrical behaviour from metallic to rectifying. Owing to the small band gap of the material, they could observe photocurrents and photo-voltages for illumination wavelengths of up to 940 nm, attractive for energy harvesting applications in the near-infrared region. Finally, an ionic gel gated phosphorene vertical p–n diode was also demonstrated [78]. They demonstrated a linear-dichroic broadband phosphorene photodetector, using the intrinsic linear dichroism arising from the in-plane optical anisotropy, which is polarisation sensitive over a broad bandwidth from 400 to 3750 nm. A perpendicular build-in electric field induced by gating in phosphorene transistors can spatially separate the photo-generated electrons and holes in the channel, effectively reducing their recombination rate, and thus enhancing the efficiency and performance for linear dichroism photodetection. Although progress has been made on phosphorene optoelectronic devices, it is still a long way to optimise and make the device really possible for applications (**Figure 15.8**).

15.5 Conclusion

The recent research studies on 2D phosphorene have attracted lots of research interest from the materials and electronics community. The high carrier mobility, a relatively narrow thickness-dependent direct band gap, anisotropy in electrical, optical, thermal and mechanical properties make it possible for

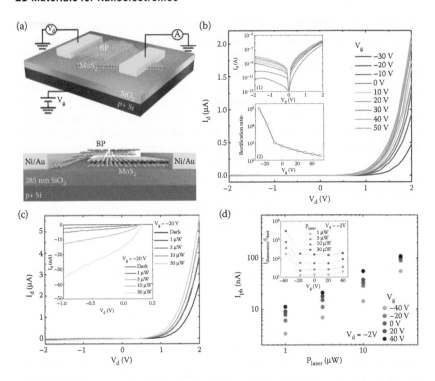

FIGURE 15.8 (a) Schematics of the device structure. A p+ silicon wafer capped with 285 nm SiO_2 is used as the global back gate and the gate dielectric. Few-layer phosphorene flakes were exfoliated onto monolayer MoS_2 in order to form a van der Waals heterojunction. (b) Gate tunable IV characteristics of the 2D p–n diode. The current increases as the back gate voltage increases. The inset (1) shows the IV characteristics under semi-log scale. The inset (2) shows the rectification ratio as a function of back gate voltage V_g. (c) IV characteristics of the p–n diode under various incident laser powers. The inset shows the details in the reverse bias region. (d) The photocurrent as a function of incident laser power. Increasing the back gate voltage can increase the photocurrent. The inset shows the ratio of Iillumination/Idark. (From Y. Deng et al., *ACS Nano* 2014, 8 (8), 8292–8299.)

applications in different fields. We especially focussed on the recent works about this novel 2D material for various electronic and optoelectronic applications. It was found that the material could be suitable for channel material for thin film transistors, but there are still many engineering problems such as reducing the contact resistance and increase the maximum on-state current. Moreover, the stability and its passivation method are still important issues. Photodetectors and p–n diodes have been demonstrated on phosphorene by different approaches, but a further understanding of the dominant mechanism in the device, and optimisation of the device structure to make the device competitive compared with other 2D optoelectronic devices, or even industrial level devices are still needed. We hope the community will keep working on it, and explore the limits of this novel 2D material.

Acknowledgements

The authors thank the U.S. National Science Foundation (ECCS-1449270) and the Army Research Office (W911NF-14-1-0572) for supporting this work.

References

1. A. Geim, K. S. Novoselov, *Nat. Mater.* 2007, 6, 183–191.
2. A. Geim, *Science* 2009, 324, 1530–1534.
3. K. S. Novoselov, V. I. Fal'ko, L. Colombo, P. R. Gellert, M. G. Schwab, K. Kim, *Nature* 2012, 490, 192–200.
4. F. Schwierz, *Nat. Nanotechnol.* 2010, 5, 487–496.
5. Q. Bao, K. P. Loh, *ACS Nano* 2012, 6, 3677–3694.
6. K. Novoselov, D. Jiang, F. Schedin, T. Booth, V. Khotkevich, S. Morozov, A. Geim, *Proc. Natl. Acad. Sci. USA.* 2005, 102, 10451–10453.
7. M. Xu, T. Liang, M. Shi, H. Chen, *Chem. Rev.* 2013, 113, 3766–3798.
8. Q. H. Wang, K. Kalantar-Zadeh, A. Kis, J. N. Coleman, M. S. Strano, *Nat. Nanotechnol* 2012, 7, 699–712.
9. D. Jariwala, V. K. Sangwan, L. J. Lauhon, T. J. Marks, M. C. Hersam, *ACS Nano* 2014, 8, 1102–1120.
10. G. Lee, Y. Yu, X. Cui, N. Petrone, C. Lee, *ACS Nano* 2013, 7, 7931–7936.
11. H. Liu, A. T. Neal, Z. Zhu, Z. Luo, X. Xu, D. Tománek, P. D. Ye, *ACS Nano* 2014, 8, 4033–4041.
12. L. Li, Y. Yu, G. J. Ye, Q. Ge, X. Ou, H. Wu, D. Feng, X. H. Chen, Y. Zhang, *Nat. Nanotechnol.* 2014, 9, 372–377.
13. F. Xia, H. Wang, Y. Jia, *Nat. Commun.* 2014, 5, 4458.
14. P. W. Bridgman, *J. Am. Chem. Soc.* 1914, 36, 1344–1363.
15. R. W. Keyes, *Phys. Rev.* 1953, 92, 580–584.
16. R. B. Jacobs, *J. Chem. Phys.* 1937, 5, 945–953.
17. H. Krebs, H. Weitz, K. H. Z. Worms, *Anorg. Allg. Chem.* 1955, 280, 119–133.
18. A. Brown, S. Rundqvist, *Acta Cryst.* 1965, 19, 684–685.
19. Y. Maruyama, T. Inabe, T. Nishii, L. He, A. J. Dann, I. Shirotani, M. R. Fahy, M. R. Willis, *Synth. Met.* 1989, 29, 213–218.
20. Y. Maruyama, T. Inabe, L. He, K. Oshima, *Synth. Met.* 1991, 43, 4067–4070.
21. S. Lange, P. Schmidt, T. Nilges, *Inorg. Chem.* 2007, 46, 4028–4035.
22. M. Köpf, N. Eckstein, D. Pfister, C. Grotz, I. Krüger, M. Greiwe, T. Hansen, H. Kohlmann, T. Nilges, *J. Cryst. Growth* 2014, 405, 6–10.
23. R. Hultgren, S. N. Gingrich, B. E. Warren, *J. Chem. Phys.* 1935, 3, 351–355.
24. J. C. Jamieson, *Science* 1963, 139, 1291–1292.
25. L. Cartz, S. R. Srinivasa, R. J. Riedner, J. D. Jorgensen, T. G. Worlton, *J. Chem. Phys.* 1979, 71, 1718–1721.
26. T. Kikegawa, H. Iwasaki, *Acta Cryst.* 1983, 39, 158–164.
27. S. M. Clark, J. M. Zaug, *Phys. Rev. B* 2010, 82, 134111.
28. J. K. Burdett, S. Lee, *J. Solid State Chem.* 1982, 44, 415–424.

29. I. Shirotani, A. Fukizawa, H. Kawamura, T. Yagi, S. Akimoto, *Solid State Physics under Pressure*, 1985, KTK, Tokyo, p. 207.
30. Y. Takao, H. Asahina, A. Morita, *J. Phys. Soc. Jpn.* 1981, 50, 3362–3369.
31. H. Asahina, K. Shindo, A. Morita, *J. Phys. Soc. Jpn.* 1982, 51, 1192–1199.
32. N. B. Goodman, L. Ley, D. W. Bullett, *Phys. Rev. B* 1983, 27, 7440–7450.
33. P. W. Bridgman, *Proc. Am. Acad. Arts Sci.* 1921, 56, 126–131.
34. D. Warschauer, *J. Appl. Phys.* 1963, 34, 1853–1860.
35. Y. Akahama, S. Endo, S. Narita, *Physica* 1986, 139–140B, 397–400.
36. M. Baba, F. Izumida, Y. Takeda, K Shibata, A. Morita, Y. Koike, T. Fukase, *J. Phys. Soc. Jpn.* 1991, 60, 3777–3783.
37. Y. Akahama, S. Endo, S. Narita, *J. Phys. Soc. Jpn.* 1983, 52, 2148–2155.
38. M. Baba, F. Izumida, A. Morita, Y. Koike, T. Fukase, *Jpn. J. Appl. Phys.* 1991, 30, 1753–1758.
39. M. Baba, Y. Nakamura, Y. Takeda, K. Shibata, A. Morita, Y. Koike, T. Fukase, *J. Phys., Condens. Matter* 1992, 4, 1535–1544.
40. K. Ito, S. Endo, *Solid State Commun.* 1980, 36, 701–702.
41. M. Okajima, S. Endo, Y. Akahama, S. Narita, *Jpn. J. Appl. Phys.* 1984, 23, 15–19.
42. A. Morita, *Appl. Phys. A* 1986, 39, 227–242.
43. J. Wittig, B. T. Matthias, *Science* 1968, 160, 994–995.
44. H. Kawamura, I. Shirotani, K. Tachikawa, *Solid State Commun.* 1984, 49, 879–881.
45. S. Zhang, J. Yang, R. Xu, F. Wang, W. Li, M. Ghufran, Y.-W. Zhang et al., *ACS Nano* 2014, 8, 9590–9596.
46. X. Wang, A. M. Jones, K. L. Seyler, V. Tran, Y. Jia, H. Zhao, H. Wang, L. Yang, X. Xu, F. *Nat. Nanotechnol.* 2015, 10, 517–521.
47. Z. Luo, J. Maassen, Y. Deng, Y. Du, M. S. Lundstrom, P. D. Ye, X. Xu. *Nat. Commun.* 2015, 6, 8572.
48. Y. Cao, A. Mishchenko, G. L. Yu, K. Khestanova, A. Rooney, E. Prestat, A. V. Kretinin et al. *Nano Lett.* 2015, 15 (8), 4914–4921.
49. A. Castellanos-Gomez, L. Vicarelli, E. Prada, J. O. Island, K. L. Narasimha-Acharya, S. I. Blanter, D. J. Groenendijk et al., *2D Mater.* 2014, 1, 025001.
50. T. Hong, B. Chamlagain, W. Lin, H.-J. Chuang, M. Pan, Z. Zhou, Y.-Q. Xu, *Nanoscale* 2014, 6, 8978.
51. X. Chen, Y. Wu, Z. Wu, S. Xu, L. Wang, Y. Han, W. Ye et al. *Nat. Commun.*, 2014, 6, Article number: 7315.
52. N. Gillgren, D. Wickramaratne, Y. Shi, T. Espiritu, J. Yang, J. Hu, J. Wei et al., *2D Mater.* 2015, 2, 011001.
53. L. Li, G. Ye, V. Tran, R. Fei, G. Chen, H. Wang, J. Wang et al. *Nat. Nanotechnol.* 2015, 10, 608–613.
54. T. Low, A. S. Rodin, A. Carvalho, Y. Jiang, H. Wang, F. Xia, A. H. Castro Neto, *Phys. Rev. B* 2014, 90, 075434.
55. T. Low, R. Roldán, H. Wang, F. Xia, P. Avouris, L. M. Moreno, F. Guinea, *Phys. Rev. Lett.* 2014, 113, 106802.
56. E. Flores, J. R. Ares, A. Castellanos-Gomez, M. Barawi, I. J. Ferrer, C. Sánchez, *Appl. Phys. Lett.* 2015, 106, 022102.
57. R. Fei, A. Faghaninia, R. Soklaski, J. Ya, C. Lo, L. Yang, *Nano Lett.* 2014, 14, 6393–6399.
58. A Jain, A. McGaughey, *Sci. Rep.* 2015, 5, 8501.
59. A. S. Rodin, A. Carvalho, A. H. Castro Neto, *Phys. Rev. Lett.* 2014, 112, 176801.
60. J. Jiang, H. S. Park, *J. Phys. D: Appl. Phys.* 2014, 47, 385304.
61. J. Jiang, H. S. Park, *Nat. Commun.* 2014, 5, 4727.

62. Y. Deng, N. J. Conrad, Z. Luo, H. Liu, X. Xu, P. D. Ye, *IEEE International Electron Devices Meeting (IEDM)*, San Francisco, CA, USA, 2014, 5.2.1–5.2.4.
63. Y. Du, H. Liu, Y. Deng, P. D. Ye, *ACS Nano* 2014, 8, 10035–10042.
64. S. Heinze, J. Tersoff, R. Martel, V. Derycke, J. Appenzeller, P. Avouris, *Phys. Rev. Lett.* 2002, 89, 106801.
65. J. Appenzeller, Y.-M. Lin, J. Knoch, P. Avouris, *Phys. Rev. Lett.* 2004, 93, 196805.
66. N. Haratipour, M. C. Robbins, S. J. Koester. *IEEE Electron Device Letters*, 2014, 36(4), 411–413.
67. H. Wang, X. Wang, F. Xia, L. Wang, H. Jiang, Q. Xia, M. L. Chin, M. Dubey, S. Han, *Nano Lett.* 2014, 14 (11), 6424–6429.
68. J. O. Island, G. A. Steele, H. S. J. van der Zant, A. Castellanos-Gomez, *2D Mater.* 2015, 2, 011002.
69. J. D. Wood, S. A. Wells, D. Jariwala, K. S. Chen, E. K. Cho, V. K. Sangwan, X. Liu, L. J. Lauhon, T. J. Marks, M. C. Hersam, *Nano Lett.* 2014, 14 (12), 6964–6970.
70. J. Kim, Y. Liu, W. Zhu, S. Kim, D. Wu, L. Tao, A. Dodabalapur, K. Lai, D. Akinwande, *Sci. Rep.* 2015, 5, 8989.
71. F. Xia, H. Wang, D. Xiao, M. Dubey, A. Ramasubramaniam, *Nat. Photonics* 2014, 8, 899–907.
72. M. Buscema, D. Groenendijk, S. Blanter, *Nano Lett.* 2014, 14 (6), 3347–3352.
73. M. Engel, M. Steiner, P. Avouris, *Nano Lett.* 2014, 4 (11), 6414–6417.
74. N. Youngblood, C. Chen, S. J. Koester, M. Li, *Nat. Photonics* 2015, 9, 247–252.
75. T. Low, M. Engel, M. Steiner, P. Avouris, *Phys. Rev. B* 2014, 90, 081408R.
76. Y. Deng, Z. Luo, N. J. Conrad, H. Liu, Y. Gong, S. Najmaei, P. M. Ajayan, J. Lou, X. Xu, P. D. Ye, *ACS Nano* 2014, 8 (8), 8292–8299.
77. M. Buscema, D. J. Groenendijk, G. A. Steele, H. S. J. van der Zant, A. Castellanos-Gomez, *Nat. Commun.* 2014, 5, 4651.
78. H. Yuan, X. E. Liu, F. Afshinmanesh, W. Li, G. Xu, J. Sun, B. Lian et al. *Nat. Nanotechnol.* 2015, 10, 707–713.

16

2D Crystal-Based Heterostructures for Nanoelectronics

Cinzia Casiraghi and Freddie Withers

Contents

2D Materials for Nanoelectronics edited by Michel Houssa, Athanasios Dimoulas and Alessandro
Molle © 2016 CRC Press/Taylor & Francis Group, LLC. ISBN: 978-1-4987-0417-5.

16.1 Introduction: A 'Legoland' of Two-Dimensional Materials

Heterostructures have already played a crucial role in technology, giving us semiconductor lasers, light-emitting diodes and fast electronic switches [1]. However, thus far the choice of materials has been extended to those that can be grown (typically by molecular beam epitaxy) one on top of another, thus limiting the types of heterostructures that can be prepared. Instead, two-dimensional (2D) crystals, characterised by out-of-plane Van der Waals (VdW) interactions, can be easily combined in one stack with atomic precision, similar to 'LEGO bricks', offering unprecedented control on the properties and functionalities of the resulting heterostructures [2,3], **Figure 16.1**. Such heterostructures do not suffer from lattice mismatch requirements because of the lack of out-of-plane covalent bonds. Interactions and transport between the layers allow one to go beyond simple incremental improvements in performance: the resulting three-dimensional (3D) structures can combine the conductivity of one 2D crystal, strength of another, chemical reactivity of the third, while the optical properties will be determined by the whole heterostructure. By carefully choosing and arranging the individual components, one can produce materials with tailored properties, determined by the design of the material itself. In addition to 2D crystals, alternative classes of nanostructures can also be incorporated into the heterostructure: from plasmonic nanostructures for improving optical absorption [4] to high quality organic crystalline films, useful for a variety of electronic and opto-electronic applications [5]. One of the major advantages of this technology is given by its compatibility with soft polymeric or plastic substrates: the 2D thickness allows for the maximum amount of mechanical flexibility [6]. Therefore, 2D crystal-based heterostructures are expected to produce a strong impact on future flexible and transparent electronics.

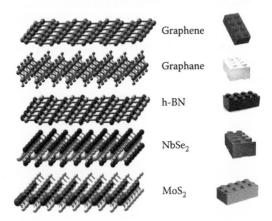

Graphene

Graphane

h-BN

NbSe$_2$

MoS$_2$

FIGURE 16.1 Schematic of the 2D crystal family and their analogy with LEGO bricks.

The class of 2D atomic crystals started with graphene – a monolayer of carbon atoms arranged into a hexagonal lattice [7]. It is a remarkable material with a set of unique properties [8–10]. It has also opened a floodgate for many other 2D crystals, including graphene derivatives [11,12], to be discovered and studied, **Figure 16.1**. Such crystals are stable, mechanically strong and carry many properties that cannot be found in their 3D counterparts. Nearly, a dozen atomically thin materials have been demonstrated so far, but the class of 2D materials is very large if one considers the existence of hundreds of layered materials such as ionic layered materials and oxides [2,13,14]. In addition, a new class of buckled graphene analogues has recently emerged and termed Xenes (e.g. silicene [15], germanene [16,17], phosphorene [18,19], etc.). Graphene stands out due to its unique electronic structure, which allows ballistic transport on a micron scale under ambient conditions [20]. Furthermore, it is the strongest material available to us, its conductivity is millions times higher than copper and it has very high thermal conductivity [2]. The 2D thickness of graphene allows for maximum electrostatic control, optical transparency, sensitivity and mechanical flexibility [6]. The other 2D crystals carry a wide range of interesting complementary properties. Of particular interest are hexagonal-boron nitride (hBN) and transition metal dichalcogenides (TMDCs). hBN is a wide band gap semiconductor (~6 eV gap) with excellent chemical and thermal stability, mechanical properties and high thermal conductivity. Currently, hBN is used as substrate in graphene-based devices because its atomically flat surface, absence of charged impurities and dangling bonds strongly reduce charge scattering in graphene, allowing for the highest mobility to be achieved in supported graphene-based devices [21]. TMDCs are structured such that each layer consists of three atomic planes: a triangular lattice of transition metal atoms sandwiched between two triangular lattices of chalcogen atoms (S, Se or Te). There exists a strong covalent bonding between the atoms within each layer and predominantly VdW bonding between adjacent layers. Those materials are structurally similar but have an array of electronic properties ranging from semiconducting to metallic, from charge density waves to superconducting, depending on their exact composition, electronic density, geometry and thickness. For example, single-layer molybdenum disulfide (MoS_2) is a semiconductor with a direct excitonic gap of approximately 1.8 eV ideal for optoelectronics; niobium diselenide ($NbSe_2$) is a superconductor with critical temperature of 7.2 K; bismuth telluride (Bi_2Te_3) is a topological insulator, etc. Few-layer and single-layer TMDCs have already displayed many novel nanoelectronic and opto-electronic phenomena such as ambipolar field-effect transistors (FETs) [22–26], photo-transistors [27], optical control of valley and spin polarisation [28–30], lateral p–n junctions and light-emitting diodes [31–34]. Recently, it has been reported that devices fabricated by encapsulating graphene with MoS_2 or WS_2 and hBN are found to exhibit consistently high carrier mobilities of about 60,000 cm^2 (V · s) [35]. Therefore, TMDCs are also excellent substrates for (encapsulated) graphene-based devices, while other materials such as atomically flat layered oxides (mica, bismuth strontium calcium copper oxide, vanadium pentoxide, etc.) give low-quality graphene devices [35]. This difference has been attributed to a self-cleansing process taking place at the interfaces between graphene, hBN and TMDCs. Surface contamination assembles into large pockets allowing the rest of the interface to become atomically clean, while this does not occur on atomically flat oxide substrates [35].

The simplest heterostructure is obtained by creating a p–n junction, which for example can be fabricated by vertically stacking a semiconductor and a metal or by stacking two semiconductors with different doping level. There are many examples of heterostructures composed of two crystals only (bilayer heterostructure), being the simplest devices to fabricate, in particular by epitaxial growth [36] or by using mechanically exfoliated crystals.

Synthesis of heterostructures is currently a major research activity as the performance of the device is strongly affected by the quality of the interfaces. Owing to the absence of broken bonds, the interfaces between two 2D crystals are expected to be pristine and with no trap states. In reality, because most of the heterostructure fabrication methods involve several transfer processes of the layers, residual contamination, strain and defects can all affect the quality of the interface. Currently, the only method available to identify the interface roughness and the electronic quality of the encapsulated components of a 2D crystal-based heterostructure is given by high-resolution cross-sectional scanning transmission electron microscopy (STEM) [37]. In case of a bilayer heterostructure containing TMDCs, Raman spectroscopy has also been suggested as qualitative technique to evaluate the quality of the contact between the two crystals [38].

Because the real potential of a heterostructure-based technology relies on the possibility to produce stacking of arbitrary complexity, in this review, we will focus on heterostructures composed by at least three layers, where every LEGO block provides a different function in the device. In the following, the heterostructure is indicated as: $Block_1/Block_2/Block_3$, etc., starting from the crystal in contact to the substrate to the crystal on the top of the heterostructure. If not specified, the substrate is highly doped silicon covered with a thin silicon oxide layer.

16.2 Handling of 2D Heterostructures: Practical Issues

Heterostructures have been initially used to study new physical phenomena such as metal–insulator transition [39] and Coulomb drag [40], and to improve the performance of graphene-based devices [41]. Those heterostructures are composed by two types of LEGO bricks only: graphene (Gr) and hBN. After hBN was demonstrated to be a suitable substrate for graphene-based devices [21], it was evident that hBN could be used also as top-gate dielectric in hBN/Gr/hBN heterostructures [41], **Figure 16.2a**. Relatively thick (~10 nm) hBN crystals were mechanically deposited on top of an oxidised Si wafer (100 nm of SiO_2). Then, sub-millimetre graphene crystallites were produced by cleavage on another substrate pre-coated with a double-layer polymer stack. The bottom polymer 'release' layer was then dissolved from the sides and the resulting film with the graphene flake was transferred on top of the chosen hBN crystal.

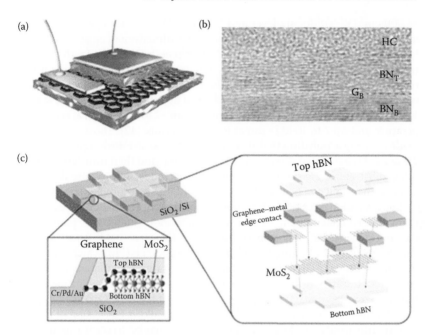

FIGURE 16.2 (a) Schematic of hBN/Gr/hBN heterostructure; (b) bright-field aberration-corrected STEM image of an enlarged region of the same heterostructure. Contamination is also visible on the top of the structure. Reproduced through/by courtesy of Sarah Haigh, University of Manchester (UK). (c) Schematic of the fabrication process used to encapsulate MoS$_2$ and to make lateral Gr/MoS$_2$ contacts. Panel (c) has been reproduced from X. Cui et al., *Nat. Nanotechnol,* 2015, doi:10.1038/nnano.2015.70 (in press) with permission from James Hone (Columbia University, USA).

In this process, it is very important not to expose the graphene surface (that goes into contact with the hBN) to any solvent, so a dry transfer technique must be used. Electron-beam lithography and oxygen plasma etching were then employed to define graphene Hall bars and finally, a second hBN crystal (~10 nm thick) was again transferred by using the same dry procedure. The top crystal was carefully aligned to encapsulate the graphene Hall bar leaving the contact regions open for depositing metal (Au/Ti) contacts. The top hBN is then used as dielectric for top gating, **Figure 16.2a**. Bright-field aberration-corrected STEM cross-sectional images show that the heterostructures produced in this way have atomically sharp interfaces, **Figure 16.2b**. The devices exhibit room-temperature ballistic transport well over a 1 μm distance and the encapsulation allows graphene to be insusceptible to the environment so that long and repeated exposure to the ambient air was found to have little effect on the device characteristics [41]. Recently, encapsulation has been extended to TMDCs [42]: hBN/MoS$_2$/hBN heterostructures, electrically contacted in a multi-terminal geometry using gate-tunable graphene electrodes (**Figure 16.2c**), show ultrahigh low-temperature mobility up to

34,000 cm^2 (V · s)$^{-1}$ for six-layer and 1000 cm^2 (V · s)$^{-1}$ for epitaxially grown monolayer MoS$_2$. In this heterostructure, one-dimensional edge contacts are fabricated by plasma etching the hBN/graphene/hBN stack. In the contact regions, graphene overlaps with MoS$_2$ and extends to the edge, where it is in turn contacted by metal electrodes, **Figure 16.2c**. Low-resistance contacts, with no thermal activation, can be achieved at sufficiently high gate voltage V_g: the contact resistance at high V_g ranges from ~2 to 20 kΩ μm at room temperature and ~0.7 to 10 kΩ · μm at low temperature [42]. Those devices were made by using a 'polydimethylsiloxane (PDMS) transfer' technique, where the 2D crystal is exfoliated directly to a PDMS stamp and then transferred to any other substrate or 2D crystal [43]. This approach does not require dissolving any polymer, so it can be used to produce contamination-free interfaces [43].

Encapsulation is particularly important for air-sensitive materials. Generally, many layered materials can be cleaved down to individual atomic planes, similar to graphene, but only a small fraction of them are stable under ambient conditions. A typical example is given by phosphorene (Ph), which reacts and decomposes in air [44]. In these cases, cleavage, transfer, alignment and encapsulation of the air-sensitive crystal into hBN has to be performed inside a glove box with controlled inert atmosphere (with level of H$_2$O and O$_2$ below 0.1 ppm) [45]. A fully motorised micro-manipulation setup allows fabrication of the heterostructure inside the glove box. This technology has been recently used to fabricate a single-layer ambipolar FET based on phosphorene. The device is stable under ambient conditions, showing that hBN is able to provide permanent protection of Ph against degradation in air, **Figure 16.3a** [45]. The efficiency of the encapsulation process can be easily tested by photoluminescence (PL) measurements: encapsulated flakes do not show any PL, in contrast to the ones exposed to air, **Figure 16.3b** and **c**. The same technology has been applied to NbSe$_2$, which is observed to remain superconducting down to the monolayer thickness [45]. This approach can significantly expand the range of experimentally accessible 2D crystals, allowing for the study of new exciting physical effects.

Note that in heterostructures made of Gr and hBN only, where the components have similar crystal structure, the possibility to tune not only the number of layers, but also the reciprocal alignment of the individual crystals, give the possibility to study new physical phenomena such as Hofstadter's butterfly [46–48]. These types of heterostructures require careful control of the reciprocal alignment between the crystals because a superlattice with a periodicity of about 15 nm is necessary to observe Hofstadter's butterfly. This condition is achieved only when Gr and hBN are aligned (mismatch angle θ < 2°) [49]. To align the crystals, the straight edges of the crystals are used as reference as they indicate principal crystallographic directions. The transfer is done under an optical microscope and the target hBN is rotated, utilising a rotation stage, relative to the graphene until their edges become parallel, **Figure 16.4a**. The alignment accuracy of this method is ~1° due to random error during transfer [47].

Raman spectroscopy has been demonstrated to be a very powerful technique in identifying aligned hBN/Gr structures (i.e. θ ≈ 0, **Figure 16.4b**): the Raman spectrum of the aligned structure is characterised by a strong broadening of the second order of the D peak (typical full width at half maximum

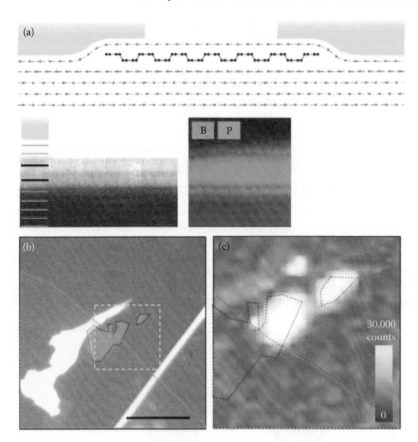

FIGURE 16.3 (a) Schematic of the hBN/Ph/hBN heterostructure made in the glove box (top); STEM image of an hBN/bilayer Ph/hBN (left) and elemental profiles superimposed to the STEM picture. (b) Image of a bilayer Ph flake (dots) partially covered with hBN (semi-transparent blue area). Scale bar: 10 μm. (c) Corresponding PL map. Reproduced through/by courtesy of Roman Gorbachev, University of Manchester (UK).

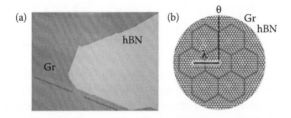

FIGURE 16.4 (a) Example of aligned hBN/Gr. (b) Because of 1.8% lattice mismatch between hBN and Gr, a superlattice with ~15 nm periodicity is formed when Gr is aligned to hBN ($\theta \approx 0$).

is ~40 cm^{-1}) [50]. This feature has been attributed to the presence of periodic strain in graphene [50]. This was confirmed experimentally by scanning probe microscopy [51]: graphene is observed to locally deform to match the crystal structure of the hBN crystal. Large graphene's areas with matching lattice constants are separated by very narrow domain walls (~2 nm) that accumulate the resulting strain. The presence of those deformations can strongly affect the electronic characteristics of hBn/Gr aligned devices [51].

16.3 Tunnel Diodes and Transistors Based on 2D Heterostructures

The first examples of a functional device based on a 2D crystal heterostructure were realised in 2011–2013, with the fabrication of (interlayer) tunnel diodes [52,53] and tunnelling transistors [54,55].

In a graphene-based tunnel diode, hBN acts as a barrier layer between two graphene electrodes, inset in **Figure 16.5a**. The I–V_b characteristics of these devices show a linear dependence at low bias and an exponential dependence at higher applied voltages, **Figure 16.5a**. The zero-bias conductivity for each type

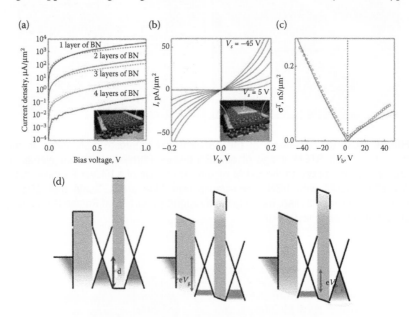

FIGURE 16.5 (a) I–V_b curve for a tunnelling diode made of different hBN layers, used as barrier. (Reproduced from L. Britnell et al., *Nano Lett.*, 12, 1707, 2012.) (b) Effect of V_g (in 10 V steps) on the I–V_b curve of a tunnelling transistor Gr/hBN/Gr. (c) Tunnelling conductivity ($V_b = 0$) as a function of V_g. Symbols are the experimental data, the solid curve is a theoretical model. (d) Band diagram of Gr/hBN/Gr tunnelling transistor under different conditions. (Reproduced by courtesy of Kostya Novoselov, University of Manchester.)

of device scales exponentially with the hBN barrier thickness and is of the order of 1 kΩ^{-1} μm^{-2} for a monolayer hBN, sandwiched between gold and graphite electrodes, decreasing to approximately 0.1 GΩ^{-1} μm^{-2} for devices with 4 hBN atomic layers [53]. This demonstrates that a single atomic layer of hBN acts as an effective tunnel barrier and that the transmission probability of the hBN barrier decreases exponentially with the number of atomic layers. On a nanometre scale, conductive atomic force microscopy was used to measure the tunnel current through the hBN layers of different thickness: the results show that the tunnelling current is spatially uniform, demonstrating that hBN is a perfect material to be used as a barrier in a tunnelling diode or transistor [52–54].

A tunnelling transistor has been fabricated by encapsulating this heterostructure in hBN, inset in **Figure 16.5b**, and placing the device on a substrate that can act as a bake gate (e.g. highly doped silicon with an oxide layer) [54]. Such a device is also called the symmetric-FET (SymFET) [56] and it exploits the unique properties of (undoped) graphene: when a small gate voltage is applied between the substrate and the bottom graphene layer (Gr$_b$), the carrier concentrations n_b and n_t in the bottom and top electrodes, respectively, is approximately the same, as shown schematically in **Figure 16.5d**. This demonstrates that Gr$_b$ does not screen out the electric field induced by the Si-gate electrode. When V_b is applied (at $V_g = 0$), the I–V_b curve is the same of a tunnelling diode (**Figure 16.5a**). When a large V_g is applied, then the tunnelling current can be modulated, **Figure 16.5b**. This clearly shows that the weak electrostatic screening (at low doping), transparency and 2D thickness of graphene allow control over charge separation and transport in the heterostructures by using external electric fields, in contrast to traditional vertical devices.

Figure 16.5c shows the low-bias tunnelling conductivity (I–V_b) as a function of V_g: the curve is very asymmetric, showing on/off ratios up to 50 for holes and ~6 for electrons [54]. The tunnelling current I depends on the density of states (DOS) of the bottom and top graphene and on the transmission probability T, which is also modulated by V_g. The application of V_g of either polarities produces a higher DOS, and leads to an increased tunnelling current, **Figure 16.5d**. The asymmetry of the curve is to relate to the barrier height (d), which is lower for holes (~1.5 eV) than electrons (~4.5 eV), **Figure 16.5d**. A method to achieve an even higher on/off ratio in a graphene-based tunnelling transistor consists in using a barrier with a smaller d (<0.5 eV), which is compatible with the achievable doping levels reached in a graphene-based field-effect device (with a thin silicon oxide layer as dielectric). Possible candidates as barrier materials are MoS_2 and WS_2. The highest on/off ratio (above 10^6 at room temperature) has been achieved by using WS_2 [55]. Being chemically stable and having only a weak impurity band, WS_2 is believed to be superior to MoS_2 [55]. In particular, WS_2 allows for switching between tunnelling and thermionic transport regimes, resulting in much better transistor characteristics and thus allowing for much higher on/off ratios and much larger ON current [55]. Furthermore, it has been demonstrated that this technology is compatible with transparent and flexible electronics by fabricating the same devices on flexible polyethylene terephthalate (PET) substrates: the WS_2-tunnelling transistor is only a few atomic layers thick and stable under bending up to 2% of strain [55].

16.4 Photodetectors Based on 2D Heterostructures

Because TMDCs are semiconductors with excitonic band gaps in the range of 1–2 eV, TMDCs-tunnelling transistors are also expected to be able to interact with visible light. Photodetectors based on this technology were reported by Britnell et al. [57]. In this work, WS_2 has been selected as active material due to its large optical absorption ($>10^7$ cm^{-1} across the visible range), its chemical stability and band gap in the visible part of the spectrum. The interlayer tunnelling geometry easily allows generating a built-in electric field to separate the photoexcited electron–hole pair by applying an external electric field (or, alternatively, by chemical doping), **Figure 16.6a**. The Gr/WS_2/Gr heterostructure is encapsulated in BN (hBN/Gr/WS_2/Gr/hBN, inset in **Figure 16.6b**) and placed on top of an oxidised silicon wafer or flexible PET film. In the case of non-flexible devices on Si/SiO_2, the doped silicon is used as a back gate and SiO_2/hBN (typically 300 nm of SiO_2 and 20 nm of hBN) can be used as the gate dielectric. The two graphene layers were connected via a 1 kΩ resistor, on which the photocurrent was measured. The current–voltage (I–V_b) characteristics strongly depends on illumination (**Figure 16.6b**) [57]. Without illumination (right axis in **Figure 16.6b**), the devices displayed strongly non-linear I–V_b curves, while under illumination, the curves are linear at low V_b (left axis in **Figure 16.6b**). At higher bias (>0.2 V), the current saturates due to charge density saturation in the photo-active region. The use of an external V_g allows for the modulation of the photocurrent by tuning the built-in electric field, although this effect is observed to saturate at a certain V_g due to charge density saturation in the photo-active region. The photo-current maps obtained by scanning photo-current microscopy shows that the photocurrent is observed only in the overlapped area of the layers composing the heterostructure, **Figure 16.6c**. In addition, **Figure 16.6c** shows that under illumination and at $V_g = V_b = 0$, finite photocurrent is measured. This can be explained by assuming the existence of a built-in electric field in the as-made heterostructure, for example, due to different doping levels between the top and bottom graphene. The highest extrinsic quantum efficiency (EQE) for these devices is ~33% (measured at $V_b = 0$, $V_g = -40$ V and ~5 μW laser power). EQE is observed not to depend on wavelength (in the visible range), as expected from the approximately constant optical absorption of WS_2 over this energy range, and to decrease with laser power density, probably due to screening of the built-in electric field by the excited electrons in the conduction band of WS_2. The strong light–matter interactions observed in devices with such thin (5–50 nm thickness) photo-active material has been attributed to Van Hove singularities in the WS_2 DOS, which lead to enhanced light absorption [57]. This feature is universal to all TMDCs, independent on their thickness [57]. Similar EQE have been reported by Yu et al. [58] for Gr/(~50 nm)MoS_2/Gr devices, although a different dependence of EQE on the wavelength was reported. Devices based on the use of GaSe as barrier have been also reported: the efficiency was found to be lower than for WS_2 but qualitatively the same behaviour was observed [57].

The possibility to control charge separation and transport in the heterostructures by using external electric fields, due to the weak electrostatic

(a)

(b)

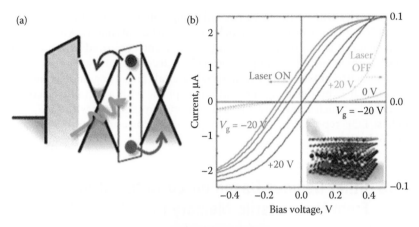

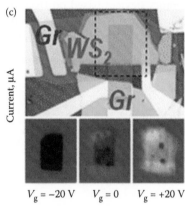

(c)

$V_g = -20\ V$ $V_g = 0$ $V_g = +20\ V$

FIGURE 16.6 (a) Band diagram of a Gr/WS$_2$/Gr heterostructure with a built-in electric field, which separates the photo-generated e–h pairs. (b) (Left axis) I–V_b curves for a device on Si/SiO$_2$ taken under illumination (2.54 eV, 10 µW power) at different V_g (in 10 V steps). (Right axis) I–V_b curves for the same device taken in the dark. (c) Optical micrograph of one of the devices. The shading of the three constituent layers denotes the regions of the respective materials – top and bottom graphene electrodes are shown in red and blue and WS$_2$ is shown in green. The dotted area is the region scanned by photocurrent, as shown in the photocurrent maps measured at $V_g = -20$ V, 0 and $+20$ V. (Adapted from L. Britnell et al., *Science*, 340, 1311, 2013.)

screening, optical transparency of graphene and the lack of Fermi-level pinning at the atomically sharp interface, allow for the realisation of such optoelectronic devices.

In devices based on Gr/TMDC/Gr heterostructures, further enhancement of the light–matter interaction can be easily obtained by using plasmonic nanostructures: for example, Reference 57 reports the use of thin gold layer deposited on the top graphene. This was annealed to produce gold nanoparticles with 5–10 nm in size, allowing enhancement of light absorption at 500–600 nm by excitation of localised surface plasmons. An enhancement factor of 10 was observed in the photocurrent, compared with the same device without metallic nanoparticles [57]. A second strategy for enhancing the photocurrent was demonstrated by Yu et al. [58] using an asymmetric device, that is by replacing the bottom graphene with a metallic electrode made by Ti, which allows to achieve an almost ohmic contact with MoS_2. The highest EQE reported in this geometry is ~55% (measured at 488 nm and ~5 μW laser power).

16.5 Other Devices Based on 2D Heterostructures: From Non-Volatile Memory to Light Emitters

Non-volatile stable memory devices based on trilayer heterostructure have been also reported [59]. A $Gr/hBN/MoS_2$ device was made on top of a Si wafer used as a back gate, **Figure 16.7a**. Here MoS_2 is used as channel, hBN

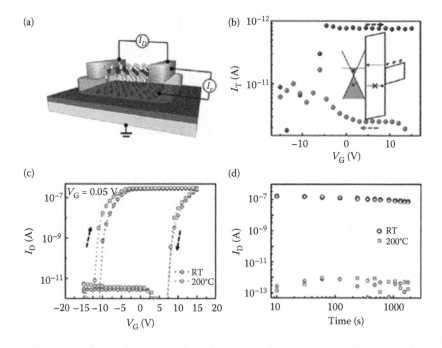

FIGURE 16.7 (a) Schematic of the $MoS_2/hBN/Gr$ heterostructure. (b) Tunnelling current (I_T) between MoS_2 and graphene through hBN as a function of V_g of forward and backward sweep directions. The inset shows the band diagram of the heterostructure. (c) Large hysteresis is observed in the I_D–V_g curve of the device. (d) Retention curve shows a high on/off ratio and nonvolatile characteristics (at room temperature and 200°C). (Reproduced from G.-H Lee et al., *APL Mat.*, 2, 092511, 2014.)

as barrier and graphene as charge trapping layer. It was observed that electrons in MoS_2 are transferred by tunnelling through hBN and trapped in floating graphene when the gate voltage is positive (forward sweep of V_G); meanwhile, the trapped electrons tunnel back to MoS_2 when V_g is negative in a releasing process (backward sweep) [59]. This behaviour is based on the observation that holes tunnelling through the hBN layer are less feasible due to relatively larger effective mass and higher barrier height for holes. As a result, the transfer curve of the MoS_2/hBN/graphene heterostructure device exhibited a large hysteresis of $\Delta V \approx 20$ V and high on/off ratio of ~10^5, which could be maintained for a long time of over 1000 s at room temperature and even at 200°C [59].

More complex and multi-functional heterostructures can be achieved by increasing the number of layers and components. Recent work performed by researchers from the University of Manchester demonstrated the possibility of fabricating efficient light-emitting heterostructure-based devices by introducing quantum wells (QWs) engineered with one atomic plane precision [60].

The device architecture is arranged as follows: graphene is used as a high transparency and yet highly conductive charge carrier injection electrode, which sandwiches the hBN tunnel barriers that in turn encase the active semiconducting layer. High quality hBN flakes are used with flake thicknesses typically ranging from one to six layers and the active semiconducting layer consists of single or bilayer semiconducting TMDC's, including WS_2, MoS_2 and WSe_2. The heterostructure devices are produced by a 'peel-lift' transfer procedure [60]. **Figure 16.8a** shows the typical schematic structure of a WS_2 single QW and **Figure 16.8b** shows the corresponding cross-sectional bright-field STEM image of a slice of the light-emitting device (LED). This demonstrates that the interfaces between neighbouring crystals are atomically flat. The 'peel-lift' procedure can be repeated multiple times to create multiple QWs, as shown in **Figure 16.8c** and **d**.

In order to understand the operation of such devices, the $I-V_b$ curve, PL and electroluminescence (EL) have been measured. **Figure 16.9a** through **c** shows the band alignment diagrams for three regimes: (a) unbiased; (b) intermediate and (c) high bias for a MoS_2 single QW. It is known from previous electron tunnelling measurements that the valence band of hBN lies 1.5 eV below the Dirac point of graphene [54], whereas the conduction band of MoS_2 is ~0.1 eV above the Dirac point of graphene (taken from the onset of screening in the $C-V_g$ data shown in Reference 35). At low V_b, there exists a small tunnel current, **Figure 16.9e**, and the PL spectrum is dominated by the neutral exciton, X^0 at 1.93 eV [61], **Figure 16.9e**. For undoped single-layer MoS_2, a sharp transition of the spectral weight from the neutral exciton to the negatively charged exciton X^- is observed as the bias is increased to $V_b = 0.5$ V; this occurs due to an increased density of electrons in the conduction band of MoS_2 [61] and simultaneously occurs with an increase of the differential conductivity; the corresponding band alignment diagram is shown in **Figure 16.9b**. As the bias is increased further to $V_b \sim 1.9$ V, then the Fermi level of the top graphene electrode is brought into the valence band of the MoS_2, which allows for an excess of electrons in the conduction band and an excess of holes in the valence band, which allows

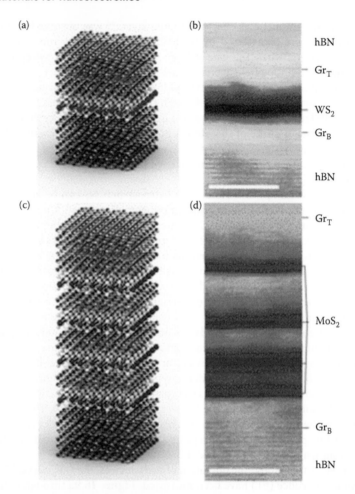

FIGURE 16.8 (a) Schematic of a WS$_2$ QW heterostructure; (b) cross-sectional bright-field STEM image of the QW LED. (c) Schematic of a multiple QW LED based on MoS$_2$; (d) cross-sectional bright-field STEM image of the heterostructure. (Reproduced from F. Withers et al., *Nat. Mater.*, 14, 301, 2015.)

for exciton formation. The exciton decay leads to the observed EL, which is facilitated by the increased lifetime of injected carriers due to confinement by the hBN tunnel barriers. The resultant decay of excitons leads to the emission of light from the QW structure as seen directly in **Figure 16.9d**.

An important quantity of any light emission device is the quantum efficiency η defined as the ratio of emitted photons, N, to the number of injected electron–hole pairs, I/e, that is, $\eta = e\,N/I$. Single QWs exhibit quantum efficiencies of approximately 1%–3%, which is an order of magnitude larger than that of light emission from in plane p–n junctions [33,34,62]. To increase the quantum efficiency, further multiple QWs are demonstrated, as shown in

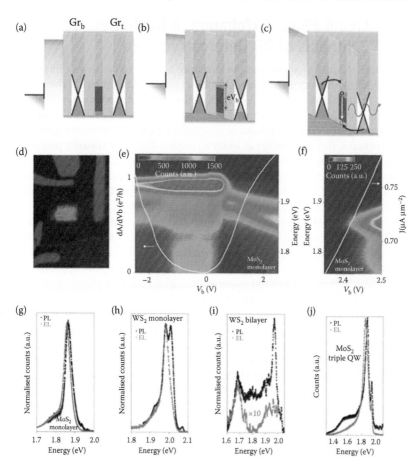

FIGURE 16.9 Band alignment diagrams for (a) unbiased, (b) applied bias and (c) large bias. (d) Optical micrograph of a single QW MoS_2 device glowing (injection current of 40 μA, V_b = 2.3 V, 300 K, area of 50 μm²; (e) PL map versus bias voltage for the device shown in (d) measured with an excitation energy of 2.33 eV at 6 K. (f) EL map versus bias and current density of the same MoS_2 device. (g–j) compares the PL and EL for QW devices consisting of different semiconducting active elements (g) MoS_2, (h) WS_2, (i) bilayer WS_2 and (j) a triple MoS_2 multiple QW. (Adapted from F. Withers et al., *Nat. Mater.*, 14, 301, 2015.)

Figure 16.8c and **d**. The EL and PL spectra from a triple MoS_2 QW are shown in **Figure 16.9j**. EL spectra of similar intensity can be collected for orders of magnitude smaller injection currents as expected for an overall increased barrier thickness, which leads to an increased confinement and can lead to QEs close to 10% in a 4QW based on MoS_2. As these LEDs comprise of only 40 atoms in thickness, they are therefore bendable and also remain highly transparent. Such structures are likely to pave the way for future transparent and flexible LEDs.

447

16.6 Liquid-Phase Exfoliation: A Cost-Effective Synthesis of 2D Heterostructures

The range of functionalities and performance is likely to be further improved by increasing the number of available 2D crystals and improving their electronic quality. However, such complex heterostructures are currently fabricated only by using mechanically exfoliated 2D crystals. Epitaxial growth is practical for fabrication of bilayer heterostructures, but for more complex heterostructures, this technology is likely to require the same transfer processes currently used for mechanically exfoliated crystals. So, alternative methods, compatible with mass production and which do not require the use of clean rooms and complex techniques should be utilised to bring the attractive qualities of 2D crystal-based heterostructures to real-life applications.

Chemical exfoliation is the easiest route for the synthesis of single layers by their separation from thicker compounds. This technique has been already used for oxide nanosheets [13], ionic layered materials [63] and it has been recently applied to the production of 2D crystals inks [64–66]. A popular approach for chemical exfoliation of layered materials is based on liquid-phase exfoliation (LPE) to produce inks of various 2D crystals [65,66]. One can then use such inks to deposit platelet layers of different materials sequentially by standard low-cost fabrication techniques (drop and spray coatings, roll-to-roll transfer, inkjet printing, etc.), **Figure 16.10**. One of the most important advantages of LPE is that the same method can be used to create inks made of nanosheets of different 2D crystals, covering a large variety of properties. Furthermore, this technique is also compatible with low-cost flexible substrates, so it is expected to produce a strong impact on the new generation of low-cost flexible devices. Although 2D crystals obtained by LPE are usually considered of 'low-quality', they have already been used to make simple devices such as sensors for hydrogen evolution reaction [67] and planar photovoltaic devices [68]. Recently, the

FIGURE 16.10 2D crystal inks-based technology allows moving from a single crystal to millions of LEGO bricks, which can be sequentially deposited in films to fabricate devices.

ink-based technology has been demonstrated to be suitable for fabrication of low-cost flexible devices based on heterostructures of arbitrary complexity [69]. In particular, the following devices were fabricated: Gr/barrier/Gr tunnelling transistors, where the tunnelling barrier (made of hBN or TMDC) is controlled by a back gate; Gr/TMDC/Gr photovoltaic devices; and in-plane transistors, where the top graphene is used as a gate and the barrier as a gate dielectric to control the in-plane current in the bottom graphene.

The devices are made in the following way [69]: the graphene ink is deposited on a Si/SiO$_2$ substrate to fabricate the bottom electrode. Then, TMDC or hBN inks are used to fabricate a thin film on top of the bottom electrode. The top graphene is usually composed of chemical vapour deposition (CVD), LPE or mechanically exfoliated graphene to ensure sufficient optical transparency and reduced screening. Note that tunnelling transistors require exactly monolayer graphene to be used as the bottom electrode in order not to screen the gate voltage. Because of the vertical geometry used, the morphology and surface properties of the film used as a barrier are crucial to establish this technology: in order to avoid a shortcut between the two graphene electrodes, the pinholes density of the barrier film has to be minimised. The presence of pinholes can immediately be detected by very low tunnelling resistance of the tunnel junctions [69]. This problem can be easily solved by using inkjet printing as this technology allows printing the patterns several times on top of each other. However, coffee stain effect and re-dispersion of the printed material may cause difficulties in controlling the surface properties of the film. In the case of drop-casting, a highly concentrated ink must be used in order to reduce the pinholes density. Amongst the deposition methods that have been tested, it has been observed that vacuum filtration allows an excellent control on the pinholes density of the TMDC barrier. The TMDC thickness of the film is controlled by using the 'fishing method' [70]: the laminate supported on the cellulose filter is vertically immersed in water, which allows a thin section of the film to delaminate and to appear as a free-standing film on the water surface. The floating TMDC film can then be collected by 'fishing' onto arbitrary substrates, including Si/SiO$_2$, plastics and quartz. This method allows controlling the thickness of the TMDC films by stacking several thin films together, for example, obtained by sequential delamination of the same laminate. The thickness of each delaminated TMDC film is around 15 nm and it contains a large proportion of holes, making a single delaminated film unsuitable for the fabrication of ink-based heterostructures [69]. However, by sequentially stacking the delaminated layers, the hole density strongly decreases since the stacked layers constitute a homogenous barrier: a tri-stacked film of about 60 nm thickness is observed to be pinhole-free [69]. Therefore, heterostructures can be easily fabricated with bi- or tri-stacked delaminated films [69]. The roughness of the uncovered TMDC film is ~24 nm and decreases to ~16 nm when the film is covered by graphene (note that such large roughness measurements are determined by a few cracks and tall inclusions) [69]. This shows that graphene partially suspends over the trenches in the LPE film, further helping preventing short circuit.

Figure 16.11a shows the strongly non-linear $I–V_b$ characteristics of a Gr/LPE WS$_2$/Gr tunnelling transistor [69]. The zero-bias conductivity goes

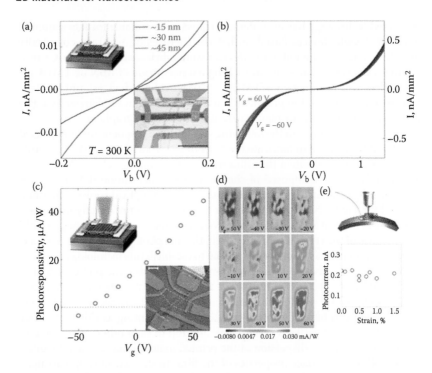

FIGURE 16.11 (a) (A) $I-V_b$ ($V_g = 0$) curves for Si/SiO$_2$/Gr/WS$_2$/Gr heterostructure (top inset) with different thickness of WS$_2$ laminate. Bottom inset: optical micrograph of one of the devices. The bottom graphene is in yellow, produced by drop coating and top graphene is in green, mechanically exfoliated few layer graphene, both marked by dashed lines. The whitish (when on Si/SiO$_2$) or reddish (when on gold contacts) area is LPE WS$_2$. Scale bar is 100 µm. (b) $I-V_b$ characteristics at different V_g (in 20 V steps). $T = 300$ K. (c) Average photoresponsivity ($V_b = 0$) of a Si/SiO$_2$/Gr/WS$_2$/Gr heterostructure (top and bottom insets, scale bar is 10 µm) as a function of V_g. (d) Photo-current maps as a function of V_g ($V_b = 0$). The photocurrent is averaged and values used in **Figure 16.11c**. (e) Top: schematic representation of the bending setup used to apply uniaxial strain. Bottom: average photocurrent obtained from the photo-current maps as a function of the applied strain. (Adapted from F. Withers et al., *Nano Lett.*, 14, 3987, 2014.)

down as the thickness of WS$_2$ layer increases and it decreases dramatically with decreasing temperature. Such a strong temperature dependence suggests an excitation mechanism for charge carrier generation, either from the graphene electrodes (in this case, the tunnelling barrier is the Schottky barrier between graphene and WS$_2$) or from the impurity band in WS$_2$ (possibly caused by the edges of the WS$_2$ nanosheets) [69]. From this point of view, the device is very different compared the one obtained by mechanically exfoliated flakes (**Figure 16.5**) as there is now a range of tunnelling barriers connected in parallel, which can contribute to the strong temperature dependence. The strong increase in the current for $V_b > 1$ V observed even at low temperatures suggests over-barrier transport between graphene and WS_2. When a V_g is applied, the zero-bias resistance is not sensitive to the back-gate voltage V_g applied, whereas the current in the non-linear region demonstrates a 30% modulation when V_g is swept between −60 and 60 V, **Figure 16.11b**. The fact that the gate voltage mostly affects the non-linear part of the $I–V_b$ dependence indicates that the changes in the current are mostly due to the changes in the relative position of the Fermi energy with respect to the top of the valence band in WS_2 and not due to the gating of WS_2. The gating of WS_2, which would result in modification of the shape of the tunnelling barrier (making it triangular), is not very efficient in LPE samples, due to the large impurity band, which screens the electric field [69]. The same heterostructure can be used for photovoltaic applications. Under illumination, the $I–V_b$ characteristics become increasingly linear, demonstrating that in this regime, the current is dominated by the photo-excited carriers. Zero-bias photo-current maps, taken at different V_g, demonstrate that the photocurrent is produced only in the regions where all three layers overlap, inset in **Figure 16.11c** and **d**. Similar to the case of the transistor, the back-gate voltage controls the value and the direction of the electric field across WS_2, and thus the magnitude and the polarity of the photocurrent (**Figure 16.11c**). For the largest electric field across WS_2 (at $V_g = 60$ V), a photoresponsivity of ~ 0.1 mA/W was recorded [69]. Although the photoresponsivity is significantly smaller than that obtained by using mechanically exfoliated flakes [57], the advantage of this technology is its compatibility with low cost, scalable methods and flexible substrates. Gr/LPE WS_2/Gr heterostructures on a flexible PET substrate show stable photocurrent (averaged on an area of 70 μm^2) up to 1.5% uniaxial strain, **Figure 16.11e**.

Although many of the ink-based devices created [68,69,71] still underperform in comparison with the benchmark structures, their versatility, low cost, simplicity of technology and unique properties (e.g. flexibility and transparency) might prove beneficial for some types of devices [72]. Furthermore, the performance of the device is expected to strongly depend on the composition of the inks: the possibility of fine-tuning the properties of the inks by varying the size and thickness of the flakes as well as the type of solvent is expected to be crucial to improve the efficiency and range of functionalities of those heterostructures.

16.7 Concluding Remarks

In conclusion, it is expected that heterostructures based on 2D crystals will allow the development of low-cost devices, able to operate at much higher speeds and frequencies, and with greatly increased multi-functionality such as personal protection combined with sensors and detectors; integrated circuits combined with solar cell batteries which provide power; paint which simultaneously provides corrosion protection and acts as antireflection coating, etc.

The compatibility of 2D crystal ink-based devices with printing technology also allows fabrication of these devices onto virtually any substrate, making it possible to integrate heterostructures not only in the printed electronics market, but also into new emerging fields such as packaging, pharmaceutics and smart textiles.

Acknowledgements

The authors acknowledge K. S. Novoselov for useful discussions and assistance with the figures. Financial support by the U.S. Army Research Office and the Royal Academy of Engineering is acknowledged.

References

1. Z. I. Alferov, *Rev. Mod. Phys.*, 73, 767, 2001.
2. A. K. Geim, I. V. Grigorieva, *Nature*, 499, 419, 2013.
3. K. S. Novoselov, A. H. Castro Neto, *Phys. Scr.*, T146, 014006, 2015, doi:10.1088/0031-8949/2012/T146/014006
4. G. Eda, S. Maier, *ACS Nano*, 7, 5660, 2013.
5. G.-H Lee et al., *APL Mat.*, 2, 092511, 2014.
6. D. Akinwande, N. Petrone, J. Hone, *Nat. Commun.*, 5, 5678, 2014.
7. K. S. Novoselov et al., *Proc. Natl. Acad. Sci. USA*, 102, 10451, 2005.
8. A. K. Geim, K. S. Novoselov, *Nat. Mater.*, 6, 183, 2007
9. A. K. Geim, *Science*, 324, 1530, 2009.
10. K. S. Novoselov, Nobel lecture: Graphene: Materials in the flatland. *Rev. Mod. Phys.* 83, 837, 2011.
11. D. C. Elias et al., *Science*, 323, 610, 2009.
12. R. R. Nair et al., *Small*, 6, 2877–2884, 2010.
13. M. Osada, T. Sasaki, *Adv. Mater.*, 24, 210, 2012.
14. S. Z. Butler et al., *ACS Nano*, 7, 2898, 2013.
15. P. Vogt et al., *Phys. Rev. Lett.* 108, 155501, 2012.
16. L. Li et al., *Adv. Mater.* 26, 4820, 2014.
17. M. E. Dávila et al. *New J. Phys.* 16, 095002, 2014.
18. L. Li et al., *Nat. Nanotechnol.* 9, 372, 2014.
19. H. Liu et al., *ACS Nano*, 8, 4033, 2014.
20. A. H. Castro Neto et al., *Rev. Mod. Phys.*, 81, 109, 2009.
21. C. R. Dean et al., *Nat. Nanotechnol*, 5, 722, 2010.
22. A. Ayari et al., *J. Appl. Phys.* 101, 014507, 2007.
23. B. Radisavljevic et al., *Nat. Nanotechnol*, 6, 147, 2011.
24. B. Radisavljevic et al., *ACS Nano*, 5, 9934, 2011.
25. R. Kappera et al., *Nat. Mater*, 13, 1128, 2014.
26. W. Bao et al., *Appl. Phys. Lett.*, 102, 042104, 2013.
27. O. Lopez-Sanchez et al., *Nat. Nanotechnol.*, 8, 497, 2013.
28. H. Zeng et al., *Nat. Nanotechnol.*, 7, 490, 2012.
29. F. K. Mak et al., *Nat. Nanotechnol.*, 7, 490, 2012.
30. T. Cao et al., *Nat. Nanotechol.*, 3, 887, 2012.
31. M. Bernardi et al., *Nano Lett.*, 13, 3664, 2013.
32. M. M. Furchi et al., *Nano Lett.*, 14, 4785, 2014.
33. B. W. H. Baugher et al., *Nat. Nanotechnol.*, 9, 262, 2014.
34. J. S. Ross et al., *Nat. Nanotechnol.*, 9, 268, 2014.

35. A. V. Kretinin et al., *Nano Lett.*, 14, 3270, 2014.
36. A. Koma et al., *J. Cryst. Growth*, 111, 1029, 1992.
37. S. J. Haigh et al., *Nat. Mater.*, 11, 764, 2012.
38. K.-G Zhou et al., *ACS Nano*, 8, 9914, 2014.
39. L. A. Ponomarenko et al., *Nat. Phys*, 7, 958, 2011.
40. R. V. Gorbachev, *Nat. Phys.*, 8, 896, 2012.
41. A. S. Mayorov et al., *Nano Lett.*, 11, 2396, 2011.
42. X. Cui et al., *Nat. Nanotechnol*, 10, 534, 2015.
43. G.-H. Lee et al., *ACS Nano*, 7, 7931, 2013.
44. A. Castellanos-Gomez et al., *2D Mater*, 1, 025001, 2014.
45. Y. Cai et al. *Nano Lett.* 15, 4914, 2015.
46. C. R. Dean et al., *Nature*, 497 598, 2013.
47. L. A. Ponomarenko et al., *Nature*, 497, 594, 2013.
48. B. Hunt et al., *Science*, 340, 1427, 2013.
49. M. Yankowitz et al., *Nat. Phys.*, 8, 382, 2012.
50. A. Eckmann et al., *Nano Lett.*, 13, 5242, 2013.
51. C. R. Woods et al., *Nat. Phys.*, 10(6), 451–456, 2014.
52. G. H. Lee et al., *Appl. Phys. Lett.*, 99, 243114, 2011.
53. L. Britnell et al., *Nano Lett.*, 12, 1707, 2012.
54. L. Britnell et al., *Science*, 335, 947, 2012.
55. T. Georgiou et al., *Nat. Nanotechnol.*, 8, 100, 2013.
56. P. Zhao et al., *IEEE Trans. Electron Devices*, 60, 951, 2013.
57. L. Britnell et al., *Science*, 340, 1311, 2013.
58. W. J. Yu et al., *Nat. Nanotechnol*, 8, 952, 2013.
59. M. S. Choi et al., *Nat. Commun*, 4, 1624, 2013.
60. F. Withers et al., *Nat. Mater.*, 14, 301, 2015.
61. K. F. Mak et al., *Nat. Mater.*, 12, 207–211, 2013.
62. A. Popischil et al., *Nat. Nanotechnol*, 9, 257–261, 2014.
63. Q. Wang, D. O'Hare, *Chem. Rev.*, 112, 4124, 2012.
64. P. Blake et al., *Nano Lett.*, 8, 1704, 2008.
65. Y. Hernandez et al., *Nat. Nanotechnol.* 3, 563, 2008.
66. J. N. Coleman et al., *Science*, 331, 568, 2011.
67. J. Yang et al., *Angew. Chem., Int. Ed.*, 52, 13751, 2013.
68. D. J. Finn et al., *J. Mater. Chem. C*, 2, 925, 2014.
69. F. Withers et al., *Nano Lett.*, 14, 3987, 2014.
70. G. Eda et al., *Nano Lett.*, 11, 5111, 2011.
71. G. Cunningham et al., *Nanoscale*, 7, 198, 2014.
72. E. B. Secor, M. C. Hersam, *J. Phys. Chem. Lett.*, 6, 620, 2015.

Index

Index

Index

Index

Index

Index

Index

Printed and bound by CPI Group (UK) Ltd, Croydon, CR0 4YY

01/11/2024

01782623-0014